Bioenergetics 4

ELSEVIER *science &*
technology books

Companion Web Site:

http://booksite.elsevier.com/9780123884251

Bioenergetics 4
David G. Nicholls and Stuart J. Ferguson

Resources for Professors:

- All figures from the book available as both PowerPoint slides and .jpeg files

- Links to web sites carefully chosen to supplement the content of the textbook

TOOLS FOR ALL YOUR TEACHING NEEDS
textbooks.elsevier.com

**ACADEMIC
PRESS**

Bioenergetics 4

DAVID G. NICHOLLS

*Buck Institute for Research on Aging,
Novato,
California, USA*

STUART J. FERGUSON

*Department of Biochemistry,
University of Oxford,
W.R. Miller Fellow of St. Edmund Hall,
Oxford, UK*

ELSEVIER

AMSTERDAM • BOSTON • HEIDELBERG • LONDON
NEW YORK • OXFORD • PARIS • SAN DIEGO
SAN FRANCISCO • SINGAPORE • SYDNEY • TOKYO
Academic Press is an imprint of Elsevier

Academic Press is an imprint of Elsevier
32 Jamestown Road, London NW1 7BY, UK
225 Wyman Street, Waltham, MA 02451, USA
525 B Street, Suite 1800, San Diego, CA 92101-4495, USA

Fourth edition 2013

Notice
No responsibility is assumed by the publisher for any injury and/or damage to persons
or property as a matter of products liability, negligence or otherwise, or from any use
or operation of any methods, products, instructions or ideas contained in the material
herein. Because of rapid advances in the medical sciences, in particular, independent
verification of diagnoses and drug dosages should be made

British Library Cataloguing-in-Publication Data
A catalogue record for this book is available from the British Library

Library of Congress Cataloging-in-Publication Data
A catalog record for this book is available from the Library of Congress

ISBN: 978-0-12-388425-1

For information on all Academic Press publications
visit our website at www.store.elsevier.com

Typeset by MPS Limited, Chennai, India
www.adi-mps.com

Printed and bound in Europe

13 14 15 16 17   10 9 8 7 6 5 4 3 2 1

CONTENTS

Preface   ix
Glossary   xi

INTRODUCTION TO PART 1   **1**

1 CHEMIOSMOTIC ENERGY TRANSDUCTION   3

1.1 The Chemiosmotic Theory: Fundamentals   3
1.2 The Basic Morphology of Energy-Transducing Membranes   7
1.3 A Brief History of Chemiosmotic Concepts   11

2 ION TRANSPORT ACROSS
   ENERGY-CONSERVING MEMBRANES   13

2.1 Introduction   13
2.2 The Classification of Ion Transport   13
2.3 Bilayer-Mediated Transport   17
2.4 Protein-Catalysed Transport   21
2.5 Swelling and the Coordinate Movement of Ions across Membranes   22

3 QUANTITATIVE BIOENERGETICS   27

3.1 Introduction   27
3.2 Gibbs Energy and Displacement from Equilibrium   30
3.3 Redox Potentials   36
3.4 Ion Electrochemical Potential Differences   43
3.5 Photons   44
3.6 Bioenergetic Interconversions and Thermodynamic Constraints on their
     Stoichiometries   45
3.7 The Equilibrium Distributions of Ions, Weak Acids and Weak Bases   47
3.8 Membrane Potentials, Diffusion Potentials, Donnan Potentials and Surface
     Potentials   50

4   THE CHEMIOSMOTIC PROTON CIRCUIT IN ISOLATED ORGANELLES   53

4.1 Introduction   53
4.2 The Proton Circuit   54
4.3 Proton Current   58
4.4 Voltage: The Measurement of Protonmotive Force Components in Isolated
     Organelles   65
4.5 Proton Conductance   73
4.6 ATP Synthase Reversal   75
4.7 Reversed Electron Transport   76
4.8 Mitochondrial Respiration Rate and Metabolic Control Analysis   77
4.9 Kinetic and Thermodynamic Competence of Δp in the Proton Circuit   83

INTRODUCTION TO PART 2   89

5   RESPIRATORY CHAINS   91

5.1 Introduction   91
5.2 Components of the Mitochondrial Respiratory Chain   91
5.3 The Sequence of Redox Carriers in the Respiratory Chain   100
5.4 Mechanisms of Electron Transfer   101
5.5 Proton Translocation by the Respiratory Chain: Loops, Conformational
     Pumps, or Both?   106
5.6 Complex I (NADH–UQ Oxidoreductase)   108
5.7 Delivering Electrons to Ubiquinone without Proton Translocation   115
5.8 Ubiquinone and Complex III   118
5.9 Interaction of Cytochrome c with Complex III and Complex IV   125
5.10 Complex IV   126
5.11 Overall Proton and Charge Movements Catalysed by the Respiratory Chain:
     Correlation with the P/O Ratio   131
5.12 The Nicotinamide Nucleotide Transhydrogenase   132
5.13 Electron Transport in Mitochondria of Non-Mammalian Cells   133
5.14 Bacterial Respiratory Chains   136

6   PHOTOSYNTHETIC GENERATORS OF PROTONMOTIVE FORCE   159

6.1 Introduction   159
6.2 The Light Reaction of Photosynthesis in *Rhodobacter Sphaeroides* and
     Related Organisms   161
6.3 The Generation by Light or Respiration of Δp in
     Photosynthetic Bacteria   172

6.4 Light-Capture and Electron Transfer Pathways in Green Plants, Algae
     and Cyanobacteria   174
6.5 Bacteriorhodopsin, Halorhodopsin and Proteorhodopsin   191

7 ATP SYNTHASES AND BACTERIAL FLAGELLA ROTARY MOTORS   197

7.1 Introduction   197
7.2 Molecular Structure   198
7.3 F_1 200
7.4 The Peripheral Stalk or Stator   209
7.5 F_o 209
7.6 The Structural Basis for H^+/ATP Stoichiometry   215
7.7 Inhibitor Proteins   216
7.8 Proton Translocation by A-Type ATPases, V-Type ATPases
     and Pyrophosphatases   217
7.9 Bacterial Flagellae   218

8 TRANSPORTERS   221

8.1 Introduction   221
8.2 The Principal Mitochondrial Transport Protein Family   222
8.3 Bacterial Transport   228

INTRODUCTION TO PART 3   253

9 CELLULAR BIOENERGETICS   255

9.1   Introduction   255
9.2   The Cytoplasmic Environment   256
9.3   Mitochondrial Monovalent Ion Transport   257
9.4   Mitochondrial Calcium Transport   259
9.5   Metabolite Communication between Matrix and Cytoplasm   268
9.6   Quantifying the Mitochondrial Proton Current in Intact Cells   276
9.7   Mitochondrial Protonmotive Force in Intact Cells   281
9.8   Permeabilised Cells   287
9.9   *In Vivo* Bioenergetics   288
9.10 Reactive Oxygen Species, 'Electron Leaks'   288
9.11 Reactive Nitrogen Species   295
9.12 Uncoupling Pathways, 'Proton Leaks'   296
9.13 The ATP Synthase Inhibitor Protein IF_1 301

10 THE CELL BIOLOGY OF THE MITOCHONDRION   303

10.1	Introduction	303
10.2	The Architecture of the Mitochondrion	303
10.3	Mitochondrial Dynamics	306
10.4	Trafficking of Mitochondria	312
10.5	Mitochondrial Biogenesis	313
10.6	Mitophagy	318
10.7	Apoptosis	321

11 SIGNALLING BETWEEN THE MITOCHONDRION AND THE CELL   327

11.1	Introduction	327
11.2	The Mitochondrial Genome	327
11.3	AMP Kinase	331
11.4	Transcription Factors and Transcriptional Coactivators in Bioenergetic Control	333
11.5	Adaptations to Hypoxia	334
11.6	Mitochondrial Protein Phosphorylation	337
11.7	mTOR	338
11.8	Sirtuins and Mitochondrial Function	340
11.9	Redox Signalling and Oxidative Stress	342

12 MITOCHONDRIA IN PHYSIOLOGY AND PATHOLOGY   345

12.1	Introduction	345
12.2	Mitochondrial Diseases	345
12.3	The Heart	350
12.4	Brown Adipose Tissue and Transcriptional Control	354
12.5	Mitochondria, the Pancreatic β Cell and Diabetes	355
12.6	Mitochondria and the Brain	361
12.7	Mitochondria and Cancer	377
12.8	Stem Cells	381
12.9	Mitochondrial Theories of Aging	383
12.10	Conclusions	386

References	387
Index	407

PREFACE

The context for the first edition of this book, published in 1982, was that Mitchell's chemiosmotic theory of energy transduction had been widely accepted, as acknowledged by the award of the Nobel Prize in 1978, yet the underpinning principles of this theory were widely misunderstood and its full scope was not appreciated. The second edition in 1992 was written against the background that, on the one hand, many general textbooks still gave too superficial a treatment to chemiosmotic mechanisms, whereas on the other hand, the high-resolution structure of a bacterial reaction centre that operates according to Mitchell's ideas had recently been reported and recognised with the Nobel Prize in 1988. The third edition in 2002 followed the 1997 Nobel Prize to Paul Boyer and John Walker for their work on the ATP synthase enzyme and marked the beginning of the explosion in mitochondrial physiology and pathology, with the calculation that 34,000 publications in the previous decade could be retrieved from PubMed with the key word 'mitochondria.' A similar calculation for the decade from June 2002 to December 2012 raises this number to an extraordinary 95,000, the large majority of which deal with aspects of what might be termed 'mitochondrial physiology'—the study of healthy and dysfunctional mitochondria in the cellular, organ and whole body context. This explosion has necessitated a change in the organisation of this fourth edition, which we have divided into three interlocking sections: basic principles; structures and mechanisms; and physiology. This last is greatly expanded from the brief chapter in the third edition. Thus, Part 1 deals with the fundamentals of bioenergetics, ion transport pathways, thermodynamics and the basis of the proton circuit. Part 2 covers the current information on the structures and mechanisms of the protein complexes of the electron transport chains in mitochondria, bacteria and photosynthetic systems, the ATP synthase and transport proteins. Part 3 covers the application of these structures and principles to physiology and pathophysiology.

It is striking that many of the misunderstandings that surrounded the chemiosmotic idea in the 1960s reoccur in the 21st century. For example, it is still common in new textbooks to see the mitochondrion depicted as having no membrane potential but just a pH gradient, while we were recently alerted to a serious error in a national biology examination, where it was incorrectly asserted that a proton-impermeable outer membrane was required for ATP synthesis. The mechanism of action of routine uncouplers such as FCCP is often erroneously described. Papers appear regularly in the most prestigious journals demonstrating a profound lack of understanding of bioenergetics—a criticism directed not only to the authors but also to deficiencies in reviewing and editing. The need to explain these fundamentals therefore continues.

Although we wish this book to continue to be considered the standard authority in the field, the present state of mitochondrial physiology poses some problems because few areas have established the same degree of precision or certainty characteristic of much of the structural and mechanistic studies. Rather than present uncritical reviews of the many physiological topics, we have imposed what we hope is a useful filter, namely that observations must be consistent with our understanding of bioenergetics at the isolated organelle and intact cellular level. There is undoubtedly a current tendency to propose an involvement of mitochondrial function and dysfunction in a large number of physiological and pathophysiological processes. Not all of these will be borne out by future experimentation. However, if only 10% prove to be verifiable in the future, this will mean that the mitochondrion will continue to play a central role in cell physiology.

The overall aim of the book is to allow readers with the standard basic knowledge of biochemistry to have an entry into the frontline research in bioenergetics, by providing a bridge to the primary research literature. To this end, we have employed a policy of focusing on the most recent authoritative reviews, with a very restricted number of key papers. Therefore, the choice of references tends to reflect the inverse of scientific priority. We trust our colleagues will recognise this limitation.

Since the last edition, there has been an explosion of crystal structures that in most cases are beginning to give mechanistic insight. The trouble with crystal structures is that they are too complex for the reader to see the woods for the trees. Consequently, we have continued our practice of using line drawings, but new for this edition is that these are usually superimposed on the actual structures. We hope this approach will be valuable. Of course, not all the mechanisms we discuss here will prove to be correct in every detail. There is a web site (http://booksite.elsevier.com/9780123884251) on which we shall post important corrections and updates in detail, and we suggest you bookmark this link.

We have been helped by generous advice and permissions to use original figures from numerous colleagues. A partial list includes Judy Armitage, Fraser Armstrong, Christopher Benz, Ben Berks, Egbert Boekema, Martin Brand, Michael Duchen, Terry Frey, Gyorgy Hajnoczky, Carola Hunte, Pankaj Kapahi, David Kelly, Edmund Kunji, John Lemasters, Zhenfeng Liu, Bernd Ludwig, Anthony Moore, Hindrik Mulder, Simon Newstead, Casey Quinlan, John Rubinstein, Guy Rutter, Leonid Sazanov, Orian Shirihai, Michael Stewart, Rolf Thauer, John Walker, Konstanze Winklhofer and Claes Wollheim. Doubtless we have overlooked some and to them we apologise whilst emphasising that we are entirely to blame for any misinterpretations and factual errors.

Writing a book such as this imposes large burdens on our families and colleagues. We thank them, in particular Solvig and Tina, very much for their forbearance.

David G. Nicholls
Stuart J. Ferguson

GLOSSARY

3-NPA	3-Nitropropionic acid
$[A]_{equil}$	Equilibrium concentration of reactant A
$[A]_{obs}$	Observed concentration of reactant A
A/B	Antiport of A against B
A:B	Symport of A and B
Ac	Acetate (ethanoate)
AcAc	Acetoacetate
AD	Alzheimer's disease
ADP/O	The number of molecules of ADP phosphorylated to ATP when two electrons are transferred from a substrate through an electron transport chain to reduce one 'O' ($\frac{1}{2}O_2$)
$ADP/2e^-$	As ADP/O, except more general because the final acceptor can be other than O_2
ANT	Adenine nucleotide translocator
AOX	Alternative oxidase
APP	Amyloid precursor protein
Bchl	Bacteriochlorophyll
bR	Bacteriorhodopsin
Bpheo	Bacteriopheophytin
BQ	Benzoquinone
BQH_2	Benzoquinol
C	Flux control coefficient
$[Ca^{2+}]_c$	Cytoplasmic free Ca^{2+} concentration
$[Ca^{2+}]_m$	Matrix free Ca^{2+} concentration
Chl	Chlorophyll
$C_M H^+$	Proton conductance (nmol H^+ min^{-1} mg^{-1} mV^{-1})
CypD	Cyclophilin D
Cyt	Cytochrome. A letter denotes the type of haem; a three-digit subscript indicates an absorbance maximum in the reduced form.
Cyt aa_3	Alternative name for complex IV (cytochrome c oxidase or cytochrome oxidase)

Cyt bc_1	Alternative name for complex III (ubiquinol–cytochrome c reductase)
dO/dt	Respiratory rate (nmol O min^{-1} mg protein^{-1})
DAD	Diaminodurane
DBMIB	2,5-Dibromo-3-methyl-6-isopropylbenzoquinone
DCCD	N,N'-dicyclohexylcarbodiimide
DCMU	3-(3,4-Dichlorophenyl)-1,1-dimethylurea
DCPIP	2,6-Dichlorophenylindophenol
E	Redox potential at any specified set of component concentrations and conditions (mV)
E_h	Actual redox potential at a defined pH (mV)
$E_{h,7}$	Actual redox potential at pH 7 (mV)
$E_{m,7}$	Standard redox potential, pH 7 (mV)
E^o	Standard redox potential
$E^{o'}$	Standard redox potential, pH specified, usually pH 7 (mV)
EP(S)R	Electron paramagnetic (spin) resonance
ER	Endoplasmic reticulum
ETF	Electron-transferring flavoprotein
F	Faraday constant ($=0.0965$ kJ mol^{-1} mV^{-1})
F_1, F_o	Matrix and membrane-located components, respectively, of the ATP synthase
FCCP	Carbonyl cyanide p-trifluoromethoxyphenylhydrazone (protonophore)
Fd	Ferredoxin
Fe–S	Iron–sulfur centre
Ferricyanide	Hexacyanoferrate (III)
Ferrocyanide	Hexacyanoferrate (II)
FTIR	Fourier transform infrared spectroscopy
FRET	Förster (or fluorescence) resonance energy transfer
G	Gibbs (free) energy (kJ)
GSH	Reduced glutathione
GSSG	Oxidised glutathione
H	Enthalpy
H$^+$/ATP	The number of protons translocated through the ATP synthase for the synthesis of 1 ATP
H$^+$/O	The number of protons translocated by the electron transport chain during the passage of two electrons to oxygen
H$^+$/2e$^-$	As H$^+$/O, but more general because the final electron acceptor need not be oxygen
h	Planck's constant
HD	Huntington's disease
$h\nu$	The energy in a photon (J)
IMM	Inner mitochondrial membrane
IMS	Intermembrane space
J_H^+	Proton current (nmol H$^+$ min^{-1} mg protein^{-1})
K	Absolute equilibrium constant

K'	Apparent equilibrium constant under defined conditions
LH1, LH2	Bacterial light-harvesting complexes 1 and 2
LHC II	A major thylakoid light-harvesting complex
MCA	Metabolic control analysis
MGD	Molybdopterin guanine dinucleotide
MPP^+	1-Methyl-4-phenyl-pyridinium ion
MPT	Mitochondrial permeability transition
MPTP	1-Methyl-4-phenyl-1,2,3,6-tetrahydropyridine
MQ	Menaquinone
MQH_2	Menaquinol
mtDNA	Mitochondrial DNA
mV	Millivolt
MV^+	Reduced methyl viologen
MV^{2+}	Oxidised methyl viologen
N-side, N-phase	Negative side of a membrane from which protons are pumped
Nbf-Cl	4-Chloro-7-nitrobenzofurazan
NMDA	N-methyl-D-aspartate
NMR	Nuclear magnetic resonance
Nuo	NADH–ubiquinone oxidoreductase
O	$\frac{1}{2}O_2$
OSCP	Oligomycin sensitivity conferring protein
OMM	Outer mitochondrial membrane
P/O ratio	As ADP/O ratio. The number of moles of ADP phosphorylated to ATP per $2e^-$ flowing through a defined segment of an electron transfer to oxygen
$P/2e^-$	As $ADP/2e^-$
P-side/ P-phase	Positive side of a membrane to which protons are pumped
P_{870}, etc.	The primary photochemically active component of a reaction centre
P_i	Phosphate anion
PC	Plastocyanin
PD	Parkinson's disease
Pheo	Pheophytin
PMF	Protonmotive force (mV)
PMS	Phenazinemethosulfate
PQ	Plastoquinone
PQH_2	Plastoquinol
PQQ	Pyrroloquinoline quinone
PSI, PSII	Photosystem I, II
PTP	Permeability transition pore
q^+/O or $q^+/2e^-$	Charge stoichiometry. The number of charges translocated across a membrane when $2e^-$ are transferred from a substrate to O_2 or another acceptor
Q_p, Q_n	Ubiquinone/ubiquinol binding sites localised toward the P- and N-sides, respectively, of the membrane in complex III
R	The gas constant ($8.3\,kJ\,mol^{-1}\,K^{-1}$)
RCR	Respiratory control ratio

RET	Reversed electron transfer
ROS	Reactive oxygen species
S	Entropy
SHAM	Salicylhydroxamic acid
SMP	Submitochondrial particle
TMPD	N,N,N',N'-tetramethyl-p-phenylenediamine
TPB^-	Tetraphenylborate anion
$TPMP^+$	Triphenylmethyl phosphonium
TPP^+	Tetraphenyl phosphonium cation
UCP	Uncoupling protein
UQ	Ubiquinone
UQH_2	Ubiquinol
$UQ^{\bullet-}$	Ubisemiquinone radical
UQ_{10}, etc.	Ubiquinone with a side chain of ten 5-carbon isoprenyl units
VDAC	Voltage-dependent anion channel
Å	Angstrom, 10^{-10} metres
Γ	Mass action ratio
Γ'	Observed mass action ratio
Δp	Protonmotive force (mV)
$\Delta\psi$	Membrane potential; that is, the electrical potential difference between two aqueous phases separated by a membrane (mV)
$\Delta\psi_m$	Mitochondrial membrane potential
$\Delta\psi_p$	Plasma membrane potential
ΔpH	pH gradient across the membrane
ΔE_h	Difference between two redox potentials (mV)
ΔG	Gibbs energy change (kJ mol^{-1})
ΔG^o	Standard Gibbs energy change
ΔG_p	Phosphorylation potential (ΔG for ATP synthesis)
ΔH	Enthalpy change
ΔS	Entropy change
$\Delta\tilde{\mu}_{H^+}$	Proton electrochemical gradient (kJ mol^{-1})
λ	Reorganisation energy (kJ mol^{-1})
ν	Frequency of radiation (s^{-1})
N	Avogadro's constant

INTRODUCTION TO PART 1

Because all biochemical reactions involve energy changes, the term *bioenergetics* could validly be applied to the whole of life sciences. However, in the field of biochemistry and biology, the term originally came to mean the study of the energy conversion processes that occur on, in, or across the inner mitochondrial membrane, the cytoplasmic membranes of bacteria and the photosynthetic thylakoid membranes that are found in the chloroplasts of plants; these comprise the so-called 'energy-conserving' membranes. Part 1 of this book focuses on fundamental bioenergetic principles. In Part 2, we review our current understanding of the molecular (and increasingly atomic) structure and mechanisms of the protein complexes catalysing these processes, whereas in Part 3 we cover the explosive growth of mammalian cellular bioenergetics, in which the knowledge gained through the study of bioenergetics has become centre stage in investigations of the physiology and pathology of the eukaryotic cell.

'Bioenergetics' originated in the quest to understand how oxidation reactions, in the form of the passage of electrons through coenzymes and proteins associated with these energy-conserving membranes, as well as photon capture in photosynthetic systems, could be coupled to the synthesis of ATP, the dominant common energy currency of the cell. The close similarity between the mechanisms and components involved in oxidative and photophosphorylation, as these processes are known respectively, allows them to be studied together. Attempts to relate the mechanism of oxidative phosphorylation to the ATP synthesising reactions in the glycolytic pathway were unsuccessful, and it turned out that this coupling was achieved via ion gradients across membranes in what became known as the chemiosmotic mechanism. It has further emerged over the years that these ion gradients, and hence the chemiosmotic mechanism, explain the coupling between a variety of other processes in cells, including how various species, from sugars and metabolites to proteins, are moved across these membranes (Figure I.1). The fundamentals of the chemiosmotic mechanism are frequently still confused today, 50

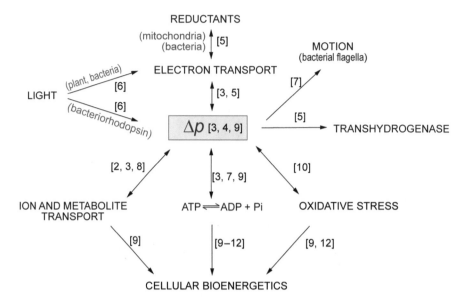

Figure I.1   Pathways of energy transduction.
The protonmotive force, Δp, interconnects multiple forms of energy. Numbers
in square brackets refer to chapters in which pathways are discussed.

years after the emergence of the theory. One of our aims here, in the fourth edition of
this book, is to continue to explain what this theory really implies as opposed to the
oversimplified and frequently misleading accounts that can be found in many textbooks.

The basic principles are wonderfully simple; indeed, the 'electrical circuit' analogy
we emphasise, with its 'voltage' and 'current' terms, continues to be useful at a research
level. We continue to emphasise the importance of understanding the basic principles
of thermodynamics applied to bioenergetic systems. We cover these in Chapter 3. It
must be remembered that although thermodynamics can never prove a mechanism, it
is ruthless in disproving energetically impossible ones. Particularly in the context of
cellular bioenergetics, failure to grasp these key concepts is still disturbingly common.
A word of caution, however: the treatment of thermodynamics in Chapter 3 is somewhat unconventional. Students preparing for examinations should check with their lec-
turers whether to adopt this simpler and more logical system, or to retain the classical
approach based on 'standard states' and 'standard free energies'.

1 CHEMIOSMOTIC ENERGY TRANSDUCTION

1.1 THE CHEMIOSMOTIC THEORY: FUNDAMENTALS

Although some ATP synthesis is catalysed by soluble enzyme systems—for example, phosphoglycerate kinase in the glycolytic pathway—the large majority is generated by membrane-bound enzyme complexes that are restricted to a particular class of membrane. These 'energy-transducing' membranes include the plasma membrane of simple prokaryotic cells such as respiratory or photosynthetic bacteria, the inner membrane of mitochondria, and the thylakoid membrane of chloroplasts (Figure 1.1). These membranes have a related evolutionary origin because chloroplasts and mitochondria are commonly thought to have evolved from a symbiotic relationship between a primitive, nonrespiring eukaryotic cell and an invading prokaryote. The membranes not only catalyse ATP synthesis but also control the transport of ions and metabolites and the oxidation state in their environment. The mechanism of ATP synthesis and ion transport associated with these diverse membranes is sufficiently related, despite the differing natures of their primary energy sources, to form the core of classical bioenergetics.

Energy-transducing membranes possess a number of distinguishing features. Each membrane has embedded within it two distinct types of proton pump. The nature of the primary proton pump depends on the energy source used by the membrane. In the case of mitochondria or respiring bacteria, an electron transfer chain (Chapter 5) catalyses the energetically 'downhill' transfer of electrons from substrates to final acceptors such as O_2 and uses this energy to generate an electrochemical gradient of protons. The term *electrochemical* is important and tells us that the energy gradient has electrical and chemical components (discussed later). Photosynthetic bacteria exploit the energy available from the absorption of quanta of visible light to generate a proton electrochemical gradient, whereas chloroplast thylakoids utilise two photon capture processes in series to generate the gradient and to drive electrons 'uphill' from water to acceptors such as $NADP^+$ (Chapter 6). The detailed topologies of the membranes differ, and to facilitate comparison it is a useful convention to define the side of the membrane to which protons are pumped as the P, or positive, side and the side from which they have originated as the N, or negative, side (Figure 1.1).

Bioenergetics. Doi: http://dx.doi.org/10.1016/B978-0-12-388425-1.00001-4

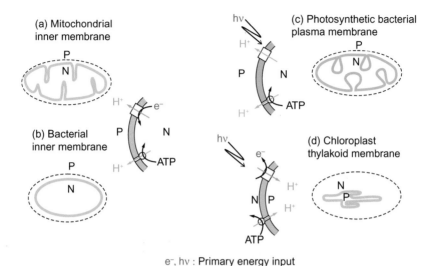

e⁻, hv : Primary energy input

Figure 1.1   Energy-transducing membranes contain pairs of proton pumps with the same orientation.

In each case, the primary pump utilising either electrons (e^-) from oxidations or driven by photons ($h\nu$) pumps protons from the N (negative) compartment to the P (positive) compartment. The photosynthetic bacterial photon-driven proton pump involves an electron transport cycle (not shown), whereas in addition to pumping protons, the chloroplast pumps electrons 'uphill.' Note that the ATP synthase in each case is shown acting in the direction of ATP hydrolysis, when it would also pump protons from the N- to the P-phase. The outer bounding membranes (dashed) do not participate in energy transduction.

In contrast to the variety of primary proton pumps, all energy-transducing membranes contain a highly conserved secondary proton pump termed the ATP synthase or the H⁺-translocating ATPase (Chapter 7). If this pump were operating in isolation in a membrane, it would hydrolyse ATP to ADP and P$_i$ (shorthand for the phosphate anion) and pump protons in the same direction as the primary pump (Figure 1.1). However, the essence of the chemiosmotic theory is that the primary proton pump generates a sufficiently large electrochemical gradient of protons to force protons back through the secondary pump so that it reverses and *synthesises* ATP from ADP and P$_i$ (Figure 1.2). Note that metabolism (i.e., electron flow or ADP phosphorylation) within both the primary pump and the secondary pump is tightly coupled to proton translocation. Thus, it is generally accepted that there is no internal 'slip' within these pumps that would allow one to occur without the other.

The proton electrochemical gradient is given the symbol $\Delta\tilde{\mu}_{H^+}$. An ion electrochemical gradient, expressed in kJ mol⁻¹, is a thermodynamic measure of the extent to which an ion gradient is removed from equilibrium (and hence capable of driving reactions) and will be discussed in Chapter 3. It is important to re-emphasise that $\Delta\tilde{\mu}_{H^+}$ has two components: one due to the concentration difference of protons across the membrane, ΔpH, and one due to the difference in electrical potential between the two aqueous phases separated by the membrane, the membrane potential, Δψ. A bioenergetic convention is to convert $\Delta\tilde{\mu}_{H^+}$ into units of electrical potential (i.e., millivolts) and to refer to this as the *protonmotive force*, or pmf, expressed by the symbol Δp.

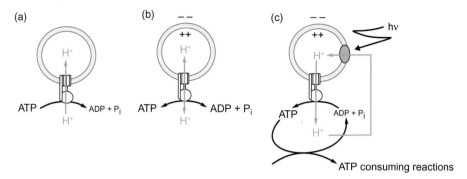

Figure 1.2 A hypothetical photosynthetic 'thylakoid' to demonstrate chemiosmotic coupling.
An ATP synthase complex is incorporated into a proton-impermeable phospholipid membrane such that the ATP binding site is on the outside. (a) ATP is added, the nucleotide starts to be hydrolysed to ADP + P$_i$, and protons are pumped into the vesicle lumen. As ATP is converted to ADP + P$_i$, the energy available from the hydrolysis steadily decreases, whereas the energy required to pump further protons against the electrochemical gradient that has already been established steadily increases. (b) Soon an equilibrium is attained. (c) If this equilibrium is now disturbed, for example, by removing ATP, the ATP synthase will reverse and attempt to re-establish the equilibrium by synthesising more ATP. Net synthesis, however, would be very small because the electrochemical gradient of protons would rapidly collapse and a new equilibrium would be established. For continuous ATP synthesis, a primary proton pump, driven in this example by photons (hv), is required to pump protons across the same membrane and replenish the gradient of protons. A proton circuit has now been established. This is what occurs across energy-conserving membranes: ATP is continuously removed for cytoplasmic ATP-consuming reactions, while the electrochemical gradient of protons, Δp, is continuously replenished by the respiratory or photosynthetic electron-transfer chains.

In only a few cases, such as the chloroplast, does Δp exist mainly as a pH difference across the energy-conserving membrane. In this example, the pH gradient, ΔpH, across the thylakoid membrane can exceed 3 units. Although the thylakoid space is therefore highly acidic, there are no enzymes in this compartment that might be compromised by the low pH. The more common situation is where $\Delta \psi$ is the dominant component and the pH gradient is small, perhaps only 0.5 pH units. This occurs, for example, in the mitochondrion, allowing enzymes in both the mitochondrial matrix and the cell cytoplasm to operate close to neutral pH. Figure 1.2 constructs a hypothetical ATP-synthesising photosynthetic organelle from first principles. A key feature is the proton circuit linking the primary pump with the ATP synthase. Under normal operating conditions, the $\Delta \tilde{\mu}_{H^+}$ generated by the primary (in this case, photosynthetic) proton pump is sufficient to drive the ATP hydrolysing proton pump in reverse—that is, as an ATP synthase.

A valuable concept that we use throughout this book is that the proton circuit (Figure 1.3) is closely analogous to an electrical circuit, and the analogy holds even when discussing detailed and complex energy flows (Chapter 4). Thus:

(a) Both circuits have generators of potential difference (the battery and the respiratory chain, respectively).

(b) Both potentials (voltage difference and Δp) can be expressed in millivolts.

(c) Both potentials can be used to perform useful work (running the light bulb and ATP synthesis, respectively).

(d) The current flowing in both circuits (amps or proton flux, J_{H^+}) is defined by Ohm's law (i.e., current = voltage/resistance, or current = voltage $\times$ conductance). Note, however, that bioenergetic proton conductances tend to increase with voltage—that is, they behave in a non-ohmic manner (see Figure 4.10).

(e) The rate of chemical conversion in the battery (or respiratory chain) is tightly linked to the current of electrons (or protons) flowing in the rest of the circuit, which in turn depends on the resistance of the circuit.

(f) Both circuits can be short-circuited (respectively by a piece of wire or a protonophore—an agent that makes membranes permeable to protons; see Chapter 2).

(g) The potentials fall as the currents drawn increase, due to 'internal resistance' in the battery or electron transport chain.

To avoid short circuits, it is evident that the membrane must be closed and possess a high resistance to protons. Protonophores, also called uncouplers, are synthetic compounds that break the energetic coupling between the primary pump and the ATP synthase. Uncouplers (Section 2.3.5) were described long before the chemiosmotic theory was propounded, and one of the most successful predictions of the theory was that they would act by increasing the proton conductance of the membrane and inducing just such a short circuit (Figure 1.3).

Mitochondrial and bacterial membranes not only have to maintain a proton circuit across their membranes but also must provide mechanisms for the uptake and excretion of ions and metabolites. It is energetically unfavourable for a negatively charged metabolite to enter the negative interior of a mitochondrion or bacterium (Chapter 3), and transport systems have evolved in which metabolites are transported together with protons or in principle by an equivalent exchange with OH^-. Alternatively, components of Δp can be exploited in other ways so as to drive transport in the desired direction (Chapter 9).

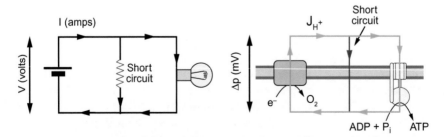

Figure 1.3   Proton circuits and electrical circuits are analogous.
A simple electrical circuit comprising battery and light bulb is analogous to a basic proton circuit. Voltage (Δp equivalent to V), current (J_{H^+} equivalent to I), and conductance $C_M H^+$ (equivalent to electrical conductance, i.e., reciprocal ohms) terms can be derived. Short circuits have similar effects, and more complex circuits with parallel batteries can be devised to mimic the multiple proton pumps in the mitochondrion (Chapter 4).

1.2 THE BASIC MORPHOLOGY OF ENERGY-TRANSDUCING MEMBRANES

1.2.1 Mitochondria and submitochondrial particles

The classical mitochondrial cross-section (Figure 1.4) is a considerable simplification of the actual structures viewed under the microscope (Section 10.2) but will serve to illustrate key features. The outer mitochondrial membrane possesses proteins, termed *porins*, that act as rather nonspecific pores for solutes of molecular weight less than 10 kDa. The outer membrane is therefore freely permeable to ions and most metabolites and plays no role in Δp generation. The mitochondrial porin, which exists in different isoforms, is also termed the *voltage-dependent anion channel* or VDAC (Section 10.2.2), although it should be emphasised that there is no potential gradient across the highly permeable outer membrane and the voltage dependency is only seen in synthetic reconstitution experiments.

The energy-transducing inner membrane has a distinct phospholipid composition, including cardiolipin but no cholesterol, reflecting its bacterial evolutionary origin. In mitochondrial preparations that have been negatively stained with phosphotungstate, it is possible to see 'knobs' on the matrix face (N-side, Figure 1.4) of the inner membrane. These are the catalytic components of the ATP synthase where adenine nucleotides and phosphate bind. The enzymes of the citric acid cycle are in the matrix, except for succinate dehydrogenase, which is bound to the N-face of the inner membrane. It must be borne in mind that the concentration of protein in the matrix can approach 500 mg/ml

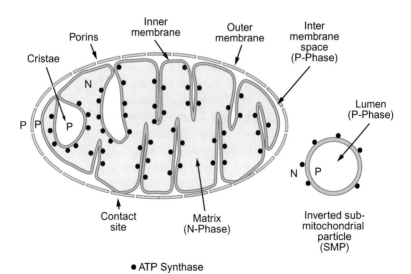

Figure 1.4   Schematic representation of a typical mitochondrion and submitochondrial particle.
P and N refer to the positive and negative compartments, respectively. Note that the shape of the cristae is highly variable and that communication between cristae and intermembrane space may be restricted. The actual range of morphologies is considerably more complex (Chapter 9).

and there may be a considerable structural organisation within this enormously concentrated solution that more closely resembles a glue than an ideal dilute medium. The matrix pools of NAD^+ and $NADP^+$ are separate from those in the cytosol, whereas ADP and ATP exchange across the inner membrane via the adenine nucleotide transporter (Chapter 8). In addition, specific carrier proteins exist for the transport of many metabolites.

Mitochondria are usually prepared by gentle homogenisation of the tissue in isotonic sucrose (for osmotic support and to minimise aggregation) followed by differential centrifugation to separate mitochondria from nuclei, cell debris, and microsomes (fragmented endoplasmic reticulum). Although this method is effective with fragile tissues such as liver, tougher tissues such as heart must either first be incubated with a protease, such as nagarse (a bacterial protease), or be exposed briefly to a blender to break the muscle fibres. Mitochondria can be prepared from yeast cells by digestion of the cell wall with snail-gut enzyme, followed by physical disruption of the resulting spheroplasts and differential centrifugation. A 'nitrogen bomb,' in which cell suspensions are saturated with the gas at high pressure so that subsequent sudden depressurisation explodes the plasma membranes, is also effective for a variety of cell types that are difficult to homogenise. Ultrasonic disintegration of mitochondria produces inverted submitochondrial particles (SMPs) (Figure 1.4). Because these have the substrate binding sites for both the respiratory chain and the ATP synthase on the outside, they have been much exploited for investigations into the mechanism of energy transduction.

Finally, increasingly sophisticated techniques are being developed to investigate mitochondrial function *in situ* within the cell, or even *in vivo*. These approaches are discussed in Part 3.

1.2.2 Respiratory bacteria and derived preparations

Energy transduction in bacteria is associated with the cytoplasmic (inner) membrane (Figure 1.5). In gram-negative bacteria (which are typically of similar size to mitochondria), this membrane is separated from a peptidoglycan layer and an outer membrane by the periplasm, which is approximately 17 nm wide. In gram-positive bacteria, the periplasm is absent and the cell wall is closely juxtaposed to the cytoplasmic membrane. Figure 1.5 is an oversimplification because in some organisms with a very high rate of respiration, there are substantial infoldings that increase the area of the cytoplasmic membrane. The archaea are an evolutionary distinct group of bacteria but which nevertheless catalyse energy transduction processes on their cytoplasmic membranes via a chemiosmotic mechanism; in general, these organisms have a single boundary membrane and thus no periplasm.

It is difficult to study energy transduction with intact bacteria because: (a) many reagents do not penetrate the outer membrane of gram-negative organisms; (b) ADP, ATP, NAD^+, and NADH do not cross the cytoplasmic membrane; (c) cells are frequently difficult to starve of endogenous substrates and thus there can be ambiguity as to the substrate that is donating electrons to a respiratory chain; and (d) the study of transport can be complicated by subsequent metabolism of the substrate.

Cell-free vesicular systems can overcome these problems. For most transport studies, right-side-out vesicles are required (Figure 1.5). These can often be obtained by

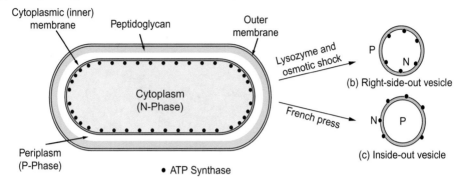

Figure 1.5   Schematic gram-negative bacterium and vesicle preparations.
P and N refer to positive and negative compartments, respectively. The periplasm is part of the P-phase, which also includes the bulk external medium because the outer membrane is freely permeable to ions. Note that gram-positive bacteria differ by lacking an outer membrane and a periplasm. Nevertheless, similar vesicle preparations can be made from these organisms as is also the case for the archaea.

weakening the cell wall, for example, with lysozyme, and then exposing the resulting spheroplasts or protoplasts to osmotic shock. Vesicles with this orientation can only oxidise substrates that have an external binding site or can permeate the cytoplasmic membrane. They cannot hydrolyse or synthesise ATP. In contrast, inside-out vesicles can frequently be prepared by extruding cells at very high pressure through an orifice in a French press (Figure 1.5). These vesicles can oxidise NADH and phosphorylate added ADP. The method of vesicle preparation varies between genera; occasionally, osmotic shock may give inside-out vesicles or a mixture of the two orientations. This last feature need not be a major problem because, for example, in a study of ATP synthesis, the reaction would be confined to the inside-out population. Nevertheless, failure to characterise the orientation of vesicles has caused confusion in the past.

Vesicle preparations have some disadvantages, such as the loss of periplasmic electron-transport or solute-binding proteins; the latter play key roles in many aspects of bacterial energy transduction (Chapters 5 and 8). Also, the membrane of a vesicle may be somewhat leaky to ions with the result that the stoichiometry of an energy transduction reaction may be adversely affected.

1.2.3 Chloroplasts and their thylakoids

Chloroplasts are plastids, organelles peculiar to plants (Figure 1.6); there may be from 1 to 100 or more chloroplasts per cell. Chloroplasts are considerably larger than the average mitochondrion, being 4–10 µm in diameter, 1–2 µm thick, and bounded by an envelope of two closely juxtaposed membranes, the matrix within the inner membrane being the stroma (Figure 1.6). Within stroma are flattened vesicles called thylakoids, the membranes of which have regions that are folded so that the contiguous membrane has

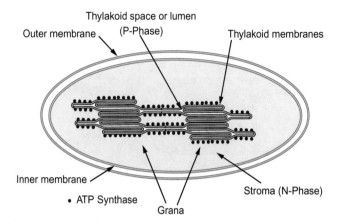

Figure 1.6   A chloroplast and its thylakoids.
Note it is probable that there is a single continuous lumen (the internal
thylakoid space). The thylakoid membrane is heterogeneous; for example, the
ATP synthase is excluded from the grana (appressed regions) where the
membrane is closely stacked (Chapter 6). Light-driven proton pumping occurs
from the N- to the P-phase (note, however, that in steady-state light the
membrane potential across a thylakoid membrane is negligible and the pH
gradient dominates; see Chapter 6).

a stacked appearance, referred to as the grana (Figure 1.6). Energy conservation occurs
across the thylakoid membranes, and light causes the translocation of protons into the
internal thylakoid space (usually called the lumen). The chloroplast ATP synthase is part
of the thylakoid membrane and is orientated with its 'knobs' on the stromal face of the
membrane. Thus, the lumen space inside the thylakoid is the P-compartment and the
stroma the N-compartment. The ATP and NADPH generated by photosynthetic phosphorylation is used by the CO_2-fixing dark reactions of the Calvin cycle located in the
stroma.

Although at first sight the structure of chloroplasts appears to be very different from
that of mitochondria, the only topological distinction is that the thylakoids, in contrast
to the mitochondria cristae, can be thought of as having become separated from the
inner membrane, with the result that the thylakoid lumen is a separate compartment,
unlike the 'cristal space,' which is continuous with the intermembrane space of mitochondria. Note, however, that even in mitochondria, there are suggestions that communication between cristal and intermembrane spaces may be restricted (Section 10.2).

Chloroplasts are prepared by gentle homogenisation in isotonic sucrose or sorbitol
of leaves (e.g., from peas, spinach, or lettuce). After removal of cell debris, the chloroplasts are sedimented by low-speed centrifugation. A rapid and careful preparation will
contain a high proportion of intact chloroplasts capable of high rates of CO_2 fixation.
Slightly harsher conditions yield 'broken chloroplasts,' which have lost the envelope
membranes and hence the stroma contents. These broken chloroplasts (thylakoid membrane preparations) do not fix CO_2 but are capable of high rates of reduction of $NADP^+$
and of photophosphorylation. They are often the choice material for bioenergetic investigations because the intact chloroplast envelope prevents access of substances such as
ADP or $NADP^+$.

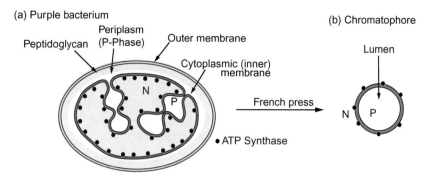

(a) Purple bacterium

Peptidoglycan   Periplasm (P-Phase)   Outer membrane

Cytoplasmic (inner) membrane

French press

• ATP Synthase

(b) Chromatophore

Lumen

Figure 1.7   Photosynthetic bacteria and chromatophores.
The cytoplasmic membrane of photosynthetically grown organisms such as *Rhodobacter sphaeroides* is highly invaginated.

1.2.4 Photosynthetic bacteria and chromatophores

Several groups of prokaryotes catalyse photosynthetic electron transfer: among these are purple bacteria, green bacteria and the cyanobacteria or blue-green algae. The purple bacteria include the (non-sulfur) Rhodospirillaceae, and the green bacteria include the Chlorobiaceae. Cyanobacteria carry out non-cyclic electron transfer (Chapter 6), use H_2O as an electron donor, and are in this respect similar to chloroplasts. Of the remaining groups, the purple bacteria, and especially the Rhodospirillaceae, have been the more intensively investigated, and several factors make them suitable for bioenergetic studies. Thus, mechanical disruption of the cells (e.g., in a French press) enables the characteristic invaginations of the cytoplasmic membrane to bud off and form isolated closed vesicles called chromatophores (Figure 1.7). Chromatophores retain the capacity for photosynthetic energy transduction and possess the same orientation as the inside-out vesicles discussed previously. Light-driven ATP synthesis can be studied, and they have been important for chemiosmotic studies, especially because they are so small (diameters in the order of 50 nm) that light scattering is negligible and suspensions are optically clear.

A further advantage of the purple bacteria is that the reaction centres (the primary photochemical complexes, Chapter 6) can be readily isolated. Finally, these organisms will grow in the dark, for example, by aerobic respiration, permitting the study of mutants defective in the photosynthetic apparatus. In addition to these bacteria, a class of archaea called halobacteria carry out a unique light-dependent energy transduction in which a single protein, bacteriorhodopsin, acts as a light-driven proton pump (Chapter 6).

1.3 A BRIEF HISTORY OF CHEMIOSMOTIC CONCEPTS

Further reading: Mitchell (1961, 2011)

As this edition is being prepared (2012), it is exactly 50 years since Peter Mitchell's first paper proposing the chemiosmotic hypothesis. To celebrate this anniversary,

Biochimica et Biophysica Acta reprinted Mitchell's privately published description of his original chemiosmotic hypothesis, "Chemiosmotic coupling in oxidative and photosynthetic phosphorylation." A brief overview of the bioenergetic climate at the time may give some background to the hypothesis. By the early 1960s, the main energetic pathways within the mitochondrion had been established, with the exception of the central 'energy-transducing' intermediate. In contrast, virtually nothing was known about the structure and mechanism of the electron transport chain and the ATP synthase. Any hypothesis had to be consistent with six basic observations:

(a) The mammalian electron transport chain is an essentially linear sequence of electron carriers with three separate regions where redox energy can be conserved in the synthesis of ATP from ADP and inorganic phosphate (P_i).

(b) Under most conditions, the rate of electron transport is controlled by the demand for ATP (respiratory control).

(c) Coupling between respiration and ATP synthesis can be disrupted by a group of agents termed *uncouplers* that abolish respiratory control and allow the mitochondrial to catalyse a rapid ATP hydrolysis.

(d) The antibiotic oligomycin inhibits both the synthesis and uncoupler-stimulated hydrolysis of ATP.

(e) The energy from electron transport can also be coupled to the energetically 'uphill' accumulation of Ca^{2+} and to the reduction of NAD^+ and $NADP^+$.

(f) The processes listed in 'e' can also be driven by the hydrolysis of ATP by anaerobic mitochondria when they can be inhibited by both oligomycin and uncouplers.

These observations (and related ones for bacteria and photosynthetic thylakoids) were consistent with pathways of energy transduction radiating from a common 'energy pool' (see Figure I.1). The only reactions in which a detailed mechanism of ATP synthesis was available were the 'substrate-level' phosphorylations in glycolysis and the citric acid cycle. It was therefore reasonable to expect that the common energy-transducing intermediate would be a chemical 'high-energy' intermediate (usually given the shorthand 'squiggle' ~). However, no squiggle was ever found.

As is so often the case in science, the solution was simple and elegant. In 1961, Mitchell, whose background was in transport rather than bioenergetics, proposed that the only intermediate was a proton electrochemical gradient across the membrane, generated by electron transport, and utilised to drive an ATP *hydrolysing* proton pump in reverse as an ATP synthase. Mitchell proposed four essential tests for the hypothesis:

(a) Respiratory and photosynthetic electron transport chains should pump protons.

(b) The ATP synthase should function as a reversible proton-translocating ATPase.

(c) Energy-transducing membranes should have a low effective proton conductance.

(d) Energy-transducing membranes should possess specific exchange carriers to permit (largely anionic) metabolites to permeate, and osmotic stability to be maintained, in the presence of a high negative inside membrane potential.

An account of Peter Mitchell and the lively debate he initiated was written by John Prebble (Prebble 2002), and an obituary by Tony Crofts is also informative (Crofts 1993).

2

ION TRANSPORT ACROSS ENERGY-CONSERVING MEMBRANES

2.1 INTRODUCTION

For an ion to be transported across a membrane, both a driving force and a pathway are required. Driving forces can be metabolic energy (e.g., ATP hydrolysis), concentration gradients, electrical potentials, or combinations of these. These forces are discussed in Chapter 3; this chapter deals with the natural and induced *pathways* across energy-conserving membranes.

2.2 THE CLASSIFICATION OF ION TRANSPORT

To reduce the complexity of membrane transport events, it is useful to classify any transport process in terms of the following four criteria (Figure 2.1):

(1) Does transport occur through the lipid bilayer or is it protein-mediated?
(2) Is transport passive or directly coupled to metabolism?
(3) Does the transport process involve a single ion or metabolite, or are fluxes of two or more species directly coupled?
(4) Does the transport process involve net charge transfer across the membrane?

2.2.1 Bilayer-mediated versus protein-catalysed transport

A consequence of the fluid-mosaic model of membrane structure is that transport can either occur through lipid bilayer regions of the membrane or be catalysed by integral, membrane-spanning proteins. The distinction between protein-catalysed transport and transport across the bilayer regions of the membrane is fundamental and is emphasised in this chapter.

Bioenergetics. Doi: http://dx.doi.org/10.1016/B978-0-12-388425-1.00002-6

(a) Does transport occur across the bilayer or is it protein-mediated?

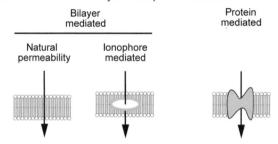

(b) Is transport passive or directly coupled to metabolism?

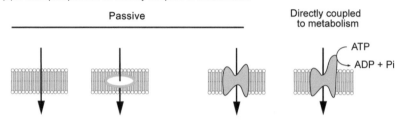

(c) Does a transport process involve a single ion or metabolite, or are fluxes of two or more species directly coupled?

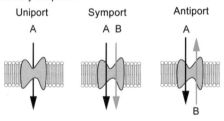

(d) Does a transport process involve net charge transfer across the membrane?

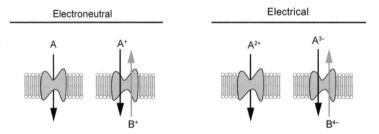

Figure 2.1   The classification of ion and metabolite transport.
(a) Transport may be bilayer-mediated (via either natural permeation across the membrane or an ionophore-induced pathway) or protein-mediated. (b) Transport by any of the three pathways in panel (a) can be passive (not directly coupled to metabolism) or, in the case of protein-catalysed transport alone, directly coupled to metabolism—for example, ATP hydrolysis. (c) Transport, by any of the pathways in panel (b), may occur as a single species, as two or more ions whose transport is tightly coupled together by symport (or co-transport), or by antiport (or exchange diffusion). Note that an antiport with OH^- cannot readily be distinguished from a symport with H^+. (d) Any of the mechanisms in panel (c) may be electroneutral or electrical (electrogenic or electrophoretic).

Although the fluid-mosaic model is sometimes visualised with protein 'icebergs' floating in a sea of lipid, the high proportion of protein in energy-conserving membranes (in the case of the mitochondrial inner membrane, 50% of the membrane is integral protein (although parts of these proteins project from the membrane), 25% peripheral protein, and 25% lipid) results in a relatively close packing of the proteins. Unlike plasma membrane proteins, there is no attachment to cytoskeletal elements.

Consistent with the proposal that mitochondria and chloroplasts evolved from respiring or photosynthetic bacteria, energy-conserving membranes tend to have distinctive lipid compositions: 10% of the mammalian mitochondrial inner membrane lipid is cardiolipin, whereas only 16% of the chloroplast thylakoid membrane lipid is phospholipid, with the remainder being galactolipids (40%), sulfolipids (4%), and photosynthetic pigments (40%). Despite this heterogeneity of lipid composition, the native and ionophore-induced permeability properties of the bilayer regions of the different membranes are sufficiently similar to justify extrapolations between energy-transducing membranes, other cell membranes, and artificial bilayer preparations. In contrast, protein-catalysed transport of a particular species can be unique not only to a given organelle but also to an individual cell type within a tissue, depending on the genes expressed in that cell. For example, the inner membrane of rat liver mitochondria possesses protein-catalysed transport properties that are absent from rat heart mitochondria.

2.2.2 Transport directly coupled to metabolism versus passive transport

A tight coupling of transport to metabolism occurs in the ion pumps, which are central to chemiosmotic energy transduction. An ion can be accumulated without direct metabolic coupling if there is an appropriate potential difference across the membrane or if its transport is coupled to the 'downhill' movement of a second ion. For example, while Ca^{2+} is accumulated into the sarcoplasmic reticulum by an ion pump (a Ca^{2+}-ATPase), the same ion is accumulated across the mitochondrial inner membrane by a uniport mechanism (Figure 2.1) driven by the membrane potential (Section 9.4.1). Only the former is strictly 'active,' because in the case of the mitochondria, Ca^{2+} accumulation occurs down the electrochemical gradient for the ion. In some texts, the terms *primary* and *secondary active transport* are used to distinguish these examples. However, confusingly, the term *active transport* is often used for any process in which a concentration gradient of a solute or ion is established across a membrane. Naturally, only protein-catalysed transport can be directly coupled to metabolism (Figure 2.1b).

2.2.3 Uniport, symport and antiport

The molecular mechanism of a transport process can involve a single ion or the tightly coupled transport of two or more species (Figure 2.1c). A transport process involving a single ion is termed a *uniport*. Examples of uniports include the uptake pathway for Ca^{2+} across the inner mitochondrial membrane (Section 9.4.1.1) and the K^+ permeability induced in bilayers by the addition of the ionophore valinomycin. A transport process involving the obligatory coupling of two or more ions in parallel is termed *symport* or *co-transport*. In this book, we use the shorthand A:B to denote symport of the species

A and B. Examples of proton symport are found at the bacterial membrane where the mechanism is used to drive the uptake of metabolites into the cell.

The equivalent tightly coupled process in which the transport of one ion is linked to the transport of another species in the opposite direction is termed *antiport* or *exchange diffusion*, represented here in the form A/B for the antiport of A against B (Figure 2.1c). Examples include the Na^+/H^+ antiport activity, which is present in the inner mitochondrial membrane (Section 9.3), and the K^+/H^+ antiport catalysed by the ionophore nigericin in bilayers. Note, however, that if one of the ions involved in a nominal symport or antiport mechanism is a proton or hydroxide ion, it is usually impossible to distinguish between the symport of a species with an H^+ and the antiport of the species with an OH^-. For example, the mitochondrial phosphate carrier (Section 9.5.1) has been represented as a P_i^-/OH^- antiport or a $H^+:P_i^-$ symport.

Closely related transport pathways exist across non-energy-conserving membranes. At the plasma membrane, the Na^+ ion can be involved in uniport (through a voltage-activated channel), symport (e.g., Na^+:glucose co-transport), and antiport (e.g., the $3Na^+/Ca^{2+}$ exchanger). More complex stoichiometries may occur; for example, some neuronal membranes possess a carrier that catalyses the co-transport of Na^+ and glutamate coupled to the antiport of a third ion, K^+.

2.2.4 Electroneutral versus electrical transport

Electroneutral transport involves no net charge transfer across the membrane. Transport may be electroneutral either because an uncharged species is transported by a uniport or as a result of the symport of a cation and an anion or the antiport of two ions of equal charge (Figure 2.1d), an example of the last being the K^+/H^+ antiport catalysed by nigericin. Electrical transport is frequently termed either *electrogenic* ('creating a potential'; e.g., proton pumping driven by ATP hydrolysis) or *electrophoretic* ('moving in response to a pre-existing potential'; e.g., Ca^{2+} uniport into mitochondria). Because these terms can refer to the same pathway observed under different conditions, the overall term *electrical* is used here.

It is important to distinguish between movement of charge at the molecular level, as discussed here, and the overall electroneutrality of the total ion movements across a given membrane. The latter follows from the impossibility of separating more than minute quantities of positive and negative charge across a membrane without building up a large membrane potential (Figure 2.3). Thus, the separation of 1 nmol of charge across the inner membranes of 1 mg of mitochondria results in the build-up of more than 200 mV of potential. In other words, a single turnover of all the electron transport components in an individual mitochondrion or bacterium will translocate sufficient charge to establish a membrane potential approaching 200 mV. The establishment of such potentials by the movement of so little charge is a consequence of the low electrical capacitance of biological membranes (typically estimated as $1\,\mu F\,cm^{-2}$). However, this property does not preclude the occurrence of steady-state charge translocations at the molecular level as long as these compensate each other. Indeed, this is the basis of the proton circuit that is the central theme of this book.

Finally, it is necessary to appreciate that the effect on an energy-transducing membrane of a tightly coupled electroneutral antiporter, such as the ionophore nigericin, which catalyses an electroneutral K^+/H^+ antiport (and is therefore unaffected by the membrane potential), is not the same as that caused by the addition of two electrical uniporters for the same ions (e.g., valinomycin plus a protonophore) when transport of both K^+ and H^+ is affected by the membrane potential. Thus, in a KCl-based medium, mitochondria tolerate low concentrations of nigericin but swell dramatically in the presence of valinomycin and depolarise in the presence of protonophores.

The four criteria discussed previously allow a comprehensive description of a transport process; for example, proton pumping by the ATP synthase is an example of a protein-catalysed, metabolism-coupled electrical uniport.

2.3 BILAYER-MEDIATED TRANSPORT

2.3.1 The natural permeability properties of bilayers

The hydrophobic core possessed by lipid bilayers creates an effective barrier to the passage of charged species. With a few important exceptions, cations and anions do not permeate bilayers. This impermeability extends to the proton, and this property is vital for energy transduction as it avoids short-circuiting the proton circuit. Not only does the bilayer have a high electrical resistance, but it can also withstand very high electrical fields. An energy-conserving membrane with a membrane potential of 200 mV across it has an electrical field in excess of $300,000 \, V \, cm^{-1}$ across its hydrophobic core.

A variety of uncharged species can cross bilayers. O_2 and CO_2 are highly permeable, as are the uncharged forms of a number of low-molecular-weight acids and bases, such as ammonia and acetic (ethanoic) acid. These last permeabilities provide a useful tool for the investigation of pH gradients across membranes (Section 3.4). The mystery of how the most polar of compounds, water, crosses membranes has been resolved by the discovery of aquaporins, a large family of water-permeating channels present in membranes and catalysing the direct transport of water (Zeuthen, 2001; Lee and Thevenod, 2006).

2.3.2 Ionophore-induced permeability properties of bilayer regions

The high activation energy required to insert an ion into a hydrophobic region accounts for the extremely low permeability of bilayer regions towards ions. It follows that if the charge can be delocalised and shielded from the bilayer, the ion permeability might be expected to increase. This is accomplished by a variety of antibiotics synthesised by some microorganisms, as well as by some synthetic compounds. These are known collectively as ionophores. These are typically compounds with a molecular weight of 500–2000 possessing a hydrophobic exterior, so as to make them lipid soluble, together with a hydrophilic interior to bind the ion. Ionophores are not natural constituents of energy-conserving membranes, but as investigative tools they are invaluable.

Ionophores can function as mobile carriers or as channel formers (Figure 2.2). Mobile carriers diffuse within the membrane and can typically catalyse the transport of approximately 1000 ions s^{-1} across the membrane. They can show an extremely high discrimination between different ions, can work across thick synthetic membranes, and

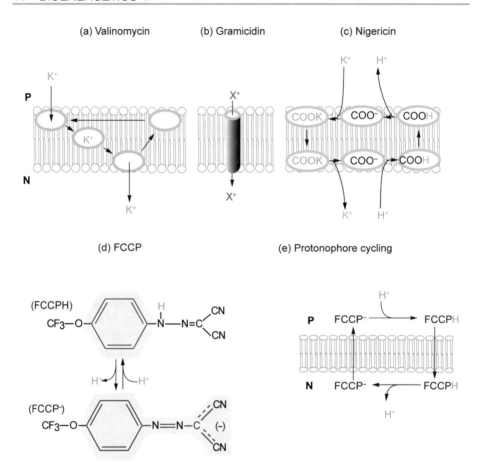

Figure 2.2   Ionophores.
Schematic function of four ionophores. (a) Valinomycin is a mobile carrier
ionophore able to cross the lipid bilayer transporting a K^+ ion. Note that the
ion's hydration sphere is lost and replaced by the ionophore. (b) Gramicidin is
a channel-forming ionophore, with less selectivity, but much higher activity,
than valinomycin. It will transport a variety of monovalent cations ('X^+'). (c)
Nigericin is a hydrophobic weak carboxylic acid permeable across lipid
bilayer regions as either the protonated acid or the neutral salt. Nigericin has a
selectivity $K^+>Rb^+>Na^+$. (d) FCCP is the most commonly employed
example of a protonophore, although many such compounds exist. The red
bonds represent the extent of the π-orbital system. If a Δp exists across the
membrane, the protonophore will cycle catalytically in an attempt to collapse
the potential. (e) $FCCP^-$ will be driven to the P-face of the membrane by the
membrane potential, whereas FCCPH will be driven towards the alkaline or
N-phase due to ΔpH. When sufficient FCCP is present (for most membranes,
10^{-9} to 10^{-5} M), the cycling can reduce both $\Delta \psi$ and ΔpH to near zero.

are affected by the fluidity of the membrane. In contrast, channel-forming ionophores
discriminate poorly between ions but can be very active, transporting up to 10^7 ions per
channel S^{-1}. Ionophores can also be categorised according to the ion transport that they
catalyse.

2.3.3 Carriers of charge but not protons

Valinomycin (Figure 2.2) is a mobile carrier ionophore that catalyses the electrical uniport of Cs^+, Rb^+, K^+, or NH_4^+. The ability to transport Na^+ is at least 10^4 less than for K^+. Valinomycin is a natural antibiotic from *Streptomyces* and is a depsipeptide—that is, it consists of alternating hydroxy and amino acids. The ions lose their water of hydration when they bind to the ionophore. Na^+ cannot be transported because the unhydrated Na^+ ion is too small to interact effectively with the inward-facing carbonyls of valinomycin, with the result that the complexation energy does not balance that required for the loss of the water of hydration. Because valinomycin is uncharged and contains no ionisable groups, it acquires the charge of the complexed ion. Both the uncomplexed and the complexed forms of valinomycin are able to diffuse across the membrane. Therefore, a catalytic amount of ionophore can induce the bulk transport of cations. It is effective in concentrations as low as 10^{-9} M in mitochondria, chloroplasts, synthetic bilayers, and, to a more limited extent, bacteria (the outer membrane can exclude it from gram-negative organisms). Other ionophores catalysing K^+ uniport include the enniatins and the nactins (nonactin, monactin, dinactin, etc., so-called from the number of ethyl groups in the structure). However, these ionophores do not have such a spectacular selectivity for K^+ over Na^+ as valinomycin. Energy-conserving membranes generally lack a native electrical K^+ permeability, and valinomycin can be exploited to induce such a permeability, in order to estimate or clamp membrane potentials, or to investigate anion transport.

Gramicidin is an ionophore that forms transient conducting dimers in the bilayer (Figure 2.2). Its properties are typical of channel-forming ionophores, with a poor selectivity between protons, monovalent cations, and NH_4^+, the ions permeating in their hydrated forms. The capacity to conduct ions is limited only by diffusion, with the result that one channel can conduct up to 10^7 ions s^{-1}.

2.3.4 Carriers of protons but not charge

Nigericin is a linear molecule with heterocyclic oxygen-containing rings together with a hydroxyl group. In the membrane, the molecule cyclises to form a structure similar to that of valinomycin, with the oxygen atoms forming a hydrophobic interior. Unlike valinomycin, nigericin loses a proton when it binds a cation, forming a neutral complex that can then diffuse across the membrane as a mobile carrier. Nigericin is also mobile in its protonated noncomplexed form, with the result that the ionophore can catalyse the overall electroneutral exchange of K^+ for H^+ (Figure 2.2). Other ionophores that catalyse a similar electroneutral exchange include X-537A, monensin and dianemycin. The latter two show a slight preference for Na^+ over K^+, whereas X-537A will complex virtually every cation, including organic amines. Nigericin has been employed to study anion transport (Section 2.5) and to modify the pH gradient across energy-conserving membranes. It is often stated that nigericin abolishes ΔpH across a membrane; in fact, the ionophore equalises the K^+ and H^+ concentration gradients, with the final ion gradients depending on the experimental conditions.

A23187 and *ionomycin* are dicarboxylic ionophores with a high specificity for divalent cations. A23187 catalyses the electroneutral exchange of Ca^{2+} or Mg^{2+} for two H^+ without disturbing monovalent ion gradients. Ionomycin has a higher selectivity for Ca^{2+} and has the additional advantage that it is nonfluorescent, allowing its use in experiments using fluorescent indicators. The ionophores are extensively used, and unfortunately often misused (Section 9.6.1), in cellular studies.

2.3.5 Carriers of protons and charge

Protonophores, also known as proton translocators or uncouplers, are synthetic chemicals that have dissociable protons and permeate bilayers both as protonated acids and as the conjugate base (Figure 2.2e). This is possible because these ionophores possess extensive π-orbital systems that so de-localise the charge of the *anionic* form that lipid solubility is retained. By shuttling across the membrane, they can catalyse the net electrical uniport of protons and increase the proton conductance of the membrane. In so doing, the proton circuit is short-circuited, allowing the process of Δp generation to be uncoupled from ATP synthesis. Uncouplers were described long before the formulation of the chemiosmotic theory. In fact, the demonstration that the majority of these compounds act by increasing the proton conductance of synthetic bilayers was important evidence in favour of the theory.

An indirect proton translocation can be induced in membranes by the combination of a uniport for an ion together with an electroneutral antiport of the same ion in exchange for a proton. For example, the combination of valinomycin and nigericin induces a net uniport for H^+, whereas K^+ cycles across the membrane. The $Ca^{2+}/2H^+$ ionophores discussed previously can also uncouple mitochondria in the presence of Ca^{2+} because a dissipative cycling is set up between the native Ca^{2+} uniporter and the ionophore.

2.3.6 The use of ionophores in intact cells

Although ionophores were introduced largely for investigations of isolated mitochondria, they have also been applied to intact eukaryotic cells in attempts to modify *in situ* mitochondrial function. However, because they display no membrane selectivity, one must be aware of the consequences of introducing these ion permeabilities into other membranes. Thus, valinomycin will hyperpolarise the plasma membrane of cells in low K^+ media by clamping the plasma membrane potential close to the K^+-diffusion potential (Section 3.8). However, at the same time the ionophore will clamp the mitochondrial $\Delta\psi$ close to zero because the K^+ concentrations in the cytoplasm and matrix are both approximately 100 mM. Nigericin can be added to cells at low concentrations with no deleterious effect except to slightly hyperpolarise the mitochondria. In contrast, protonophores have multiple effects on cells, acidifying the cytoplasm and depleting the cytoplasm of ATP by allowing the ATP synthase to reverse and hydrolyse glycolytically generated ATP. The $Ca^{2+}/2H^+$ ionophores A23187 and ionomycin have been used in more than 20,000 studies. However, it is seldom recognised that they will intercalate into the inner membrane of the *in situ* mitochondria, setting up a dissipative cycling with the native Ca^{2+} uniporter and effectively creating a protonophoric uncoupling that can readily de-energise and kill the cell. In Chapter 9, we emphasise the importance of monitoring mitochondrial function in intact cells exposed to any agent that could conceivably affect bioenergetic function. Finally, it is important to appreciate that ionophores such as valinomycin may fail to act on intact bacteria due to their absorption to cell walls.

2.3.7 **Lipophilic cations and anions**

The ability of π-orbital systems to shield charge and enhance lipid solubility has been exploited in the synthesis of a number of cations and anions that are capable of being transported across bilayer membranes even though they carry charge. Examples include the tetraphenyl phosphonium cation (TPP$^+$) and the tetraphenylborate anion TPB$^-$ (see Figure 4.10). These ions are not strictly ionophores because they do not act catalytically but are instead accumulated in response to $\Delta\psi$. Lipophilic cations and anions were of value historically in demonstrations of their energy-dependent accumulation in mitochondria and inverted submitochondrial particles, respectively. These experiments eliminated the possibility of specific cation pumps driven by chemical intermediates. Subsequently, the cations have been employed for the estimation of $\Delta\psi$ (Section 4.4.3). Fluorescent lipophilic cations are employed to monitor changes in $\Delta\psi$ both in isolated mitochondria (Section 4.4.6) and in mitochondria *in situ* within intact cells (Section 9.7.1).

2.4 **PROTEIN-CATALYSED TRANSPORT**

The characteristics of protein-catalysed transport across energy-conserving membranes are usually sufficiently distinct from those of bilayer-dependent transport, whether in the absence or the presence of ionophores, to make the correct assignment straightforward. The mechanisms of transport proteins of the mitochondrial inner membrane and bacterial cytoplasmic membrane are discussed in detail in Chapter 8; here, we merely summarise the distinctions between protein-catalysed and bilayer-mediated transport.

Transport proteins share the features of other enzymes; they can display stereospecificity, can frequently be inhibited specifically, and are genetically determined. This last feature means that it is not possible to make the same kinds of generalisations as for bilayer transport. For example, if FCCP induces proton permeability in mitochondria, it can generally be assumed that the effect will be similar in chloroplasts, bacteria and synthetic bilayers. In contrast, a transport protein may not only be specific to a given organelle but also may be restricted to the organelle from one tissue. Thus, the citrate carrier is present in liver mitochondria (Section 9.5.3), where it is involved in the export of intermediates for fatty acid synthesis, but it is absent from heart mitochondria. The well-known lactose:H$^+$ symporter (lac permease) is restricted to *Escherichia coli* and related genera. The strongest evidence for the involvement of a protein in a transport process is often the existence of specific inhibitors. For example, whereas pyruvate was for many years considered to permeate into mitochondria through the bilayer, which is feasible because it is a monocarboxylic weak acid, it was later found that cyanohydroxycinnamate was a specific transport inhibitor. This provided the first firm evidence for a transport protein for this substrate, although the molecular identification of the pyruvate carrier has only recently been accomplished (Section 9.5.3).

Transport proteins have been studied using many approaches, and this has led to a plethora of names, including carriers, permeases, porters and translocases, all of which are synonyms for transport protein. Although the term *carrier* is most commonly used, it is inappropriate because there is no evidence that any protein functions by the carrier type of mechanism exemplified by valinomycin. Instead mitochondrial and bacterial transporters undergo subtle conformational changes to expose their binding site alternately to N- and P-phases (Chapter 8).

2.5 SWELLING AND THE COORDINATE MOVEMENT OF IONS ACROSS MEMBRANES

The driving forces for the movement of ions across membranes will be derived quantitatively in the next chapter. Here, we discuss qualitatively how the movement of ions on different carriers within the same membrane may be coupled to each other.

The overriding principle of bulk ion movement across a closed membrane is that there must at no time be more than a slight charge imbalance across the membrane. We have seen that the electrical capacity of a mitochondrion or bacterium is tiny, and thus that the uncompensated movement of less than 1 nmol of a charged ion per milligram protein is sufficient to build up a $\Delta\psi$ of greater than 200 mV. Thus, during the operation of a proton circuit, the charge imbalance would never exceed this amount, even though the proton current might exceed 1000 nmol H^+ min^{-1} (mg protein)$^{-1}$.

In order to illustrate the coordinate movement of ions, we first discuss a simple but powerful technique that was much used to establish the pathways and mechanisms of ion transport across the mitochondrial inner membrane: osmotic swelling. Swelling (i.e., the net inward transport of water and solutes) is possible because mitochondria, in common with many membranes, possess aquaporins that create selective channels for water, and perhaps CO_2, across the inner membrane (Section 2.3.1).

Mitochondria will swell and ultimately burst unless they are suspended in a medium that is isotonic with the matrix and impermeant across the inner membrane. Swelling does not mean that the inner membrane stretches like a balloon but, rather, that the inner membrane unfolds as the matrix volume increases, ultimately rupturing the outer membrane. Mitochondria will also swell in an isotonic solution in which the principal solute is permeable across the inner membrane. To observe osmotic swelling of mitochondria in ionic media, both the cation and the anion of the major osmotic component of the medium must be permeable, and the requirement for overall charge balance across the membrane must be respected (Figure 2.3).

Suspensions of mitochondria are turbid and scatter light due to the difference in refractive index between matrix contents and the medium, and any process that decreases this difference will decrease the scattered light. Thus, paradoxically, an increase in matrix volume due to the influx of a permeable solute results in a decrease in the light scattered as the matrix refractive index approaches that of the medium. This provides a very simple semiquantitative method for the study of solute fluxes across the mitochondrial inner membrane by monitoring the decrease in light scattered in the 90° geometry of a fluorimeter, or the increase in transmitted light in a normal spectrophotometer, during the swelling process. Swelling can proceed sufficiently to rupture the outer membrane and release adenylate kinase and cytochrome c, which are located in the intermembrane space.

The simplest case to consider is that of mitochondria where both the respiratory chain and the ATP synthase are inhibited. In the examples shown in Figure 2.3, the swelling of

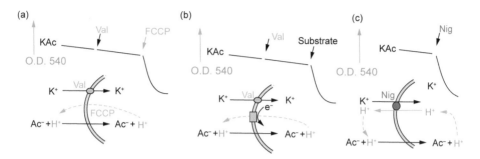

Figure 2.3   Ion transport across the inner mitochondrial membrane studied by osmotic swelling.
Bulk solute entry and osmotic swelling of mitochondria in ionic media requires both charge and pH balancing. To illustrate this, suppose that nonrespiring mitochondria are suspended in iso-osmotic K-acetate (KAc). Acetate is freely permeable as the neutral HAc, shown as $Ac^- + H^+$ (but not as the acetate anion Ac^-) and ionises in the matrix. (a) No swelling occurs because K^+ is poorly permeable; addition of valinomycin (Val) to induce electrical K^+ permeability does not allow swelling because there is a charge and pH imbalance that is overcome by addition of the protonophore FCCP. (b) Alternatively, swelling in the presence of valinomycin can be induced by initiating respiration. (c) No imbalance exists when K^+ permeability is induced by the electroneutral K^+/H^+ antiport ionophore nigericin (Nig), and swelling occurs on ionophore addition.

rat liver mitochondria suspended in isotonic (120 mM) concentrations of K-acetate illustrates the need for charge balance. Nigericin supports swelling in K-acetate, whereas valinomycin is ineffective. The reason for the difference is the need for charge balance. K^+ entry catalysed by valinomycin is electrical, whereas acetate permeates the bilayer as the neutral protonated acid. Therefore, in the presence of valinomycin and potassium acetate, a membrane potential (positive inside) rapidly builds up, preventing further K^+ entry. The permeation of acetic acid also ceases, as dissociation of the co-transported proton within the matrix builds up a proton concentration gradient (acid in the matrix), which opposes further acetic acid entry (see Section 3.5 for an explanation). These problems are not encountered with potassium acetate in the presence of nigericin because cation and anion entry are both now electroneutral (Figure 2.3c). The proton entering with acetic acid is re-exported by the ionophore in exchange for K^+. It is therefore possible to use swelling not only to determine if a species is permeable but also to determine the mode of entry. Thus, with FCCP present, swelling occurs in the presence of potassium acetate plus valinomycin. The ion fluxes that lead to accumulation of potassium acetate in the matrix are shown in Figure 2.3.

Matrix volume changes occurring in respiring mitochondria have to take account of the contribution of the protons pumped across the membrane by the respiratory chain. Respiration-dependent swelling occurs in the presence of an electrically permeant cation (which is accumulated due to the membrane potential) and an electroneutrally permeant weak acid (accumulated due to ΔpH; Section 3.5), as shown for K-acetate plus valinomycin in Figure 2.3. Conversely, a rapid contraction of pre-swollen mitochondria will occur on initiation of respiration when the matrix contains an electroneutrally permeant cation (expelled by ΔpH) and an electrically permeant anion (expelled by $\Delta\psi$).

Swelling induced under conditions of Ca^{2+} overload (the mitochondrial permeability transition; Section 9.4.1.5) is generally considered to be due to modification of the inner membrane adenine nucleotide transporter to produce a nonselective pore permeable to cations and anions up to 1.4 kDa.

Consideration of charge balance is also necessary in reconstitution experiments. For example, if a pH indicator is trapped inside a phospholipid vesicle with an acidic lumen and the external pH is subsequently raised into the alkaline range, the indicator trapped inside will continue to indicate an acid pH even if the protonophore FCCP is added. Significant proton efflux is not possible because efflux of a tiny quantity of protons generates a membrane potential, positive outside. However, if external K^+ is available and valinomycin is added, protons can efflux via FCCP because the requirement for charge balance is satisfied by the influx of K^+. The internal indicator then signals an alkaline pH.

2.5.1 The ammonium swelling technique for the detection of mitochondrial anion carriers

This technique is of historic interest because it provided the first demonstration of the major metabolite carriers and their functioning as linked systems of antiporters. When nonrespiring mitochondria are suspended in an isotonic solution of the ammonium salt of a bilayer-permeant weak acid such as acetate, the mitochondria undergo very rapid osmotic swelling, followed by a decrease in the light scattered as the matrix refractive index decreases to that of the medium. Swelling can occur because NH_4^+ crosses as the neutral NH_3 and acetate$^-$ crosses as the protonated acetic acid, both passive diffusion processes that result in the NH_4^+ and CH_3COO^- concentrations in the matrix becoming equal to those in the suspending medium. There is thus no charge or pH imbalance during transport, and massive swelling can occur as the matrix expands to regain osmotic equilibrium with the suspending medium.

In the case of ammonium phosphate, swelling is again observed, indicating that the phosphate anion crosses as the neutral species, which can be visualised as a $H^+{:}H_2PO_4^-$ symport or the indistinguishable P_i^-/OH^- antiport (Figure 2.4). However, when mitochondria are suspended in isotonic ammonium malate, no swelling occurs until a low concentration of P_i is added because malate permeates the inner membrane on a carrier protein in exchange for P_i. Phosphate can thus cycle across the membrane between the malate and phosphate carriers. The situation with a tricarboxylic acid such as citrate is still more complex: low concentrations of both phosphate and malate are required because the citrate/isocitrate carrier exchanges with malate, which in turn exchanges with P_i (Figure 2.4).

Careful examination of the stoichiometries in Figure 2.4 reveals that the net accumulation of $H_2PO_4^-$ is accompanied by the entry of one proton, that malate uptake is accompanied by two protons and that three protons enter with each citrate. In thermodynamic terms, this means that the phosphate gradient in/out can in theory equal the proton concentration gradient out/in (membrane potential has no effect), that the malate gradient can equal the square of the proton concentration gradient and that citrate or

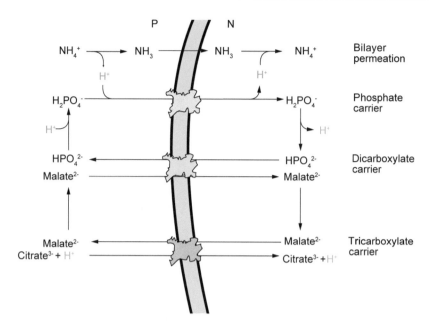

Figure 2.4   Ammonium swelling and mitochondrial metabolite carriers.
Nonrespiring mitochondria suspended in isotonic ammonium salts of
transportable metabolites swell osmotically if the metabolite is permeant. The
ammonium ion donates a proton to the P-phase and takes up a proton in the
N-phase and thus mimics respiration. Phosphate is transported as the fully
protonated H_3PO_4 (or an indistinguishable $H_2PO_4^-/OH^-$ exchange), malate
exchanges with phosphate and citrate exchanges with malate. Note that
respectively one, two and three protons are taken up and that this implies that
the equilibrium distributions of phosphate, malate and citrate across the
membrane equate to the first, second and third powers of the pH gradient,
respectively (see text).

isocitrate can be accumulated up to the cube of the proton concentration gradient. Note
that each carrier catalyses electroneutral transport so that membrane potential does not
affect the distribution. This means that with a ΔpH of -0.5 units (i.e., matrix alkaline),
the equilibrium concentrations of $H_2PO_4^-$, malate^{2-} and citrate^{3-} in the matrix can be
respectively 3, 10 and 30 times higher than in the cytoplasm. These quantitative aspects
are discussed in Chapter 3.

QUANTITATIVE BIOENERGETICS

The Measurement of Driving Forces

3.1 INTRODUCTION

Thermodynamics has a fearsome reputation among biologists, but the reader is strongly advised to follow through the derivations, if only to exorcise the idea, which amazingly still exists in many general biochemistry textbooks, that ATP is a 'high-energy' compound. Thermodynamic ignorance is also responsible for some extraordinary errors found in the current literature, particularly in the field of mitochondrial physiology (see Part 3). A thermodynamically impossible reaction is just that—impossible.

3.1.1 Systems

In thermodynamics, three types of system are studied. *Isolated* (or adiabatic) systems are completely autonomous, exchanging neither material nor energy with their surroundings (e.g., a closed, perfectly insulated Dewar (thermos) flask) (Figure 3.1). *Closed* systems are materially self-contained but can exchange energy across their boundaries (e.g., a hot water bottle). *Open* systems exchange both energy and material with their environment (e.g., all living organisms). The complexity of the thermodynamic treatment of these systems increases as their isolation decreases. Open systems strictly require a non-equilibrium thermodynamic treatment; classical equilibrium thermodynamics cannot be applied precisely to open systems because the flow of matter across their boundaries precludes the establishment of a true equilibrium.

The most significant contribution of equilibrium thermodynamics to bioenergetics comes from considering individual reactions or groups of reactions as closed systems, asking questions about the nature of the equilibrium state for that reaction, and establishing how far removed the actual reaction is from that equilibrium state. Despite this

Bioenergetics. Doi: http://dx.doi.org/10.1016/B978-0-12-388425-1.00003-8

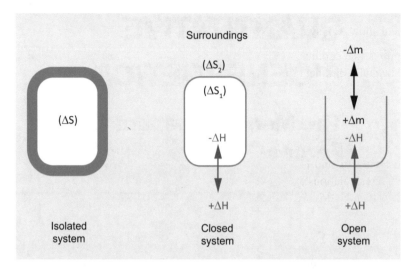

Figure 3.1   Gibbs energy and entropy changes in isolated, closed and open systems.
A process can only occur if it results in an increase in the entropy (ΔS) of the system plus its surroundings—that is, $\Delta S > 0$. An isolated system exchanges neither enthalpy nor material with its surroundings. Closed systems (e.g., closed containers with conducting walls) exchange enthalpy (ΔH) but not material. For a closed system at constant temperature and pressure, the enthalpy exchange can be equated with the heat flow across the walls. The change in entropy of the surroundings, ΔS_2, is equal to $\Delta H/T$ under these conditions, where T is the absolute temperature. Gibbs energy change, ΔG, is defined as being equal to $-T\Delta S$, which equals $T\Delta S_2 - T\Delta S_1$ in the above example. Therefore, $\Delta G = \Delta H - T\Delta S_1$. This means that under these conditions, ΔG can be determined using only those parameters that refer to the system itself (the entropy change of the system and the enthalpy change), whereas entropy changes in the surroundings need not be determined. A decrease in ΔG implies an increase in the entropy of the universe and so can occur spontaneously, if a pathway exists. Open systems exchange both enthalpy and material, Δm, and require more complex non-equilibrium thermodynamic treatments.

limitation, equilibrium thermodynamics is immensely powerful in bioenergetics because it can be applied to the following problems:

(a) Calculating the conditions required for equilibrium in an energy transduction, such as the utilisation of the protonmotive force to produce ATP, and by extension determining how far such a reaction is displaced from equilibrium under the actual experimental conditions. It is this displacement from equilibrium that defines the capacity of the reaction to perform useful work.

(b) Eliminating thermodynamically 'impossible' reactions. Although no thermodynamic treatment can prove the existence of a given mechanism, equilibrium thermodynamics can readily disprove any proposed mechanism that disobeys its laws. For example, if reliable values are available for the protonmotive force and the free energy for ATP synthesis (concepts that will be developed quantitatively later), it is possible to state unambiguously the lowest value of the H^+/ATP stoichiometry (the number of H^+ that must enter through the ATP synthase to make an ATP), which would allow ATP synthesis to occur.

Remember that equilibrium thermodynamics gives no clue as to the rate of a process. That is the realm of kinetics, which depends on the properties of the enzymes catalysing the reaction.

3.1.2 Entropy and Gibbs energy change

Why do biologists and biochemists use free energy rather than entropy in their thermodynamic calculations? The universe is by definition an isolated system, and in an isolated system the driving force for a reaction is an increase in entropy, which may be broadly equated to the degree of disorder of the system. In a closed system, a process can occur spontaneously (in the presence of a catalyst) if the entropy of the system *plus its surroundings* increases. Although it is not possible to measure directly the entropy changes in the rest of the universe caused by the energy flow across the boundary of the system, this parameter can be calculated under conditions of constant temperature and pressure (and therefore applicable to most biological systems) from the flow of enthalpy (heat) across the boundaries of the system (Figure 3.1). The thermodynamic function that takes account of this enthalpy flow is the *Gibbs energy change*, ΔG, which is the quantitative measure of the net driving force (at constant temperature and pressure):

$$\Delta G = \Delta H - T\Delta S \qquad [3.1]$$

where ΔH is the enthalpy change of the system, and ΔS is the entropy change, again of the system. This is the Gibbs–Helmholtz equation. Thus, ΔG can be determined using only those parameters that refer to the system, and entropy changes in the surroundings need not be determined.

A process that results in a decrease in Gibbs energy ($\Delta G < 0$) is one that causes a net increase in the entropy of the system plus surroundings and is therefore able to occur spontaneously *if* a mechanism is available.

Gibbs energy changes (also termed free energy changes) occur in bioenergetics in four different guises; indeed, the subject might well be defined as the study of the mechanisms by which the different manifestations of Gibbs energy are interconverted:

(1) Gibbs energy changes themselves are used in the description of substrate reactions feeding into the respiratory chain and of the ATP that is ultimately synthesised.
(2) The oxidation-reduction (redox) reactions occurring in the electron transfer pathways in respiration and photosynthesis can validly be quantified in terms of Gibbs energy changes but are usually referred to in terms of closely derived redox potential changes.
(3) The available energy in a gradient of ions is quantified by a further variant of the Gibbs energy change, namely the ion electrochemical gradient.
(4) In photosynthetic systems, the Gibbs energy available from the absorption of quanta of light can be compared directly with the other Gibbs energy functions within the reaction centre or the cell.

It should be emphasised that these different conventions merely reflect the diverse historical background of the topics that are brought together in chemiosmotic energy transduction.

3.2 GIBBS ENERGY AND DISPLACEMENT FROM EQUILIBRIUM

Since the first edition, this book has taken a rather unorthodox approach to Gibbs energy changes. Whereas the classical physical chemistry approach emphasises standard free energies (a hypothetical condition where all components are present at unit activity), we prefer to discuss reactions purely in terms of displacement from equilibrium. This leads to considerably simpler, more intuitive and more symmetrical equations without sacrificing precision. We strongly recommend taking the time to follow through this section.

Consider a simple reversible reaction [A] $\rightleftharpoons$ [B], occurring in a closed system. By observing and measuring the concentration of reactant A ($[A]_{obs}$) and product B ($[B]_{obs}$), we can calculate the *observed mass action ratio* Γ (capital gamma), equal to $[B]_{obs}/[A]_{obs}$. If the mixture of reactant and product happens to be at equilibrium, the mass action ratio of these equilibrium concentrations $[B]_{equil}/[A]_{equil}$ is termed the *equilibrium constant K*.

The absolute value of the Gibbs energy (G) increases the further Γ is displaced from K—that is, the further the reaction is from equilibrium. When G is plotted as a function of the logarithm of Γ/K (Figure 3.2), a parabola is obtained. The curve shows the following features:

(a) The Gibbs energy content G is at a minimum when the reaction is at equilibrium. Thus, any change in Γ away from the equilibrium ratio requires an increase in the Gibbs energy content of the system and so cannot occur spontaneously.

(b) The slope of the curve is zero at equilibrium. This means that a conversion of A to B that occurs at equilibrium without changing the mass action ratio Γ (e.g., by replacing the reacted A and removing excess B as it is formed) would cause no change in the Gibbs energy *content*. Another way of saying this is that the slope ΔG (i.e., the Gibbs energy *change* in units of kJ mol^{-1}) is zero at equilibrium.

(c) When the reaction A→B has not yet proceeded as far as equilibrium, a conversion of A to B without changing the mass action ratio Γ results in a decrease in G—that is, the slope ΔG is negative. This implies that such an interconversion can occur spontaneously, *provided that a mechanism exists*.

(d) The slope of the curve decreases as equilibrium is approached. Thus, ΔG decreases the closer the reaction is to equilibrium. Note that ΔG does *not* equal the Gibbs energy that would be available if the reaction were allowed to run down to equilibrium but, rather, gives the Gibbs energy that would be liberated per mole if the reaction proceeded with no change in substrate and product concentrations. This closely reflects the conditions prevailing *in vivo* where substrates are continuously supplied and products removed.

(e) For the reaction to proceed beyond the equilibrium point would require an input of Gibbs energy; this therefore cannot occur spontaneously.

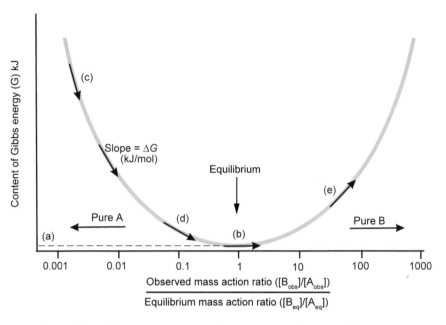

Figure 3.2   Gibbs energy content of a reaction as a function of its displacement from equilibrium.
Consider a closed system containing components A and B at concentrations [A] and [B], which can be interconverted by a reaction [A] ⇌ [B]. The reaction is at equilibrium when the mass action ratio [B]/[A] = K, the equilibrium constant. The curve shows qualitatively how the Gibbs energy content (G) of the system varies when the total [A]+[B] is held constant but the mass action ratio is varied away from equilibrium. The slope of the tangential arrows represents schematically the Gibbs energy change, ΔG, for an interconversion of A to B occurring at different displacements from equilibrium, without changing the mass action ratio (e.g., by continuously supplying substrate and removing product). For details of (a)–(e), see text.

The discussion may be generalised and placed on a quantitative footing by considering the reaction where a moles of A and b moles of B react to give c moles of C and d moles of D; that is,

$$aA + bB \rightleftharpoons cC + dD \tag{3.2}$$

The equilibrium constant K for the reaction is defined as follows:

$$K = \frac{[C]_{eq}^{c}[D]_{eq}^{d}}{[A]_{eq}^{a}[B]_{eq}^{b}} \, \text{Molar}^{(c+d-a-b)} \tag{3.3}$$

where the equilibrium concentration of each component is inserted into the equation to obtain an equilibrium mass action ratio. Note that the equilibrium constant has units unless the number of product molecules equals those of the substrates.

This complicated-looking equation will not be around for long, however, because we can now define the observed mass action ratio Γ when the reaction is held away from equilibrium by:

$$\Gamma = \frac{[C]_{obs}^c [D]_{obs}^d}{[A]_{obs}^a [B]_{obs}^b} \text{Molar}^{(c+d-a-b)} \qquad [3.4]$$

Note the symmetry between Eqs. 3.3 and 3.4. We now state, without deriving it, the key equation that relates the Gibbs energy change, ΔG, for the generalised reaction given in Eq. 3.2 to its equilibrium constant and observed mass action ratio given in Eqs. 3.3 and 3.4, respectively:

$$\Delta G = -2.3RT \log_{10} \left[\frac{K}{\Gamma} \right] \qquad [3.5]$$

where the factor 2.3 comes from the conversion from natural logarithms, R is the gas constant, and T is the absolute temperature.

This key equation tells us the following:

(a) ΔG has a value that is a function of the displacement from equilibrium. The numerical value of the factor 2.3 RT means that at 37°C a reaction that is maintained one order of magnitude away from equilibrium possesses a ΔG of 5.9 kJ mol^{-1}.
(b) ΔG is negative if $\Gamma < K$ and positive if $\Gamma > K$.

Note again that ΔG is a differential; that is, it measures the change in Gibbs energy that would occur if 1 mol of substrate were converted to product without changing the mass action ratio Γ (e.g., by continuously replenishing substrate and removing product). It does not answer the question, "How much energy is available from running down this reaction to equilibrium?"

3.2.1 ΔG for the ATP hydrolysis reaction

The consequences of Eq. 3.5 may be illustrated by reference to the hydrolysis of ATP to ADP and P$_i$. At pH 7.0, and in the presence of an approximately physiological 10^{-2} M Mg^{2+}, this reaction has an apparent equilibrium constant K' of approximately 10^5 M. By *apparent* equilibrium constant, we mean that obtained by putting the total chemical concentrations of reactants and products into the equation without considering the concentration of water, the pH and the effect of the ionisation state of each component. That such a surprising oversimplification is possible will be explained later. The equation for the equilibrium of the ATP hydrolysis reaction is

$$K' = \frac{[\Sigma ADP][\Sigma P_i]}{[\Sigma ATP]} M = 10^5 M \qquad [3.6]$$

where each concentration represents the total sum of the concentrations of the different ionised species of each component, including that complexed to Mg^{2+} (discussed later).

Table 3.1   The Gibbs energy change for the hydrolysis of ATP to ADP + P_i as a function of the displacement from equilibrium

for $K' = 10^5$ M, pH 7, 10 mM Mg^{2+}, 10 mM P_i

Γ' (M)	K'/Γ'	ΔG (kJ/mol^{-1})	[ATP]/[ADP]	Relevant condition
10^5	1	0	0.0000001	Equilibrium
10^3	10^2	-11	0.00001	
1	10^5	-28	0.01	
10^{-1}	10^6	-34	0.1	
10^{-3}	10^8	-46	10	Matrix
10^{-5}	10^{10}	-57	1000	Cytoplasm

Because equilibrium is attained when the apparent mass action ratio Γ', obtained using exactly the same simplifications as for K', is 10^5 M, the equilibrium concentration of ATP in the presence of 10^{-2} M P_i and 10^{-3} M ADP (which are approximate figures for the cytoplasm) would be only 10^{-10} M, or approximately 1 part per 10 million of the total adenine nucleotide pool!

The variations of ΔG with the displacement of the ATP hydrolysis mass action ratio from equilibrium are shown in Table 3.1. Mitochondria are able to maintain a mass action ratio in the incubation medium or cytoplasm that is as low as 10^{-5} M, ten orders of magnitude away from equilibrium. Under these conditions, the incubation might contain 10^{-2} M P_i, 10^{-2} M ATP, and only 10^{-5} ADP. To synthesise ATP under these conditions requires an input of Gibbs energy of 59 kJ per mole of ATP produced. The reason for a lower ΔG for the ATP/ADP pool in the mitochondrial matrix is discussed in Chapter 8.

Note that ΔG for ATP synthesis (sometimes referred to as the 'phosphorylation potential,' ΔG_p) is obtained from the corresponding value for ATP hydrolysis by simply changing the sign.

3.2.2 The uses and pitfalls of standard Gibbs energy, $\Delta G°$

A special case of the general equation for ΔG (Eq. 3.5) occurs under the totally hypothetical condition when the concentration of all reactants and products are in their 'standard states'—that is, 1 M for solutes, a pure liquid such as water, or a pure gas at 1 atm. These conditions define the standard Gibbs energy change $\Delta G°$.

Considering again our generalised reaction in Eq. 3.2, under these 'standard' conditions Γ has a value of $1\,M^{(c + d - a - b)}$ and Eq. 3.5 reduces to:

$$\Delta G° = -2.3RT\log_{10}K \qquad [3.7]$$

This equation, found in many biochemistry and physical chemistry textbooks without being derived from an equation similar to Eq. 3.5, creates an apparent mathematical problem that is a frequent cause of confusion. If the number of product and substrate species differs, then K will not be dimensionless. However, it is not possible to take the logarithm of a dimensional value (for example '10 M^{-1}'). The above derivation explains that in this equation the term $\log_{10}K$ is dimensionless because the units of K have been cancelled by those of Γ.

Equation 3.7 is frequently misunderstood. It is important to appreciate that ΔG° is simply related to the logarithm of the equilibrium constant and as such gives no information whatsoever concerning the Gibbs energy of the reaction in the cell. It is therefore absolutely incorrect to use ΔG° values to predict whether a reaction can occur spontaneously or to estimate the Gibbs energy available from a reaction.

Equation 3.7 can, however, be used to derive the more commonly used form of the Gibbs energy equation in which the equilibrium constant is substituted by ΔG°. If we take Eq. 3.5 and divide both K and Γ by the standard state concentrations to make them dimensionless, and then rearrange the equation, we get:

$$\Delta G = -2.3RT\log_{10}K + 2.3RT\log_{10}\Gamma \qquad [3.8]$$

Combining with Eq. 3.6 and eliminating K gives:

$$\Delta G = \Delta G^{\circ} + 2.3RT\log_{10}\Gamma$$

or as usually written,

$$\Delta G = \Delta G^{\circ} + 2.3RT\log_{10}\left\{\frac{[C]^c_{obs}[D]^d_{obs}}{[A]^a_{obs}[B]^b_{obs}}\right\} \qquad [3.9]$$

Equation 3.9 is the most common form of the Gibbs energy equation and the one found in most textbooks. Just as Eq. 3.5 has terms for Γ and K, so Eq. 3.9 has terms for Γ and ΔG°. Note that Eq. 3.9 reverts to Eq. 3.7 at equilibrium when $\Delta G = 0$ and, of course, $\Gamma = K$.

Both Eq. 3.5 and Eq. 3.9 can be used correctly to calculate ΔG; however, Eq. 3.5 is more intuitive because it emphasises the fact that ΔG is a function of the extent to which a reaction is removed from equilibrium. In addition, it is not immediately evident from Eq. 3.9 that ΔG° and $2.3RT\log_{10}\Gamma$ are dimensionally homogeneous terms or why apparent equilibrium constants and apparent mass action ratios (discussed next) can be used that make simplifying assumptions about the states of ionisation of reactants and products, the pH, etc.

3.2.3 Absolute and apparent equilibrium constants and mass action ratios

To avoid confusion or ambiguity in the derivation of equilibrium constants, and hence Gibbs energy changes, a number of conventions have been adopted. Those most relevant to bioenergetics are the following:

(a) True thermodynamic equilibrium constants (K) are defined in terms of the chemical activities rather than the concentrations of the reactants and products. Generally, in biochemical systems it is not possible to determine the activities of all the components, and so equilibrium constants are calculated from concentrations. This introduces no error as long as the observed mass action ratio and the equilibrium constants are calculated under comparable conditions (remember that ΔG is calculated from the ratio of Γ and K under identical conditions).

(b) When water appears as either a reactant or a product in dilute solutions, its concentration will be the same under both equilibrium and observed conditions. This means that the water term can be omitted from both the equilibrium and the observed mass action ratio equations (again, ΔG is calculated from the ratio of Γ and K'; see Eq. 3.5).

(c) If one or more of the reactants or products are ionisable, or can chelate a cation, there is an ambiguity as to whether the equilibrium constant should be calculated from the total sum of the concentrations of the different forms of a compound or just from the concentration of that form which is believed to participate in the reaction. The hydrolysis of ATP to ADP and P_i is a particularly complicated case: not only are all the reactants and products partially ionised at physiological pH but also Mg^{2+}, if present, chelates ATP and ADP with different affinities. Thus, ATP can exist at pH 7 in the following forms:

$$[\Sigma ATP] = [ATP^{4-}] + [ATP^{3-}] + [ATP \cdot Mg]^{2-} + [ATP \cdot Mg]^- \quad [3.10]$$

If it were known that the true reaction was

$$ATP \cdot Mg^{2-} + H_2O \rightleftharpoons ADP \cdot Mg^- + HPO_4^{2-} + H^+ \quad [3.11]$$

then the true equilibrium constant would be

$$K = \frac{[Mg \cdot ADP^-][HPO_4^{2-}][H^+]}{[Mg \cdot ATP^{2-}]} M^2 \quad [3.12]$$

This equilibrium constant would be independent of pH or Mg^{2+} because changes in these factors are allowed for in the equation. However, the reacting species are not known unambiguously, and even if they were, their concentrations would be difficult to assay because enzymatic or chemical assay determines the total concentration of each compound (e.g., ΣATP). Therefore, in practice, an apparent equilibrium constant, K', is employed, calculated from the total concentrations of each reactant and product, ignoring water and any effects of ionisation or chelation and omitting any protons that are involved (see Eq. 3.6).

The most important limitation of the apparent equilibrium constant is that K' is not a universal constant but, rather, depends on all those factors that are omitted from the equation, such as pH and cation concentration. K' is thus only valid for a given pH and cation concentration, and it must be qualified by information about these conditions. Because the standard Gibbs energy change is derived directly from the apparent equilibrium constant, this parameter must be similarly qualified. Finally, and most important, the apparent mass action ratio, Γ', must be calculated under exactly the same set of assumptions; if this is done, when the ratio K'/Γ' is calculated for Eq. 3.5, all the assumptions cancel out and a true and meaningful ΔG is obtained. In biochemistry, the terms $\Delta G^{\circ\prime}$ and K' are frequently used to specify that a $[H^+]$ of 10^{-7} M is being considered, but in principle these parameters can be specified for any condition of pH, ionic strength, temperature, $[Mg^{2+}]$ etc., that is convenient—as long as Γ' is always calculated under exactly the same set of conditions.

3.2.4 **The myth of the 'high-energy phosphate bond'**

It is still possible to come across statements to the effect that the phosphate anhydride bonds of ATP are 'high-energy' bonds capable of storing energy and driving reactions in otherwise unfavourable directions. However, it should be clear from Table 3.1 that it is the extent to which the observed mass action ratio is displaced from equilibrium which defines the capacity of the reactants to do work, rather than any attribute of a single component of the reaction. A hypothetical cell could utilise any reaction to transduce energy from the mitochondrion. For example, if the glucose 6-phosphatase reaction were maintained 10 orders of magnitude away from equilibrium, then glucose 6-phosphate hydrolysis would be thermodynamically just as capable of doing work in the cell as is ATP. Conversely, the Pacific Ocean could be filled with an equilibrium mixture of ATP, ADP and P_i, but the ATP would have no capacity to do work. It *is* important that the equilibrium constant for the ATP hydrolysis reaction has about the value it does, because it means that even in the cytoplasm under conditions in which ΔG for ATP synthesis is almost $60\,kJ\,mol^{-1}$, providing a sufficient driving force for many energetically uphill cellular processes, there is still a sufficient concentration of ADP to bind to the adenine nucleotide translocator responsible for translocating the nucleotide into the matrix (Chapter 8).

3.3 **REDOX POTENTIALS**

Further reading: Schafer and Buettner (2001), Price *et al.* (2001)

3.3.1 **Redox couples**

Both the mitochondrial and the photosynthetic electron transfer chains operate as a sequence of reactions in which electrons are transferred from one component to another. Although many of these components simply gain one or more electrons in going from the oxidised to the reduced form, in others the gain of electrons induces an increase in the pK of one or more ionisable groups on the molecule, with the result that reduction is followed immediately by the gain of one or more protons.

Cytochrome c undergoes a $1e^-$ reduction:

$$Fe^{3+}.cyt\,c + 1e^- \rightleftharpoons Fe^{2+}.cyt\,c \qquad [3.13]$$

NAD^+ undergoes a $2e^-$ reduction and gains one H^+:

$$NAD^+ + 2e^- + H^+ \rightleftharpoons NADH \qquad [3.14]$$

whereas ubiquinone undergoes a $2e^-$ reduction followed by the addition of $2H^+$:

$$UQ + 2e^- + 2H^+ \rightleftharpoons UQH_2 \qquad [3.15]$$

This last equation is sometimes incorrectly referred to as a 'hydrogen transfer.'

Redox reactions are not restricted to the electron transport chain. For example, lactate dehydrogenase also catalyses a redox reaction:

$$\text{Pyruvate} + \text{NADH} + \text{H}^+ \rightleftharpoons \text{Lactate} + \text{NAD}^+ \qquad [3.16]$$

Whereas all redox reactions can quite properly be described in thermodynamic terms by their Gibbs energy changes, electrochemical parameters can be employed because the reactions involve the transfer of electrons. Although the thermodynamic principles are the same as for the Gibbs energy change, the origins of redox potentials in electrochemistry sometimes obscure this relationship.

The additional facility afforded by an electrochemical treatment of a redox reaction is the ability to dissect the overall electron transfer into two half-reactions involving the donation and acceptance of electrons, respectively. Thus, Eq. 3.16 can be considered as the sum of two half-reactions:

$$\text{NAD}^+ + \text{H}^+ + 2\text{e}^- \rightleftharpoons \text{NADH} \qquad [3.17]$$

and

$$\text{Pyruvate} + 2\text{H}^+ + 2\text{e}^- \rightleftharpoons \text{Lactate} \qquad [3.18]$$

Note that these two equations can be combined to regenerate Eq. 3.16. An oxidised–reduced pair such as NAD^+/NADH or pyruvate/ lactate is termed a *redox couple*.

3.3.2 Determination of redox potentials

Each of the half-reactions described previously (Eqs. 3.17 and 3.18) is reversible and so can in theory be described by an equilibrium constant. However, it is not immediately apparent how to treat the electrons, which have no independent existence in solution. A similar problem is encountered in electrochemistry when investigating the equilibrium between a metal (i.e., the reduced form) and a solution of its salt (i.e., the oxidised form). In this case, the tendency of the couple to donate electrons is quantified by forming an electrical cell from two half-cells, each consisting of a metal electrode in equilibrium with a 1 M solution of its salt. An electrical circuit is completed by a bridge that links the solutions without allowing them to mix. The electrical potential difference between the electrodes may then be determined experimentally.

To facilitate comparison, electrode potentials are expressed in relation to the *standard hydrogen electrode*:

$$2\text{H}^+ + 2\text{e}^- \rightleftharpoons \text{H}_2 \qquad [3.19]$$

Hydrogen gas at 1 atm is bubbled over the surface of a platinum electrode, which has been coated with the finely divided metal to increase the surface area. When this electrode is immersed in 1 M H^+, the absolute potential of the electrode is defined as zero (at 25°C). The standard electrode potential of any metal/salt couple may now be determined by forming a cell comprising the unknown couple together with the standard hydrogen electrode, or more conveniently with secondary standard electrodes whose electrode potentials are known.

A similar approach has been adopted for biochemical redox couples. As with the hydrogen electrode, it is not feasible to construct an electrode out of the reduced component of the couple, so a Pt electrode is employed. The oxidised and reduced components of the couple (e.g., the oxidised and reduced states of a cytochrome) rarely equilibrate with the Pt electrode sufficiently rapidly for a stable potential to be registered because the polypeptide chain hinders electron transfer to and from the redox centre. Therefore, a low concentration of a second redox couple, capable of reacting with both the cytochrome redox couple and the Pt electrode, is added to act as a redox mediator. As will be shown later, the redox mediator is oxidised or reduced until it exhibits the same redox potential as the cytochrome redox couple. As long as the concentration relationships of the latter are not disturbed, the electrode can therefore register the cytochrome's redox potential. The use of one or more mediating couples is of particular importance when the redox potentials of membrane-bound components are being investigated (Chapter 5).

Unlike metal/salt and H^+/H_2 couples, both components of a biochemical couple can generally exist in aqueous solution. Standard conditions are defined in which both the oxidised and the reduced components are present at unit activity, or 1 M in concentration terms, and pH = 0. Note that these unphysiological conditions parallel those for $\Delta G°$. The experimentally observed potential relative to the hydrogen electrode is termed the *standard redox potential*, $E°$.

3.3.3 Redox potentials and the [oxidised]/[reduced] ratio

Just as the standard Gibbs energy change, $\Delta G°$, does not reflect the actual conditions existing in the cell, the standard redox potential, $E°$, must be qualified to take account of the relative concentrations of the oxidised and reduced species.

The actual redox potential E at pH = 0 (i.e., 1 M H^+) for the redox couple:

$$Oxidised + ne^- \rightleftharpoons Reduced$$

is given by the relationship:

$$E = E° + 2.3\frac{RT}{nF}\log_{10}\left\{\frac{[\text{oxidised}]}{[\text{reduced}]}\right\} \qquad [3.20]$$

where R is the gas constant, and F the Faraday constant. Note that this equation is closely analogous to the 'conventional' equation involving standard Gibbs energy changes (Eq. 3.8). Remember that a pH of zero is not physiological.

3.3.4 Redox potential and pH

In many cases (e.g., Eqs. 3.13 and 3.14), protons are involved in the redox reaction, in which case the generalised half-reaction becomes:

$$Oxidised + ne^- + mH^+ \rightleftharpoons Reduced$$

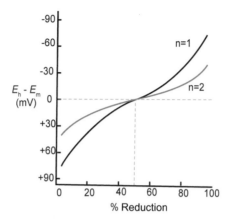

Figure 3.3   The variation of E_h with the extent of reduction of a redox couple M_{red}/M_{ox}.
$n = 1$ and $n = 2$ refer to one- and two-electron oxido-reductions, respectively. For example, if the $2e^-$ redox couple $NAD^+/NADH$ is exactly 50% reduced, $E_h = E_m = -320\,mV$. In the mitochondrial matrix, it is usually approximately 10% reduced, so E_h is approximately $-290\,mV$.

The standard redox potential at a pH other than zero is more negative than $E°$, the difference being $2.3RT/F.(m/n)$ mV per pH unit. This corresponds to $-61\,mV/pH$ at 37°C when $m = n$ (e.g., the fumarate/succinate couple) and $-30\,mV/pH$ when $m = 1$ and $n = 2$ ($NAD^+/NADH$; see Table 3.2). Note that the potentials are still calculated relative to the standard hydrogen electrode at pH $= 0$.

The usual biochemical convention is to define redox potentials for pH 7. The standard redox potential under these conditions is given the symbol $E°'$; it is also sometimes referred to as the midpoint potential $E_{m,7}$ because it is the potential, at pH 7, where the concentrations of the oxidised and reduced forms are equal (Figure 3.3).

Note that although the potential of the standard hydrogen electrode (at pH $= 0$) always remains zero, $E_{m,7}$ for the $H^+/\frac{1}{2}H_2$ couple is $7 \times (-61) = -427\,mV$—that is, much more reducing—because the concentration of the oxidised component of the couple, H^+, is present at such low concentration ($10^{-7}\,M$).

So far, we have concentrated on standard or midpoint potentials. However, as for the case of Gibbs energy changes, it is the actual concentrations in the cell that define the redox potential. The actual redox potential at a pH of 'x' ($E_{h,pH = x}$) is related to the midpoint potential at that pH by the following relationship:

$$E_{h,pH=x} = E_{m,pH=x} + \frac{2.3RT}{nF} \log_{10} \left\{ \frac{[oxidised]}{[reduced]} \right\} \qquad [3.21]$$

Thus, for example, in the mitochondrial matrix (pH 7.0), the $NAD^+/NADH$ pool might be 10% reduced so that $E_{h,\,pH = 7}$ will be $-290\,mV$, whereas the 99% reduced matrix $NADP^+/NADPH$ pool will have a $E_{h,\,pH = 7}$ of $-380\,mV$ (Table 3.2), the difference in potential being due to the operation of the NADP-linked isocitrate dehydrogenase and nicotinamide nucleotide transhydrogenase (Section 5.11).

Table 3.2   Some midpoint potentials and examples of actual redox potentials

Oxidised $+ ne^- + mH^+ =$ reduced

	n	m	$E_{m,7}$ (mV)	ΔE_m per pH	Typical ox/ red ratio	$E_{h,,7}$ (mV)[a]
Ferredoxin oxidised/reduced	1	0	-450	0		
$H^+/\frac{1}{2}H_2$ (at 1 atm)	1	1	-420	-60		
O_2 (1 atm[b])/(superoxide)	1	0	-330	0	10^{-5}	-30
$NAD^+/NADH$	2	1	-320	-30	10	-290
$NADP^+/NADPH$	2	1	-320	-30	0.01	-380
Menaquinone/menaquinol	2	2	-74	-60		
Glutathione oxidised/reduced (when GSH + GSSG = 10 mM)	2	2	-172	-60	0.01	-240^c
Fumarate/succinate	2	2	$+30$	-60		
Ubiquinone/ubiquinol	2	2	4	-60		
Ascorbate oxidised/reduced	2	1	$+60$	-30		
Cyt c oxidised/reduced	1	0	$+220$	0		
Ferricyanide oxidised/reduced	1	0	$+420$	0		
O_2 (1 atm[b])/2 H_2O (55 M)	4	4	$+820$	-60		

[a] Approximate values for mitochondrial matrix under typical conditions.
[b] 1 atm oxygen = 1.25 mM.
[c] See Eq. 3.23.

3.3.5 The special case of glutathione

Further reading: Schafer and Buettner (2001)

The glutathione couple:

$$GSSG + 2H^+ + 2e^- \rightleftharpoons 2GSH \qquad [3.22]$$

where GSH is the monomeric reduced form and GSSG the dimeric oxidised form, is one of the most important eukaryotic reactions defining the redox status of the cytoplasm and mitochondrial matrix. Because there are unequal numbers of substrate and product molecules, the absolute concentrations of these species, rather than merely their ratios, affect the redox potential.

In the general case, because $n = 2$,

$$E_{h,pH7} = E^{o\prime} + 30\log_{10}\left\{\frac{[GSSG]}{[GSH]^2}\right\} mV \qquad [3.23]$$

Note that we must perform the same device of dividing the concentration terms by standard conditions as in Eq. 3.8 in order for the logarithmic term to be dimensionless. Under standard conditions (1 M GSSG and 1 M GSH), $E^{o\prime} = -240$ mV at 25°C and pH 7.0. However, when the total glutathione pool concentration (defined as [GSH] + 2[GSSG]) is 10 mM, typical for the cytoplasm, a thiol redox potential of -200 mV implies a concentration of [GSSG] of 1.2 mM. If the pool size is lowered to 1 mM, then [GSSG] must be decreased 60-fold to only 20 μM to maintain the redox potential. This could impose kinetic problems for glutathione reductase, the enzyme

Table 3.3 Interconversion between redox potential differences and Gibbs energy change for one-electron and two-electron transfers

ΔE_h (mV)	ΔG (kJ mol^{-1})	
	$n = 1$	$n = 2$
0	0	0
+100	−9.6	−19.3
+200	−19.3	−38.6
+500	−48.2	−96.5
+1000	−96.5	−193
+1200	−116	−231

reducing GSSG to GSH, and helps to explain why a maintained pool size of glutathione is critical for preventing oxidative stress in the mitochondria.

3.3.6 Redox potential difference and the relation to ΔG

The Gibbs energy that is available from a redox reaction is a function of the difference in the actual redox potentials ΔE_h between the donor and acceptor redox couples. (Note that the difference in redox potential between two couples [redox potentials $E_{(A)}$ and $E_{(B)}$] is written in most books as E, but we believe that use of ΔE_h clarifies that a *difference* between two couples, or a redox span in an electron transport system, is being considered.) In general terms for the redox couples A and B:

$$\Delta E_h = E_{h(A)} - E_{h(B)} \qquad [3.24]$$

There is a simple and direct relationship between the redox potential difference of two couples, ΔE_h, and the Gibbs energy change ΔG accompanying the transfer of electrons between the couples:

$$\Delta G = -nF\Delta E_h \qquad [3.25]$$

where n is the number of electrons transferred, and F is the Faraday constant. From this, it is apparent that an oxido-reduction reaction is at equilibrium when $\Delta E_h = 0$. Table 3.3 relates redox potential differences and Gibbs energy changes.

In the case of the mitochondrion, $E_{h,7}$ for the NAD$^+$/NADH couple is approximately −280 mV and $E_{h,7}$ for the O$_2$/H$_2$O couple is +780 mV (note that these values differ from the midpoint potentials shown in Table 3.2 because the NAD$^+$/NADH ratio in the matrix is only approximately 0.1 and because oxygen comprises only 20% of air). The redox potential difference $\Delta E_{h,7}$ of 1.16 V is the measure of the thermodynamic disequilibrium between the couples. Applying Eq. 3.24 to the transfer of 2 mol of electrons from NADH to O$_2$ yields:

$$\Delta G = -2 \times 96.5 \times 1.16 = -224 \text{kJ} \cdot \text{mol}^{-1} \qquad [3.26]$$

One complication arises where the donor and acceptor redox couples are on the opposite sides of a membrane across which an electrical potential exists (Figure 3.4).

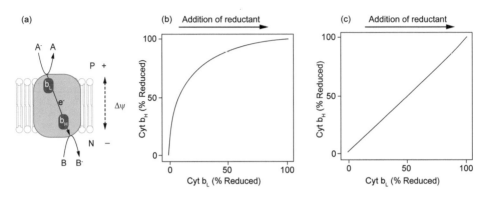

Figure 3.4 ΔE_h **and** ΔG **for an electron transfer between redox couples located on opposite sides of a membrane sustaining a membrane potential.**
(a) Single electrons are transferred from a redox couple on the P-side of the membrane (A/A$^-$) to one on the N-side (B/B$^-$). In the absence of a $\Delta\psi$, the redox potential difference is given by $\Delta E_b = (E_{bA} - E_{bB})$ and the free energy by $\Delta G = -nF\Delta E_b$, where n represents the number of electrons transferred ($n = 1$ in this example). The presence of a membrane potential opposes the electron transfer and ΔG is reduced; now, $\Delta G = -nF(\Delta E_h + \Delta\psi)$. To a detection system in the P-phase, it appears that the midpoint potential of B$^-$/B has become more negative. (b) Depolarised rat skeletal mitochondria are titrated with reductant (succinate) in the presence of antimycin A. The N-side located cyt b$_H$ ($E_{m,7}$ + 50 mV) is more readily reduced than the P-side located b$_L$ ($E_{m,7}$ −100 mV), consistent with the midpoint potentials of the two cytochromes. (c) Addition of ATP allows the ATP synthase to run in reverse, generating a $\Delta\psi$. The membrane potential redistributes electrons towards b$_L$, equalising their degree of reduction. Thus, there is an apparent change in midpoint potential of b$_H$.
Source: Data redrawn from Quinlan et al. (2011).

Examples of this occur in the respiratory chain (Chapter 5). If an electron enters a protein via a centre at the N-side of the membrane and is transferred to a centre at the P-side, then it is intuitive that the membrane potential contributes to the free energy change of the process. The value of the membrane potential must be added to the redox potential difference to calculate the effective Gibbs energy change:

$$\Delta G = -nF(\Delta E_h + \Delta\psi) \quad [3.27]$$

Conversely, if the electron travels from the P-side to the N-side, then $\Delta\psi$ must be subtracted; thus,

$$\Delta G = -nF(\Delta E_h - \Delta\psi) \quad [3.28]$$

Mitochondrial complex III (Section 5.8) possesses two b-type cytochromes on opposite sides of the hydrophobic core of the membrane: cyt b$_L$ (b$_{566}$) is toward the P-side of the membrane, and b$_H$ (b$_{560}$) is toward the N-side. The two cytochromes exchange electrons, and in the absence of a membrane potential b$_H$ ($E_{m,7}$ +50 mV) will be more reduced than b$_L$ ($E_{m,7}$ −100 mV) (Figure 3.4). However, when a $\Delta\psi$ is imposed, electrons will be driven toward cyt b$_L$ on the P-side of the membrane and the effect is to equalise the degree of reduction of the two cytochromes. Another example of a $\Delta\psi$ driving electrons towards a more negative redox couple is discussed for the bacterium *Nitrobacter* in Chapter 5.

A final note about nomenclature: a powerful *oxidising* agent is an oxidised component of a redox couple with a relatively positive E_m (e.g., $\frac{1}{2}O_2/H_2O$), whereas a powerful *reducing* agent is the reduced component of a redox couple with a relatively negative E_m (e.g., $NAD^+/NADH$). Note that these terms are all relative; for example, NAD^+ can act as an oxidant for ferredoxins with a more negative E_m value, whereas H_2O is a reductant for the enormously electropositive oxygen-evolving complex in the green plant photosystem II (Section 6.4.2.1).

3.4 ION ELECTROCHEMICAL POTENTIAL DIFFERENCES

We have tried to emphasise in this chapter that the Gibbs energy change is a function of displacement from equilibrium. The disequilibrium of an ion or metabolite across a membrane can be subjected to the same quantitative treatment. As before, the derivation is not only valid for energy-transducing membranes but also has equal applicability to all membrane transport processes.

There are two forces acting on an ion gradient across a membrane, one due to the concentration gradient of the ion and one due to the electrical potential difference between the aqueous phases separated by the membrane (the 'membrane potential' $\Delta\psi$). These can initially be considered separately. It is important to remember that the term *membrane potential* is shorthand for 'the difference in electrical potential between two aqueous compartments separated by a membrane' and says nothing about the nature of the membrane itself, the nature of any ions transported across the membrane, or any charge on its surface.

Consider the Gibbs energy change for the transfer of 1 mol of solute across a membrane from a concentration $[X]_A$ to a concentration $[X]_B$, where the volumes of the two compartments are sufficiently large that the concentrations do not change significantly.

In the absence of a membrane potential, ΔG is given by:

$$\Delta G = 2.3RT\log_{10}\frac{[X]_B}{[X]_A} \qquad [3.29]$$

Note that this equation is closely analogous to that for scalar reactions (Eq. 3.5). In particular, ΔG in both cases is $5.9\,kJ\,mol^{-1}$ at 37°C for each 10-fold displacement from equilibrium.

The second special case is for the transfer of a charged species driven by a membrane potential in the absence of a concentration gradient. In this case, the Gibbs energy change when 1 mol of cation X^{m+} is transported down an electrical potential of $\Delta\psi$ mV is given by:

$$\Delta G = -mF\Delta\psi \qquad [3.30]$$

In the general case, the ion will be affected by both concentrative and electrical gradients, and the net ΔG when 1 mol of X^{m+} is transported down an electrical potential of

$\Delta\psi$ mV from a concentration of $[X^{m+}]_A$ to $[X^{m+}]_B$ is given by the general electrochemical equation:

$$\Delta G = -mF\Delta\psi + 2.3RT\log_{10}\left\{\frac{[X^{m+}]_B}{[X^{m+}]_A}\right\} \qquad [3.31]$$

ΔG in this equation is often expressed as the ion electrochemical gradient $\Delta\tilde{\mu}_X^{m+}$ (kJ mol^{-1}).

In the specific case of the proton electrochemical gradient, $\Delta\tilde{\mu}_{H^+}$, Eq. 3.31 can be considerably simplified because pH is a logarithmic function of $[H^+]$:

$$\Delta\tilde{\mu}_{H^+} = -F\Delta\psi + 2.3RT\Delta pH \qquad [3.32]$$

where ΔpH is defined as the pH in the P-phase (e.g., cytoplasmic) minus the pH in the N-phase (e.g., matrix). Note that this means that in a respiring mitochondrion, ΔpH is usually negative. $\Delta\psi$ is defined as P-phase minus N-phase and is usually positive.

Mitchell defined the term protonmotive force (PMF or Δp) in units of voltage, where:

$$\Delta p(mv) = -\frac{\Delta\tilde{\mu}_{H^+}}{F} \qquad [3.33]$$

This facilitates comparison with redox potential differences in the electron transfer chain complexes, which generate the proton electrochemical gradient, as well as emphasising that we are dealing with a *potential* driving a proton circuit.

A $\Delta\tilde{\mu}_{H^+}$ of 1 kJ mol^{-1} corresponds to a Δp of 10.4 mV. Conversely, a Δp of 200 mV corresponds to a $\Delta\tilde{\mu}_{H^+}$ of 19 kJ mol^{-1}. Using Δp and substituting values for R and T at 37°C, the final equation is:

$$\Delta p(mv) = \Delta\psi - 61\Delta pH \qquad [3.34]$$

3.5 PHOTONS

In photosynthetic systems, the primary source of Gibbs energy is the quantum of electromagnetic energy, or photon, which is absorbed by the photosynthetic pigments. The energy in a single photon is given by $h\nu$, where h is Planck's constant (6.62×10^{-34} J s), and ν is the frequency of the radiation (s^{-1}). One photon interacts with one molecule, and therefore N photons, where N is Avogadro's constant, will interact with 1 mol.

The energy in 1 mol (or einstein) of photons is therefore:

$$\Delta G = Nh\nu = \frac{Nhc}{\lambda} = \frac{120000}{\lambda}\,kJ\,einstein^{-1} \qquad [3.35]$$

where c is the velocity of light, and λ is the wavelength in nm. Thus, even the absorption of an einstein of red light (600 nm) makes available 200 kJ mol^{-1}, which compares favourably with the Gibbs energy changes encountered in bioenergetics.

3.6 BIOENERGETIC INTERCONVERSIONS AND THERMODYNAMIC CONSTRAINTS ON THEIR STOICHIOMETRIES

The critical stages of chemiosmotic energy transduction involve the interconversions of ΔG between the different forms discussed in the previous sections. In the case of the mitochondrion, these are redox potential difference (ΔE_h) to protonmotive force (Δp) to ΔG for ATP synthesis. Whereas bioenergetic systems operate under non-equilibrium conditions *in vivo* (i.e., they are 'open'), with isolated preparations it is frequently possible to allow a given interconversion to achieve a true equilibrium by the simple expedient of inhibiting subsequent steps. For example, isolated mitochondria can achieve an equilibrium between the protonmotive force and ATP synthesis if reactions that hydrolyse ATP are absent.

In some cases, a true equilibrium is not achievable in practice. For example, because of the inherent proton permeability of the mitochondrial membrane (Section 4.6), there is always some net leakage of protons across the membrane that results in the steady-state value of Δp lying below its equilibrium value with ΔE_h. Consequently, there is always some flux of electrons from NADH to oxygen. Under these conditions, it is valid to obtain a number of values for different flux rates and to extrapolate back to the static head condition of zero flux.

A test of whether an interconversion is at equilibrium is to establish whether a slight displacement in conditions will cause the reaction to run towards more reactant or product, depending on the nature of the displacement. In respiring mitochondria, this test can be fulfilled by the ATP synthase and by two of the three respiratory chain proton pumps (complexes I and III; see Chapter 5). A process is of course at equilibrium when the overall ΔG is zero.

3.6.1 Proton pumping by respiratory chain complexes

If two electrons falling through a redox span of ΔE_h mV within the respiratory chain pump n protons across the membrane against a protonmotive force of Δp, then equilibrium would be attained when:

$$n\Delta p = 2\Delta E_h \qquad [3.36]$$

Thus, the higher the H^+/O stoichiometry (n) of a respiratory chain complex with a particular value of ΔE_h, the lower the equilibrium Δp that can be attained, just as a bicycle has less ability to climb a hill in high rather than low gear.

Note that Eq. 3.36 will only hold if the electrons enter and leave the redox span on the same side of the membrane. If, as in the case of electron transfer from succinate dehydrogenase (on the matrix face) to cytochrome c (on the cytoplasmic face), the electrons effectively cross the membrane, they will be aided by the membrane potential (Figure 3.4) and the relationship becomes:

$$n\Delta p = 2(\Delta E_h + \Delta\psi) \qquad [3.37]$$

3.6.2 Proton pumping by the ATP synthase

The equilibrium relationship between the protonmotive force and the Gibbs energy change ($\Delta G_{p,matrix}$) for the ATP synthase reaction in the mitochondrial matrix is given by:

$$\Delta G_{p,matrix} = x'F\Delta p \qquad [3.38]$$

where x' is the H^+/ATP stoichiometry. Note that the higher the H^+/ATP stoichiometry, the higher the $\Delta G_{p,\ matrix}$ that can be attained at equilibrium with a given Δp. The same equation applies to the bacterial cytoplasm and the chloroplast stroma.

Earlier methods to estimate x' by direct measurement of proton extrusion or by determining thermodynamic equilibrium can now be supplemented by direct structural investigation of the membrane-located proton turbine F_o component of the ATP synthase, by counting the number of c-subunits in the rotor (see Chapter 7). In vertebrates, this number appears to be 8, giving a value of x' of 8/3, or 2.7, whereas fungi, eubacteria and chloroplasts may have 10–15 c-subunits, implying the requirement of between 3.3 and $5\,H^+$ for each ATP. The corollary is that at higher values of x' a lower Δp is required to generate a given ΔG_p.

Because one additional proton is expended in the overall transport of P_i and ADP into, and of ATP out of, the mitochondrial matrix (Section 8.5), the relationship for the eukaryotic cytoplasmic ATP/ADP + P_i pool becomes:

$$\Delta G_{p,cytoplasm} = (x' + 1)F\Delta p \qquad [3.39]$$

This means that a substantial proportion of the Gibbs energy for the cytoplasmic ATP/ADP + P_i pool comes from the transport step rather than the ATP synthase itself. Naturally, this occurs at a cost: the overall H^+/ATP stoichiometry is increased by 1 to account for the additional proton.

3.6.3 Thermodynamic constraints on stoichiometries

Because the previous equations contain a term for the stoichiometry, it is possible to determine the thermodynamic parameters at equilibrium, substitute these values into the equations and hence calculate the stoichiometry term (H^+/2e⁻, H^+/ATP, etc.) without actually measuring the movement of protons across the membrane. This is known as the thermodynamic stoichiometry. Naturally, such calculations are only as accurate as the determination of the thermodynamic parameters, but it does offer an alternative approach to the non-steady-state technique that will be discussed in Chapter 4.

3.6.4 The 'efficiency' of oxidative phosphorylation

A statement of the type "Oxidation of NADH by O_2 has a $\Delta G°$ of $-220\,\text{kJ}\,\text{mol}^{-1}$ while ATP synthesis has a $\Delta G°$ of $+31\,\text{kJ}\,\text{mol}^{-1}$. Thus if three ATP molecules are synthesised for each NADH oxidised, mitochondrial oxidative phosphorylation traps approximately $93\,\text{kJ}\,\text{mol}^{-1}$ of the energy available from NADH oxidation, an efficiency of 42%" used to appear in many textbooks of biochemistry but is now mercifully rare. This analysis has at least two shortcomings. First, it refers to standard conditions that are not found in cells and are merely restatements of the equilibrium constant (Section 3.2.2). Second, there is no basis in physical chemistry for dividing an output ΔG ($93\,kJ\,mol^{-1}$ in this case) by the input ΔG ($220\,kJ\,mol^{-1}$) to calculate an efficiency. This will now be explained.

Under cellular conditions, $2\,mol$ of electrons flowing from the $NAD^+/NADH$ couple to oxygen liberate approximately $220\,kJ$. In the ideal case in which there is no proton leak across the membrane, this would be conserved in the generation of a Δp of approximately $200\,mV$, while approximately $10\,mol$ of H^+ is pumped across the membrane. The energy initially conserved in the proton electrochemical gradient is thus approximately $10 \times 200 \times F = 200\,kJ$. If $3\,mol$ of ATP were synthesised per pair of electrons passing down the respiratory chain and the ATP is subsequently exported to the cytoplasm at a ΔG of approximately $60\,kJ/mol$, then $180\,kJ$ would be conserved, showing that the oxidative phosphorylation machinery can closely approach equilibrium and that there are no large energy losses between electron transport and ATP synthesis. In this sense, the machine can be regarded as highly efficient. However, it is important to realise that as ATP turnover increases (e.g., in an exercising muscle), the ΔG of the cytoplasmic ATP/ADP + P_i pool will be significantly lower than $60\,kJ\,mol^{-1}$. Under these conditions, the overall 'efficiency' falls as the inevitable price of running a reaction away from close-to-equilibrium conditions. This will be discussed in Section 4.8.

Comparison of oxidative phosphorylation in mitochondria with that in *Escherichia coli* (Section 5.14.2) shows that in the latter, NADH oxidation is coupled to the synthesis of fewer ATP molecules than in mitochondria. The reason for this appears to be that fewer protons are pumped for each pair of electrons flowing from NADH to oxygen. In addition, the number of c-subunits in the F_0 rotor of the ATP synthase may be higher than in mitochondria (Section 7.5). Because all other energetic parameters (ΔG and Δp) are similar, it could be said that oxidative phosphorylation is less 'efficient' in the bacterium due to failure to conserve fully the energy from respiration in the form of the Δp.

In practice, all energy-transducing membranes have a significant proton leak and thus the actual output is reduced so that equilibrium between ATP synthesis and respiration is not reached. Irreversible thermodynamics, which is beyond the scope of this book, is able to calculate that the true efficiency (i.e., power output divided by power input) is optimal when the mitochondria are synthesising ATP rapidly, because the proton leak is greatly decreased (Section 4.7) and most proton flux is directed through the ATP synthase.

3.7 THE EQUILIBRIUM DISTRIBUTIONS OF IONS, WEAK ACIDS AND WEAK BASES

The membrane potential and pH gradients across the inner mitochondrial membrane affect the equilibrium distribution of permeant ions and species with dissociable protons. These driving forces are of importance in controlling transport between the mitochondrion and its cytoplasmic environment. In addition, the equilibrium distribution of synthetic cations and dissociable species provides the basis for the experimental determination of $\Delta\psi$ and ΔpH across the inner membrane of both isolated and *in situ* mitochondria.

3.7.1 Charged species and $\Delta\psi$

As with all Gibbs energy changes, an ion distribution is at equilibrium across a membrane when ΔG, and hence $\Delta\tilde{\mu}$, for the ion transport process is zero. At equilibrium, the ion electrochemical equation (Eq. 3.26) becomes:

$$\Delta G = 0 = -mF\Delta\psi + 2.3RT\log_{10}\left\{\frac{[X^{m+}]_B}{[X^{m+}]_A}\right\} \qquad [3.40]$$

This rearranges to give the equilibrium Nernst equation, relating the equilibrium distribution of an ion to the membrane potential:

$$\Delta\psi = 2.3\frac{RT}{mF}\log_{10}\left\{\frac{[X^{m+}]_B}{[X^{m+}]_A}\right\} \qquad [3.41]$$

An ion can thus come to electrochemical equilibrium when its concentration is unequal on the two sides of the membrane. The Nernst potential is the value of $\Delta\psi$ at which an ion gradient is at equilibrium as calculated from Eq. 3.41. This is the diffusion potential condition (Section 3.8).

A membrane potential is a delocalised parameter for any given membrane and acts on all ions distributed across on a membrane. It therefore follows that a membrane potential generated by the translocation of one ion will affect the electrochemical equilibrium of all ions distributed across the membrane. The membrane potential generated, for example, by proton translocation will therefore affect the distribution of a second ion. If the second ion only permeates by a simple electrical uniport, it will redistribute until its electrochemical equilibrium is regained, and the resulting ion distribution will enable the membrane potential to be estimated from Eq. 3.41. The mitochondrial membrane potential, for example, will not be appreciably perturbed by the distribution of the second ion provided the latter is present at low concentration. This is because there is steady-state proton translocation, and any transient drop in membrane potential following redistribution of the second ion is compensated by the proton pumping.

This is the principle for most determinations of $\Delta\psi$ across energy-transducing membranes (Section 4.2.1). The equilibrium ion distribution varies with $\Delta\psi$ as shown in Table 3.4. Note the following:

(a) Anions are excluded from a negative compartment (e.g., the mitochondrial matrix).
(b) Cation accumulation is an exponential function of $\Delta\psi$.
(c) Divalent cations can be accumulated to much higher extents than monovalent cations.
(d) An ion will not distribute according to Eq. 3.41 if it can be metabolised or if it can be additionally transported by any mechanism, such as an ion pump, other than by a bilayer-mediated uniport (Section 2.3.1).

3.7.2 Weak acids, weak bases and ΔpH

An electroneutrally permeant species will be unaffected by $\Delta\psi$ and will come to equilibrium when its concentration gradient is unity (Table 3.4). Weak acids and bases (i.e.,

Table 3.4   The equilibrium distribution of ions permeable by passive uniport across a membrane

$\Delta\psi$ (mV)	$[X]_{in}/[X]_{out}$			
	Charge on ion			
	−1 (e.g., SCN^-)	0	1 (e.g., K^+/ valinomycin)	2 (e.g., Ca^{2+})
30	0.3	1	3	10
60	0.1	1	10	100
90	0.03	1	30	1,000
120	0.01	1	100	10,000
150	0.003	1	300	100,000
180	0.001	1	1000	1,000,000

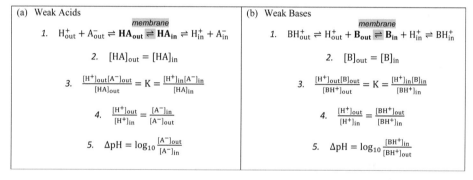

Figure 3.5   The equilibrium distribution of electroneutrally permeant weak acids and bases as a function of ΔpH.
1. Many weak acids and weak bases can cross membranes (shaded) only as the uncharged species (HA and B, respectively). *2.* At equilibrium, the concentrations of the uncharged species are the same on both sides of the membrane. *3.* If the pK is the same in both compartments, the Henderson–Hasselbalch equations for the two compartments are related. *4.* It follows that the proton gradient is inversely proportional to the gradient of the weak acid anion and is proportional to the weak base cation gradient. *5.* The pH gradient can be calculated. Note that the acid anion is concentrated in the alkaline compartment, whereas the base cation is concentrated in the acidic compartment. Thus, weak acids are used to monitor the alkaline-inside matrix pH gradient and weak bases for the acidic lumen of, for example, thylakoids.

those with a pK between 3 and 11) can often permeate in the uncharged form across bilayer regions of the membrane (Chapter 2), whereas the ionised form remains impermeant, even though it may be present in great excess over the neutral species. As a result, the *neutral* species (protonated acid or deprotonated base) equilibrates without regard to $\Delta\psi$. However, if there is a ΔpH, the Henderson–Hasselbalch equation requires that the concentration of the *ionised* species on the two sides of the membrane must differ (Figure 3.5). Weak acids accumulate in alkaline compartments (e.g., the mitochondrial matrix), whereas weak bases accumulate in acidic compartments. If the equilibrium gradient can be measured, for example, radioisotopically, then ΔpH can be estimated. This principle is widely used to determine ΔpH across energy-transducing membranes.

3.8 MEMBRANE POTENTIALS, DIFFUSION POTENTIALS, DONNAN POTENTIALS AND SURFACE POTENTIALS

There are two ways in which a true, bulk-phase membrane potential (i.e., transmembrane electrical potential difference) may be generated. The first is by the operation of an electrogenic ion pump such as operates in energy-transducing membranes. The second is by the addition to one side of a membrane of a salt, the cation and anion of which have unequal permeabilities. The more permeant species will tend to diffuse through the membrane ahead of the other ion and thus create a *diffusion potential*. Diffusion potentials may be created across energy-transducing membranes, for example, by the addition of external KCl in the presence of valinomycin, which provides permeability only for K^+, thus generating a $\Delta\psi$, positive inside. Under ideal conditions, the magnitude of the diffusion potential can be calculated from the Nernst equation (Eq. 3.41). Such diffusion potentials are usually transient because the other ion, in this case Cl^-, will slowly permeate across the membrane and so eventually (after minutes to hours) the KCl concentration will become equal on the two sides of the membrane.

3.8.1 Eukaryotic plasma membrane potentials

Diffusion potentials across inner mitochondrial and related membranes tend to be transient due to the rapid movement of counter-ions and high surface-to-volume ratio and are not generally physiologically significant. This is in contrast to eukaryotic plasma membranes where the generally slow transport processes enable potentials sometimes to be sustained for several hours. In this case, the diffusion potentials due to the maintained concentration gradients across the plasma membrane play the dominant role in determining the membrane potential. In the case in which K^+, Na^+, and Cl^- gradients exist across the membrane, the membrane potential is a function of the ion gradients weighted by their permeabilities, and it is given by the Goldman equation:

$$\Delta\psi = 2.3 \frac{RT}{F} \log_{10} \left\{ \frac{P_{Na}[Na^+]_{out} + P_K[K]_{out} + P_{Cl}[Cl^-]_{in}}{P_{Na}[Na^+]_{in} + P_K[K]_{in} + P_{Cl}[Cl^-]_{out}} \right\} \qquad [3.42]$$

Note that if only a single ion is permeant, this equation reduces to the Nernst equation (Eq. 3.41).

3.8.2 Donnan potentials

The limiting case of a diffusion potential occurs when the counter-ion is completely impermeant. This condition pertains in mitochondria due to the 'fixed' negative charges of the internal proteins and phospholipids. As a result, when the organelles are suspended in a medium of low ionic strength, such as sucrose, and an ionophore such as valinomycin is added, the more mobile cations attempt to leave the organelle until the induced potential balances the cation concentration gradient. This is a stable Donnan potential.

3.8.3 Surface potentials

Surface potentials are quite distinct from the above. Due to the presence of fixed negative charges on the surfaces of energy-transducing membranes, the proton concentration in the immediate vicinity of the membrane is higher than in the bulk phase. However, Δp is not affected because the increased proton concentration is balanced by a decreased electrical potential. The proton electrochemical potential difference across the membrane, Δp, is thus unaffected by the presence of surface potentials, although membrane-bound indicators of $\Delta\psi$, such as the carotenoids of photosynthetic membranes (Chapter 6), might be influenced. Surface potentials can also control the access of ions to transport pathways; thus, Ca^{2+} uptake by the mitochondrial Ca^{2+} uniporter (Section 9.4.1.1) is inhibited by Mg^{2+} and a range of polycations such as La^{3+} and the hexavalent ruthenium red.

4 THE CHEMIOSMOTIC PROTON CIRCUIT IN ISOLATED ORGANELLES

Theory and Practice

4.1 INTRODUCTION

Further reading: Brand and Nicholls (2011)

Chapter 3 focused on the thermodynamic 'voltage' terms of bioenergetic energy transduction. In the present chapter, we deal with the 'current' terms, centred on the proton circuit linking electron transport to the ATP synthase and a range of other functions that utilise the proton current. The molecular architecture and mechanism of the many protein complexes associated with the proton circuit are dealt with in Part 2 of the book, whereas discussion of mitochondrial function in the cellular and physiological context is reserved for Part 3.

4.1.1 Isolated mitochondria or intact cells?

Although this chapter is mainly concerned with isolated mitochondria, much of the material is applicable to other energy-transducing membranes. Increasingly techniques are being developed to allow mitochondrial bioenergetics to be determined in the intact cell with a precision approaching that for the isolated organelle (Chapter 9). It is perhaps appropriate at this stage to review some of the advantages and limitations of the two classes of preparation.

Isolated mitochondria are relatively simple and can be prepared from a range of tissues from animals of different ages. A wide range of substrates and reagents can be added directly, and one has close control over the environment of the experiment. They are generally appropriate for studies of electron transport mechanisms, mitochondrial metabolic pathways, ion transport, etc. The major limitation is that the preparation lacks the more physiological cellular context, and it is difficult to eliminate the possibility that the act of isolation alters or damages their function, particularly when small amounts of material are available. Some tissues, notably brain, are so complex that a

Bioenergetics. Doi: http://dx.doi.org/10.1016/B978-0-12-388425-1.00004-X

mitochondrial preparation from even a limited region comes from a highly heteroge-neous mixture of cells.

Intact cells retain interactions with the cytoplasm and the other organelles, avoid artefacts of cell fractionation, and allow the bioenergetic consequences of any cell biological or genetic manipulation to be investigated. However, their complexity, limited access to substrates and reagents and, frequently, the necessity to work with neonatal tissue all produce limitations.

In the continuum of experimental design, biophysical investigations gravitate towards the isolated organelle, whereas increasingly what can be termed 'mitochondrial physiology' exploits intact cell preparations. The latter will provide the focus of Part 3 of this book.

4.2 THE PROTON CIRCUIT

Further reading: Nicholls (2008b), Mitchell (2011)

The close analogy between the proton circuit and the equivalent electrical circuit (see Figure 1.3) will be emphasised not only as a simple model but also because similar laws govern the flow of energy around both circuits. In an electrical circuit, the two fundamental parameters are potential difference (in volts) and current (in amps). From measurements of these functions, other factors may be derived, such as the rate of energy transmission (in watts) or the resistance of components in the circuit (in ohms). As will be discussed in this chapter, it is possible, with reasonable precision, to quantify both the voltage (Δp) and cur-rent (proton current) components of the proton circuit in a population of mitochondria and to apply Ohm's law ($I = V/R$) to derive a further important parameter, the proton conduc-tance, C_MH^+ (reciprocal of resistance), of the mitochondrial inner membrane. Note, how-ever, that C_MH^+ is not constant but, rather, varies with Δp (Section 4.5.1).

In Figure 4.1, three extreme states of a simple electrical circuit are shown, together with highly simplified analogous proton circuits across the mitochondrial inner mem-brane (the circuit operating across a photosynthetic or bacterial membrane would be closely similar). In an electrical open circuit (Figure 4.1a), electrical potential (voltage) is maximal, but no current flows because the reduction-oxidation (redox) potential dif-ference generated by the chemical redox reactions within the battery is precisely bal-anced by the back pressure of the electrical potential. The tight coupling of the reactions within the battery to electron flow prevents any net chemical reaction. In the case of an ideal mitochondrion with no proton leak across the inner membrane, the proton circuit is open-circuited when there is no pathway for the protons extruded by the respiratory chain to re-enter the matrix (e.g., when the ATP synthase is inhibited or when there is no turnover of ATP). As with the electrical circuit, the voltage (i.e., the protonmotive force, Δp) across the membrane is maximal under these conditions. As the redox reactions are tightly coupled to proton extrusion, there would be no respiration in this limiting condi-tion, and a thermodynamic equilibrium would exist between Δp and at least one of the proton-translocating regions of the respiratory chain (Section 3.6.1).

In Figure 4.1b, the electrical and proton circuits are shown operating normally and performing useful work. Both potentials are slightly less than under open-circuit con-ditions because it is the slight disequilibrium between the redox potential difference

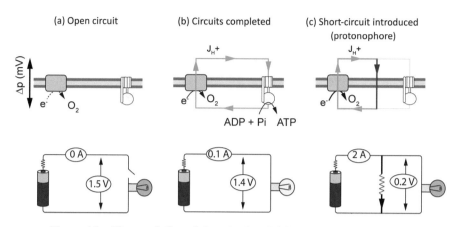

Figure 4.1   The regulation of the mitochondrial proton circuit by analogy to an electrical circuit.

The electron transport chain is simplified to a single block (see Figure 4.2). An illustrative electrical circuit (with arbitrary values) is shown above, where the protonmotive force is represented by the voltage across the circuit, and the proton current is represented by the electrical current. Voltage and current are related by Ohm's law, $I=V/R$. (a) Open circuit, no ATP synthesis, no proton leak; therefore, zero current (no respiration), potential (Δp) maximal. The battery circuit is broken by the open switch: no current flows and voltage is maximal. (b) Circuits completed, current flows (respiration occurs), useful work is done (ATP is synthesised), potential (Δp) decreases slightly due to 'internal resistance.' The light bulb has a resistance of $14\,\Omega$ and passes a current of $0.1\,A$. The voltage drops to $1.4\,V$ because of the battery's internal resistance. (c) Low-resistance short circuit introduced by addition of a high concentration of protonophore. Energy is dissipated, potentials are low, current (respiration) is high. The battery is short-circuited by a low-resistance ($0.1\,\Omega$) wire. The maximal current is drawn from the battery, but the voltage drops close to zero.

available and the back-potential in the circuit that provides the net driving force enabling the battery or respiratory chain to operate. The 'internal resistance' of the battery may be calculated from the drop in potential required to sustain a given current. Analogously, the 'internal resistance' of the respiratory chain may be estimated, and it is found to be rather low (see Figure 9.17).

An electrical circuit may be shorted by introducing a low resistance pathway in parallel with the existing circuit—for example, putting a copper wire across the battery terminals (Figure 4.1c). Current can now flow from the battery without having to do 'useful' work. Current flow is maximal, the voltage (Δp) is low, and much heat can be evolved. This 'uncoupling' can be accomplished in the proton circuit by the addition of protonophores (Section 2.3.5), enabling respiration to occur without stoichiometric ATP synthesis; a specialised class of mitochondria, in brown adipose tissue, possess a unique proton conductance pathway that performs an analogous function (Sections 9.12 and 12.4).

In practice, the proton circuits are slightly more complicated because 'useful' (ATP-synthesising) and 'wasteful' (proton leak) pathways tend to operate in parallel, and multiple proton pumps feed into the proton circuit. Using the electrical analogy, these pumps act in parallel with respect to the proton circuit and in series with respect to the pathway of electron transfer (Figure 4.2). This means that the proton current is the sum of the

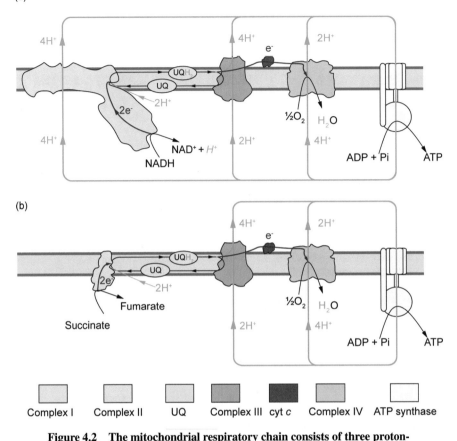

(a)

(b)

Complex I   Complex II   UQ   Complex III cyt *c*   Complex IV   ATP synthase

Figure 4.2   The mitochondrial respiratory chain consists of three proton-translocating complexes (I, III and IV) that act in parallel with respect to the proton circuit and in series with respect to the electron flow.
(a) Pathway of electron transfer for NADH-linked substrates: Complex I translocates $4H^+/2e^-$. Note the distance between the site(s) of proton translocation and the pathway of electron transport. The matrix proton (*italicised*) released in the N-phase when NADH is oxidised does not count towards the stoichiometry of proton translocation because an equivalent proton reassociates with NAD^+ when it is re-reduced to NADH by NAD^+-linked dehydrogenases. The ubiquinone (UQ) pool is mobile within the hydrophobic core of the membrane. The two protons required for the reduction of UQ to ubiquinol (UQH_2) come from the N-phase but are released to the P-phase when UQH_2 is reoxidised by complex III. The overall stoichiometry of proton translocation by the UQ pool and complex III is $4H^+/2e^-$ but $2q^+/2e^-$. Note that the pathway of electrons within complex III is simplified in this figure (see Chapter 5). Cytochrome *c* (cyt *c*) is loosely associated with the P-face of the IMM and shuttles single electrons from complex III to complex IV. Complex IV translocates $2H^+/2e^-$ but $4q^+/2e^-$, the two additional H^+ from the N-phase being required to reduce $\frac{1}{2}O_2$ to H_2O.
(b) The desaturation of a C–C bond, for example, in succinate, delivers electrons at a redox potential close to 0 mV. This type of reaction is therefore thermodynamically incapable of reducing NAD^+ and thus complex I is not involved. Succinate dehydrogenase (often called complex II) reduces UQ to UQH_2 without translocating protons. The H^+/ATP stoichiometry of the ATP synthase is discussed in Section 7.6.

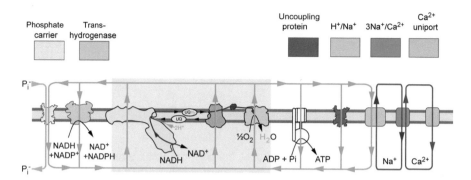

Figure 4.3   Secondary transport processes coupled to the proton circuit.
The proton current generated by the three primary proton pumps (complexes
I, III and IV) is utilised by several processes in addition to the ATP synthase,
including metabolite carriers (Section 9.5.3), the energy-linked
transhydrogenase (Section 5.12), the uncoupling protein (Section 9.12.2) and/
or proton leaks (not shown here), and Na^+ and Ca^{2+} cycling across the
membrane.

currents generated by each of the three complexes, but that each complex generates the
full protonmotive force. The pumps, which are described in more detail in Chapters 5
and 6, include complexes I, III and IV and the ATP synthase (sometimes termed
complex V). Electrons at a redox potential of approximately $-300\,mV$ can be fed into
complex I of the electron transport chain from NADH (which is oxidised to NAD^+).
Less 'energetic' electrons (at $\sim 0\,mV$) from complex II (succinate dehydrogenase), glycerol phosphate dehydrogenase, or the electron-transferring flavoprotein of fatty acid
β-oxidation enter the chain by reducing the ubiquinone (UQ) pool that links complexes I
and III, thus bypassing complex I. In addition, under artificial conditions it is possible to
introduce or remove electrons at the interfaces of an individual proton pump. For example, the redox dye TMPD can shuttle electrons from external ascorbate to cyt c, allowing
complex IV to be studied by itself in the intact mitochondrion.

   In addition to the ATP synthase and proton leak, several secondary processes can utilise the proton current (Figure 4.3), for example, to transfer electrons from NADH to
$NADP^+$ (the nicotinamide nucleotide transhydrogenase), to cycle Na^+ and Ca^{2+} across
the inner membrane (Section 9.4), and to transport metabolites such as P_i (Section 9.5).
It is notable that the proton circuit shown in Figure 4.3 is virtually identical to that originally proposed by Mitchell in 1966.

4.2.1 What do voltage and current measurements tell us?

Continuing the electrical analogy, the voltage (protonmotive force in millivolts) and current (proton current in e.g. nmol H^+ min^{-1} mg $protein^{-1}$) components of the proton circuit
are determined together or separately in parallel experiments. Proton current is generally
more informative because correctly designed experiments can give information on rates
of substrate transport, metabolism, ATP synthesis, proton leak, respiratory capacity, etc.
'Voltage' determinations (Δp or $\Delta \psi$) can readily distinguish between a 'healthy' polarised

and 'damaged' depolarised population, and more subtle measurements can detect the small changes as energy supply and demand are varied. We first discuss the proton current.

4.3 PROTON CURRENT

Further reading: Brand and Nicholls (2011)

At first sight, experimental determination of the proton current seems inherently impossible: how can one quantify the minute ion currents flowing across the inner membranes of mitochondria in an incubation? In practice, it is simple. Because there is a tight coupling between electron transport and proton extrusion in mitochondria and aerobic bacteria, for a given substrate the total proton current flowing around proton circuits in the incubation will be proportional to the rate of oxygen utilisation. The current of protons flowing around the proton circuit (J_{H^+}) may thus be readily calculated knowing the rate of respiration and the H^+/O stoichiometry:

$$J_{H^+} = (\delta O/\delta t) \times (H^+/O) \qquad [4.1]$$

For a given substrate, therefore, proton current and respiratory rate vary in parallel, and thus an oxygen electrode (see Figure 4.5) is an effective way of monitoring J_{H^+} as long as the mitochondrial H^+/O stoichiometry is known. Before discussing the proton current, it is necessary to determine the stoichiometric relationships between electron transport and proton extrusion.

4.3.1 The stoichiometry of proton extrusion by the respiratory chain

An H^+/O stoichiometry determined by any method has to satisfy the constraints imposed by thermodynamics. In other words, the energy conserved in the proton electrochemical potential has to lie within the limits imposed by the redox span of the proton-translocating region. In addition, the proton-translocating regions of complexes I and III (which are described in Chapter 5) are known to be in near-equilibrium because they can be readily reversed. Therefore, an approximate stoichiometry for these regions can be deduced on purely thermodynamic grounds, knowing ΔE_h (Section 3.3) and the components of Δp. Accurate stoichiometries have been notoriously difficult to obtain: the thermodynamic approach of equating the Gibbs energy change in reversible regions of the respiratory chain with the magnitude of the Δp under near-equilibrium conditions is beset with problems in the accurate determination of the latter parameter (as discussed later). Non-steady-state determinations of H^+ extrusion (discussed later) are also subject to controversy: it is difficult to eliminate the movement of compensatory ions masking the pH change, and some of the few protons that appear per respiratory chain complex in these experiments could conceivably originate from pK changes on the protein itself rather than as a result of transmembrane translocation. Nevertheless, most investigators agree that the NADH/O and succinate/O reactions translocate $10H^+/2e^-$ and $6H^+/2e^-$, respectively. Remember the bioenergetic convention of referring to 'O', which is equivalent to $\frac{1}{2}O_2$.

4.3.2 Experimental determination of H^+/O

Further reading: Mitchell and Moyle (1967)

The total proton current in a population of mitochondria can be determined by measuring the rate of oxygen uptake and multiplying by the H^+/O stoichiometry of proton extrusion by the electron transport chain. The stoichiometry cannot be determined directly under steady-state conditions because there is no means of quantifying the flux of protons when the rate of H^+ efflux exactly balances that of re-entry. However, it is possible to measure the initial ejection of protons that accompanies the onset of respiration before H^+ re-entry has become established.

Clearly, the demonstration of electron transport-coupled proton extrusion was central to the establishment of the chemiosmotic theory, and Mitchell demonstrated that, by making a small precise addition of O_2 to an anaerobic mitochondrial suspension in the presence of substrate and monitoring the extent of H^+ extrusion with a rapidly responding pH-electrode, thus providing a value for the H^+/O stoichiometry of the segment of the respiratory chain between the substrate and O_2. The practical details are shown in Figure 4.4. A number of precautions are necessary. First, an electrical cation permeability has to be introduced to allow charge compensation of the proton extrusion, which would otherwise be limited by the rapid build-up of $\Delta\psi$. Second, the pulse of O_2 must be small enough to prevent ΔpH from reaching its maximum value when the rate of proton leakage back into the matrix equals the rate of extrusion and no net extrusion can be detected. Even with submaximal O_2 pulses, some protons will leak back across the membrane (and thus be undetected) before the burst of respiration is completed. These protons must be allowed for. The problem is enhanced if P_i (which is nearly always present in mitochondrial preparations) is allowed to re-enter the mitochondrion during the O_2 pulse. $H^+:P_i^-$ symport (remember that this is functionally indistinguishable from OH^-/P_i antiport) is extremely active in most mitochondria, and the ΔpH-induced P_i uptake results in movement of protons into the matrix and hence an underestimate of H^+/O stoichiometry. Thus, inhibition of the phosphate symport by N-ethylmaleimide significantly increases the apparent H^+/O ratio (without affecting the real value). An alternative method for determining H^+/O ratios is based on the measurement of the initial rates of respiration and proton extrusion when substrate is added to the substrate-depleted mitochondria. This approach tends to give higher values than the oxygen pulse procedure.

4.3.3 $H^+/2e^-$ and $q^+/2e^-$ ratios for individual complexes

Electron acceptors other than O_2 can be used to select limited regions of the mitochondrial respiratory chain, allowing an $H^+/2e^-$ ratio for the span to be obtained. In addition, the charge stoichiometry (q^+/O or $q^+/2e^-$) may be determined experimentally by quantifying the compensatory movement of K^+ in the presence of valinomycin or by structural analysis of the sites of electron entry and exit. Charge and proton stoichiometry are not necessarily synonymous because electrons can enter and leave respiratory complexes on opposite sides of the membrane (Figures 3.4 and 4.2). This will become clearer in Chapter 5.

The currently accepted values for complex I are 4 for both $H^+/2e^-$ and $q^+/2e^-$ ratios. Electrons enter the complex from the matrix via NADH oxidation to NAD^+ and leave on the same side by reduction of UQ to UQH_2 (Figure 4.2a). However, complex III (UQH_2–cyt c)

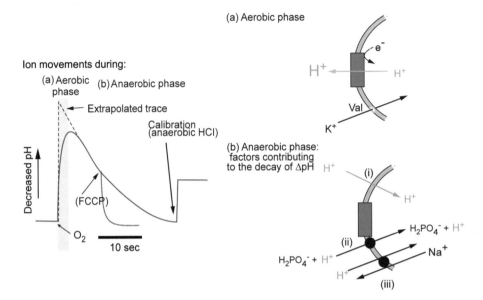

Figure 4.4   Determination of mitochondrial H$^+$/O ratios by the 'oxygen pulse' technique.
A concentrated mitochondrial suspension is incubated anaerobically in a lightly buffered medium containing substrate (e.g., succinate or β-hydroxybutryate from which matrix NADH can be generated), valinomycin and a high concentration of KCl. The pH of the suspension is continuously monitored with a fast-responding pH electrode. To initiate a transient burst of respiration, a small aliquot of air-saturated medium, containing approximately 5 nmol of O per mg protein, is rapidly injected (note that some textbooks erroneously state that hydrogen peroxide solution is added). There is a transient acidification of the medium as the respiratory chain functions for 2 or 3 s while using up the added O$_2$. Valinomycin and K$^+$ are necessary to discharge any $\Delta\psi$ that would limit proton extrusion. When O$_2$ is exhausted, the pH transient decays as protons leak back into the matrix. This decay can be due to: (i) proton permeability of the membrane (note that FCCP accelerates the decay); (ii) the action of the endogenous Na$^+$/H$^+$ antiport; or (iii) electroneutral P$_i$ entry. The trace must then be corrected by extrapolation to allow for proton re-entry which occurred before the oxygen was exhausted (Mitchell and Moyle, 1967).

has an H$^+$/2e$^-$ ratio of 4, but because electrons enter the complex from the matrix side (via UQH$_2$) and leave from the cytoplasmic face (to cyt c), the charge/2e$^-$ (q$^+$/2e$^-$) ratio is only 2. Finally, complex IV (cyt c–O$_2$) has an H$^+$/2e$^-$ ratio of 2 but a q$^+$/2e$^-$ ratio of 4.

4.3.4 The oxygen electrode: monitoring proton current

Armed with the stoichiometries given in Section 4.3.3, it is now possible to quantify the total proton current in the population by measuring the rate of oxygen consumption—that is, with some form of oxygen-sensitive electrode. After correction for nonmitochondrial respiration, it is equally valid to use the respiration of a population of intact cells to quantify the total proton current. This powerful technique for monitoring *in situ* mitochondrial function is developed in Part 3.

The Clark oxygen electrode (Figure 4.5a) has long been the most versatile tool for investigating the mitochondrial proton circuit. Although the electrode only determines directly the rate of a single reaction, the final transfer of electrons to O_2, information on many other mitochondrial processes can be obtained simply by arranging the incubation conditions so that the desired process becomes a significant step in determining the overall rate. Several such steps may be investigated, including substrate transport across the membrane, substrate dehydrogenase activity, respiratory chain activity, adenine nucleotide transport across the membrane, ATP synthase activity, and H^+ permeability of the membrane.

Three basic states of the proton circuit were shown in Figure 4.1: open circuit, where there is no evident means of proton re-entry into the matrix; a circuit completed by proton re-entry coupled to ATP synthesis; and a circuit completed by a proton leak not coupled to ATP synthesis. These states can readily be created in the O_2-electrode chamber (Figure 4.5b).

When the oxygen electrode was first being applied to mitochondrial studies, Chance and Williams (1955) proposed a convention following the typical order of addition of agents during an experiment:

State 1: Mitochondria alone (in the presence of P_i)
State 2: ADP added, respiration low due to lack of exogenous substrate
State 3: ADP and substrate present, rapid respiration
State 4: All ADP converted to ATP, respiration slows
State 5: Anoxia

In previous editions, we defined state 2 differently as respiration in the presence of substrate but in the absence of ADP. We continue to do so in this edition. However, only states 3 and 4 are in general use, and these can be further subdivided into state 3_{ADP} (substrate plus ADP), state 3_u (substrate plus sufficient protonophore to allow uncontrolled respiration), state 4 (as above), and state 4_{oligo} (substrate plus ADP plus oligomycin to inhibit the ATP synthase).

Although mitochondria contain adenine nucleotides within their matrices, the amount is relatively small (~ 10 nmol mg protein^{-1}), and when the mitochondria are introduced into the incubation, this internal pool will be very rapidly phosphorylated until it achieves equilibrium with Δp. That there is any respiration at all after this is only due to a proton leak across the inner membrane that allows some proton re-entry in the absence of net ATP synthesis. The leak is due to an endogenous "non-ohmic" proton leak that becomes most apparent at very high Δp (Section 4.5.1) or in brown adipose tissue in the presence of a specific 'uncoupling protein' (Section 9.12.2). One of the secondary factors that contribute to this proton leak is the slow cycling of Ca^{2+} across the membrane (Section 9.4.1).

4.3.5 Practical determination of proton current

Further reading: Brand and Nicholls (2011)

A range of incubation media are used for isolated mitochondria. Essential features are osmotic support (200–300 mOsM), preferably with KCl but often with sucrose or

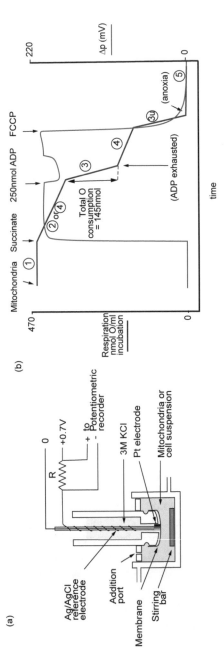

Figure 4.5 The Clark oxygen electrode.

(a) O_2 is reduced to H_2O at the Pt electrode, which is maintained 0.7V negative with respect to the Ag/AgCl reference electrode, and a current flows that is proportional to the O_2 concentration in the medium. A thin O_2-permeable membrane prevents the incubation from making direct contact with the electrodes. Because the electrode slowly consumes O_2, the incubation must be continuously stirred to prevent a depletion layer forming at the membrane. The chamber is sealed except for a small addition port. The electrode is calibrated with both air-saturated medium and, under anoxic conditions, following dithionite addition. (b) In this experiment, mitochondria were added to an oxygen electrode chamber, followed by succinate as substrate. P_i is present. Respiration is slow as the proton circuit is not completed by H^+ re-entry through the ATP synthase. That there is any respiration at all is because of a slow proton leak across the membrane. A limited amount of ADP is added, allowing the ATP synthase to synthesise ATP coupled to proton re-entry across the membrane, 'state 3.' When this is exhausted, respiration slows. The circled numbers refer to the respiratory 'states' (Section 4.3.4). If the amount of ADP is known, the oxygen uptake during the accelerated state 3 respiration can be quantified, allowing a P/O ratio to be calculated (moles ATP synthesised per mole O). Because the proton leak is greatly reduced in state 3 (Section 4.5.1), almost all the oxygen uptake during state 3 is effectively used for ATP synthesis. In this example, the ADP/O ratio for the substrate is found to be $250/145 = 1.72$. Note the bioenergetic convention of referring to 'O' (i.e., $\frac{1}{2}O_2$), which is equivalent to $2e^-$. Also, the controlled respiration prior to addition of ADP, which is strictly termed 'state 2,' is functionally similar to state 4, and the latter term is usually used for both states. The blue trace reports the values of the Δp during the experiment (Section 4.4.2).

sucrose-mannitol; a sub-micromolar Ca^{2+} concentration (sometimes with the addition of EGTA); a pH close to that of the cytoplasm (~7.0); and 2–10 mM P_i to allow ATP synthesis on addition of ADP. Albumin is frequently added if residual fatty acids are problematic.

The classical oxygen electrode chamber (Figure 4.5b) has a volume of 0.2–2 ml. The electrode is mounted either in the base of the chamber or in the lid, which is designed to be moved down until all air is expelled through the small hole that serves to make additions during the experiment. The Clark type O_2 electrode (Figure 4.5) records the current flowing between a platinum electrode and a silver reference electrode held 0.7 V positive. A thin O_2-permeable membrane prevents the incubation from making direct contact with the metal electrodes. The current is proportional to the oxygen tension, and because the electrode slowly consumes O_2, it is necessary to stir the incubation to prevent a depletion layer forming at the membrane. The electrode is calibrated with air-saturated medium and under anoxic conditions following dithionite addition. Similar measurements can be made with bacterial cells or cell-free membrane vesicles, while light-dependent oxygen evolution can be monitored with thylakoids.

Alternative multi-well plate-based equipment, using proprietary solid O_2 sensors whose fluorescence is quenched by O_2, can also be used for isolated mitochondria, although their major use is with intact cells (Section 9.6). In one device, the Seahorse extracellular flux analyser, the mitochondria are attached to the bottom of the wells by centrifugation and incubated in a volume of approximately 1 ml. Periodically, a piston descends to isolate a volume of approximately 10 μl, and the sensor records the drop in O_2 tension in this small volume as the mitochondria respire (see Figure 9.6). The piston then rises, allowing the bulk medium to re-equilibrate the O_2 tension. This cycle can be repeated indefinitely, facilitating long-term experiments that are not feasible with the Clark-type assembly.

4.3.6 Design and interpretation of oxygen electrode experiments

Let's consider a classic experiment similar to that originally performed by Chance and Williams (Figure 4.5). Mitochondria are added to an incubation containing an oxidisable substrate and P_i. A steady respiration rate is attained (our 'state 2'). ADP is then added (state 3_{ADP}), and this is followed by oligomycin to inhibit the ATP synthase so that respiration slows (state 4_{oligo}). Alternatively, if a limited amount of ADP was added, respiration will slow once the ADP is phosphorylated to ATP (state 4). Finally, sufficient protonophore is added to achieve uncontrolled respiration (state 3_u). The information content of this simple experiment is surprisingly high, revealing many aspects of mitochondrial bioenergetic function, although a full analysis must wait until protonmotive force measurements have been discussed because ultimately it is the level of Δp that largely controls respiration in state 4.

For a typical 'healthy' mitochondrion, state 2 respiration is largely controlled by the activity of the proton leak (see Figure 4.12). State 3_{ADP} is controlled to similar extents by ATP turnover (involving the ATP synthase, adenine nucleotide translocator, and phosphate transporter) and factors upstream of the proton circuit, including substrate supply and oxidation (involving substrate transport, metabolism and electron transport complexes).

State 4_{oligo} is largely controlled by the proton leak, as in state 2, and is a means to detect any abnormal 'uncoupling' in the preparation. However, state 4_{ADP}, achieved by allowing a limited ADP addition to be phosphorylated, may show a prolonged higher respiration rate if the mitochondrial preparation is contaminated with any extracellular ATPases (e.g., myosin ATPase in skeletal muscle preparations) that can regenerate ADP.

State 3_u is entirely controlled by the upstream events of substrate supply, metabolism and electron transport and is thus a powerful means to detect any upstream deficiencies. An inhibited state 3_{ADP} accompanied by a normal state 3_U is indicative of a limitation in ATP synthesis or export. For example, brown fat mitochondria (Section 9.12.2) have a high physiological proton leak pathway and a limited ATP synthase. Thus, state 3_u can be much higher than state 3_{ADP}.

A measure of 'coupling' that retains utility is the respiratory control ratio (RCR). This is an empirical parameter frequently used for assessing the integrity of a mitochondria preparation. It is based on the observation that damaged mitochondrial preparations tend to show an increased proton leak above and beyond that which is now known to be an inherent physiological property of the mitochondrion. The RCR (usually defined as state 3_{ADP}/state 4) must be qualified by the substrate used. Succinate and glycerophosphate feed electrons to the UQ pool and therefore translocate fewer protons per electron than NADH-linked substrates so that a given respiration rate generates a lower proton current. Although this does not affect the RCR directly, these substrates tend to generate a higher Δp than NADH-linked substrates. Because the inherent proton leak is highly voltage dependent (Section 4.5.1), this will increase the proton current through the leak, even though the properties of the leak are not changed. There is no 'correct' value for the RCR; depending on the tissue and the substrate, values can vary from 3 to 15 or more for well-prepared mitochondria, although an unexpectedly low RCR is an indication that the preparation may be damaged.

Brown fat mitochondria, which possess an additional regulated proton leak mediated by uncoupling protein 1 (UCP1; Section 9.12.2), generally show no respiratory control unless UCP1 is inhibited. In addition, these mitochondria have an unusually low ATP synthase activity combined with a high respiratory capacity, restricting the RCR (state 3_{ADP}/state 4). The alternative RCR (state 3_u/state 4_{oligo}) with UCP1 inhibited can be derived.

A more detailed description of respiratory control must wait until protonmotive forces are discussed. Note that most bacterial cells do not show significant respiratory control owing, presumably, to the continuous activity of the ATP synthase and other protonmotive force-driven reactions. Respiratory control can be observed in some inside-out vesicle preparations. The rate of electron transport in isolated thylakoids accelerates when ATP synthesis is occurring.

4.3.7 P/O and P/2e⁻ ratios

The P/O ratio is the number of moles of ADP phosphorylated to ATP per $2e^-$ flowing through a defined segment of an electron transfer chain to oxygen. If the terminal acceptor is not oxygen, then the term $P/2e^-$ ratio is used. Its origin lies in the era when ATP synthesis was believed to occur by a chemical coupling mechanism at three sites (two for succinate) in the electron transport chain, resulting in predicted integral P/O ratios of 3 and 2, respectively. In chemiosmotic terms, a 'pure' P/O ratio would equal the ratio between the H^+/O ratio for the electron transport chain with the substrate in question and the H^+/ATP ratio for the ATP synthase including the 'extra proton' required for the transport of adenine nucleotides and P_i across the inner membrane (see Figure 9.4) and is thus a fixed value for a given substrate. This means that the P/O ratio is of little value in assessing mitochondrial function. In practice, the presence of the parallel proton leak pathway is a complicating factor, and reported changes in P/O in preparations simply reflect changes in leak.

Historically, P/O ratios were estimated in oxygen electrode experiments from the extent of the burst of accelerated state 3 respiration obtained when a small measured aliquot of ADP is added to mitochondria respiring in state 4. Almost all the added ADP is phosphorylated to ATP, with the ATP:ADP ratio typically being at least 100:1 when state 4 is regained, and the ratio 'moles of ADP added/moles of O consumed' can be calculated. In order to correct for the proton leak, it is the convention to assume that the leak ceases during state 3 respiration, which is largely valid, due to its 'non-ohmic' nature (Section 4.5.1).

4.4 VOLTAGE: THE MEASUREMENT OF PROTONMOTIVE FORCE COMPONENTS IN ISOLATED ORGANELLES

In the electrical circuit model of the proton circuit (Figure 4.1), the oxygen electrode corresponds to the 'ammeter.' To more fully characterise the circuit, it is necessary to provide a 'voltmeter' to monitor the protonmotive force (Δp). Measurement of Δp (or, more commonly, changes in Δp (discussed later)) has played a number of roles. First, in the early days of the chemiosmotic theory, quantification of Δp was vital to establish the thermodynamic feasibility of the theory. Second, a qualitative assessment of $\Delta \psi$, the major component of Δp, has been extensively employed as a crude live/dead assay, mostly in cellular applications. Third, in combination with respiration, application of Ohm's law ($I = V/R$) has enabled the proton conductance of the inner membrane to be quantified. Finally, more sophisticated semiquantitative techniques have been employed to determine whether an intervention results in a subtle increase or decrease in Δp, as a means of complementing respiratory experiments. Under physiologically relevant conditions, Δp can decrease by 10–20% during the state 4–state 3 transition (Figure 4.5) or when Ca^{2+} is being accumulated, and it can rise slightly when substrate supply is enhanced. Parallel determination of respiration with Δp (or, more commonly, $\Delta \psi$) provides the most comprehensive information on the proton circuit, for example, allowing a distinction to be made between an increase in state 4 respiration being due to downstream utilisation of the circuit by slight uncoupling (depolarising; i.e., lowering Δp) and an increase being due to an increased upstream substrate availability (hyperpolarising; i.e., increasing Δp).

4.4.1 Early estimates of Δp

Further reading: Mitchell and Moyle (1969), Nicholls (1974)

The first determination of Δp in mitochondria dates back to 1969, when Mitchell and Moyle employed pH- and K^+-selective electrodes in an initially anaerobic, low K^+

incubation. Valinomycin was present to allow K^+ to equilibrate, and $\Delta\psi$ was calculated from the K^+ uptake. ΔpH was calculated by monitoring proton extrusion and calculating the accompanying internal acidification from the buffering capacity of the matrix. A value of 228 mV was obtained for Δp for mitochondria respiring under 'open circuit' conditions in the absence of ATP synthesis (state 4), but this is an overestimate because too low a value was taken for the matrix volume, thus exaggerating the gradients. The technique was modified for radioactive assay by substituting the β-emitter ^{86}Rb for K^+ and by using ^3H-labelled weak acids (e.g., acetate) and bases (e.g., methylamine) to determine ΔpH (Figure 3.5). The use of valinomycin in the previous experiments has a major disadvantage in that $\Delta\psi$ is artifactually clamped at a value corresponding to the Nernst equilibrium for the pre-existing K^+ gradient across the membrane. Most of the change in Δp in different states is reflected in changes in ΔpH, and it is essential that this parameter is also measured to give a meaningful estimate of Δp.

4.4.2 Estimation of membrane potential ($\Delta\psi$) by permeant ion distribution

Under most conditions, the membrane potential component of Δp is dominant, accounting for approximately 150 mV, or 80% of the total Δp. A typical ΔpH of 0.5 pH units alkaline in the matrix will add 30 mV to Δp (Eq. 3.34). However, under extreme conditions (e.g., Ca^{2+} accumulation by P_i-depleted mitochondria), ΔpH can account for up to 50% of the total Δp. Although the large majority of studies focus on monitoring changes in $\Delta\psi$, it must be remembered that this does not represent the full thermodynamic driving force for ATP synthesis. However, under most conditions in the presence of P_i, changes in $\Delta\psi$ tend to parallel similar but smaller changes in ΔpH. The pH component can be minimised by the inclusion of high P_i concentrations or addition of nigericin (in KCl-based media); the approximation is then made that $\Delta\psi$ accounts for the entire Δp.

Although there are one or two electrophysiological studies in which mitoplasts (swollen mitochondria lacking an intact outer membrane) have been patched in attempts to detect and quantify $\Delta\psi$, indirect techniques are almost universally employed, based on the determination of the equilibrium concentration gradient of a membrane-permeant monovalent cation across the inner membrane and substituting these values into the Nernst equation (Eq. 3.41). Use of these cations (or anions if the membrane potential is positive inside) is preferable to the earlier use of valinomycin plus K^+ or $^{86}Rb^+$ (Section 4.4.1), where $\Delta\psi$ is essentially clamped by the K^+ gradient across the membrane (Tedeschi, 2005; Nicholls, 2005a), because very low concentrations of permeant cation can be added, minimising perturbation to the potential. A suitable cation should be permeant across lipid bilayer regions by passive electrical uniport (Figure 2.1) and not be transported by any inner membrane carrier. This requires a structure that is predominantly hydrophobic and where a fixed positive (or negative) charge is delocalised around the molecule by a π-orbital system similar to that seen in the anionic form of the protonophore FCCP (see Figure 2.2). Examples in common use include the phosphonium cations triphenylmethyl phosphonium (TPMP$^+$) and tetraphenyl phosphonium (TPP$^+$) (Section 4.4.3). In each case, their positive charges are sufficiently shielded or delocalised to allow the cations to cross lipid bilayer regions of the inner mitochondrial membrane (IMM). The thiocyanate anion, SCN$^-$, has been used when $\Delta\psi$ is negative (positive inside).

In order to estimate $\Delta\psi$, it is necessary to calculate the ratio of the concentration of the cation in the internal compartment (e.g., matrix) to that in the external medium and to input this value into the Nernst equation (Eq. 3.41). The logarithmic relationship between $\Delta\psi$ and the ion gradient means that the technique is able to detect small changes at high potentials, but it is increasingly prone to error at low potentials.

4.4.3 Phosphonium cations

Further reading: Isaev *et al.* (1970), Serviddio and Sastre (2010)

Two approaches have been used to estimate $\Delta\psi$ from the distribution of phosphonium cations, employing radioisotopes or external macroelectrodes. In either case, it is necessary to calculate the total matrix volume by quantifying the ^3H_2O-permeable/^{14}C-sucrose-impermeable space in the mitochondrial pellet (Figure 4.6). Typical values range from 0.5 to 1 µl/mg mitochondrial protein. It is also necessary to apply a correction for the activity coefficients of the cation, particularly because they tend to be rather hydrophobic and to partition into the membrane. Activity coefficients can be estimated by comparing the gradients of K^+ and the cation across the inner membrane in the presence of the K^+ ionophore valinomycin and varying low K^+ concentrations.

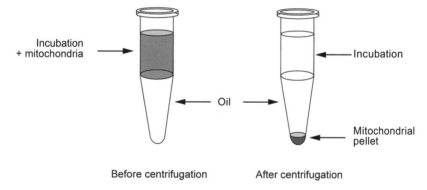

Before centrifugation   After centrifugation

Figure 4.6   Silicone-oil centrifugation of mitochondria for the determination of matrix volume.
Mitochondria are incubated under the desired conditions in media containing ^{14}C-sucrose and ^3H_2O. An aliquot of the suspension is added to an Eppendorf tube containing silicone oil with a density slightly greater than the incubation medium and centrifuged for approximately 1 min at 10,000 g. The mitochondria form a pellet under the oil and can be solubilised for liquid scintillation counting. An aliquot of the supernatant is also counted. The ^{14}C-sucrose in the incubation allows the sucrose-permeable spaces, V_s, in the pellets to be calculated. This gives the extramatrix contamination of the pellet with incubation medium (i.e., extramitochondrial space plus intermembrane space). The difference between the ^3H_2O-pemeable space in the pellet (V_h) and V_s gives the sucrose-impermeable space, which is taken to represent the matrix volume (because water but not sucrose can permeate the inner membrane). The technique can be used for additional experiments in which it is necessary to determine the proportion of a reagent or metabolite accumulated in the matrix.

Isotopic determinations using ^{3}H-TPP$^+$ or ^{3}H-TPMP$^+$ require that the mitochondria are rapidly separated from the medium by filtration or centrifugation through silicone oil (as for the matrix volume determination). Contamination of the pellet with medium is corrected for by including ^{14}C-sucrose (which does not cross the inner membrane).

Rather than quantifying the internal accumulation of the phosphonium cations directly, gradients can be calculated from the fall in external concentration as they are accumulated into the matrix. This does not require labelled cations. A simple macro-electrode can be constructed that is selectively permeable to the phosphonium cations and so can detect the fall in external concentration as the cation is accumulated into the mitochondrion. Several examples exploiting this technique are shown later, and the powerful combination in the same chamber of a TPP$^+$ electrode to monitor $\Delta\psi$ plus an oxygen electrode to monitor proton current (Figure 4.7) allows the calculation of proton

(a)

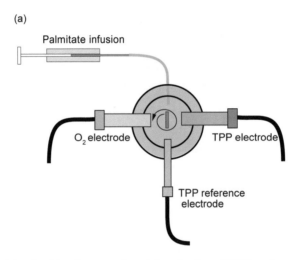

Figure 4.7   Combination tetraphenylphosphonium (TPP$^+$) and oxygen electrodes for the continuous monitoring of $\Delta\psi$, respiration, J_{H^+} and C_mH^+. Top view of a closed mitochondrial incubation chamber fitted with an O$_2$ electrode and a TPP electrode with its reference, widely used to monitor $\Delta\psi$ and respiration simultaneously, and hence to calculate C_mH^+ by Ohm's law. The TPP electrode measures the fall in external TPP$^+$ concentration as it is accumulated into the mitochondria. Knowing the matrix volume (Figure 4.6), the TPP$^+$ gradient across the IMM can be calculated, and hence the Nernst equation can be used to calculate $\Delta\psi$. Here, the chamber is equipped with the means to slowly infuse palmitate. In this experiment, the reversible activation of uncoupling protein 1 (UCP1) by palmitate is studied in brown adipose tissue mitochondria. The mitochondria are initially incubated in the presence of pyruvate, ATP, CoA and carnitine and maintain a high $\Delta\psi$ and low respiration because UCP1 is inhibited. Infusion of palmitate activates the uncoupling protein, increasing C_mH^+. At the same time, the palmitate is activated to palmitoyl CoA and then palmitoyl carnitine, supplying additional substrate to the mitochondrion. At the conclusion of the infusion, the mitochondria automatically recouple as the palmitate is activated and oxidised, allowing the fatty acid to leave its binding site on UCP1. *Source: Data from Locke et al. (1982). Continued →*

conductance (Section 4.5) and facilitates modular kinetic analysis (Section 4.8.1). Although much less employed, the anionic tetraphenyl boron or thiocyanate anion can be used with a suitable selective electrode to monitor $\Delta\psi$ where the P-side is internal (e.g., chromatophores or submitochondrial particles) (Chapter 1).

4.4.4 Extrinsic optical indicators of $\Delta\psi$

Further reading: Scaduto and Grotyohann (1999)

Discussion here is limited to investigation of isolated organelles; the complexity intro-duced by monitoring organelle function in intact cells is reserved for Part 3. In addition to the phosphonium ions discussed previously, other lipophilic cations and anions with extensive π-orbital systems that allow charge to be delocalised throughout the structure, and hence are membrane permeant, can achieve a Nernst equilibrium across energy-transducing membranes and can thus be used to monitor $\Delta\psi$. Many of these compounds have characteristic absorption spectra in the visible region, and their planar structure allows them to aggregate and form stacks when at high concentrations, such as when

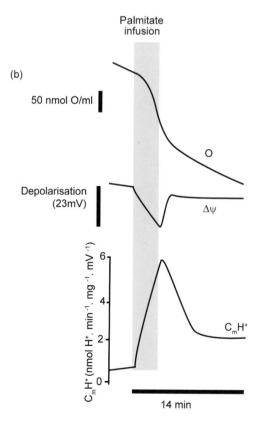

Figure 4.7 *Continued*

accumulated into the matrix, thus reducing their ability to absorb light. This phenomenon ('quenching') can be exploited to monitor the uptake of the probes by mitochondria in suspension in a cuvette from the decrease in total absorbance (or emission if the probe is fluorescent) without separating mitochondria from the incubation (Figure 4.8). Great care must be taken both in the selection of the probe and regarding the incubation conditions in order to obtain a reproducible signal because many factors other than potential can interfere with the signal. Calibration is frequently performed with reference to diffusion potentials obtained with valinomycin in the presence of varying external KCl concentrations. A particular problem is that some probes are mitochondrial inhibitors; control experiments with an oxygen electrode to establish that the highest concentration to be used does not restrict maximal respiration are therefore important when establishing conditions. No probes are perfect, but rhodamine-123 and tetramethylrhodamine methyl or ethyl esters show low toxicity.

Membrane-permeant anions are accumulated by chromatophores and sonicated submitochondrial particles with positive inside membrane potentials. The anionic bisoxonols are very sensitive, with a large fluorescent yield, but respond only slowly to changes in $\Delta\psi$, requiring up to 1 min to re-equilibrate. With intact eukaryotic cells, bisoxonols may be used in combination with fluorescent cations to monitor both mitochondrial and plasma membrane potentials.

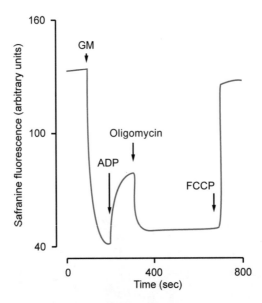

Figure 4.8   Monitoring $\Delta\psi$ in isolated mitochondria by safranine fluorescence.
Brain mitochondria were incubated in the presence of safranine and fluorescence monitored at 495 nm excitation and 585 nm emission. Where indicated, glutamate + malate (GM) ADP, oligomycin and FCCP were added. Note the partial depolarisation going from state 4 to state 3 on addition of ADP.
Source: Data from Komary et al. (2010).

4.4.5 Intrinsic optical indicators of $\Delta\psi$

Further reading: Smith (1990)

A $\Delta\psi$ of approximately $200\,\mathrm{mV}$ corresponds to an electrical field across the energy-transducing membrane in excess of $300,000\,\mathrm{V\,cm^{-1}}$. It is not surprising, therefore, that certain integral membrane constituents respond to the electrical field by altering their spectral properties. Such *electrochromism* is due to the effect of the imposed field on the energy levels of the electrons in a molecule. The most widely studied of these intrinsic probes of $\Delta\psi$ are the carotenoids of photosynthetic energy-transducing membranes. Carotenoids are a heterogeneous class of long-chain, predominantly aliphatic pigments that are found in both chloroplasts and photosynthetic bacteria. Their roles include light harvesting (Chapter 6) and protection against oxidative damage of the photosynthetic apparatus because they can trap reactive excited states of oxygen molecules. A common feature of carotenoids is a central hydrophobic region with conjugated double bonds allowing delocalisation of electrons and giving carotenoids their characteristic visible spectrum. The shifts in their absorption spectra in response to the membrane potentials experienced by energy-transducing membranes are only a few nanometres and so the signal is usually detected by dual-wavelength spectroscopy (Chapter 5). The carotenoids respond with extreme rapidity (nanoseconds or less), allowing primary electrogenic events in the photosynthetic apparatus to be followed (Chapter 6). The carotenoid band shift can be calibrated, especially in the case of chromatophores, by monitoring the band shift imposed in the dark with valinomycin and varying external KCl to generate defined potassium diffusion potential concentrations calculated from the Nernst equation (Eq. 3.41). One limitation of the technique, however, is that because carotenoids are integral membrane components, they only detect the field in their immediate environment, which need not correspond to the bulk-phase membrane potential difference measured by distribution techniques, particularly because surface potential effects could be significant. In this context, it is generally found that in chromatophore membranes the carotenoid shift gives much larger values of $\Delta\psi$ than ion distribution methods.

4.4.6 Extrinsic indicators of ΔpH

As discussed in Chapter 3, weak acids are accumulated by 10-fold per pH unit into acidic compartments and weak bases by 10-fold per pH into acidic compartments. Using similar separation techniques as for the labelled cations discussed previously to determine the gradient of radiolabelled acetate and methylamine allows negative (alkaline inside) and positive (acidic inside) pH gradients to be quantified (Figure 3.5). ΔpH can also be estimated from the fluorescence quenching of certain acridine dyes, which are weak bases and so will tend to accumulate on the acidic side of a membrane where their fluorescence may be quenched. There are often problems of quantifying such quenching in terms of a pH gradient, but they can be useful qualitative probes because small amounts of membranes may be assayed without any requirement to separate the membranes from the suspending medium.

Green fluorescent protein variants that respond to pH have been targeted to the mitochondrial matrix and cytoplasm, and their use is discussed in the cellular context in Part 3.

4.4.7 Factors controlling the contribution of $\Delta\psi$ and ΔpH to Δp

Some of the events that regulate the partition of Δp between $\Delta\psi$ and ΔpH are summarised in Figure 4.9. The mitochondrion is used as an example, but the discussion is equally valid for other energy-transducing systems. Starting from a 'de-energised' state of zero proton-motive force, the operation of an H^+ pump in isolation leads to the establishment of a Δp in which the dominating component is $\Delta\psi$ (Figure 4.9a). The electrical capacity of the membrane is so low that the net transfer of 1 nmol of H^+ mg protein^{-1} across the membrane establishes a $\Delta\psi$ of approximately 200 mV. The pH buffering capacity of the matrix is approximately 20 nmol of H^+ mg protein^{-1} per pH unit, and the loss of 1 nmol of H^+ will only increase the matrix pH by 0.05 units (i.e., equivalent to 3 mV). Δp will thus be approximately 99% in the form of a membrane potential (Figure 4.9b). In the absence of a significant flow of other ions, it will stay this way in the steady-state.

If an electrically permeant ion such as Ca^{2+}, or K^+ plus valinomycin, is now added (Figure 4.9c), its accumulation in response to the high $\Delta\psi$ will tend to dissipate the membrane potential and hence lower Δp. The respiratory chain responds to the lowered Δp by a further net extrusion of protons, thus restoring the protonmotive force. Because

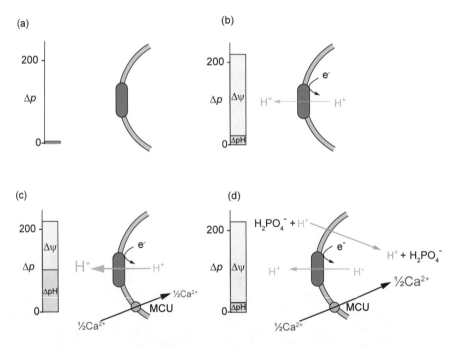

Figure 4.9   Factors controlling the partition of Δp between $\Delta\psi$ and ΔpH.
Starting from a non-respiring mitochondrion (a), the initiation of respiration (b) leads to a high $\Delta\psi$ and low ΔpH because the electrical capacitance of the membrane is very low. (c) Addition of a permeant cation such as Ca^{2+} (which is transported through the mitochondrial Ca^{2+} uniporter (MCU; Section 9.4.1) with two charges) starts to collapse $\Delta\psi$; this allows further proton extrusion leading to a build-up of ΔpH. If a permeant weak acid such as P_i is additionally present, it will be accumulated with a proton, driven by ΔpH (d). This will tend to neutralise the ΔpH, allowing $\Delta\psi$ to be re-established, thus allowing further Ca^{2+} uptake.

of the pH buffering capacity of the matrix, the uptake of 10 nmol of Ca^{2+} (or 20 nmol of K^+), balanced by the extrusion of 20 nmol of H^+, will lead to the establishment of a ΔpH of approximately -1 unit (equivalent to 60 mV). Because the respiratory chain can only achieve the same total Δp as before, this means that the final $\Delta\psi$ must be nearly 60 mV lower than before uptake of the cation. Thus, cation uptake in the absence of anion transport leads to a redistribution from $\Delta\psi$ to ΔpH. The lowered $\Delta\psi$ means that cation uptake under these conditions becomes self-limiting, as the driving force steadily decreases until equilibrium is attained. For example, the uptake of Ca^{2+} by mitochondria in exchange for extruded protons is limited to approximately 20 nmol mg protein^{-1} in the strict absence of permeant weak acids, by which time $\Delta\psi$ has decreased (and -60ΔpH increased) by approximately 120 mV.

The third event that can influence the relative contributions of $\Delta\psi$ and ΔpH is the redistribution of electroneutrally permeant weak acids and bases (Figure 4.9d). Uptake of a weak acid in response to the ΔpH created by prior cation accumulation dissipates the pH gradient and allows the respiratory chain to restore $\Delta\psi$. However, the net accumulation of both cation and anion can result in osmotic swelling of the mitochondrial matrix (Figure 2.3b). This does not occur when the ions are Ca^{2+} and P_i because formation of an osmotically inactive calcium phosphate complex prevents an increase in internal osmotic pressure (Section 9.4.1).

It is clear from the previous discussion that $\Delta\psi$ and ΔpH indicators—being ions, weak acids, or weak bases—can disturb the very gradients to be measured unless care is taken. This is particularly true in the presence of valinomycin because the ionophore brings into play the high endogenous K^+ of the matrix, with the result that $\Delta\psi$ will become clamped at the value given by the initial K^+ gradient. This risk is less apparent with cations such as TPP^+ that can be employed at very low concentrations.

It is a common misconception even nowadays that the mitochondrial protonmotive force is essentially just a pH gradient. We stress again that due to the low electrical capacitance of membranes, the default position is that the membrane potential should be the dominant contributor to the protonmotive force (Section 4.4.7). The same applies to most bacterial species. Possibly the confusion stems in part from the situation with the thylakoids of chloroplasts where indeed the pH gradient is the dominant term. However, this is the anomaly and is a consequence of Cl^- and Mg^{2+} movements across the thylakoid membrane, which allow substantial net proton accumulation in the interior. Note that the thylakoid lumen, with a pH of approximately 4, does not contain any enzymes, whereas the cytoplasm and mitochondrial matrix both contain multiple pH-sensitive metabolic pathways evolved to function over a narrow pH range.

4.5 PROTON CONDUCTANCE

In an electrical circuit, the conductance of a component is calculated from the current flowing per unit potential difference. A similar calculation for the proton circuit enables the effective proton conductance of the membrane $(C_M H^+)$ to be calculated:

$$C_M H^+ = J_{H^+} / \Delta p \qquad [4.2]$$

where J_{H^+} is the proton current per milligram of mitochondrial protein. This can be illustrated by a simple example:

Liver mitochondria oxidising succinate in 'state 4' might typically respire at 15 nmol O min^{-1}mg^{-1} and maintain a Δp of 220 mV. If the H$^+$/O ratio for the span succinate-oxygen is 6, then

$$J_{H^+} = (\delta O/\delta t) \times (H^+/O) = 15 \times 6 = 90 \text{ nmol H}^+ \text{ min}^{-1} \text{ mg}^{-1}$$
$$C_M H^+ = J_{H^+}/\Delta p = 90/220 = 0.4 \text{ nmol H}^+ \text{ min}^{-1} \text{ mg}^{-1} \text{ mV}^{-1}$$

If now the protonophore FCCP is added and Δp drops to 40 mV while respiration increases to 100 nmol O min^{-1}mg^{-1}, then the new values are as follows:

$$J_{H^+} = (\delta O/\delta t) \times (H^+/O) = 100 \times 6 = 600 \text{ nmol H}^+ \text{ min}^{-1} \text{ mg}^{-1}$$
$$C_M H^+ = J_{H^+}/\Delta p = 600/40 = 15 \text{ nmol H}^+ \text{ min}^{-1} \text{ mg}^{-1} \text{ mV}^{-1}$$

The magnitude of the endogenous proton conductance of the membrane is a central parameter underlying the bioenergetic behaviour of a given preparation of mitochondria and other energy-transducing membranes. Evidently, for an efficient transduction of energy, $C_M H^+$ should be as low as possible. Note that the respiratory chain does not distinguish between a Δp that is lowered by ATP synthesis and one that is lowered to the same extent by the addition of a proton translocator.

The combination of an oxygen electrode and a TPP$^+$ electrode (Figure 4.7) recording from a mitochondrial suspension is a powerful tool for continuously monitoring the proton circuit. When conditions are optimised to minimise the ΔpH component of Δp (phosphate present or nigericin addition in a KCl-based medium), $\Delta\psi$ monitored by the TPP$^+$ electrode will approximate Δp and together with the rate of respiration will allow J_{H^+} and $C_M H^+$ to be calculated. In Section 9.12, we illustrate this approach with reference to highly specialised mitochondria in brown adipose tissue that demonstrate a regulated proton leak pathway.

4.5.1 The basal proton leak

Further reading: Brand *et al.* (1994)

A proton leak exists in all mitochondria, including those that do not express uncoupling proteins. This basal leak is responsible for the state 4 respiration seen in even the most carefully prepared mitochondria and accounts for a high proportion of the basal metabolic rate in tissues such as skeletal muscle. The molecular basis of the leak is unclear; it does not correlate with the phospholipid composition of the bilayer, but it has been suggested that it may correlate with the level of the adenine nucleotide translocator, protons leaking across the membrane at the junction between protein and lipid. Within the context of electrical analogy for the proton circuit, the basal leak displays a non-ohmic current/voltage relationship, which can be investigated by progressively restricting respiration. Proton conductance is most apparent at very high Δp and decreases more than proportionately with Δp (Figure 4.10). This 'non-ohmic' I/V relationship suggests that

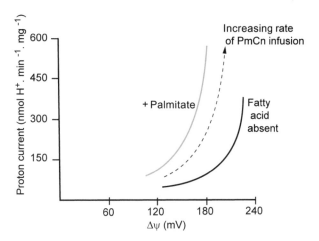

Figure 4.10   Determining the current voltage (*I/V*) relationship for the brown adipose tissue UCPI for BAT mitochondria in the presence or absence of fatty acids.

Brown adipose tissue mitochondria were incubated in state 4 in the presence of excess ATP to inhibit the uncoupling protein. Palmitoyl carnitine (PmCn) was infused as substrate at varying rates, and respiration and $\Delta\psi$ were determined as in Figure 4.7 to generate an *I/V* curve for the inherent inner membrane proton leak in the absence of the uncoupling protein UCP1. The experiment was then repeated in the presence of 1.3 µM palmitate to activate the uncoupling protein. Note that the conductances are 'non-ohmic' and that fatty acid lowers $\Delta\psi$. *Source: Redrawn from Rial et al. (1983).*

the endogenous leak may have evolved to limit the maximal value of Δp attainable by mitochondria or other membranes. Because the production of reactive oxygen species by mitochondria is highly dependent on Δp (Section 9.10), proton leaks, whether endogenous or uncoupler protein mediated, may serve the purpose of restricting oxidative damage (but see Section 9.12.1). The voltage gating means that under conditions of intense ATP synthesis (i.e., state 3), the accompanying relatively small drop in Δp means that the leak pathway will be inhibited and almost all the proton current will be directed through the ATP synthase.

4.6 ATP SYNTHASE REVERSAL

The ATP synthase is reversible and is only constrained to run in the direction of net ATP synthesis by the continual regeneration of Δp and the use of ATP by the cell. If the respiratory chain is inhibited and ATP is supplied to the mitochondrion, or if sufficient Ca^{2+} is added to depress Δp below that for thermodynamic equilibrium with the ATP synthase reaction, the enzyme complex functions as an ATPase, generating a Δp comparable to that produced by the respiratory chain. The proton circuit generated by ATP hydrolysis must be completed by a means of proton re-entry into the matrix. Proton translocators therefore accelerate the rate of ATP hydrolysis, just as they accelerate the rate of respiration; this is the 'uncoupler-stimulated ATPase activity,' which is of particular importance in the cellular context, when mitochondrial dysfunction can cause ATP synthase reversal and drain glycolytically generated ATP. In some circumstances, an 'inhibitor protein,' IF1, can limit this reversal (Sections 7.7 and 9.13).

The classic means of discriminating whether a mitochondrial energy-dependent process is driven directly by Δp or indirectly via ATP is to investigate the sensitivity of the process to the ATP synthase inhibitor oligomycin. A Δp-driven event would be insensitive to oligomycin when the potential was generated by respiration, but it would be sensitive when Δp was produced by ATP hydrolysis. The converse would be true of an ATP-dependent event. If Δp or $\Delta \psi$ is being monitored, mitochondria (isolated or *in situ*), which are net generators of ATP, will hyperpolarise (i.e., Δp will increase) on addition of oligomycin, whereas those whose Δp is supported by ATP hydrolysis will depolarise. This 'null-point' assay is a simple way of monitoring mitochondrial function within cells (Figure 12.2).

Under physiological conditions, the mitochondrial ATP synthase will not normally be called upon to act as a proton-translocating ATPase, except possibly during periods of anoxia when glycolytic ATP could be utilised to maintain the mitochondrial Δp. However, some bacteria, such as *Streptococcus faecalis* when grown on glucose, lack a functional respiratory chain and rely entirely on hydrolysis of glycolytic ATP to generate a Δp across their membrane and enable them to transport metabolites.

4.7 REVERSED ELECTRON TRANSPORT

The near equilibrium in state 4 between Δp and the redox spans of complexes I and III suggests that conditions could be devised in which these segments of the respiratory chain could be induced to run backwards, driven by the inward flux of protons. Note that this does not apply to complex IV, which is thermodynamically irreversible. Reversed electron transfer (RET) may be induced in two ways, either through generating a Δp by ATP hydrolysis or by using the flow of electrons from succinate or cyt c to O_2 to reverse electron transfer through complexes I or I and III, respectively (Figure 4.11). RET can most clearly be demonstrated in inverted submitochondrial particles where the NAD^+ binding site of complex I is accessible to added NAD^+. Such flow of electrons (e.g., from succinate) involves the majority of the electron flux passing to O_2 and thereby generating Δp, whereas a minority is driven energetically uphill to reduce NAD^+ at the expense of Δp. Note, however, that an actual reversal will only be transient in intact mitochondria due to the limited NAD^+ pool. What is commonly referred to as reversed electron transport is an increased steady-state reduction level of the NAD^+ pool in the presence of a substrate feeding into the UQ pool, relative to an NAD^+-linked substrate, due to the higher Δp that can be generated by bypassing complex I. This in turn occurs because complex I is thermodynamically 'weaker' than complex III; that is, it reaches equilibrium between electron flux and proton pumping at a slightly lower Δp than does complex III. Thus, forward electron flow through Complex II (e.g., from succinate) can drive reverse electron flow through complex I. Note also that RET does not occur in eukaryotic cells because succinate is generated by NAD-dependent dehydrogenases in the tricarboxylic acid cycle and the net electron flux through complex I

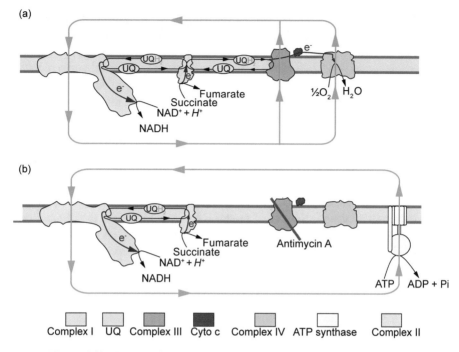

Figure 4.11   Reversed electron transfer (RET) in the mitochondrial respiratory chain.
Proton circuits for submitochondrial particles incubated in the presence of NAD^+. (a) A Δp is generated by succinate oxidation. Most of the electrons pass through complexes III and IV to generate a large Δp that drives complex I in reverse, causing NAD^+ reduction; that is, succinate acts as donor of electrons not only for conventional forward electron transfer through complexes III and IV but also for reversed electron transfer through complex I. Note that although conditions can be designed for RET with submitochondrial particles and isolated mitochondria, RET does not occur in intact mammalian cells. (b) Complex III is inhibited by antimycin A and Δp is generated by ATP hydrolysis. Succinate merely donates electrons for reversed electron transfer through complex I. Note the unconventional directions of the arrows.

is always forward. In contrast, bacteria are more versatile (Chapter 5). Reversed electron transport driven by Δp generated through respiration is an essential process in some bacterial species (Chapter 5).

4.8 MITOCHONDRIAL RESPIRATION RATE AND METABOLIC CONTROL ANALYSIS

Respiratory control is a fundamental property of mitochondria. That is, their respiration is usually controlled by the rate of re-entry of protons through the ATP synthase (plus the endogenous leak and any other proton-consuming pathways) (Figure 4.3). The activity of the ATP synthase is in turn linked to the rate of ATP synthesis, which in the cell is governed by the rate of ATP turnover in the cytoplasm. So how does the respiratory chain know how fast to operate? The control of respiration is a complex function shared between different bioenergetic steps, and the control exerted by these steps can differ between mitochondria and under different metabolic conditions for the same mitochondria. The quantitative analysis of these fluxes is termed *metabolic control analysis* and will be discussed later. Here, we first give a simplified explanation based on oxygen electrode experiments of the type depicted in Figure 4.5 to discuss respiratory control.

A fundamental factor that controls the rate of respiration is the thermodynamic disequilibrium between the redox potential spans across the proton-translocating regions of the respiratory chain and Δp. In the absence of ATP synthesis, respiration is automatically regulated so that the rate of proton extrusion by the respiratory chain precisely balances the rate of proton leak back across the membrane. If proton extrusion were momentarily to exceed the rate of re-entry, Δp would increase, the disequilibrium between the respiratory chain and Δp would in turn decrease, and respiratory chain activity would decrease, restoring the steady state. Again, the electrical circuit analogy is useful here.

In the example shown in Figure 4.5, respiration is disturbed by the addition of exogenous ADP, mimicking an extramitochondrial hydrolysis of ATP such as would occur in an intact cell. The added ADP exchanges with matrix ATP via the adenine nucleotide translocator, and as a result, the ΔG_p for the ATP synthesis reaction in the matrix is lowered, disturbing the ATP synthase equilibrium. The following events then occur sequentially (but note that the gaps between them would be on the millisecond timescale):

(a) The ATP synthase operates in the direction of ATP synthesis and proton re-entry to attempt to restore ΔG_p.
(b) The proton re-entry lowers Δp.
(c) The thermodynamic disequilibrium between the respiratory chain and Δp increases.
(d) The proton current, and hence respiration, increases.

This accelerated state 3 respiration is once more self-regulating so that the rate of proton extrusion balances the (increased) rate of proton re-entry across the membrane. Net ATP synthesis, and hence state 3 respiration, may be terminated in three ways:

(a) When sufficient ADP is phosphorylated to ATP for thermodynamic equilibrium between the respiratory chain and Δp to be regained.
(b) By preventing adenine nucleotide exchange across the membrane with an inhibitor such as atractyloside (also called atractylate).
(c) By inhibiting the ATP synthase, for example, by the addition of oligomycin.

Energy transduction between the respiratory chain and the protonmotive force is extremely well regulated in that a small thermodynamic disequilibrium between the two can result in a considerable energy flux. Δp drops by less than 20% when ADP is added to induce maximal state 3 respiration, and then it recovers as ATP synthesis proceeds to completion (Figure 4.5). The actual disequilibrium between the respiratory chain and Δp can be even less because the ΔE values across proton translocation segments of the respiratory chain may also decrease in state 3. Effective energy transduction during state 3 is also apparent at the ATP synthase. A high rate of ATP synthesis can be maintained with only a slight thermodynamic disequilibrium between Δp and ΔG_p.

Protonophores uncouple oxidative phosphorylation by inducing an artificial proton permeability in bilayer regions of the membrane. They may thus be used to override the inhibition of proton re-entry that results from an inhibition of net ATP synthesis. As a consequence, proton translocators such as FCCP can induce rapid respiration, regardless of the presence of oligomycin or atractylate or the absence of ADP.

4.8.1 Metabolic control analysis

Further reading: Brand (1997), Brand and Nicholls (2011)

In Section 4.8, we explained a simplified model of the connections between mito-chondrial respiration, Δp and $C_M H^+$. However, in practice, there are many more steps interacting with the proton circuit, including the supply and transport across the inner membrane of substrate; the supply of electrons to the respiratory chain via the metab-olite dehydrogenases; and, at the other end of the circuit, the activity of the adenine nucleotide translocator and the rate of ATP turnover.

One approach to this complexity is to invoke non-equilibrium thermodynamics where fluxes are described in terms of the net thermodynamic driving forces under near-equilibrium conditions, but the most useful technique is to apply quantitative metabolic control analysis (MCA) to provide a simple description of how control is distributed between multiple steps.

'Control' has a precise meaning in MCA. Consider a simple metabolic pathway com-prising two enzymes, E_1 and E_2, where the overall flux through the pathway in steady state is J; that is,

$$A \xrightarrow{E_1} B \xrightarrow{E_2} C \longrightarrow J \qquad [4.3]$$

The *flux control coefficient* C relates changes in the overall flux through the pathway to changes in the activity of an enzyme or transport process. Strictly, it is defined as the frac-tional change in flux divided by the fractional change in the amount of the enzyme as the change tends to zero; that is, for E_1 in the previous example, the control coefficient $C_{E_1}^J$ equals

$$C_{E_1}^J = \lim_{\delta E \to 0} \frac{\delta J/J}{\delta E_1/E_1} \qquad [4.4]$$

We can illustrate this with a simple example. Consider a mitochondrion respiring in state 3. If we deliberately alter the activity of a single step in the overall sequence, such as the adenine nucleotide translocator, by a small fraction, say 1%, what effect does this have on the overall respiration rate? Two extreme results are possible in this type of experiment. First, the change in flux through the entire pathway may be the same per-centage as the change in activity of the single step (i.e., 1%). In this case, the flux con-trol coefficient of the adenine nucleotide translocator would be said to be 1. The second extreme would be when a 1% change of the translocator activity had no effect on the overall flux. In this case, the step would have a flux control coefficient of 0.

In practice, flux control coefficients of 1 are rare; the idea of a single rate-determining step, to which a flux control coefficient of 1 corresponds, although often encountered in chemical reactions, rarely applies to metabolic sequences. Instead, there is an interplay between many steps, each of which may have significant flux control coefficients, with values in nonbranched pathways between 0 and 1. The *summation theorem* states that in any pathway, the sum of all the individual flux control coefficients is always 1; that is, for the pathway in Eq. 4.3,

$$C_{E_1}^J + C_{E_2}^J = 1 \qquad [4.5]$$

The *elasticity coefficient* ε is the fractional change in activity of an enzyme or transport process in response to a small change in its substrates, products, or other effectors. For the example in Eq. 4.3, consider how the activity V_{E_2} of enzyme E_2 responds to changes in concentration of its substrate B:

$$\varepsilon_B^{E_2} = \lim_{\delta B \to 0} \frac{\delta V_{E_2}/V_{E_2}}{\delta B/B} \qquad [4.6]$$

Finally, the *connectivity theorem* states that the products of the flux control coefficient and the elasticity to a given substrate for all enzymes connected by that substrate add up to 0.

4.8.2 Bottom-up analysis

If all this seems a little dry and theoretical, we now apply MCA to mitochondrial oxidative phosphorylation, which is especially suited to this type of analysis. For an isolated mitochondrion, some processes that can be analysed are summarised in Figure 4.12. Two approaches are possible—bottom-up and top-down. Bottom-up analysis examines the effects of titrating specific mitochondrial enzymes and transporters with irreversible inhibitors and determining the effects on respiratory rate and ATP synthesis.

As will be discussed in Chapter 12, even modest restrictions in respiratory chain capacity *in vivo* greatly sensitise neurons to damaging stimuli. Indeed, chronic complex I restriction in animal models can simulate the neurodegenerative characteristics of Parkinson's disease, whereas complex II inhibition reproduces the damage to the striatum found in Huntington's disease. A bottom-up approach has been made for isolated brain mitochondria by titrating mitochondria with high-affinity inhibitors acting on individual complexes. For example, flux control coefficients of 0.29, 0.2, and 0.13 were determined in state 3 for complexes I, III and IV, respectively, of presynaptic mitochondria (Davey *et al.*, 1998). The high coefficient seen at complex I is consistent with the sensitivity of the neuron to even slight inhibition of this complex (Chapter 12).

4.8.3 Top-down (modular) analysis

A limitation of the bottom-up technique is the requirement for irreversible inhibitors and the need to know each system component, which is difficult, especially for more complex systems such as cells or tissues. One solution is to simplify the mitochondrion by grouping processes into just three 'modules' linked by the common intermediate, Δp (Figure 4.12). The modules are substrate oxidation (substrate transport, metabolism and

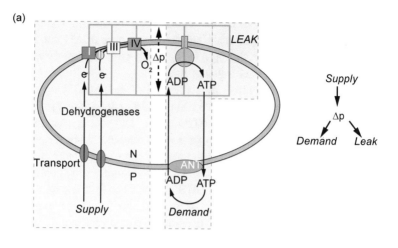

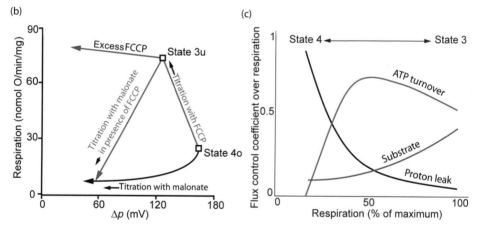

Figure 4.12 Modules for metabolic control analysis.
(a) The complexity of the bioenergetic pathways may be lessened for metabolic control analysis by grouping processes together as three modules ('supply,' 'demand' and 'leak'), which are linked by a single common intermediate (here Δp). ANT, adenine nucleotide translocator. (b) Modular kinetic analysis of the above. Kinetic responses of the three modules to Δp for isolated mitochondria oxidising succinate. Black line, response of proton leak to Δp by titrating state 4 respiration with malonate; red line, response of substrate supply (oxidation rate) to Δp by titrating state 4 respiration with FCCP; blue line, response of ATP turnover (plus proton leak) to Δp by titrating state 3_u respiration with malonate. (c) Flux control coefficients of the three modules over respiration rate when rates of ATP turnover are varied by titration with hexokinase in the presence of glucose and ATP.
Source: Adapted from Brand and Nicholls (2011).

electron transport), ATP turnover (ATP synthesis, transport and turnover), and proton leak (all other reactions that utilise Δp) (Figure 4.3). The next stage is to determine the kinetic response of each module to a change in Δp. This is done by titrating another module with appropriate activators or inhibitors. The response of substrate oxidation to Δp can be determined by titrating the proton leak with a protonophore (Figure 4.12).

The response of proton leak to Δp can be followed by titrating respiration with inhibitors in the absence of ATP turnover, whereas the response of ATP turnover to Δp can be found by titrating respiration during ATP synthesis with respiratory inhibitors and subtracting the proton leak for each value of Δp.

These titrations allow the elasticity coefficient of each block toward $\Delta \psi$ to be calculated. Finally, the connectivity theorem can be used to determine the flux control coefficients of the blocks over respiration, ATP turnover and proton leak. Figure 4.12c shows the control exerted by the three blocks in isolated mitochondria over respiration. Control is shared between multiple steps and the distribution changes with metabolic state during a transition from state 4 to state 3. The mitochondria are supplied with succinate as respiratory substrate together with ADP and P_i. The initial state 4 is attained when the net conversion of ADP and P_i to ATP ceases. As intuitively expected, the overall flux control coefficient of the set of reactions—adenine nucleotide translocation, ATP synthesis and consumption of ATP ('ATP turnover')—is 0 in state 4. In the previous sections, we made the simplification that the proton leak across the mitochondrial membrane completely controls the respiration in state 4 (i.e., has a flux control coefficient of 1). However, more careful analysis shows that although the proton leak is indeed dominant (flux control coefficient of 0.9), there is also significant control (coefficient of 0.1) in the set of reactions catalysing transport of succinate into the mitochondrion and its oxidation by the electron transport chain.

If glucose and incremental amounts of hexokinase are now added, to accelerate ATP turnover, respiration will steadily increase until the rate of ATP synthesis reaches a maximum and the mitochondria are in state 3. The first additions of hexokinase each cause a marked increase in the respiration rate and thus the flux control coefficient of the 'ATP turnover' reactions is high, corresponding to the classic respiratory control. As further hexokinase is added, other components of the ATP turnover reactions, particularly the adenine nucleotide translocator, assume an increasing share of the control. Concomitantly, the control by hexokinase becomes a progressively smaller component of the control exerted by the ATP turnover reactions. At the limit of state 3 respiration, further additions of hexokinase are without effect on the respiration rate, and thus its flux control coefficient falls to 0, the classic 'uncontrolled respiration' (state 3).

In state 3, control is shared almost equally (Figure 4.12c) between ATP turnover reactions and those of 'succinate utilisation.' More detailed analysis shows that it is distributed between the adenine nucleotide translocator, the dicarboxylate translocator, the cytochrome bc_1 complex and cytochrome oxidase. As the respiration rate alters between the extremes of state 3 and state 4, the quantitative contribution of each of these components varies; for example, control due to the adenine nucleotide translocator rises to its greatest flux control strength at 75% of the maximum respiration rate. *The important outcome of this analysis is that neither in state 3 nor in state 4 is a single step responsible for the control of the mitochondrial respiration rate.* Traditional attempts to correlate respiratory control with the [ATP]/[ADP][P_i] ratio or a single irreversible step in the electron transport chain (e.g., a step in the cytochrome oxidase reaction) are thus not tenable.

Top-down analysis can be used for more complex systems such as intact cells and can include cytoplasmic metabolic blocks such as glycolysis and cellular ATP turnover. In an intact cell, the factors controlling mitochondrial respiration rate will be more varied and complex than those considered in the previous paragraphs. The major respiratory substrate will not be succinate but, rather, NADH generated in the matrix. There will also be important differences between mitochondria from different cell types. Mitochondria in a liver cell respire at a rate intermediate between state 3 and state 4. Control analysis shows that this rate is controlled by processes (e.g., glycolysis, fatty acid oxidation and the tricarboxylic acid cycle) that supply mitochondrial NADH (flux control coefficient of 0.15 to 0.3) by the proton leak (flux control coefficient of 0.2) and by the ATP turnover reactions (flux control coefficient of 0.5). Oxidation of NADH is less important with a flux control coefficient between 0 and 0.15. Fluctuations in rate can be caused by hormones or increases in cytoplasmic and matrix Ca^{2+} via three separate effects: alteration of either ATP turnover, NADH supply, or proton leak. Each of these effects may be important. Muscle mitochondria can experience periods of resting activity when they may be close to the state 4 respiration rate but upon initiation of contraction the ATP demand and raised Ca^{2+} may be such as to cause transition to state 3. If anaerobiosis approaches, then the rate of respiration could conceivably pass transiently through a stage where cytochrome oxidase has a higher flux control coefficient due to restriction on the supply of oxygen.

4.9 KINETIC AND THERMODYNAMIC COMPETENCE OF Δp IN THE PROTON CIRCUIT

Although bioenergetics has for many years been grounded on the validity of the basic principles of the chemiosmotic theory, it is instructive to review some of the key experiments that provided a stringent test of the theory, in particular supporting the concept that the generation of Δp was a primary event rather than a secondary consequence of an alternative coupling mechanism.

4.9.1 ATP synthesis driven by an artificial protonmotive force

Further reading: Jagendorf (2002)

An artificially generated Δp must be able to cause the net synthesis of ATP in any energy-transducing membrane with a functional ATP synthase. The first demonstration that this was so came from chloroplasts after equilibration in the dark at acid pH. They could be induced to synthesise ATP when the external pH was suddenly increased from 4 to 8, creating a transitory pH gradient of 4 units across the membrane, the *acid bath experiment* (Figure 4.13). For many years, this experiment has been interpreted on the basis that thylakoids normally operate with ΔpH as the main component of Δp due to the ease with which Cl^- redistributes across the thylakoid membrane to collapse $\Delta\psi$ (Chapter 6).

An important corollary of this experiment is that the ATP synthase can be driven by ΔpH alone. There is no thermodynamic objection to this, but it is currently being argued (Chapter 7) that mechanistically a $\Delta\psi$ is required. In this context, it has been recently proposed that the acid bath experiment should be reinterpreted. It is argued that the transition to higher external pH was accompanied by the efflux of the succinate monoanion (towards

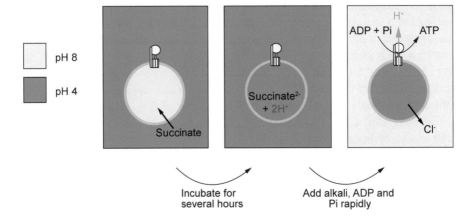

Figure 4.13 The 'acid bath' experiment: a ΔpH can generate ATP.
Thylakoid membranes were incubated in the dark at pH 4 in the presence of electron transport inhibitors in a medium containing succinate, which slowly permeated into the thylakoid space, liberating protons and lowering the internal pH to approximately 4. The external pH was then suddenly raised to 8, creating a ΔpH of 4 units across the membrane. A Δψ may also be induced (see text). ADP and P_i were simultaneously added, and proton efflux through the ATP synthase led to the synthesis of approximately 100 mol of ATP per mole of synthase. Cl^- efflux and other ion movements occurred for charge balance. Protonophores such as FCCP inhibited the ATP production.

which the membrane is claimed to be permeable), thus generating a diffusion potential, positive inside, for this ion. Thus, an induced Δψ would be at least part of the driving force for the ATP synthesis seen in this type of experiment. It remains to be finally decided if this reinterpretation is valid, but it is important to understand that it makes no difference to the validity of the experimental approach for showing that an imposed protonmotive force, independent of electron transport reactions, can drive ATP synthesis.

For an analogous acid bath experiment with mitochondria or bacteria, an ionophore such as valinomycin is needed to allow movement of compensating charge. Submitochondrial particles, which are inverted relative to intact mitochondria, are treated with valinomycin to render them permeable to K^+, incubated at low pH in the absence of K^+ to acidify the interior, and then transferred to a medium of higher pH containing K^+ along with ADP and P_i. K^+ entry creates a diffusion potential, positive inside, and this, together with the artificial ΔpH that has just been created, generates a short-lived Δp. Protons exit through the ATP synthase, generating a small amount of ATP. K^+ enters on valinomycin to maintain charge balance. Eventually, the K^+ and H^+ gradients run down to the extent that ATP synthesis ceases. An analogous approach has been used to demonstrate Δp-driven secondary active transport (Chapter 8).

4.9.2 Kinetics of proton utilisation

If Δp is the intermediate between electron transport and ATP synthesis, then the sudden imposition of an artificial Δp of comparable magnitude to that normally produced by the respiratory chain should lead to ATP synthesis with minimal delay and at an initial rate comparable to that seen in the natural process. In other words, the proton circuit requires a cause-and-effect relationship. Tests of kinetic competence have been made for both the thylakoid and the submitochondrial particle systems as described in Section 4.9.1, except that the protonmotive force was imposed by rapid mixing. The subsequent reaction period can be altered by varying the length of tubing between the mixing and quenching points (where the reaction is terminated by concentrated acid). In this way, ATP synthesis on the millisecond timescale can be followed. In both preparations, ATP synthesis was initiated with no significant lag and at initial rates comparable to those seen for energy transduction using light (thylakoids) or respiration (submitochondrial particles). Indeed, in the case of the submitochondrial particles, the onset of ATP synthesis was more rapid than following initiation of respiration.

4.9.3 Kinetics of charge movements driven by electron transport

Although the experiments described previously are clearly consistent with the kinetic competence of Δp as the intermediate, an important complementary experiment would be to show that the generation of Δp by electron transport preceded ATP synthesis. This requires a method with a high time resolution for detection of Δp. The carotenoid band shift, an indicator of membrane potential in thylakoid membranes and bacterial chromatophores (Figure 4.8), has an almost instant response to an imposed membrane potential, and responded within microseconds to the initiation of light-driven electron transport initiated by a laser flash. Furthermore, the subsequent decay of the membrane potential was accelerated by the presence of ADP and P_i. Because the increased decay is due to the passage of protons through the ATP synthase to make ATP, it follows that ATP synthesis occurs after the formation of $\Delta \psi$.

4.9.4 Light-dependent ATP synthesis by bovine heart ATP synthase

An essential feature of the chemiosmotic theory is that the primary and secondary proton pumps should be functionally and structurally separable. In order to observe proton translocation, the purification of proton-translocating complexes must be followed by their reincorporation into synthetic, closed membranes that have low permeabilities to ions. Historically, such 'reconstitutions' allowed aspects of the chemiosmotic theory to be tested, such as whether each complex was capable of pumping protons as an autonomous unit. Reconstitution has been an important technique to investigate mechanism, co-factors, etc. for respiratory chain complexes and metabolite transporters.

Membrane proteins are generally purified following solubilisation of the membrane with a detergent (usually non-ionic) that disrupts protein–lipid but not protein–protein interactions. Once purified, there are two principal ways in which they can be reconstituted into a membrane structure. The first is to mix with phospholipids the purified protein dispersed in a suitable detergent, preferably one with a high critical micellar concentration (the concentration at which micelles form from monomers in solution), and then to allow the concentration of detergent to fall slowly either by dialysis or by gel filtration. Under optimal conditions, this can lead to the formation of unilamellar phospholipid vesicles.

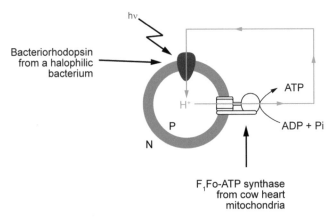

Figure 4.14   A proton circuit between a light-driven proton pump (bacteriorhodopsin) and ATP synthase from mitochondria.
The establishment of the proton circuit depends on the majority of the bacteriorhodopsin molecules adopting (for poorly understood reasons) the orientation in which they pump protons inwards. Similarly, the ATP synthase had to incorporate predominantly with the topology shown. Opposite orientations of both bacteriorhodopsin and ATP synthase would in principle also have permitted an H^+ circuit, in the opposite direction, to be established; in practice, this would have meant that added ADP and P_i (both membrane impermeant) would not have been able to reach the active site of the ATP synthase.

The protein can in principle be oriented in either of two ways. If the protein uses a substrate, such as ATP, to which the phospholipid bilayer is impermeable, then mixed orientation is not a problem because only those molecules with their catalytic site facing outwards will be accessible to the substrate. On the other hand, if the protein is a photosynthetic system, then asymmetry can obviously not be imposed in this way. Fortunately, proteins frequently orient asymmetrically because the differences in radius of curvature for the two sides of the vesicle may be an important factor. A more demanding type of reconstitution is when the presence of two different proteins (e.g., a primary and secondary pump) is required in the same membrane. The problem here is to ensure not only that at least a majority of vesicles contain both proteins but also that the relative orientations of the two proteins allow coupling between them via the proton circuit.

A second procedure for reconstitution is to incorporate the purified protein into a planar bilayer that can be formed over a tiny orifice that separates two reaction chambers. The insertion of protein is frequently achieved by fusing phospholipid vesicles containing the protein of interest with the planar bilayer. Alternatively, in some cases it has been possible to form the bilayer directly by application of a protein–phospholipid mixture in a suitable volatile solvent to the aperture. The amount of enzyme incorporated into such bilayers is usually so small that biochemical or chemical assays of activity are not possible. However, the crucial advantage of this type of system is that macroscopic electrodes can be inserted into the two chambers and thus direct electric measurements (either current or voltage) of any ion or electron movements driven by the reconstituted protein can be made.

An important qualitative demonstration of the proton circuit comes from an instructive reconstitution experiment. The ATP synthase of mitochondria ought to be able to drive ATP synthesis if it is incorporated in a phospholipid vesicle, with the correct relative orientation, along with another protein that generates protonmotive force of the correct polarity. A dramatic substantiation of this point was achieved when this experiment was done using the light-driven proton pump, bacteriorhodospin (Chapter 6), from a halophilic bacterium (Figure 4.14). An important point about this experiment is that the ATP synthase and the bacteriorhodopsin originate from such disparate sources; it is inconceivable that the coupling between them occurred through any other mechanism than the proton circuit shown in Figure 4.14.

INTRODUCTION TO PART 2

Having covered the basic universal features of bioenergetic energy transduction in Part 1, Part 2 delves into the molecular, and where possible atomic, mechanisms that underlie the action of the proton pumps that transduce redox energy and photon energy into proton electrochemical potential, the ATP synthase that utilises this energy for the phosphorylation of ADP to ATP, and the mitochondrial and bacterial transporters. One chapter is devoted to each topic. The major advance that has occurred since the third edition has been in the proliferation and refinement of crystal structures at atomic resolution of representative examples of each class of pump or transporter, such that we are now approaching the stage at which a precise mechanism can be proposed for these types of energy-transducing steps. We start in Chapter 5 with the general principles of electron transfer and then apply these to mitochondrial and bacterial electron transfer pathways. In Chapter 6, we review photosynthetic systems from bacteria and green plants; again, enormous advances have been made in determining the crystal structures of key complexes, including the oxygen evolving centre of photosystem II that extracts electrons from water. We also review the unique light-driven proton pump that is bacteriorhodopsin. Chapter 7 focuses on what must be the most remarkably biological assemblies on Earth: the linked proton turbine and ATP generator that comprise the ATP synthase and the bacterial flagellum. In Chapter 8, we discuss the recent advances in our understanding of the structure and function of mitochondrial and bacterial transporters.

5

RESPIRATORY CHAINS

Further reading: Rich and Marechal (2012)

5.1 INTRODUCTION

This chapter describes our knowledge of the respiratory chains of mitochondria and selected species of bacteria, and provides a brief outline of some of the approaches that have been taken to investigate these systems. The respiratory chain of mammalian mitochondria is an assembly of more than 20 discrete carriers of electrons that are mainly grouped into several multi-polypeptide complexes (Figure 5.1). Three of these complexes (I, III and IV) act as oxidation–reduction-driven proton pumps. There are now detailed crystal structures for each of these complexes, and the sequences of all the constituent polypeptides are available. This information has advanced functional understanding considerably, but many aspects remain to be understood at the molecular level. We illustrate methods for studying electron transport by reference to mitochondria, although comparable approaches are applied to bacteria and photosynthetic organisms.

5.2 COMPONENTS OF THE MITOCHONDRIAL RESPIRATORY CHAIN

The mitochondrial respiratory chain transfers electrons through a redox potential span of 1.1 V, from the $NAD^+/NADH$ couple to the $O_2/2H_2O$ couple. It is important to emphasise that the term *low potential* refers to a redox couple with a relatively negative redox potential. Conversely, *high potential* couples have a relatively positive redox potential. Confusingly, the redox span to the $O_2/2H_2O$ couple from low redox potential couples is relatively large while that from high redox potential couples is relatively small. We have to blame sign conventions for this.

With the exception of complex IV, much of the respiratory chain is reversible (Section 3.6.4), and to catalyse both the forward and reverse reactions it is necessary for the redox

Bioenergetics. Doi: http://dx.doi.org/10.1016/B978-0-12-388425-1.00005-1

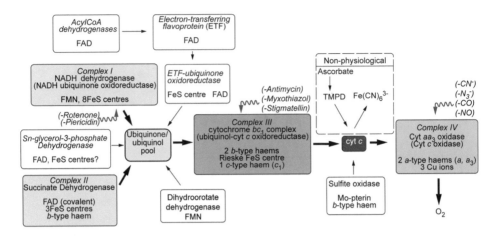

Figure 5.1   An overview of the redox carriers in the mitochondrial respiratory chain and their relation to the four respiratory chain complexes.
A wavy arrow indicates a site of action of an inhibitor. Sulfite oxidase and cyt c are in the intermembrane space. The active site of glycerol-3-phosphate dehydrogenase faces the inner membrane space; the other dehydrogenases act on substrates supplied directly from the matrix side. Note that components are not all present at equal stoichiometry (see text).

components to operate under conditions in which both the oxidised and reduced forms exist at appreciable concentrations. In other words, the operating redox potential of a couple, E_h (Section 3.3.3), should not be far removed from the midpoint potential of the couple, E_m. As shown later (Section 5.4.1), this constraint is generally obeyed, and this in turn gives some rationale to the selection of redox carriers within the respiratory chain.

The initial transfer of electrons from the soluble dehydrogenases of the citric acid cycle requires a cofactor that has midpoint potential in the region of $-300\,mV$ and is sufficiently mobile to shuttle between the matrix dehydrogenases and the membrane-bound respiratory chain. This function is filled by the $NAD^+/NADH$ couple, which has an $E_{m,7}$ of $-320\,mV$.

Although the majority of electrons are transferred to the respiratory chain in this way, a group of enzymes catalyse dehydrogenations where the midpoint potential of the substrate couple is close to $0\,mV$, and thus these reactions are not thermodynamically capable of reducing NAD^+. These feed electrons into the UQ/UQH_2 pool bypassing complex I and are reviewed in Section 5.7. A third site of electron entry from sulfite oxidase (the vital final step of degradation of sulfur-containing amino acids in liver and which occurs at a very low rate compared with other inputs) is at cyt c, which is also where electrons can be donated in *in vitro* experiments from chemicals such as ascorbate, mediated by tetramethyl-p-phenylenediamine (TMPD).

The redox carriers within the respiratory chain consist of *flavoproteins*, which contain tightly bound FAD or FMN as prosthetic groups (note that unlike $NAD^+/NADH$, these flavins do not diffuse from one enzyme to another) and undergo a $(2H^+ + 2e^-)$ reduction; *cytochromes*, with haem prosthetic groups undergoing a $1e^-$ reduction; *iron–sulfur* (non-haem iron) proteins, which possess Fe–S centres, also reduced in a $1e^-$ step; *ubiquinone*, which is a free, lipid-soluble cofactor reduced by $(2H^+ + 2e^-)$; and *protein-bound Cu*, reversibly reducible from Cu^{II} to Cu^I.

Cytochromes are classified according to the structure of their haem prosthetic group, which in a *b*-type cytochrome is the same as in haemoglobin. In a *c*-type cytochrome, the haem is covalently attached to the polypeptide chain via two cysteine groups. It is not known what advantage *c*-type cytochromes gain by undergoing this post-translational modification to generate the covalent attachment, but in some multi-haem proteins it allows for the haems to be packed closely together with relatively few amino acids per haem in the structures. The haem in an *a*-type cytochrome differs from a *b*-type by having a 15-carbon atom farnesyl side chain instead of a vinyl group at carbon atom 2 and by replacement of the methyl group on carbon atom 8 by a formyl group. Again, the rationale for these modifications is surprisingly unclear. In bacterial respiratory chains, other types of haem groups are encountered, notably those called *d*, d_1, and *o*. The former two involve quite significant modification to the standard haem structure, whereas the *o* type is intermediate between the *b* and *a* types, possessing the farnesyl group but lacking the formyl group. Again, the consequences of these modifications are still not always understood.

5.2.1 Fractionation, reconstitution and organisation of mitochondrial respiratory chain complexes

Of the approximately 20 discrete electron carriers in the mitochondrial electron transport chain, the only mobile components are ubiquinone (and its reduced form ubiquinol, UQH_2), which can diffuse within the hydrophobic core of the membrane, and the water-soluble cytochrome *c*, which is located on the P-side of the membrane. Ubiquinone (also called coenzyme Q, UQ, or simply Q) is found in mammalian mitochondria as UQ_{10}—that is, with a side chain of ten 5-carbon isoprene units (see Figure 5.4). The remaining redox carriers are components of the electron transport chain complexes.

Certain non-ionic detergents at low concentrations disrupt lipid–protein interactions in membranes, leaving protein–protein associations intact. Using these, the mitochondrial respiratory chain can be fractionated into four complexes, termed complex I (or NADH-UQ oxidoreductase), complex II (succinate dehydrogenase), complex III (UQH_2–cyt *c* oxidoreductase, or bc_1 complex) and complex IV (cytochrome *c* oxidase). Complex V is another name for the ATP synthase (Chapter 7). It is a source of confusion that *s,n*-glycerophosphate dehydrogenase and ETF–ubiquinone oxidoreductases do not have the 'complex' nomenclature, even though they are connected to the respiratory chain in a similar manner to complexes I and II.

The electron transfer activity of each complex is retained during this solubilisation, and when complexes I, III, or IV are reconstituted into artificial bilayer membranes, their ability to translocate protons is restored. Fractionation and reconstitution of the complexes has served a number of purposes:

(1) The complexity of the intact mitochondrion is reduced.
(2) It is possible to establish the minimum number of components that are required for function.
(3) During the period in which the chemiosmotic theory was being tested, reconstitution proved one of the most persuasive techniques for eliminating the necessity of a direct chemical or structural link between the respiratory chain and the ATP synthase.

For example, it proved possible to 'reconstitute' ATP synthesis by combining complex IV and the bovine heart mitochondrial ATP synthase in bilayer vesicles. However, the technical problems surrounding reconstitution are considerable. First, the complex must be incorporated into a bilayer in a way that retains catalytic activity. Second, allowance has to be made for the possibility of a random orientation of the reconstituted proton pumps that could prevent the detection of net transport. Incorporation into vesicles is normally performed by dissolving the proteins in a detergent together with phospholipid and then slowly removing the detergent by dialysis. Alternatively, the proteins can be sonicated together with phospholipid. An example of a reconstituted system was given in Section 4.14.

The complexes and other components, such as ETF–ubiquinone oxidoreductase, are not equimolar in the membrane. Thus, in cow heart mitochondria, the complex I:II:III:IV ratio is estimated to be 1:1.3:3:6, but different relative stoichiometries may apply to other mitochondria. In addition, there is a considerable molar excess of the ubiquinone/ubiquinol pool (Figure 5.1), consistent with its role as a diffusing connector in the respiratory chain. The amount of cyt c is usually in modest excess over complex IV, again consistent with its role diffusing between complex III and complex IV.

The isolated complexes readily reassociate functionally: for example, complex I and complex III assemble spontaneously in the presence of phospholipid and UQ_{10} to reconstitute NADH–cyt c oxidoreductase activity. Such reconstitution experiments provided an important approach to establishing the order of the components in the chain. For many years, it has been accepted that each of the complexes shown in Figure 5.1 independently diffused in the inner mitochondrial membrane and that the more mobile ubiquinone/ubiquinol could transfer electrons from any complex I, complex II, glycerophosphate dehydrogenase, or ETF–ubiquinone oxidoreductase to any cyt bc_1 complex in the same membrane. Similarly, the relatively mobile cyt c would connect any complex III to any complex IV. However, it is likely that the diffusing UQH_2 or cyt c will interact with one of the nearest complexes III or IV, respectively. There is experimental support for such a model, including estimates of diffusion coefficients for the components.

Recently, there has been increasing support for the idea that the complexes may exist in the membrane as super-complexes, such that electrons from complex I might flow to complex III within a super-complex, drastically decreasing the diffusion path for UQH_2. This approximates to a 'fixed wire' for electrons, with some copies of complex III attached to complex I, some to complex II and some to ETF–ubiquinone oxidoreductase, etc. Such an organisation lessens the flexibility of the respiratory chain system to respond to differences in substrate supply and at the limit would suggest, contrary to most observations, that rates of succinate and NADH oxidation should be mutually independent because separate 'wires' of electron transfer proteins would catalyse the reactions. However, this is not consistent with reversed electron transport (Section 4.7), where electrons from succinate can flow both down the electron transport chain to generate a Δp and back up through complex I, demonstrating that complexes I, II and III must share a common UQ/UQH_2 pool. Extrapolation of the wire model to bacterial electron transfer systems is even more problematic because there are many more electron donors (with their specific proteins) and acceptors than in the case of mitochondria.

The functional significance of super-complexes remains under debate. One study showed that the kinetics of cyt c oxidation in yeast mitochondria are not consistent with a fraction of the total cyt c being permanently sandwiched between complexes III and IV (Trouillard et al., 2011).

5.2.2 Methods of detection of redox centres

5.2.2.1 Cytochromes

The cytochromes were the first components to be detected, owing to their distinctive, redox-sensitive, visible spectra. An individual cytochrome exhibits one major absorption band in its oxidised form, whereas most cytochromes show three absorption bands when reduced. Absolute spectra, however, are of limited use when studying cytochromes in intact mitochondria or bacteria, due to the high nonspecific absorption and light scattering of the organelles or cells. For this reason, cytochrome spectra are studied using a sensitive differential, or split-beam, spectroscopy in which light from a wavelength scan is divided between two cuvettes containing incubations of mitochondria identical in all respects except that an addition is made to one cuvette to create a differential reduction of the cytochromes (Figure 5.2a). The output from the reference cuvette is then automatically subtracted from that of the sample cuvette to eliminate nonspecific absorption. Figure 5.2d shows the reduced, oxidised and reduced-minus-oxidised spectra for isolated cyt c, together with the complex reduced-minus-oxidised difference spectra obtained with sub-mitochondrial particles, in which the peaks of all the cytochromes are superimposed.

The individual cytochromes may most readily be resolved on the basis of their α-absorption bands in the 550- to 610-nm region (Figure 5.2d). The sharpness of the spectral bands can be enhanced by running spectra at liquid N_2 temperatures (77°K) due to a decrease in line broadening resulting from molecular motion and to an increased effective light path through the sample resulting from multiple internal reflections from the ice crystals.

Room-temperature difference spectroscopy can only clearly distinguish single a-,b-, and c-type cytochromes. However, each is now known to comprise two spectrally distinct components. The a-type cytochromes can be resolved into a and a_3 (the origins of the subscript 3 lie in the distant past, but its use continues to indicate that the two a-type haems behave differently) in the presence of CO, which combines specifically with a_3. a and a_3 are chemically identical but are in different environments. The b-cytochromes consist of two components with different E_m values: high, b_H ($E_{m,7} = +50\,mV$) and low, b_L ($E_{m,7} = -100\,mV$). These components respond differently when a Δp is established across the membrane (see Figure 3.4). It is now clear (Section 5.8) that the two components reflect the presence on one polypeptide chain of two b-type haems; the different local environments provided by the polypeptide chain account for the differences in spectral and redox properties.

The two c-type cytochromes, cyt c and cyt c_1, can be resolved spectrally at low temperatures. Cyt c_1 is an integral protein within complex III (Section 5.8), whereas cyt c is a peripheral protein on the P-face of the membrane and links complex III with complex IV.

(a) Split-beam spectrophotometer

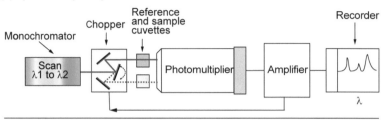

(b) Dual-wavelength spectrophotometer

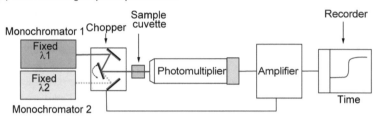

Figure 5.2 Spectroscopic techniques for the study of the respiratory chain.
(a) The split-beam spectrophotometer uses a single monochromator, the output from which is directed alternately (by means of a chopper oscillating at ~300 Hz) into reference and sample cuvettes. A single large photomultiplier is used, and the alternating signal is amplified and decoded so that the output from the amplifier is proportional to the difference in absorption between the two cuvettes. If the monochromator wavelength is scanned, a difference spectrum is obtained. The split beam is therefore used to plot difference spectra that do not change with time. (b) The dual-wavelength spectrophotometer uses two monochromators, one of which is set at a wavelength optimal for the change in absorbance of the species under study and one set for a nearby isosbestic wavelength at which no change is expected. Light from the two wavelengths is sent alternately through a single cuvette. The output plots the difference in absorbance at the two wavelengths as a function of time and is therefore used to follow the kinetics or steady-state changes in the absorbance of a given spectral component, particularly with turbid suspensions. *Continued →*

5.2.2.2 Fe–S centres

Although their distinctive visible spectra aided the early identification and investigation of the cytochromes, the other major class of electron carriers, the iron–sulfur (Fe–S) proteins (Figure 5.3), have ill-defined visible spectra but characteristic electron spin resonance spectra (ESR or EPR). The unpaired electron, which may be present in either the oxidised or the reduced form of different Fe–S proteins, produces the ESR signal. Each Fe–S group detectable by ESR is termed a centre or cluster. A single polypeptide may contain more than one centre. Complexes I–III between them have 11 such centres.

Fe–S proteins contain Fe atoms covalently bound to the apoprotein via cysteine sulfurs and bound to other Fe atoms via acid labile sulfur bridges (Figure 5.3). Fe–S centres may contain two or four Fe atoms, even though each centre only acts as a $1e^-$ carrier.

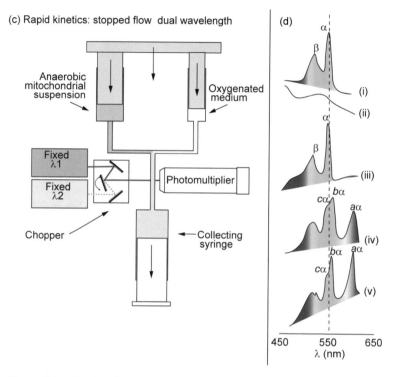

Figure 5.2 *Continued*
(c) To improve the time resolution of the dual-wavelength spectrophotometer, a
rapid-mixing device can be added. The syringes are driven at a constant speed,
and the 'age' of the mixture will depend on the length of tubing between the
mixing chamber and the cuvette. When the flow is stopped, the transient will
decay and this can be followed. (d) The absolute oxidised (i) and reduced (ii)
spectra were obtained with purified cyt *c* in a split-beam spectrophotometer
with water in the reference cuvette. The reduced minus oxidised spectrum (iii)
was obtained with reduced cyt *c* in one cuvette and oxidised cyt *c* in the other.
(iv) The reduced (with dithionite) minus oxidised (with ferricyanide) spectrum
from bovine heart SMPs. (v) The scan was repeated at 77°K; note the greater
sharpness of the α-bands.

Fe–S proteins are widely distributed among energy-transducing electron transfer chains
and can have widely different $E_{m,7}$ values from as low as $-530\,mV$ for chloroplast ferredoxin (Section 6.4) to $+360\,mV$ for a bacterial periplasmic HiPIP ('high-potential iron–
sulfur protein'). This emphasises the general point that the redox potential of a particular
type of centre can be considerably 'tuned' by the environment provided by the protein.

5.2.2.3 Flavins, quinones and quinols

Flavin mononucleotide (FMN) is a component of complex I, whereas flavin adenine
dinucleotide (FAD) is present in complex II, ETF and α-glycerophosphate dehydrogenase. FAD is additionally present in a number of enzymes, including pyruvate and

(a)   (b)

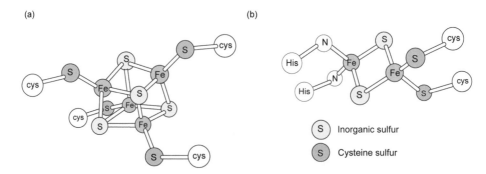

Figure 5.3   Iron–sulfur centres.
(a) A centre with four Fe and four acid-labile sulfurs is shown. On treatment
with acid, these sulfurs (yellow) are liberated as H_2S. Although there are four
Fe atoms, the entire centre undergoes only a $1e^-$ oxido-reduction. (b) The
structure of the 2Fe–S centre in complex III with two histidine ligands is
shown; other 2Fe–S structures will have four cysteine ligands to the Fe.

α-ketoglutarate dehydrogenases. The oxidised flavin moiety can gain an electron and
a proton, forming an intermediate free radical ($FMNH^•$ or $FADH^•$) detectable by EPR,
followed by a second electron plus another proton forming the fully reduced state
($FMNH_2$ or $FADH_2$) (Figure 5.4b). The oxidised and fully reduced forms are difficult to
detect optically, especially if other redox centres are present. The flavin is well adapted
to undergo both a one-electron oxidation/reduction step (e.g., when one electron at a
time is passed to Fe–S centres) and a two-electron reduction (e.g., accepting electrons
from NADH in complex I).

The 50-carbon hydrocarbon side chain of ubiquinone renders UQ_{10} highly hydrophobic (Figure 5.4). UQ undergoes an overall $2H^+ + 2e^-$ reduction to form UQH_2 (ubiquinol), although in general the reaction will take place in two one-electron steps (Figure
5.4); the partially reduced free radical form $UQ^{•-}$ (ubisemiquinone) plays a defined role
in both the photosynthetic reaction centre (Chapter 6) and the cyt bc_1 complex, where it
is stabilised by binding sites in the proteins. Ubiquinone reduction and ubiquinol oxidation will always occur at catalytic sites provided by the membrane proteins for which
they are substrates; the (de)protonation and oxidation/reduction steps are believed to
always proceed as shown in Figure 5.4b.

The radical form can be detected by its ESR spectrum or a characteristic absorption
band in the visible region, but ubiquinone and ubiquinol are more difficult to detect
because in common with proteins, they absorb at approximately 280 nm, although the
absorbance of the oxidised and reduced forms differs.

The simplest postulate for the role of UQ is as a mobile redox carrier linking complexes I and II (and the other flavin-linked dehydrogenases; Figure 5.1) with complex III,
although the 'Q-cycle' of electron transfer in complex III involves a more sophisticated
and integral role (Section 5.8). Although UQ_{10} is the physiological mediator, its hydrophobic nature makes it difficult to handle, and ubiquinones with shorter side chains,
and consequently greater water solubility, are usually employed *in vitro*. Some anaerobic bacterial respiratory chains employ menaquinone in place of UQ (Section 5.13.2),

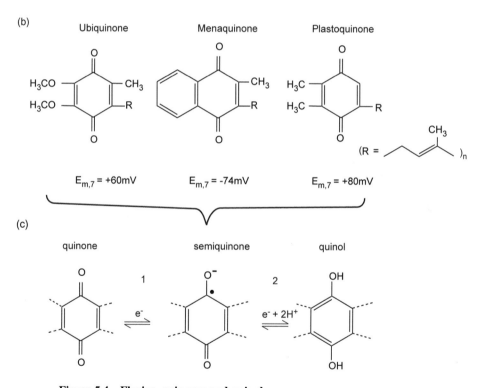

Figure 5.4 Flavins, quinones and quinols.
(a) Structure of the flavin ring in its three states fully oxidised, semi-reduced (a radical), and fully reduced. (b) Structures of the common quinone and quinols found in energy-transducing membranes. The length of the side chain can vary; for example, in ubiquinone $n=10$ in mammals but $n=6$ in yeast. (c) The two steps of quinol oxidation/quinone reduction. Note that $E^{0'}$ for step 1 is usually approximately $100\,mV$ more positive than $E^{0'}$ for step 2; that is, $Q\bullet^-$ is a stronger reductant than QH_2.

whereas in the chloroplast the corresponding redox carrier (Section 6.4.3) is plastoquinone (Figure 5.4).

5.3 THE SEQUENCE OF REDOX CARRIERS IN THE RESPIRATORY CHAIN

The sequence of electron carriers in the mitochondrial respiratory chain (Figure 5.1) was largely established by the early 1960s through a combination of oxygen electrode and spectroscopic techniques. This work was greatly facilitated by the ability to feed in and extract electrons at a number of locations along the respiratory chain, corresponding to the junctions between the respiratory complexes. Thus, NADH reduces complex I, succinate reduces complex II, and TMPD reduces complex IV (cytochrome oxidase) via cyt c. In this last case, ascorbate is usually added as the reductant to regenerate TMPD from its oxidised form, known as Wurster's blue. Ferricyanide (hexacyanoferrate (III)) is a nonspecific, but impermeant, electron acceptor and can be used not only to dissect out regions of the respiratory chain but also to provide information on the orientation of the components within the membrane. The reconstitution approach also showed that the complexes could not interact randomly; for example, complex I could only transfer electrons to complex III and the transfer depended on the presence of ubiquinone.

It should be emphasised that it is now clear that the electron carriers do not operate in a simple linear sequence but, rather, that electrons may divide between carriers in parallel (as happens in complex III) and that there have to be mechanisms for permitting a switch from one to two electron steps.

The discovery of electron transfer inhibitors acting at specific sites enabled the relative positions of electron entry and inhibitor action to be determined (Figure 5.1). Armed with this information, it was possible to proceed to a spectral analysis of the location of each redox carrier relative to these sites. Briefly, a component will become oxidised if it is located downstream (in the sense of nearer the 'oxygen end' of the chain) of the inhibitor site, and it will become more reduced if upstream of the site. An independent approach to the ordering of the redox components came with the development of techniques for studying their kinetics of oxidation following the addition of oxygen to an anaerobic suspension of mitochondria (Figure 5.2). The sequence with which the components become oxidised can reflect their proximity to the terminal oxidase and also whether they are kinetically competent to function in the main pathway of electron transfer. The rapidity of the oxidations observed under these conditions requires the use of stopped-flow techniques (Figure 5.2c).

The carriers in the respiratory chain must be ordered in such a way that their operating redox potentials, E_h (Chapter 3), form a rough sequence from NADH to O_2. E_h is determined from the midpoint potential, E_m, and the extent of reduction (Eq. 3.20). Although the extent of reduction of a component in the respiratory chain can be measured spectroscopically, indirect methods are needed to measure the midpoint potential *in situ* (Figure 5.6). Note that the midpoint potential of a component in the respiratory chain can be different from that of the purified, solubilised component.

5.4 MECHANISMS OF ELECTRON TRANSFER

Further reading: Moser *et al.* (2006)

A fundamental process in bioenergetics is the transfer of electrons between redox components that may be metallic (e.g., Fe) or organic (e.g., flavin). We know from many protein structures that these centres are rarely directly adjacent to each other but are usually separated by protein components (e.g., peptide bonds or side chains) or water. As discussed later in this Chapter and in Chapter 6, protein crystal structures have shown that electrons must be capable of being transferred over distances of up to 14 Å. The electron passes from one centre to another by a process known as *tunnelling*, a prediction of quantum mechanics. Specifically, the wave function for an electron held in an energy well on a donor shows that there is a finite, if very low, probability that the electron will be found at an acceptor some distance away. In effect, the electron can tunnel through a barrier. An insensitivity of electron transfer rate to temperature, even down to liquid helium temperatures, is diagnostic of tunnelling, and such behaviour has been observed in proteins. Development of the theory of electron transfer within and between proteins indicates that proteins present a rather uniform barrier through which the electron tunnels and that three factors influence the rate:

(a)   The distance between the electron donor centre and the acceptor is the most important factor. Distance is relatively easy to define if the transfer is between two ions (e.g., two Cu ions) but more difficult to define if, for example, the donor and acceptors are both haems, in which case the distance is generally taken to be from haem edge to edge and not iron to iron.

(b)   The size of the free energy (or redox potential) difference between the electron residing on the donor and the acceptor.

(c)   The response of the donor and acceptor, and their environments, to the increased positive charge on the donor and the increased negative charge on the acceptor that follow the transfer of the electron. The latter term is the 'reorganisation energy,' usually given the Greek letter λ. It is important to note that this does not include what might be termed chemical events, such as a concomitant transfer of a proton or dissociation of a ligand from one or both of the centres between which electrons are exchanged: such events can limit, or 'gate,' the rate of an electron transfer process.

Understanding rates of electron transfer reactions requires the grasping of some difficult concepts different from those that apply to chemical reactions, where bonds are broken and made. In the general case for transfer of single outer shell electrons, the Gibbs activation energy, ΔG^* (defined as the Gibbs free energy change for the formation of the transition state), is related to the standard free energy change ΔG° and the standard reorganisation energy λ_o by the following relationship, derived by R. A. Marcus:

$$\Delta G^* = (\Delta G^\circ + \lambda_o)^2 / 4\lambda_o \qquad [5.1]$$

Remember that ΔG° is a restatement of the equilibrium constant (Eq. 3.7). The transition state also requires some explanation. The electron is either on the donor or on the acceptor, governed by quantum mechanics, and never halfway between. However, it is possible to conceive of an intermediate configuration and polarisation of the molecule and the surrounding solvent, and the activation energy of the transition state largely reflects this.

The rate equation for such an electron transfer reflects the magnitude of the activation energy:

$$k_{et} = A \cdot e^{-\frac{\Delta G^*}{RT}} \qquad [5.2]$$

where the factor A includes probability terms that will be discussed later.

To take a simple example, in a solution containing a mixture of Fe^{2+} and Fe^{3+} ions, exchange of electrons between the two will involve solvent polarisation and hence reorganisation energy. In this case, forward and reverse rates are equal, and the free energy change is zero. From Eq. 5.1, therefore, the activation free energy is $0.25\lambda_o$.

Let us now consider a more complex case: electron transfer from a molecule of reduced cyt c to the inorganic chemical hexacyanoferrate (VI) (also known as ferricyanide), which is often used experimentally as a nonphysiological electron acceptor. In this case, there is a 200-mV difference in midpoint potentials (Table 3.2), so ΔG^o is negative by $-19\,kJ\,mol^{-1}$ (see Eq. 3.25). The ligands to the iron in cyt c and hexacyannoferrate do not change as the former undergoes oxidation and the latter reduction, but there will be changes in ligand to Fe bond lengths. The reorganisation energy is the energy required to move various nuclei in reduced cyt c to the positions they will adopt in the oxidised cytochrome immediately before the electron is transferred to ferricyanide, where reorganisation may also be required. Substitution of Eq. 5.1 into Eq. 5.2 shows that the electron transfer rate constant will depend on the relative magnitudes of ΔG^o and λ_o.

In most circumstances, ΔG^o will be negative, whereas λ_o is always positive and relatively invariant for the types of biological reactions of interest here. As the magnitude of ΔG^o increases from a very small value, ΔG^* will at first move toward a minimum (actually zero) when $-\Delta G^o = \lambda_o$, and under this condition the rate of electron transfer is at a maximum. Because the square term in Eq. 5.1 cannot have a negative value, as ΔG^o becomes still more negative (i.e., as the thermodynamic driving force becomes ever larger) ΔG^* will rise, meaning that the electron transfer rate (Eq. 5.2) becomes ever slower. This counterintuitive behaviour is called the inverted region and has been demonstrated using chemical systems with different sets of electron donors and acceptors at a fixed separation distance. This inverted region is thought to be of considerable importance in photosynthesis (see Chapter 6).

We must now discuss the factor A in Eq. 5.2. One component, often termed β, relates to the parameters of electron tunnelling from a donor to an acceptor; this is a complex term but is related to the probability that the wave function of the donor can penetrate through a barrier to reach a potential well on an acceptor. In biology, the barrier will almost always be protein; experimental analysis suggests that all proteins present much the same value of β, a critical point in favour of electrons tunnelling through a protein from a donor to an acceptor. If individual proteins had evolved distinct pathways via defined routes of covalent bonds to catalyse the electron transfer, one might expect variable values of β. Although this issue is controversial, it is assumed here that such defined routes do not exist in proteins and the former description is correct. Despite this, there are rare instances (Sections 5.12 and 6.4.2) in which the side chain of an amino acid participates as an oxidation/reduction centre. This requires a very electropositive redox centre to be the acceptor of an electron from the amino acid because side chains are thermodynamically difficult to oxidise. However, these are the exceptions, and in general the idea of a 'best pathway' for electrons through particular bonds or amino acids that happen to lie between redox centres is incorrect; even bound water molecules within proteins can support electron tunnelling.

The second important factor defining A is the distance between donor and acceptor. Experiments show that the rate constant for electron transfer (k_{et}) is an exponential function of the separation distance (this is because all wave functions decay exponentially at long range). This is intuitive; the greater the distance, the slower the rate. It turns out for normally encountered values for ΔG^o and λ_o that once the distance exceeds approximately 14Å, then the rate of electron transfer becomes too slow to match the millisecond turnover time characteristic of enzymes. Indeed, rates of $10^6 s^{-1}$ are frequently encountered for overall reactions in biological electron transfer processes. For most individual steps of biological electron transfer, ΔG^o is between 0 and $-10 kJ\,mol^{-1}$ (i.e., ΔE_h is between 0 and 100mV), and λ_o varies between proteins by only a factor of two or three. It is in this context that distance between centres correlates closely with rate; edge-to-edge distances of 3, 10 and 14Å are predicted to allow rates of approximately 10^{10}, 10^7 and $10^4 s^{-1}$ with a driving force of $-10 kJ\,mol^{-1}$. At 25Å, electron transfer would take hours.

It is striking that for all the known protein structures with more than one redox group, the edge-to-edge distance between centres that exchange electrons is always below 14Å in the conformation where transfer occurs. Change of distance caused by conformational change can be a means of controlling electron transfer as exemplified by the cyt bc_1 complex (Section 5.8.6). Thus, specificity and directionality in electron transfer is achieved as a consequence of the spatial relationships of the participating groups; transfer of an electron from a donor to the 'wrong' acceptor will occur so slowly as to be insignificant.

Electron transfer occurs not only within a protein, or proteins, that form a permanently associated complex but also between proteins that interact transiently—for example, transfer from cyt c_2 to haem 4 (h4) in Figure 5.5 or from cyt c to cytochrome oxidase. The requirement is that a transiently formed complex places the redox groups of the donor and acceptor proteins at an edge-to-edge distance no more than 14Å, meaning that the redox groups of the two proteins cannot be deeply buried.

5.4.1 Midpoint potentials are not always in sequence

A puzzling feature of many proteins involved in biological electron transfer is that they often have at least one cofactor for which the midpoint redox potential is dissimilar to the others. For example, Figure 5.5 shows a sequence of electron carriers within the reaction centre of *Rhodopseudomonas viridis* (see Figure 6.6). The initial electron donor is cyt c_2 ($E_h +240 mV$) and the final acceptor, via four haems (h4–h1), is the 'special pair' of bacteriochlorophyll molecules (Bchl)$_2$ ($E_h +500 mV$) (see Figure 6.6). The negative midpoint potentials of h4 and h2 at first sight suggest that they might not be involved in the overall electron transfer process. However, the crystal structure of the reaction centre (see Figure 6.6) and the limit of 14Å between haem edges show that h4

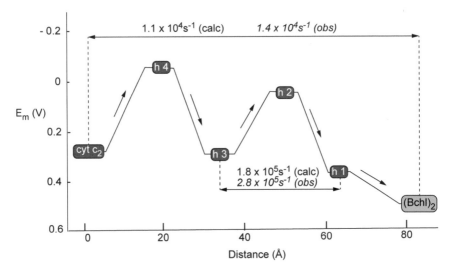

Figure 5.5 Electron transfer over large distances in proteins is catalysed by chains of redox centres (haems (h) 1–4) with both uphill and downhill steps.
This energy diagram corresponds to part of the reaction centre of *Rhodopseudomonas viridis* (Section 6.2.3). Note the good agreement between measured rates and those calculated according to current electron transfer theory (see text).
(Adapted from Page et al. (1999), in which further details of the calculation procedures can be found).

and h2 must be on the path of electron transfer. The overall translocation of an electron from cyt c_2 to h3 is thermodynamically favourable, although in the steady state only a very small fraction of h4 can be in the reduced state. Similar arguments apply to the role of h2 in electron transfer from h3 to h1. Although electron transfer involving h4 and h2 will not be as fast as it would be if their redox potentials were more positive, edge-to-edge distance is the dominant criterion and the overall rate of electron transfer will still be fast enough to more than match the rates of the chemical steps at the beginning or end of any chain of electron transfer reactions.

Nevertheless, it is still puzzling why 'downstream' redox centres with midpoint potentials more negative than their donors should occur, particularly because this implies a high activation energy (Eq. 5.1). There may be two reasons. First, the reactions could be a locus of control, although there is currently no evidence for such an effect. More simply, many electron transfer proteins are evolutionarily related to one another and, if the kinetics are adequate, the recruitment of a particular redox centre for a new reaction would not have exerted any pressure for a change in its redox potential.

5.4.2 Redox potentiometry

Redox potentiometry combines dual-wavelength spectroscopy with redox potential determinations, the latter using immersed Pt and reference electrodes (Figure 5.6). For most biological couples, it is necessary to add a low concentration of an intermediate

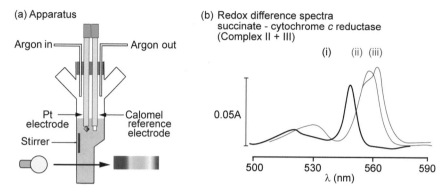

Figure 5.6 Redox potentiometry of respiratory chain components.
(a) Apparatus for the simultaneous determination of redox potential and
absorbance. (b) Difference spectra obtained with a suspension of succinate-
cytochrome c reductase (i.e., complexes II + III). The complex, held in
solution by a low concentration of detergent, was added to an anaerobic
incubation containing redox mediators. The ambient redox potential was
varied by the addition of ferricyanide. (i) Reference scan (baseline) at
+280 mV (all cytochromes oxidised), second scan at +145 mV (cyt c_1 now
reduced). (ii) Baseline at +145 mV (cyt c_1 reduced), second scan at −10 mV
(cyt b_L additionally reduced). (iii) Baseline at −10 mV (c_1 and b_L reduced),
second scan at −100 mV (b_H additionally reduced).

redox mediator to speed the equilibrium between the Pt electrode and the redox centres.
For example, the haems of complex III are buried between transmembrane helices and
are thus inaccessible without a secondary mediator. Because mediators will only function effectively in the region of their midpoint potentials (so that there are appreciable
concentrations of both their oxidised and reduced forms), a set of mediators is required
to cover the whole span of the respiratory chain, with midpoint potentials spaced at
intervals of approximately 100 mV. Mediators are usually employed at concentrations
of 10^{-6} to 10^{-4} M. Many mediators are autoxidisable, and the incubation has to be maintained anaerobic both for this reason and to prevent a net flux through the respiratory
chain from upsetting the equilibrium.

A second requirement for membrane-bound systems is that mediators must be able
to permeate the membrane in order to equilibrate with all the components. This introduces a considerable complication if the mitochondria are studied in the presence of a
membrane potential, as $\Delta\psi$ (or indeed ΔpH) will almost certainly affect the distribution
of the oxidised and reduced forms of the mediators across the membrane differentially
(Chapter 4) so that the redox potential of the mediator at the platinum electrode will differ from that at the cytochrome or other component. The simplest redox potentiometry
is therefore performed with mitochondria or submitochondrial particles at zero Δp.

The practical determination of the E_m of a respiratory chain component (Figure 5.6)
involves incubating mitochondria anaerobically in the presence of the secondary mediators. The state of reduction of the relevant component is monitored by dual-wavelength
spectrophotometry (Figure 5.2b), whereas the ambient redox potential is monitored by
a Pt or Au electrode relative to a calomel reference. The electrode allows the secondary mediators and the respiratory chain components all to equilibrate to the same redox

potential. This potential can then be made more electronegative (i.e., reducing; by the addition of ascorbate, NADH, or dithionite) or more electropositive (i.e., oxidising; by the addition of ferricyanide). The degree of reduction of the component is monitored simultaneously by spectrophotometry, and in this way a redox titration for the component can be established.

Considerable information can be gathered from such a titration. In addition to E_h, the slope of $\log_{10}[ox]/[red]$ establishes whether the component is a $1e^-$ carrier (60 mV per decade) or a $2e^-$ carrier (30 mV per decade) (see Table 3.2). By repeating the titration at different pH values, it can be seen whether the midpoint potential is pH dependent, implying that the component is a $(H^+ + e^-)$ carrier. Finally, the technique frequently allows the resolution of a single spectral peak into two or more components based on differences in E_m. In this case, the basic redox plot (see Figure 3.3) is distorted, being the sum of two plots with differing E_m values that can then be resolved. One of the most interesting findings with this technique was that cyt b in complex III can be resolved into two components (Section 5.8). Redox potentiometry can also be employed for Fe–S proteins, in which case the redox state of the components is monitored by ESR.

5.4.3 E_h values for respiratory chain components fall into isopotential groups separated by regions where redox potential is coupled to proton translocation

The midpoint potentials of some of the identifiable components of the mitochondrial respiratory chain for mitochondria respiring in state 4 are depicted in Figure 5.7. Once the E_m values have been established for nonrespiring mitochondria, an E_h for a component can be assigned to any component in respiring mitochondria simply by determining the degree of reduction. The oxido-reduction components fall into four roughly equipotential groups, the gaps between which correspond to the regions where proton translocation occurs. Thus, the drop in E_h of the electrons across these gaps is conserved in Δp.

5.5 PROTON TRANSLOCATION BY THE RESPIRATORY CHAIN: LOOPS, CONFORMATIONAL PUMPS, OR BOTH?

Mitchell's original formulation of the chemiosmotic mechanism envisaged that electron transport chains would pump protons by alternating 'hydrogen' (i.e., $2e^- + 2H^+$) and pure electron carriers looped across the membrane in such a way that the 'hydrogen' carrier (B in Figure 5.8a) accepts $2e^- + 2H^+$ from a reduced cofactor on the N-face of the membrane, forming BH_2. BH_2 would be reoxidised by releasing its protons to the P-phase and transferring its electrons to a sequence of pure electron carriers reducing a $2e^- + 2H^+$ acceptor (C in Figure 5.8a) taking up protons from the matrix. It was sometimes postulated that the 'loop' hypothesis requires that the loops should span the membrane and that the appropriate redox centres should be located at the two sides (P and N) of the membrane. However, all that is required for a functional loop is that there should be a means for taking up and releasing the protons at the two sides. Thus, specific pathways through a protein from a site of oxidation or reduction to a surface can contribute

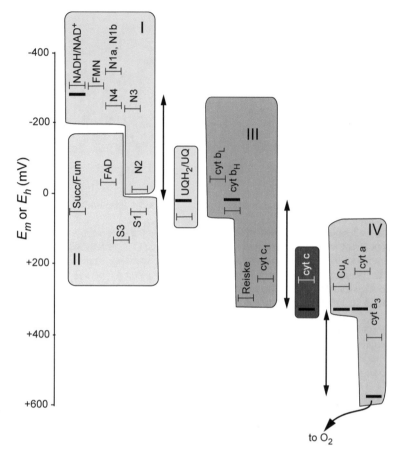

Figure 5.7 E_m values for components of the mitochondrial respiratory chain and E_h values for mitochondria respiring in state 4.
Values are consensus values for mammalian mitochondria. (⊢⊣), $E_{m,7}$ values obtained with de-energised (that is zero Δp) mitochondria; (—) $E_{h,7}$ values for mitochondria in state 4.

to the operation of a loop mechanism, and such pathways clearly occur, for example, in complex IV (Section 5.9). A pure 'loop' mechanism implies a fixed stoichiometry of $1H^+$ translocated per electron for a single loop such as in Figure 5.8a.

A loop mechanism does operate in several bacterial electron transport reactions (Section 5.13), and elements of a loop mechanism are used in the mitochondrial electron transport system, as discussed later in this chapter. However, it is also now clear that *direct* proton pumping across a membrane also occurs in many systems, including complexes I and IV (Sections 5.6 and 5.9). This means that the oxidation/reduction reaction at one or more centres in the protein is linked with conformational changes that result in net proton translocation across the membrane. This type of mechanism places no limits on the stoichiometry of proton translocation, within the thermodynamic constraints imposed by the relationship between the available driving force and the size of Δp (Section 3.6).

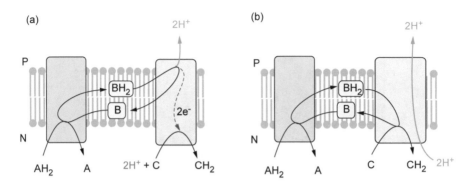

Figure 5.8   Net proton translocation by (a) a redox loop and (b) a proton pump.
In the loop model (a), most of the charge transfer across the membrane will be due to the inward movements of electrons to reduce C, but there will be some contribution from the outward movement of H^+, depending on the depth in the membrane of the site of BH_2 reoxidation. In the pump model (b), all the charge translocation is due to the pumping of protons. Note an important rule for calculating proton translocation stoichiometries: protons that are taken up and released in the matrix by the reduction and oxidation of $NAD^+/NADH$ or $FAD/FADH_2$ do *not* count because at steady state these redox mediators continuously cycle between the two redox states and there is no net production or consumption of protons. On the other hand, protons that are continuously consumed, as in the final transfer of electrons to O_2 to make H_2O in complex IV, or continuously generated, as in the water-splitting reaction in photosynthesis, must be taken into account, as must the protons that are taken up from the matrix by UQ reduction in complex I and released to the P-phase by UQH_2 oxidation in complex III.

In a purely *conformational pump* model, a redox carrier anchored within a flexible protein is proposed to undergo redox-induced changes in pK_a, the directionality of proton transport being assured by coordinate conformational changes that make the redox site alternately accessible from either side of the membrane (Figure 5.9). A conformational redox pump must coordinate a redox change, a conformational change, and a protonation change. The reversibly protonated site in this model need not necessarily be limited to a conventional redox centre; it is equally possible that a redox-induced conformational change alters the pK_A of an amino acid side chain. The closest approach to understanding such a scheme in molecular detail comes from bacteriorhodopsin (Section 6.5.1).

5.6   COMPLEX I (NADH–UQ OXIDOREDUCTASE)

Further reading: Sazanov and Hinchliffe (2006), Berrisford and Sazanov (2009), Hirst (2010), Efremov and Sazanov (2011), Baradan *et al.* (2013)

Mitochondrial complex I is—given its function of coupling the passage of two electrons from NADH to ubiquinone to the translocation of four protons—a surprisingly

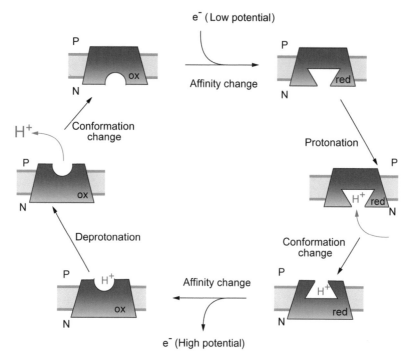

Figure 5.9   A model for a redox-driven proton pump.
A hypothetical model is shown for a redox pump with a stoichiometry
of $1H^+/e^-$. Reduction by an electronegative electron causes the proton binding
site to adopt a high-affinity state, sufficient to bind a proton from the alkaline
N-phase. A spontaneous conformation change now makes the binding site
accessible to the P-phase (note: the protein itself does not rotate in the
membrane). Loss of the electron (at a more positive potential) causes the
binding site to adopt a low-affinity state, so the proton is released into the
acidic environment of the P-phase. The cycle is completed by a conformational
change, again making the binding site accessible to the N-phase.

large and complicated protein, with a molecular weight in the order of 980 kDa and
as many as 44 subunits. Progress has been made in understanding the structure of the
mitochondrial enzyme, to which we return later, but as with many of the respiratory
chain complexes, bacterial analogues exist that have significantly fewer polypeptides
(14 core subunits) but similar redox and cofactor group content. Structural information about the bacterial enzyme has therefore provided considerable insight into the
structure–function relationship of the complex.

In this section, we use a bacterial convention for the naming of the 14 core subunits
of complex I: NuoA–NuoM. An alternative bacterial nomenclature terms the subunits
Nqo1–Nqo14, but in a different order. The mitochondrial complex I, with its 14 core
and approximately 30 accessory subunits, has a separate nomenclature system.

The bacterial protein from *Thermus thermophilus* has two clearly demarcated
domains. A hydrophobic proton translocating module in the membrane comprises subunits NuoAJKLMN, of which NuoLMN (ND5, ND4, and ND3 are the equivalent mitochondrial subunits; Figure 5.10) are each clearly related to sodium/proton antiporters

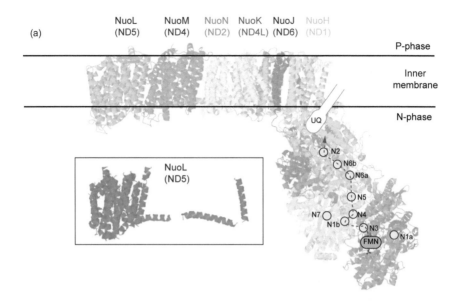

Figure 5.10   The overall structure of NADH–ubiquinone oxidoreductase from *T. thermophilus* (a model for mitochondrial complex I) and its relationship to conformational changes that are proposed to be linked to proton translocation.

(a) A crystal structure at 3.3 Å viewed side on to the membrane. The globular NuoBCDEFGI subunits bind nine Fe–S centres (the approximate locations of the Fe–S centres are shown by the yellow circles), seven of which lie on the pathway of electron transfer from $FMNH_2$ (formed by reduction of FMN by NADH), to the ubiquinone binding site, which is a narrow chamber leading from the membrane bilayer. This chamber allows the quinone head group to approach within 10 Å of the N-2 Fe–S centre, which itself is 25–30 Å above the phospholipid head group region on the N-side of the membrane. The eight transmembrane subunits are NuoL, NuoM, NuoN, NuoK, NuoJ, and NuoH plus NuoA, which is adjacent to NuoH but not visible in the figure. The insert shows the NuoL subunit in isolation to emphasise the 'horizontal' long HL helix, but note that the apparent break in the middle of the helix is not real but a consequence of the relative tilt at which it is observed. *Continued →*

found, for example, in the bacterium *Bacillus subtilis*. A hydrophilic module containing seven polypeptides (NuoBCDEFGI), to which are bound the FMN and all the Fe–S centres, projects into the bacterial cytoplasm (equivalent to the mitochondrial matrix), making an angle of approximately 120° with the membrane (Figure 5.10). The site of NADH oxidation by the tightly bound FMN is located towards the furthest extremity of the hydrophilic domain. NuoEFG are regarded as the NADH-oxidising domain, and NuoBCDI are regarded as the quinone module. These domains are clearly related to structures found separately in various types of bacterial hydrogenases, which is suggestive of an evolutionary route to complex I. The NuoH subunit is at the junction between the hydrophilic domain and the hydrophobic domain (Figure 5.10) and is related to a distinct kind of hydrogenase, where it has been implicated in ion translocation. There are 64 transmembrane helices in the *T. thermophilus* enzyme.

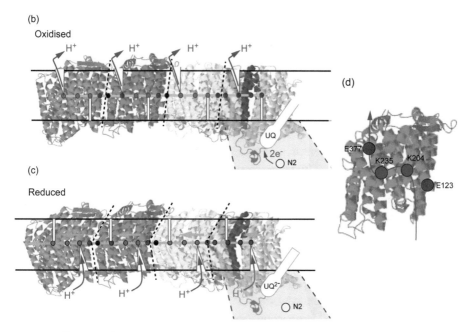

Figure 5.10 *Continued*
(b) Schematic of proposed conformational state of the protein following the oxidation of ubiquinol and release of protons to the P-side of the membrane. Notice the four half channels that are open to the P-side of the membrane and closed to the N-side. (c) Proposed conformation state with the unprotonated dianionic ubiquinol species in its binding site. The half channels are closed to the P-side and open to the N-side. Note that in comparing panels b and c, the helices have moved, as has the hydrophilic axis, which is green in panel a and orange in panel c (the conformational change is schematic). Charged residues on this axis are shown in green or red. Note that the globular domain has been truncated, but the delivery of electrons (one at a time but schematically shown as $2e^-$ in panel d) from the N2 centre is shown. The dotted lines indicate approximate boundaries between the four proton pumping segments of the enzyme. (d) The putative proton-translocation channel in one (subunit NuoM) of the Na^+/H^+ -like subunits. The proposed approximate pathway (arrow) for H^+ movement involves E123 (helix 5) and K204 (helix 7) in the N-side half channel and then K235 (helix 8) and E 377 (helix 12). A bacterial nomenclature is used for the subunits, with the human mitochondrial equivalent in brackets. For a complete list of nomenclature variants, see Hirst (2010). *The structure is adapted from Baradaran et al. (2013) PDB 3M9S.*

A crucial and surprising feature of complex I is the total lack of redox groups in the hydrophobic domain, contrary to expectations over many years and diagrams presented in many books (including our third edition). This means that conformational changes in the hydrophilic domain induced by redox changes must in some way be transmitted to the hydrophobic domain to drive conformational changes there, leading to proton pumping.

5.6.1 The hydrophilic domain of the bacterial enzyme

The action of complex I starts with the oxidation of NADH by the FMN moiety at the extremity of the hydrophilic module, approximately 50Å from the membrane surface. FMN is reduced to $FMNH_2$ by a $2e^- + 1H^+$ transfer from NADH plus gain of an additional H^+. $FMNH_2$ is then reoxidised in two $1e^- + 1H^+$ steps, involving transient formation of the $FMNH^•$ radical (Figure 5.4), by a 'wire' of seven Fe–S clusters. Sequence analysis shows that these clusters are also present in mitochondrial complex I. (Note that *T. thermophilus* has two extra Fe–S centres relative to the mitochondrial enzyme, but these do not seem fundamental to the function of complex I.) The final Fe–S centre, N2, has the least negative E_h value of the seven (Figure 5.7). It lies approximately 30Å from the surface of the membrane and approximately 12Å from the supposed UQ binding site, which itself is approximately 20Å from the plane of the membrane (Figure 5.10). N2 is bound to the NuoI subunit, and an unusual feature of this Fe–S centre (but conserved in all complex I molecules) is that there are two consecutive cysteines in the sequence of the subunit that provide sulfurs for the Fe–S centre when complex I is in the oxidised state, introducing strain into the structure. Strikingly, when N2 is reduced, one of these cysteines (which one depends on the exact condition) dissociates from the cluster and may take up a proton as a consequence. Both this and the very small concomitant movement of helices at the end of the globular domain closest to the interface with the membrane domain may be crucial for the functioning of the complex. Overall, coupling of electron transport to movement of protons depends on the significantly exergonic reduction of the N2 centre by the roughly isopotential wire of Fe–S centres that delivers electrons from NADH; the large $E_{m,7}$ span from N3 to N2 (Figure 5.7) must provide the majority of the driving force for the proton pumping events associated with reduction and reoxidation of N2.

The N2 centre passes electrons on to UQ, whose binding site is long and narrow (Figure 5.10) and can be regarded as a chamber provided largely by the NuoB and D subunits. This is notable because the binding site is many angstroms away from the hydrophobic core of the membrane, where the bulk UQ/UQH_2 pool is located. Although UQ has not been directly observed bound in this region, analogs have been observed in crystal structures. Mutation of a conserved tyrosine residue in NuoD severely compromises reduction of ubiquinone, supporting this location for the site. Furthermore, the inhibitors rotenone and piericidin bind at, or close to, this site. This quinone site forces the UQ side chain to be extended and contrasts with the geometry of other quinone sites as seen, for example, in complex III. Sequential electron transfer from centre N2 first gives ubisemiquinone, $UQ^{•-}$, and, after N2 has been re-reduced by a second electron delivered by the Fe–S wire, transfer of this second electron will produce UQ^{2-}.

The NuoH subunit provides the connection to the hydrophobic domain, which we now review.

5.6.2 The hydrophobic domain of the bacterial enzyme

A striking feature of complex I is that three major peptides of the hydrophobic domain—NuoL, -M and -N—are each related to proteins known to have Na^+/H^+ antiport function. NuoM and NuoN each have 14 structurally conserved transmembrane helices. NuoL has an equivalent set of 14 helices but has two additional transmembrane helices that are connected by a long helix (called HL) that runs horizontally along the N-side of the membrane and is approximately 110Å long (Figure 5.10, insert).

The sets of 14 transmembrane helices found in NuoLMN contain a novel fold. NuoLMN each have two helices that are kinked (or broken) in the middle of the membrane by π bulges formed where two peptide carbonyls are hydrogen bonded to water, disrupting the α-helix. There are conserved lysine and glutamate residues associated with the kinks in the helices, and mutagenesis studies suggest that these are involved in proton translocation. Lysine has not previously been considered a likely side chain to be involved in proton translocation because in aqueous solution the side chain pK_a is approximately 10. However, in a protein, the local environment can perturb pK_a values by as much as 4 units, making a protonation/deprotonation cycle associated with the lysine side chain feasible. Such a mechanism has been suggested for proteorhodopsin (Section 6.5.3).

The structural analysis of NuoLMN suggests that the three subunits may work in parallel as proton pumps, with conformational changes in each, driving protons via the breaks in the helices from half channels at the N-side of the membrane to half channels at the P-side. For example, it is plausible that conformational changes could cause transfer of a proton from the lysines on helix 5 to the lysine (NuoLN) or glutamate (NuoM) on helix 12. This would transfer a proton from a half channel facing the N-side of the membrane to one open to the other side (Figure 5.10).

A reasonable extrapolation from the structure is that the parallel operation of these channels in the three subunits could explain the translocation of $3H^+$ per $2e^-$ transferred from NADH to ubiquinone. However, the generally accepted proton translocation stoichiometry for complex I is $4H^+/2e^-$. The structure suggests that NuoH, NuoJ and NuoK together could provide a fourth pair of half channels, closely resembling the three pairs proposed in NuoLMN. NuoH has eight helices with an unusual fold, with most of them tilted relative to the membrane normal. Remarkably, given that there is no clue from their sequence, helices 2–6 can be superimposed onto either helices 4–8 or helices 9–13 of the NuoL M or N subunits, with charged residues, glutamates, or lysines of NuoH located in similar positions to their counterparts in NuoLMN. NuoH contains several more charged residues, many glutamates, in the membrane, which could provide a funnel from the quinone binding site to a network of four interacting carboxylates deep in the membrane, three being contributed by NuoH and one by NuoA. Subunits NuoJ and NuoK provide the other half channel to work alongside that provided by NuoH. Thus, the set of subunits NuoAHJK could provide the fourth route for protons to be pumped across the membrane. At this point, we recall that the accepted $H^+/2e^-$ stoichiometry for NADH ubiquinone oxidoreductase is 4, consistent with the structure.

5.6.3 How does electron transport drive proton pumping?

Although we can model hypothetical proton pumps in the hydrophobic domain, we have still to explain how the substantial amount of free energy required to pump these protons against the protonmotive force is transduced from the hydrophilic domain. The presence of the unprecedented horizontal helix HL in NuoL originally led to the suggestion that the helix could move 'horizontally' like a piston to drive conformational changes in, and proton translocation by, the NuoLMN subunits. However, the subsequent realisation that NuoH may have similarity to NuoLMN, and the recognition that the 16th helix of NuoL (the final 'vertical' one in Figure 5.10, insert) is still some distance (~30Å) from either centre N2 or the UQ binding site, calls this hypothesis into question. An alternative role for the N-side-located long HL helix, in conjunction with a series of β hairpins joining helices at the opposite P-side of the membrane, is that of a supporting framework (in effect, a stator to anchor ends of the helices while still permitting movement in the middle regions). According to this view, conformational changes, particularly associated with the π-bulge kinks in the mid-membrane parts of the NuoHJKLMN subunits, are transmitted laterally (think of a series of elbow nudges between people standing side-by-side) from NuoH to NuoN to NuoM and finally to NuoL (Figure 5.10). Although still just as speculative as the probably incorrect piston mechanism, this mechanism would involve conformational changes initiated by the redox change at N2 and the UQ binding site being transmitted to the relatively close NuoH, from which they could be laterally transmitted to NuoN, NuoM, and finally NuoL. The nature of the quinone site suggests that the energy released by formation of the dianion form of ubiquinol could interact electrostatically with a linear sequence of four carboxylates provided by NuoH and NuoA (Section 5.6.2). Energy provided by this interaction, probably reinforced by cysteine movement at Fe–S N2 being transmitted by helices on NouD and NuoB, could drive the conformational changes to pump a proton through the two half channels provided by NuoHJK.

Although this provides a plausible mechanism for the pumping of the first proton, how could these events also drive the additional proton pumping by NuoLMN? The structure shows that there is a continuous hydrophilic axis, comprising interacting charged and polar residues with surrounding water molecules, running through the centre of the whole length of the membrane domain (Figure 5.10). In this model, concerted conformational changes, initiated by events at Fe–S cluster N2 and in the quinone chamber as two electrons sequentially arrive from N2, would be transmitted along the entire hydrophilic axis of the hydrophobic domain, creating conformational changes driving proton pumping at all four pump domains. An implication of this mechanism is that mutations preventing conformational change in the first proton translocation pathway provided by NuoHJK should prevent proton pumping by the other three channels. Observations from mutagenesis studies support this notion. In contrast, the structure does not readily suggest how any conformational changes at N2 and the UQ site could be transmitted directly to the HL helix for the 'piston effect' that was initially proposed for the long HL helix.

Although the overall bioenergetics allows the redox potential gap from N3 via N2 to UQ to be utilised to pump $4H^+/2e^-$ against a Δp of approximately $180\,mV$ (Section 3.6.1), the thermodynamic consequences of this mechanism remain to be fully developed. The $E_{m,7}$ for the UQ/UQ^{2-} couple at the UQ site in the oxidised conformation of the complex (Figure 5.10b) will likely be similar to that for N2. If electrostatic repulsion does provide the driving force to generate the reduced conformation (Figure 5.10c), then the energy would be provided by a large positive shift in the UQ/UQ^{2-} midpoint potential as the conformation changes. In other words, the $E_{m,7}$ for the UQ/UQ^{2-} couple becomes more positive as the conformation changes, thus providing the required energy change. Note that it is essential that the two N-phase protons required to protonate UQ^{2-} and produce UQH_2 do not have access to the UQ chamber until this conformational change has taken place. In addition, it has to be established that essentially all this redox energy can be transmitted along the hydrophilic axis to drive the four parallel proton pumps, with the necessary cyclic changes in pK of the protonatable groups responsible for binding protons from the

N-phase and releasing protons to the P-phase (Section 5.5). Note that complex I operates close to equilibrium and can reverse (reversed electron transport; Section 4.7), meaning that a high Δp generated by succinate oxidation or ATP synthase reversal must be able to reverse the entire sequence, lowering the redox potential of the bound UQ^{2-} sufficiently for it to transfer electrons to the N2 centre.

5.6.4 Mitochondrial complex I

Further reading: Hirst *et al.* (2003), Janssen *et al.* (2006), Hunte *et al.* (2010)

Progress has been made in determining a structure for a mitochondrial complex I using protein from the yeast *Yarrowia lipolytica* (not all species of yeasts possess a proton-translocating complex I). The resolution is only 6Å, but this is sufficient to show that the horizontal HL helix and the general structural features are very similar to the bacterial enzyme. Mutagenesis studies on this mitochondrial enzyme also support the importance of many of the residues implicated in the bacterial enzyme. A notable feature of the mitochondrial enzyme is that the core hydrophobic domain subunits—ND1, -2, -3, -4, -4L, -5 and -6 (corresponding to the bacterial NuoH, N, A, M, K, L and J subunits, respectively)—are all coded for by mitochondrial DNA, so that biosynthesis of complex I involves the coordination of expression and assembly from the nuclear and mitochondrial genome (Section 10.5.2). The hydrophilic domain, with 7 of its 8 Fe-S centres proving the electron wire, comprises seven nuclear-encoded core subunits, which for the human complex are termed NDUFV1, -V2, -S1, -S2, -S3, -S7, and -S8 (corresponding to the bacterial NuoF, -E, -G, -C, -D, -B, and -I subunits).

In addition to the 14 core subunits common to the bacterial complex, the mitochondrial complex I contains no less than 31 or 32 additional accessory or supernumerary subunits, whose functions are less well understood but probably include protection against oxidative damage, binding of the complex to the membrane, structure stabilisation, and complex assembly. The MWFE subunit (NDUFA1) is a phosphorylatable integral membrane protein that has been proposed to be required for the synthesis and assembly of the core mitochondria-encoded subunits of the hydrophobic domain. NDUFA13 is the only hydrophobic subunit in the hydrophilic domain; in addition to being a possible structural component, this subunit has been implicated in apoptosis. NDUFA11 has homology to inner membrane protein import components (Section 10.5.1).

5.7 DELIVERING ELECTRONS TO UBIQUINONE WITHOUT PROTON TRANSLOCATION

In addition to complex I, at least four other enzymes in mammalian mitochondria feed electrons to the UQ pool (Figure 5.1). Succinate dehydrogenase, or complex II, transfers electrons from succinate as part of the tricarboxylic acid (TCA) cycle. This is the only membrane-bound member of the cycle, and its succinate binding site faces the mitochondrial matrix. The second enzyme, also located on the matrix face of the membrane, is the ETF–ubiquinone oxidoreductase; the third is dihydroorotate dehydrogenase (an enzyme involved in pyrimidine biosynthesis); and the fourth (not found in all types of mitochondria) is *s,n*-glycerophosphate dehydrogenase. The latter two enzymes bind their substrate from the outer (P-side) of the membrane. All four are flavoproteins transferring electrons from substrate couples with midpoint potentials close to 0 mV directly to the UQ/UQH_2 pool, bypassing complex I. This direct transfer requires these enzymes to be membrane bound. As would be expected on thermodynamic grounds, none is proton translocating. The feeding into the respiratory chain, via ETF, of electrons from the flavin-linked step in fatty acid oxidation is often overlooked but is functionally very significant in many mammalian cells.

5.7.1　Complex II (succinate dehydrogenase)

Further reading: Maklashina and Cecchini (2010), Iverson (2012)

Figure 5.11 shows that four polypeptides play a functional role in succinate dehydrogenase. That furthest from the membrane contains the covalently bound FAD from which electrons pass sequentially into the membrane sector of the enzyme via a 40Å chain of three Fe–S centres located in the second peripheral subunit. The two integral membrane polypeptides contain one haem group sandwiched in between transmembrane helices. It is not understood exactly how the quinone is reduced, but the terminal 3Fe–4S centre is within 8Å of the quinone binding site. The role of the single haem is not clear, but it cannot be truly essential because it is absent, for example, in the *Escherichia coli* enzyme. The quinone binding site receives both protons and electrons from the N-phase (Figure 5.11); thus, the reduction of UQ by succinate is not associated with any charge movement across the membrane. The redox potential of the 4Fe–4S centre is much lower than those of the adjacent centres. As explained in Section 5.4.1, this is not a reason for excluding its role in the linear chain of centres.

Complex II could be redesigned as a Δp consumer by taking the protons for quinone reduction from the P-phase. Although this would make no sense in mitochondria, in *B. subtilis* a transfer of electrons occurs from succinate to menaquinone, which has a midpoint potential 115 mV more negative than ubiquinone. Such a transfer is thermodynamically unfavourable, and to overcome the energetic barrier, this organism has a succinate dehydrogenase with two haems, one at each side of the membrane, and the quinone reduction site at the P side. This succinate dehydrogenase thus consumes Δp as an electron has to move to the P-side from the catalytic centre at the N-side (Figure 5.11), although there are suggestions that in the *B. subtilis* enzyme some proton movement in the same direction as the electron movement might partially compensate for the direction of electron movement. It is clear that evolution has found a way to modify the bioenergetics of the succinate dehydrogenase reaction by switching between a one- and a two-haem succinate dehydrogenase.

5.7.2　Electron-transferring flavoprotein–ubiquinone oxidoreductase

Further reading: Watmough and Frerman (2010)

The water-soluble ETF is located in the mitochondrial matrix, contains one molecule of FAD, and accepts electrons from several flavin-containing dehydrogenases, including

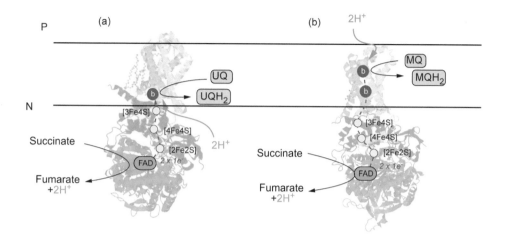

Figure 5.11   (a) Crystal structure of mitochondrial succinate dehydrogenase and (b) schematic model of *B. subtilis* succinate dehydrogenase.
The enzymes have four subunits, with that furthest from the membrane having a covalently bound FAD. A second peripheral subunit contains three Fe–S centres—S-1 [2Fe–2S], S-2 [4Fe–4S], and S-3 [3Fe–4S]—with respective $E^{0\prime}$ values of 0, −260, and 60 mV; see Section 5.4.1 for a discussion of the consequences of the negative potential of S-2. The cyt *b* has $E^{0\prime} = -185$ mV. For the mitochondrial enzyme (a), UQ reduction is believed to be on the N-side of the membrane, close to the haem group. For the *B. subtilis* enzyme (b), the site of menaquinone reduction is believed to be at the P-side of the membrane from where protons are taken. The two haems have $E^{0\prime}$ values of −95 and 65 mV, being respectively located toward the P- and N-sides of the membrane, overcoming the 160-mV difference between the haems (see Figure 3.4). Thus, the membrane potential will act as a driving force for the movement of electrons from the Fe–S centres to the site of menaquinone reduction at the P-side. Note how the organisation of the protein allows the $FADH_2$ to be converted back to FAD at the active site. Structures based on (a) PDB 1Z0Y, Sun *et al.* (2005) and (b) fumarate reductase of *W. succinogenes*. PDB 2BS4, Lancaster *et al.* (2005).

enzymes that catalyse the C–C bond unsaturation step in fatty acid oxidation or various steps in amino acid and choline catabolism. The resulting $FADH_2$ of ETF is oxidised by the ETF–ubiquinone oxidoreductase, which contains an FAD, an Fe–S centre, and a UQ binding site. It is not certain whether the FAD or the Fe–S centre is the immediate electron acceptor for ETF, but the structure suggests that $FADH_2$ could transfer its electrons to bound UQ. The protein is mainly globular, with 10 α-helices and 21 β-strands.

ETF–ubiquinone oxidoreductase does not contain any transmembrane helices, and the association with the membrane can be attributed to a series of hydrophobic residues that contribute to an α-helix and a β-sheet that are adjacent to the hydrophobic ubiquinone binding pocket (Figure 5.12). The structure of this protein shows that quinones can be bound by proteins that are largely globular (rather than transmembranous) but provide a surface that can dip into the membrane sufficiently for UQ and UQH_2 to be able to exchange directly with the core of the lipid bilayer. Further examples of such monotopic membrane proteins are discussed later (Section 5.13).

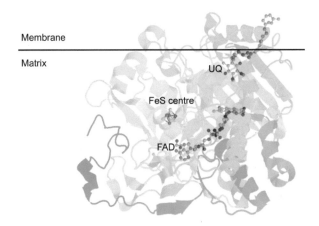

Membrane

Matrix

UQ

FeS centre

FAD

Figure 5.12   ETF–ubiquinone oxidoreductase.
The enzyme is an example of a monotopic protein that does not have any
transmembrane helices but instead associates with the membrane by a parallel
hydrophobic α-helix and β-sheet. Structure based on PDB 2GMH (Zhang
et al., 2006).

5.7.3 *s,n*-Glycerophosphate dehydrogenase and dihydroorooatate dehydrogenase

The enzyme that oxidises *s,n*-glycerophosphate at the outer surface (P-side) of the inner
mitochondrial membrane, contains FAD and at least one Fe–S centre. It is presumably
organised similarly to the ETF–ubiquinone oxidoreductase such that it can receive UQ
from, and deliver UQH_2 to, the hydrophobic core of the membrane. The gene for this
enzyme is found on genomes from those of humans to yeast, but in higher eukaryotes its
expression may differ significantly between cell types, making difficult generalisations
about the role of *s,n*-glycerophosphate oxidation in cellular bioenergetics. When present, it
plays an important role in the transfer of electrons from cytoplasmic NADH to the UQ pool
(Section 9.5.2). Dihydroorooatate dehydrogenase has a similar orientation.

5.8 UBIQUINONE AND COMPLEX III

Further reading: Crofts (2004), Swierczek *et al.* (2010)

The transfer of electrons from UQH_2 to cyt *c*, and the associated proton translocation,
is catalysed by complex III, also termed the cyt bc_1 complex, or more precisely ubiquinol–cytochrome *c* oxidoreductase. This complex is also found in many species of
bacteria (Section 5.15) and is similar in many respects to the cyt b_6f complex of thylakoids (Section 6.4.3). The redox groups in cyt bc_1 comprise a 2Fe–2S centre, located
on the Rieske protein, the haem of cyt c_1, and two *b*-type haems located on the same
polypeptide. One of these, with an $E_{m,7}$ of approximately $-100\,mV$ and thus known as

b_L (sometimes called b_{566} because of its α-band absorption maximum), is located toward the P- (cytoplasmic) side of the mitochondrial membrane. The second haem, b_H ($E_{m,7}$ +50 mV; sometimes called b_{560} because of its α-band at ~560 nm), is positioned toward the N- (matrix) side of the membrane. There are two binding sites for ubiquinone/ubiquinol—one, known as Q_p (also termed Q_o or Q_z), close to the P-face of the membrane, and the other, known as Q_n (also termed Q_i or Q_z), nearer the N-side of the membrane.

We first discuss the mechanism of the complex and then relate this to the structure. The pathway of electron flow, often called the Q-cycle, is at first sight convoluted, and we describe it in detail (Figure 5.13). The discussion is for the mitochondrial case; the bacterial bc_1 complexes are very similar.

5.8.1 Stage 1: UQH₂ oxidation at Q_p

A pool of ubiquinone and ubiquinol exists in the inner mitochondrial membrane in significant molar excess over the other components of the respiratory chain. The midpoint potential, $E_{m,7}$, for the UQH$_2$/UQ couple is +60 mV (Figure 5.4), whereas the actual $E_{h,7}$, taking into account the UQH$_2$/UQ ratio, is close to 0 mV. A molecule of UQH$_2$ from the pool diffuses to a binding site Q_p, which is close enough to the Rieske protein for a hydroxyl group of the ubiquinol to hydrogen bond to one of the histidines that coordinates the Fe–S centre (Figure 5.12). What then happens is that the oxidation of UQH$_2$ to UQ takes place in two stages:

(1) The first electron is transferred from UQH$_2$ to the Rieske protein (Figure 5.13a), releasing two protons to the cytoplasm and leaving the free radical semiquinone anion species UQ$^{\bullet-}$ at the Q_p site.
(2) The second electron is transferred to the b_L haem, which is also close to the P-face.

The $E_{m,7}$ for the UQ$^{\bullet-}$/UQH$_2$ couple is approximately +280 mV, close to that for the Rieske protein, and 220 mV more *positive* than the $E_{m,7}$ of the two-electron oxidation (+60 mV; see above). This implies that the second stage of the oxidation will be energetically favourable as the semiquinone seeks to lose the second electron, and this is reflected in the E_m for the UQ/ UQ$^{\bullet-}$/ couple, which at −160 mV is 220 mV more *negative* than that of the two-electron oxidation. The first one-electron oxidation step has thus generated the semiquinone anion, which is a very strong reductant. Under certain conditions, it can be detected by ESR. UQ$^{\bullet-}$ is also an intermediate in the bacterial photosynthetic reaction centre (Section 6.2.2.6). As discussed in Section 9.10.2, the unpaired electron on UQ$^{\bullet-}$ can in some circumstances be transferred directly to oxygen, thus generating the dangerous superoxide anion.

The electron received by the Rieske protein passes down the chain to cyt c_1, cyt c, and cytochrome oxidase (Figure 5.12a). Note that we still do not fully understand the events at the Q_p site. Attempts to create mutations that send both electrons from the oxidation of ubiquinol to the Fe–S centre have failed for the bc_1 complex, but in the case of the analogous $b_6 f$ complex in thylakoids (Chapter 6), a 'broken' Q-cycle has been achieved by mutations causing loss of the haem corresponding to b_H (Malnoe et al., 2011). Even in this mutation, there is only a very slow oxidation of the semiquinone by the Fe–S centre. The driving force for this reaction is very large, and it may be that this retards the electron transfer rate for the reasons discussed in Section 5.4.

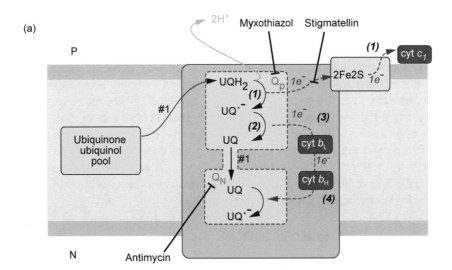

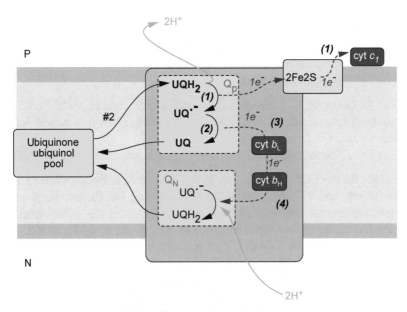

Figure 5.13   The Q-cycle in mitochondria and bacteria.
(a) This illustrates the electron transfer events that follow the oxidation of a ubiquinol (#1) at the P-side of the inner mitochondrial membrane under conditions in which the quinone binding site at the N-side is initially either vacant or occupied by a ubiquinone molecule. There is evidence that an internal channel allows UQ to move from the Q_p to the Q_n site without equilibrating with the bulk pool (Section 5.8.6). (b) This illustrates the electron transfer events that follow the oxidation of a second ubiquinol (#2) at the P-side of the membrane when the Q_n is occupied by a ubisemiquinone radical. Note that the Q_p site has also been termed the Q_i or Q_c site ('c' indicates the cytoplasmic side of the membrane in bacterial cytochrome bc_1 complexes) in various systems and the Q_p site is also known as the Q_o or Q_z site. The inhibitory sites of action of myxothiazol, stigmatellin and antimycin are also shown.

5.8.2 Stage 2: UQ reduction to UQ$^{\bullet-}$ at Q$_n$

The electron on b_L ($E_{m,7}$ −100 mV) now passes to the other haem, b_H ($E_{m,7}$ +50 mV). Note that these redox potentials were measured in the absence of a $\Delta\psi$, which affects the distribution of electrons between them, as discussed in Section 3.3.6 (see also Figure 3.4). The presence of a physiological $\Delta\psi$ of approximately 150 mV opposes the electron transfer from the b_L on the P-side of the hydrophobic core of the membrane to the b_H on the N-side. Thus, the electron retains its original energy on passing from b_L to b_H because the drop to a more positive redox potential is compensated by the energetically unfavourable electron transfer from the P-side to the N-side of the membrane. This organisation also implies that very high $\Delta\psi$ will retard electron transfer between the b-type haems, leading to a prolongation of the occupancy of the Q$_p$ site by UQ$^{\bullet-}$ enhancing the chances of O$_2^{\bullet-}$ production (Section 9.10.2).

UQH$_2$ and UQ can in principle migrate freely from one side of the hydrophobic core to the other, regardless of $\Delta\psi$, because these hydrophobic carriers are uncharged. The second quinone binding site, Q$_n$, in the close vicinity of b_H, binds UQ and allows the transfer of the electron from the reduced b_H with the formation of UQ$^{\bullet-}$ (Figure 5.12). At first glance, this seems thermodynamically unlikely because the $E_{m,7}$ for the UQ/UQ$^{\bullet-}$/ couple in free solution is −160 mV, whereas that for b_H is +50 mV. If, however, Q$_n$ were to bind the semiquinone much more strongly than UQ, this would have the effect of shifting the $E_{m,7}$ to a more positive value—that is, making the UQ more readily reducible. A 10-fold difference in the binding of the semiquinone relative to UQ shifts $E_{m,7}$ 60 mV more positive than if the reaction occurred in free solution. A 300-fold stronger binding of the semiquinone would thus make the $E_{m,7}$ 150 mV more positive.

We have not cheated the first law of thermodynamics here because the energy required for the addition of a second electron to generate unbound UQH$_2$ (discussed later) is proportionately *increased*; that is, the E_m for the couple UQH$_{2\,free}$/UQ$^{\bullet-}_{bound}$ is made proportionately more negative. This is confirmed by actual measurements of the two $E_{m,7}$ values, using ESR to detect the semiquinone. We will come across this concept of driving an apparently unfavourable reaction by making a product very tightly bound again in Chapter 7 when we discuss the ATP synthase. In fact, it is worth noting that binding sites within proteins can in general stabilise oxidation and protonation states that would not be stable in homogeneous solutions and hence there are many substantially perturbed pKa values for amino acid side chains within proteins.

5.8.3 Stage 3: UQ$^{\bullet-}$ reduction to UQH$_2$ at Q$_n$

We now have a semiquinone firmly bound to Q$_n$. In the next part of the cycle (Figure 5.13b), a second molecule of UQH$_2$ is oxidised at Q$_p$ in a repeat of stage 1—one electron passing to cyt c_1 and the other via b_L to b_H. This second electron now completes the reduction of UQ$^{\bullet-}$ to UQH$_2$, the two protons required for this being taken up from the matrix (Figure 5.14b). The UQH$_2$ returns to the bulk pool and the cycle is completed. Q$_n$ and Q$_p$ are not equivalent in this model: Only Q$_n$ stabilises the semiquinone through strong binding—supported by its detection by ESR. At Q$_p$, the redox potentials of the two steps are widely separated and the semiquinone has only a transient existence.

5.8.4 **The thermodynamics of the Q-cycle**

The overall reaction catalysed by the bc_1 complex involves the *net* oxidation of 1 UQH_2 to UQ (two UQH_2 oxidised in stage 1 and one UQ reduced in stage 3), together with the reduction of two cyt c_1, the release of $4H^+$ at one side of the membrane and the uptake of $2H^+$ from the other. The imbalance in the proton stoichiometry is only apparent because the regeneration of UQH_2 from UQ at the acceptor site of complexes I and II involves the uptake from the matrix of a further $2H^+$ (see Figure 4.2).

In the model we have discussed, the main *charge* transfer across the membrane is the movement of the two electrons between the b haems on opposite sides of the hydrophobic barrier. Q_p is close to the cytoplasmic face, so the two additional electrons transferred via the Fe–S centre to cyt c_1 will not significantly decrease the transmembrane charge transfer, and release of protons at Q_p will contribute little to the charge displacement. Q_n is more deeply buried into the matrix side of the membrane such that the entry of protons from the matrix contributes partially to the charge movement across the membrane as a whole. However, this does not affect the charge calculation because the charge displacement for $2H^+$ moving from the N-phase to Q_n plus $2e^-$ moving from Q_p to Q_n is independent of the exact position of the latter in the membrane.

The overall action of the bc_1 complex is the oxidation of one ubiquinol by two molecules of cytochrome c, an energetically downhill reaction, coupled to the uphill translocation of protons. To understand how it can function as the equivalent of a proton pump (it is not a proton pump in the sense that complex I and bacteriorhodopsin, Section 6.5, are proton pumps), removing protons from the matrix at low electrochemical potential and releasing them in the cytoplasm at a Δp approximately 180 mV higher, we consider two conditions: (1) where Δp is present purely as a membrane potential (approximating to the condition in the respiratory chain) and; (2) where Δp is present purely as a ΔpH (as would occur in a thylakoid membrane, where a closely analogous cycle probably operates, Section 6.3). In the first case, the protons are present at equal concentrations on both sides of the membrane and the work that must be done is to push $2e^-$ from the cytoplasmic to the matrix side of the membrane against a high membrane potential. As stated previously, this is energetically possible because the electrons are transferred from a negative (low) potential haem to a positive (high) potential haem. In the case of a pure ΔpH, the electrons would flood from b_L to b_H because they have no $\Delta \psi$ to push against. This would drive UQH_2 oxidation at Q_p and UQ reduction at Q_n, enabling protons to be translocated against a high ΔpH.

5.8.5 **Inhibitors of the Q-cycle**

Antimycin, myxothiazol, and stigmatellin inhibit the mitochondrial bc_1 complex. Antimycin acts at Q_n, preventing the formation of the relatively stable $UQ^{\bullet -}$. If oxygen is added to an anaerobic suspension in the presence of this inhibitor, an additional *reduction* of the b cytochromes occurs. This paradoxical *oxidant-induced reduction* occurs because UQH_2 bound at the Q_p site can now transfer an electron down through complex IV via the Rieske proton and cyt c_1, and the resulting $UQ^{\bullet -}$ can transfer a further electron to any b haems not previously reduced. Myxothiazol blocks events at Q_p, while stigmatellin inhibits electron transfer to the Rieske protein. It should be clear from inspection of Figure 5.13 that oxidant-induced reduction of the b cytochromes does not occur in the presence of these inhibitors. The effect of these inhibitors on the leakage of electrons to form superoxide is discussed in Section 9.10.2.

5.8.6 The structure of complex III

Further reading: Crofts (2004), Crofts *et al.* (2008)

The functional information about complex III summarised previously was obtained before detailed crystal structures were available. For the most part, these fully support the earlier biochemical investigations, but two features in particular substantially add to our understanding. One of the most important insights relates to how the electron transport pathway is bifurcated at the Q_p site; in other words, why does one electron transfer to the Rieske Fe–S centre and thereafter to cyt c_1, while the second passes to cyt b_L and then to cyt b_H?

The Rieske protein 2Fe–2S cluster is attached to the polypeptide by chelation of one Fe to two cysteines and the other to two histidine residues. The polypeptide chain is folded as a globular structure extending into the aqueous P-phase, incorporating the 2Fe–2S centre close to its surface and anchored to the membrane via a hydrophobic N-terminal helix (Figure 5.14). Crystal structures obtained under different conditions show that the globular head of the Rieske protein can reposition itself. When its Fe–S centre is reduced (Figure 5.14C,a), the globular head docks onto cyt c_1, which has a globular domain and hydrophobic anchor similar to the Rieske protein, except it is the C-terminus that provides the anchor. Two exposed histidine ligands of the Fe–S centre insert into the c_1 rather like the prongs of an electrical plug. This allows electron transfer to take place (Figure 5.14C,b). The Fe–S centre, now in its Fe^{3+} state, has a diminished affinity for the cyt c_1, and the globular head repositions itself close to the Q_p centre, stabilised by electrostatic interactions with amino acid residues (Figure 5.14C,c). In this conformation, the Fe–S centre accepts an electron from UQH_2 at the Q_p site, with the formation of $UQ^{\bullet-}$ releasing $2H^+$ to the P-phase (Figure 5.14C,d). The movement, which is approximately 20Å, can occur on the sub-millisecond timescale and thus can match the required electron transfer rates. Because 14Å is the limit for rapid electron transfer between two redox centres, when the Fe–S centre is bound to cyt c_1 it is too far from the Q_p site to accept an electron. Thus, the second electron, released from $UQ^{\bullet-}$, has no alternative but to transfer to the b_L haem (Figure 5.14C,e).

The scheme in Figure 5.14 is a simplification because crystal structures show that complex III is a dimer in which the two monomeric units do not function independently; the globular domain of the Rieske protein of one monomer interacts with the Q_p site and the cyt c_1 in the other. The two monomers are packed together to form two separate cavities, the walls of which contain the quinone binding sites. This provides a second example of cooperation between monomers because in any given one cavity there is a Q_p site provided by one monomer and a Q_n site provided by the other. In principle, this means that the UQ molecule produced by oxidation of UQH_2 at a Q_p site can then diffuse to the Q_n site within the cavity without having to equilibrate with the bulk quinone pool, contributing to catalytic efficiency. A further consequence of the dimeric structure

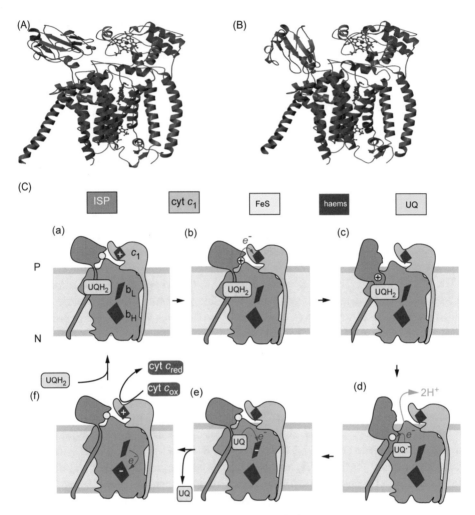

Figure 5.14   Conformational changes in the cytochrome bc_1 complex.
(A and B) Structure of the bc_1 complex. Only cyt b, cyt c_1, and the Rieske iron–
sulfur protein (ISP) are shown. Note that the functional complex III is a dimer and
includes several so-called core subunits (which are actually peripheral in the
N-phase). There is evidence that the ISP of one monomer transfers electrons for
the second monomer. In panel A, ISP is in its C-conformation and in panel B it is
in the ISP_B conformation. Structures based on PDB 1BE3 and 1BGY (Iwata
et al., 1998). (C) Conformational changes in the globular head group of the
Rieske ISP during UQH_2 oxidation at the Q_p site. (a) The C-conformation, with
its (reduced) 2Fe–2S centre close to the haem of oxidised cyt c_1; this allows
electron transfer to occur to the latter, leaving the ISP 2Fe–2S centre oxidised. (b)
The ISP spontaneously changes conformation to the ISP_B conformation, bringing
the 2Fe–2S centre close to the Q_p site. (c) Transfer of $1e^-$ from UQH_2 bound at
the Q_p site to the 2Fe–2S centre of ISP neutralises the latter. Ubisemiquinone
($UQ^{•-}$) is formed at Q_p after two protons are released to the P-phase. (d) The
electron from $UQ^{•-}$ is then transferred across the membrane through the b haems
and UQ is released. (e and f) Finally, an electron is transferred from the c_1 haem
to cyt c and a new molecule of UQH_2 binds to complete the cycle.
Adapted from Crofts et al. (1999).

is that electrons may exchange between the haems on one cytochrome b subunit with those on the other subunit. This 'H-shaped' electron transfer system, which distributes electrons within the millisecond timescale of turnover between four quinone oxidation/reduction sites (two Q_p and two Q_n), has been termed a 'bus bar' by analogy with components found in electronic circuits.

The crystal structures of the mitochondrial cyt bc_1 complexes show the positions of as many as 11 subunits in each monomer. Eight of these have no catalytic role in the oxidation of ubiquinol. One may assist assembly of subunits from the cytoplasm, whereas two others may catalyse removal of targeting sequences. As in the case of complex I, bacterial cyt bc_1 complexes have far fewer subunits, sometimes just the three that bind the redox groups.

5.9 INTERACTION OF CYTOCHROME C WITH COMPLEX III AND COMPLEX IV

Further reading: Lange and Hunte (2002)

Cytochrome c is a 12-kDa water-soluble protein that is located on the P-face of the inner membrane. Mitochondrial cyt c accepts an electron from cyt c_1 and donates it to the copper A (Cu_A) centre of complex IV (Section 5.10). Thus, after dissociation in its ferrous state from the bc_1 complex, cyt c must rapidly diffuse to its other main partner, complex IV. The asymmetrically positioned haem group of cyt c has one edge within 5Å of the surface of the protein where a patch of lysines are found. These have been implicated in the interaction of cyt c with at least one of its partners because their chemical modification inhibits electron transfer, both from complex III and to complex IV. In addition, association of cyt c with either complex protects against such chemical modification. A crystal structure for complex III with bound cyt c has recently shown exactly how close the two haems are in the complex, with the edge-to-edge distance between the haems of cyt c_1 and cyt c being only 9Å, which would facilitate very rapid electron transfer. The interaction between the cytochromes must be transient and yet specific, which is compatible with it being mediated by nonpolar forces and a cation–π interaction (between a cation and a π-orbital system, such as an aromatic side chain) as seen in the crystal structure. The lysine patch on cyt c is not directly involved in the complex and may be involved in steering the cyt c onto cyt c_1. The types of interactions seen in the cyt c_1/cyt c complex appear to be the dominant features of transient electron transfer complexes and are also observed for the interface of the bacterial reaction centre cytochrome c_2 complexes. Curiously, the crystal structure of the complex between the dimeric bc_1 complex and cyt c shows that cyt c is bound to only one monomer.

The initial electron acceptor of complex IV, the Cu_A centre, is close to the surface of the protein and also has anionic sites suitable for docking with the lysine patch on cyt c, but there is currently no structure to confirm this. Involvement of the same region of the cyt c surface for interaction with both its electron donor and acceptor indicates a single route for electron transfer into and out of its haem. This means that a cyt c molecule could not remain sandwiched between cyt c and cyt c_1 as demanded by a 'wire' model of the electron transfer chain. The rest of the surface of the 30-Å-diameter cytochrome may function to insulate the haem from adventitious electron transfer. As with most other electron transfers between centres, the electron transfer involves electrons tunnelling through the relatively uniform protein dielectric and does not require specifically positioned amino acid side chains.

Clearly, if a single patch on the surface of cyt c is responsible for interaction with both redox partners, it follows that after reduction by complex III, the cytochrome must dissociate from the complex before associating with complex IV to pass the electron to the Cu_A. This is in accord with the current view that the integral complexes III and IV are thought to diffuse relatively slowly in the plane of the membrane while the peripheral protein, cyt c, undergoes more rapid lateral diffusion along the surface.

5.10 COMPLEX IV

Further reading: Kaila *et al.* (2010)

The final step in the electron transport chain of mitochondria and certain species of respiratory bacteria operating under aerobic conditions is the sequential transfer of four electrons from the reduced cyt c pool to O_2, forming $2H_2O$ in a $4e^-$ reaction catalysed by a cytochrome c oxidase:

$$O_2 + 4e^- + 4H^+ \rightarrow 2H_2O$$

The names cytochrome c oxidase and ferrocytochrome c:O_2 oxidoreductase refer to the catalysis of oxidation of cyt c by oxygen. Complex IV is an often-used alternative name, whereas earlier nomenclature such as cytochrome oxidase and cytochrome $aa3$ oxidase are also used. This protein complex is a member of the haem–copper oxidase family, being only one example of a variety of proteins that occur in bacteria (discussed later).

The protons required for the reduction of oxygen are taken from the N-side of the membrane, whereas the electrons from the oxidation of reduced cyt c come from the P-phase. Regardless of the exact location of the water-forming reaction in the membrane, this is equivalent to the translocation of two positive charges from the N- to P-phase per $\frac{1}{2}O_2$ reduced. Thus, the reduction of oxygen to water automatically generates a Δp. Note that protons used in the reduction of water are not translocated all the way across the membrane by this process; how far they move can be deduced from the structure (Figure 5.15). This imbalance between proton uptake and release at the two sides of the membrane disappears when the overall oxidation of ubiquinol by oxygen is considered. In addition, complex IV is a proton pump, translocating two additional protons per $2e^-$ from the N- to the P-phase.

Cytochrome oxidase poses several challenging problems, including (1) how the protons required for oxygen reduction are taken from the N-side of the membrane; (2) the mechanism of oxygen reduction to water, and (3) the mechanism by which the reduction of oxygen is coupled to the pumping of protons across the membrane. Understanding

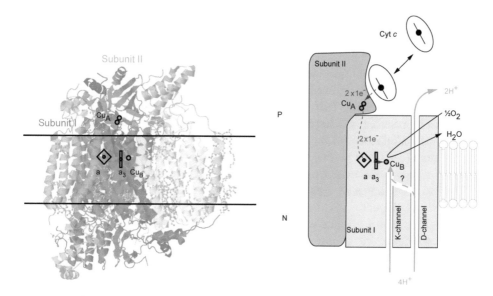

Figure 5.15   Schematic representation of subunits I and II of bovine heart mitochondrial complex IV.
The crystal structure (PDB 3ASO, Suga *et al.*, 2011) has established the relative positions of the *a*- and *a₃*-type haems and the two copper centres (copper atoms are represented by open spheres). An approximate deduced site of cyt *c* docking onto subunit II is shown; this site has to allow the haem of cyt *c* to come within at least 14 Å of the Cu_A site in order to facilitate sufficiently rapid electron transfer. The reaction is shown in terms of two electrons, and thus consumption of half an oxygen molecule, so as to facilitate comparison with the operation of the electron transport chain as a whole, which is traditionally analysed in terms of two electrons. Note that a cyt *c* molecule loses one electron upon oxidation and thus two molecules must dock sequentially in order to transfer two electrons into cytochrome oxidase. For each two electrons reaching an oxygen atom from cyt *c*, four positive charges are moved through the oxidase and thus across the membrane; two of these can be regarded as pumped all the way across the membrane, but the other two charge movements result from the movement of two electrons from the P-side to meet two protons coming from the N-side. Currently, there is uncertainty regarding the pathways for proton movement from the N-side; the tentatively accepted contributions of the D and K channels (see text) are shown. All stoichiometries shown should be multiplied by 2 in order to account for reduction of one oxygen molecule (O_2).

of all these issues has been advanced by the availability of crystal structures for the enzyme, but is far from complete.

5.10.1 Structure of complex IV

Further reading: Yoshikawa (1999), Saraste (1999), Abramson *et al.* (2001)

High-resolution structures of the molecule, in both fully oxidised and fully reduced states, have been obtained from both a mitochondrial and a bacterial (*P. denitrificans*) source (e.g., Figure 5.15). The key catalytic functions are found in each case in subunits I and II. Other subunits (as many as 11 for the mitochondrial enzyme) are not relevant to catalysis and are not considered here. The mitochondrial protein is a dimer, but unlike complex III, this has no functional significance. The structures show that subunit II, in addition to two transmembrane helices, has a globular domain, folded as a β barrel, that projects into the P-phase. This is the location of the Cu_A centre, which has two copper atoms in a cluster with sulfur atoms. The binuclear copper centre undergoes a one-electron oxidation/reduction reaction, but why this is advantageous relative to a single copper atom (e.g., as in plastocyanin, Section 6.4.3) is not clear. A function of the Cu_A centre is to receive electrons, one at a time, from cyt c, which must bind to subunit II.

The two haem groups of complex IV, located approximately $15\,\text{Å}$ below the P-surface of the bilayer (Figure 5.15), are sandwiched between some of the 12 transmembrane α-helices of subunit I. Haem a is slightly closer to the Cu_A centre than the second haem, a_3, and accepts electrons from Cu_A. Haem a_3 is within $5\,\text{Å}$ of haem a. The two haems are chemically identical but quite distinct spectrally, principally because one axial coordination position to the a_3 haem iron is not occupied by an amino acid side chain. This is the position where oxygen binds before its reduction to water, and it is also the site of binding of several inhibitors, including cyanide, azide, nitric oxide, and carbon monoxide.

Immediately adjacent to haem a_3 is a third copper atom, known as Cu_B; it has three histidine ligands, suggesting that a fourth coordination position may be occupied by a reaction product during some stages of the oxygen reduction reaction. The crystal structure surprisingly indicated that one of these histidine ligands was cross-linked through a covalent bond to a nearby tyrosine residue, a feature that was confirmed by analysis of peptides.

Plausible channels, containing bound water molecules and located between some of the helices, can be identified. These could provide a pathway for conducting protons from the N-phase to the site of oxygen reduction. These channels are known as D and K after a conserved glutamate and conserved lysine projecting into these channels. There are no such obvious channels linking the haem a_3 to the P-side of the membrane, perhaps explaining the failure of protons from the P-phase to be recruited for the reduction of oxygen to water.

The structure does not explain how the additional protons can also be pumped across the membrane. As discussed previously (Figure 5.9), a proton-pumping mechanism requires a redox-driven change of proton binding affinity and conformation to ensure that protons move unidirectionally across the membrane. Comparison of the fully oxidised and fully reduced structures of the bacterial enzyme has not detected any significant conformational changes. In contrast, comparison of the structures of the mitochondrial enzyme in these two oxidation states has revealed a conformational change, associated with an aspartate residue, near the P-side of the membrane and some distance from the haems and Cu_B. This has led to a proposal for a proton-pumping mechanism, involving a so-called 'H-channel,' whereby the translocated protons move through a hydrogen-bonded system that is independent of the D and K channels, and conformational energy is transduced from the site of oxygen reduction.

A difficulty with this H-channel mechanism is that the key residue that undergoes the conformational change is not found in any of the proton-pumping bacterial complex IVs, nor indeed in all the mitochondrial enzymes, while the lack of effect of mutagenesis in the bacterial enzymes of counterparts to the implicated residues eliminates such a pathway for the bacterial enzymes. Although it is possible that the mitochondrial and bacterial complexes actually differ, this runs counter to evolutionary and mechanistically unifying principles. Most investigators favour a mechanism in which the route of proton pumping passes close to the haem a_3/Cu_B. However, the temptation to imagine that the extended side chain of the a-type haem is important for proton translocation must be resisted because there are relatives of cytochrome c oxidase in bacteria that still pump protons despite having the standard haem b in both the haem binding sites. Nevertheless, there is increasing evidence that one or more propionates of the haem groups may play a role in the proton-pumping mechanism.

The fact that two putative proton channels, D and K, can be identified in cytochrome oxidase has led to attempts, by mutagenising key residues, to determine whether one of these channels provides the protons required for reduction of oxygen while the second is concerned with the proton-pumping process. Initial studies did suggest such a division of function, but currently the situation is thought to be more complex. For each eight protons needed when one O_2 is reduced (four for water formation and four pumped), approximately six are thought to travel via the D channel with only one or two via the K channel. Thus, no clear distinction between pathways for 'chemical' and 'translocated' protons can be drawn. If the D channel conducts proton both for water formation and for pumping, some distribution mechanism is needed. A glutamate residue equidistant from the a and a_3 haems is a candidate, undergoing protonation/deprotonation cycles and conformational changes linked to the different redox states of the a_3/Cu_B, although additional components remain to be identified. It is possible that movements of bound water molecules are important, as is the case for bacteriorhodopsin (Section 6.5.1).

5.10.2 Electron transfer and the reduction of oxygen

Complex IV must reduce one molecule of oxygen to two molecules of water while not releasing any reactive oxygen species, such as superoxide. One provisionally accepted reaction scheme is shown in Figure 5.16, which shows a cycle of catalytic activity starting with the two centres (a_3 and Cu_B) at the catalytic site reduced. Note that this scheme does not involve the other two centres, haem a and Cu_A. The complete cycle has removed four protons from the N-phase. At the catalytic site, these have met four electrons from the P-phase. Therefore, the P-phase has lost four negative charges and the N-phase has lost four positive charges: thus, a $\Delta\psi$ is generated. It is important that the scheme does not involve the generation of any potentially toxic free radical oxygen species. Aspects of this scheme are contentious: How many protons are released at each step? When exactly are the water molecules released?

Any scheme has to satisfy thermodynamic constraints. In the model, the ΔG from the transfer of four electrons to an oxygen molecule is roughly balanced by that resulting from the formation of the highly oxidising $a_3^{4+} = O$ and tyrosine radical species. Electrons enter the complex from cyt c on the P-side at an $E_{h,7}$ of approximately +290 mV for mitochondria in state 4 and are ultimately transferred to the $\frac{1}{2}O_2$/H_2O couple with an $E_{h,7}$ in air-saturated medium at approximately +800 mV. However, because the equivalent of four negative charges cross from the P to the N faces of the membrane

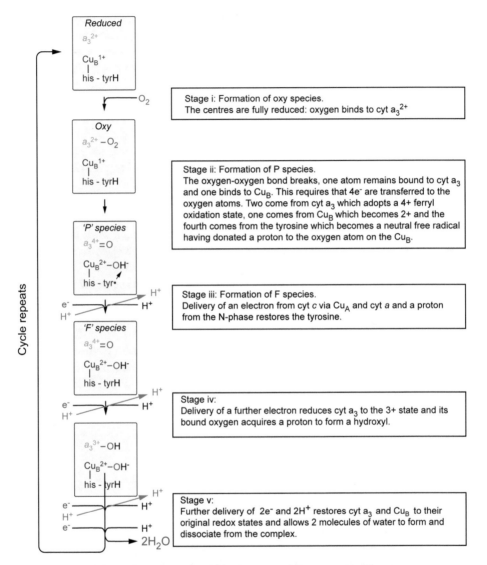

Figure 5.16 Simplified scheme for the reaction between cytochrome oxidase and oxygen and the connection to its proton-pumping activity.
Only cyt a_3, Cu_B, and the tyrosine 244 side chain (which is cross-linked to histidine 240, a ligand to the Cu_B) are considered in this scheme. The tyrosine is shown protonated. Four electrons are delivered from cyt c via the Cu_A and cyt a. The four black protons are required for the formation of $2\,H_2O$. The blue protons are pumped across the membrane; three sites are shown, although the stoichiometry is $4H^+$ pumped per $2\,H_2O$ because it is thought that one site is responsible for pumping two protons. For further details, see text.

against a $\Delta\psi$ of approximately 180 mV, this reduces the available energy (just as the reverse process in the bc_1 complex increased the energy). The effective redox span in state 4 is therefore slightly more than 300 mV (i.e., +800−290−180 mV). Four electrons falling through this potential would be sufficient to translocate up to six protons across the membrane against a Δp of approximately 200 mV. However, unlike the remainder of the respiratory chain, complex IV is irreversible. The actual H^+/O stoichiometry for the proton-pumping activity alone is lower, $4H^+/4e^-$, reflecting this lack of reversibility. Recall that the combination of this proton pumping and the meeting of protons and electrons within the protein (Figure 5.15) results in an overall charge movement of $8q^+/4e^-$.

5.11 OVERALL PROTON AND CHARGE MOVEMENTS CATALYSED BY THE RESPIRATORY CHAIN: CORRELATION WITH THE P/O RATIO

For every two electrons passing down the mitochondrial electron transport chain, complex I pumps four protons and additionally takes up $2H^+$ from the matrix to reduce UQ to UQH_2 (Figure 5.17). As the electrons pass from UQH_2 to cyt c, complex III releases four protons at the P-side and takes up two from the N-side. Note that the UQ/UQH_2 couple linking complex I to complex III acts as part of a classical 'Mitchell loop' (Section 5.5). Finally, complex IV pumps $2H^+$ across the membrane and additionally takes up $2H^+$ from the matrix for the formation of water. Note that we ignore the proton liberated when NADH is oxidised to $NAD^+ + H^+$ because the coenzyme recycles continuously and an equivalent matrix proton is taken up when NAD^+ is reduced by a dehydrogenase. We would similarly ignore the matrix protons liberated when $FADH_2$ is oxidised to FAD when succinate or similar substrates are used. In contrast, the two protons that are removed from the matrix to form each water molecule must be included. The overall result is that the matrix loses $12H^+$ and the P-phase gains $10H^+$ for every pair of electrons passing through the entire electron transport chain. This apparent lack of balance disappears if we include the upstream dehydrogenases responsible for reducing NAD^+: These liberate $2H^+$ into the matrix for each molecule oxidised—for example, malate = oxaloacetate + $2H^+ + 2e^-$.

The electrons passing through complex I do not cross the membrane and so do not contribute to charge translocation. However, the electrons moving from Q_P to Q_N in complex III and the electrons translocated to the water formation site in complex IV do contribute. Remember that the positive charges moved from the N-phase to cyt a_3 by the protons required for water formation are equivalent in charge terms to negative charges moving from a_3 to the matrix. Electron translocation in complex III and complex IV therefore each contribute two charges/$2e^-$, whereas proton pumping by complex I and complex IV contributes $4q^+/2e^-$ and $2q^+/2e^-$, respectively. Overall, this gives $4q^+/2e^-$ for complexes I and IV and $2q^+/2e^-$ for complex III. There is of course no creation of net positive or negative charges during this process: the $2e^-$ fed into complex I balance the $2H^+$ used to form H_2O:

$$2e^- + 2H^+ + \tfrac{1}{2}O_2 \rightarrow H_2O$$

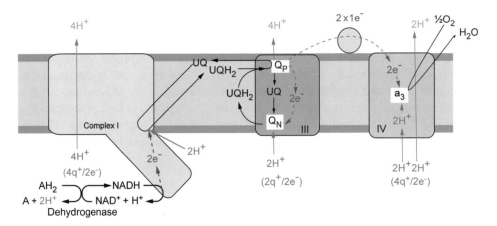

Figure 5.17 Schematic representation of the movement of protons and electrons by the mitochondrial electron transport chain.

The overall proton (and charge) translocation stoichiometry for the transfer of $2e^-$ from succinate to oxygen is $6/2e^-$. If the electrons originate from a NAD-linked substrate, thus passing through complex I, then the overall proton and charge stoichiometry is $10/2e^-$. Current structural information for the animal mitochondrial ATP synthase (Section 7.6) indicates that 2.6 (8/3) H^+ are required to synthesise 1 ATP in the matrix, with a further proton required for transport of ADP, Pi and ATP (Section 9.5.1). In the strict absence of a proton leak, the maximal P/O ratio (Section 4.3.7) would be 2.8 for a NAD-linked substrate and 1.7 for succinate.

Because the complexes act in parallel with respect to the proton circuit (see Figure 4.2), each must generate the identical Δp. The lower charge translocation stoichiometry for complex III relative to complex IV thus accords with its smaller redox span, and hence lower ΔG available for proton translocation. It is a frequently encountered misunderstanding to envisage that complex III makes 'less protonmotive force' than complexes I or IV.

5.12 THE NICOTINAMIDE NUCLEOTIDE TRANSHYDROGENASE

Further reading: Shimomura *et al.* (2009), Jackson (2012)

Although the midpoint potentials for the $NAD^+/NADH$ and $NADP^+/NADPH$ couples are the same (Table 3.2), the ratio $NADPH/NADP^+$ is much greater than $NADH/NAD^+$ in the mitochondrial matrix. One process (others include an NADP-linked isocitrate dehydrogenase) maintaining this disequilibrium is the protonmotive force-dependent transhydrogenase, which catalyses the following reaction:

$$NADP^+ + NADH + nH^+_{\text{P-phase}} = NADPH + NAD^+ + nH^+_{\text{N-phase}}$$

where n is almost certainly 1. The observed mass action ratio (Section 3.2) may exceed 500. Note that the transhydrogenase uses, rather than generates, protonmotive force. It is also found in the cytoplasmic membranes of many bacterial species, where it may play a role in providing NADPH for reductive biosynthesis—a role that can be deduced from experiments in which overexpression of the transhydrogenase results in boosted production of amino acids in industrial processes. In mammalian mitochondria, NADPH is required for reduction of glutathione, but other roles have not been excluded. Loss of transhydrogenase in a mouse strain results in the appearance of diabetes (Section 12.5.2.2), suggesting that the NADPH generated by transhydrogenase has a role in insulin signalling. The transhydrogenase is not present in all eukaryotes, but there is no rationale for its variable occurrence. In some parasites, the transhydrogenase is argued to function as a protonmotive force generator—that is, the reaction runs from right to left in the previous equation.

The mitochondrial transhydrogenase has a single 110-kDa polypeptide with a central hydrophobic region predicted to contain 14 transmembrane helices, a 40-kDa N-terminal globular domain that binds NAD(H), and a 20-kDa C-terminal globular domain that binds NADP(H), with both binding sites being exposed to the matrix. The enzyme functions as a dimer. Direct $2e^- + 1H^+$ (so-called 'hydride') transfer from NADH to NADP$^+$ requires the direct juxtaposition of binding sites for the two substrates. It is envisaged that at any one time, this can occur only between one N-terminal/C-terminal domain pairing, with the other pair within the dimer being in a conformationally distinct state. This would accommodate an alternating site mechanism (Chapter 7).

The structural changes accompanying the coordinated interconversions between the two conformations of each of the two component pairs in the dimeric enzyme are envisaged to be coupled to proton translocation through the transmembrane domain. A conformational change would be coupled, via a set of protein conformational changes, to proton transfer through the transmembrane domain without any need for the translocated protons to pass through the NAD(H) and NADP(H) binding sites.

The transhydrogenase is an interesting exception to the rule that there should be a difference in standard midpoint potential across an energy-transducing step (Figure 5.7).

5.13 ELECTRON TRANSPORT IN MITOCHONDRIA OF NON-MAMMALIAN CELLS

Further reading: Muller *et al.* (2012), Iwata *et al.* (2012), Shiba *et al.* (2013)

The electron transport systems of mitochondria from mammalian sources have been the most studied at the biochemical level, in part because large-scale preparation of mitochondria from these sources has been relatively easy. This emphasis tends to obscure the significant differences that are found in mitochondria from other sources. As examples, we give a brief overview of mitochondrial electron transport in plants, fungi and parasites.

A feature found in plant mitochondria that distinguishes them from their mammalian counterparts is the frequent presence of an electron transport pathway from ubiquinol

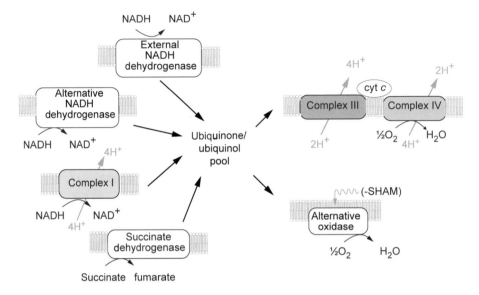

Figure 5.18   General features of the organisation of the electron transport system of plant mitochondria.
The role of a ubiquinone pool in connecting the various pathways of electron transport is evident.

to oxygen that is independent of complexes III and IV. This pathway is characteristically inhibited by salicylhydroxamic acid (usually called SHAM) but not by antimycin or myxothiazol (Figure 5.18). It is regarded as an 'uncoupled' pathway because no proton translocation occurs and the protein responsible is generally called the alternative oxidase or AOX. AOX is present in many species of fungi and also in a quite distinct group of eukaryotes that include trypanosomes; the latter are causative agents for sleeping sickness among other ailments. Because AOX does not occur in animals, a drug targeting AOX might prove helpful in treatments for sleeping sickness, which is currently difficult to treat.

AOX is a monotopic membrane protein without transmembrane helices, but with helices running parallel to the membrane, that have hydrophobic surfaces to interact with the lipid bilayer (Figure 5.19). There is a plausible binding site for UQH_2 that can be oxidised by a nearby $Fe–OH^-–Fe$ centre (such centres occur elsewhere; e.g., in methane monooxygenase), which is the site of oxygen reduction. Because this protein is located on the N-side of the membrane and has no transmembrane helices, it is clear that protons released in ubiquinol oxidation can only be delivered to the N-side of the membrane and cannot contribute to the generation of a protonmotive force. In any case, there is no net release of protons because an equal number are required for water formation. The organisation of AOX is analogous to that seen in another monotopic membrane protein, ETF–ubiquinone oxidoreductase (Section 5.7.2).

What is the physiological function of the alternative oxidase? In the case of mitochondria from the plant Arum (common name Lords & Ladies), it is to generate heat that will volatilise insect attractants to aid pollination. There are striking examples of

Inter-membrane space

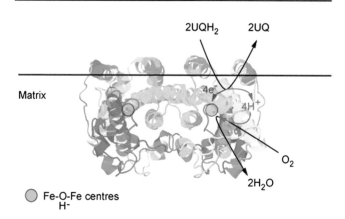

Matrix

2UQH$_2$ 2UQ

4e$^-$

4H$^+$

O$_2$

2H$_2$O

Fe-O-Fe centres
H$^-$

Figure 5.19   Structure of the alternative oxidase.
The alternative oxidase (in this case, the enzyme from the parasite *Trypanosoma brucei*) is a dimeric monotopic integral membrane protein and can be deduced from sequence analysis to have a very similar structure in plants, fungi and other protists. The Fe–OH$^-$–Fe centres at which reduction of oxygen to water occurs are shown in brown. Note how amphiphilic helices allow this protein to interact with the membrane to a depth of approximately 10 Å. Ubiquinol is thought to bind at approximately 5 Å from the iron centre and will approach the enzyme from the bilayer. Protons from ubiquinol oxidation will be released to the matrix side of the membrane; hence, no protonmotive force can be generated. The figure shows the oxidation of ubiquinol and reduction of oxygen on one monomer; whether the protons from ubiquinol oxidation are delivered directly to the site of oxygen reduction remains to be clarified.
Structure from PDB 3VV9 courtesy of Prof. A. L. Moore.

this behaviour; for example, a plant called the Titan arum (*Amorphophallus titanium*) releases a nauseous cocktail of volatiles (grown in all botanical gardens throughout the world but indigenous only to Sumatra and flowering only every 7 years). On the other hand, in the American and Asian skunk cabbage, heat production is directed towards permitting growth in subzero temperatures. This thermogenic mechanism offers a striking contrast with that evolved by mammalian brown fat mitochondria (Section 12.4), in which proton translocation is normal but a dissipative proton re-entry pathway exists. It is important to appreciate that the extent of activity of the alternative oxidase pathway varies between different plants. In potato, for example, it is present in only low levels, in contrast to the plants discussed previously, in which it is the dominant pathway. In plants and fungi, AOX is generally considered to be a stress protein that is expressed when the main respiratory chain is impaired by factors such as growth inhibition or the plant is exposed to wounding, drought, or adverse salinity.

A third rationale for the alternative oxidase pathway is that it provides a mechanism for oxidative metabolism in the absence of ATP synthesis. This may be the reason that free-living trypanosomes make use of the standard complexes III and IV but the bloodstream form, in which rich nutrients are available, relies on the AOX. The mechanism whereby electrons are distributed between the two electron transport pathways to oxygen is not fully understood, but it may depend on the UQH_2/UQ ratio.

Many plant mitochondria possess an inner membrane rotenone-insensitive NADH dehydrogenase in which the active site is exposed to the cytoplasm. Transfer of electrons from this enzyme to ubiquinone is not associated with the translocation of charge across the membrane.

Yeast mitochondria can have two NADH dehydrogenases that are not found in mammalian mitochondria. One, known as Ndi1, catalyses oxidation of matrix NADH by ubiquinone without proton translocation; in some yeasts, such as *Saccharomyces cerevisiae*, complex I is absent and Ndi1 acts alone. The second, known as Nde, has its active site facing the cytoplasm; both Ndi1 and Nde have counterparts in many types of plant mitochondria. The structure of a yeast Ndi1 shows that it is, like the AOX, a monotopic membrane protein in which dimerisation between two momomers creates an ampiphilic domain that provides anchoring to the membrane. Because Ndi1 does not traverse the membrane, it cannot be proton pumping, and unlike complex I, the NADH and ubiquinone binding sites are adjacent, with a single FAD aiding the electron transfer. In common with AOX, there is interest in using Ndi1 to bypass a defective human mitochondrial electron transfer complex and hence ameliorate a disease.

Other proteins associated with the respiratory chain in yeast mitochondria include an L-lactate:cyt *c* oxidoreductase and a cyt *c* peroxidase, both located in the intermembrane space. In the oxidoreductase, electron transfer occurs from lactate via FMN and from *b*-type haem to cyt *c*, which can transfer electrons to either cyt aa_3 or the peroxidase; the latter enables yeast to reduce hydrogen peroxide to water as a terminal step of the electron transport system. The peroxidase contains *b*-type haem, but a second redox active group is a specific tryptophan side chain. This is one of the relatively rare instances of an amino acid side chain undergoing an oxidation–reduction reaction as part of an electron transport process; another example is given in Section 6.4.2.

Importantly, there are examples of eukaryotes in which the mitochondrial electron transport chain is able to reduce fumarate or nitrate and nitrite, reactions previously thought to be found only in bacteria. There are evolutionary implications. There are also anaerobes in which the mitochondria can generate hydrogen and an aerobe in which the electron transport chain can oxidise sulfide. Among the relatively unknown groups of eukaryotic anaerobes are parasites that can be harmful to human health.

5.14 BACTERIAL RESPIRATORY CHAINS

Further reading: Swierczek *et al.* (2010)

Oxidative phosphorylation is vital for many bacteria that cannot exist by fermentation alone. The availability of cell-free vesicular systems has been important in the investigation of the multiple donors, pathways, and acceptors of electrons that not only vary by organism but also can vary within the same organism depending on the growth conditions. We restrict our discussion to a limited number of bacterial electron transfer chains that have either been intensively investigated or that provide novel mechanistic insights.

5.14.1 *Paracoccus denitrificans*

Further reading: van Spanning *et al.* (2012)

This soil organism allows us to start on familiar territory because many features of its electron transport system are similar, whatever its growth mode, to their mitochondrial counterparts (Figure 5.20). However, complexes I, III and IV (the last two are usually referred to as bc_1 and cytochrome aa_3, respectively) all contain fewer polypeptide chains than their mitochondrial counterparts, facilitating their structure–function analysis.

Two *c*-type cytochromes, cyt c_{550} and cyt c_{552}, seem to be able to act as alternates in the electron transport pathway (Figure 5.20); such degeneracy of components appears quite common in bacteria, but the advantage conferred by this feature is not clear. Cyt c_{552} is membrane anchored, whereas cyt c_{550} is closely related to mitochondrial cyt *c* in terms of both structure and redox potential. Deletion of the cyt c_{550} gene does not stop electron transfer to cyt aa_3. A branched aerobic electron transport chain is a very common feature among bacteria, but the reasons for it, the control of expression of the different components and the regulation of the distribution of electrons between the branches are not understood in detail.

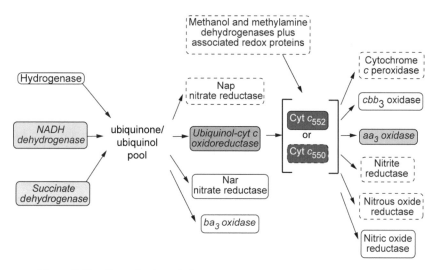

Figure 5.20   Organisation of electron transport components in *P. denitrificans*.
Only the components in italics are thought to be constitutive. The other components are induced by appropriate growth conditions and are unlikely to be all present at once. NADH dehydrogenase, succinate dehydrogenase, ubiquinol cytochrome *c* oxidoreductase, and aa_3 oxidase correspond to mitochondrial complexes I–IV. Continuous boxes indicate integral membrane components; dashed lines represent periplasmic components. Further details of methanol and methylamine oxidation are given in Figure 5.21 and of nitrate respiration in Figure 5.22.

Paracoccus denitrificans has three routes by which electrons can be transferred to oxygen. In addition to a cyt aa_3 similar to mitochondrial complex IV, a cbb_3 oxidase can accept electrons from the c-type cytochromes, whereas a cyt ba_3 (ba_3 oxidase) can bypass bc_1 and accept electrons directly from UQH_2 (Figure 5.20).

The set of electron carriers in ba_3 oxidase differs from mammalian complex IV (Cu_A, cyt a, cyt $a3$, Cu_B) by a UQH_2 binding site replacing Cu_A and a b-type cytochrome in place of cyt a. The oxidase pumps protons by a similar mechanism to complex IV, but because the bc_1 complex is bypassed, the overall $H^+/2e^-$ stoichiometry from UQH_2 to oxygen will be lower. Cyt ba_3 is very similar to the cyt bo_3 of *E. coli* (Section 5.13.2); both are members of the superfamily of terminal oxidases known as the haem–copper oxidases containing a binuclear active site consisting of a high-spin haem (a_3, b_3, or o_3) and a closely associated Cu_B.

The cbb_3 oxidase (Figure 5.20) is also a member of this family. A c-type cytochrome replaces the Cu_A of cyt aa_3 and transfers electrons through two b-type haems, one designated b_3, to Cu_B. These changes result in an oxidase with much higher affinity for oxygen than aa_3 while retaining the capacity for pumping $2H^+$ per $2e^-$, even though it functions at very low oxygen concentrations, which decreases the driving force for formation of water. A structure of the cbb_3 oxidase from another organism shows that unlike the aa_3 type, it has only one proton access route from the N-side of the membrane but has broad similarity with aa_3 oxidase.

Paracoccus denitrificans can use final electron acceptors other than oxygen (Figure 5.20). Among these is H_2O_2, commonly found in soil. Reduction of H_2O_2 is catalysed by a periplasmic cyt c peroxidase, which is dihaem c-type cytochrome. Anaerobic electron acceptors in *P. denitrificans* are described in Section 5.13.3 after we discuss how this organism is able to oxidise compounds that have only one carbon atom.

5.14.1.1 Oxidation of compounds with one carbon atom

Paracoccus denitrificans can grow on methanol or methylamine as the sole carbon source. The respective dehydrogenases (Figure 5.21) are found in the periplasm. That for methanol contains pyrroloquinoline quinone (PQQ) as a cofactor. Electrons are transferred from reduced PQQ to c-type cytochromes, probably including cyt c_{550}, that feed into aa_3 oxidase (Figure 5.21). Δp is established by the inward movement of charge and outward pumping of protons through cyt aa_3 together with the release and uptake of protons at the two sides of the membrane associated with methanol oxidation and oxygen reduction. Methylamine dehydrogenase contains a novel type of redox centre, a tryptophyl-tryptophan involving a covalent bond between two tryptophan side chains. Electrons pass from the redox centre of trimethylamine dehydrogenase, via a $1e^-$ carrier copper protein, amicyanin, to the c-type cytochromes (Figure 5.21). The formaldehyde produced by either oxidation is oxidised by cytoplasmic (N-phase) enzymes to CO_2 with concomitant generation of NADH. The CO_2 thus produced is refixed into cell material. In other organisms that grow on methanol, some of the formaldehyde can be directly incorporated into cell material.

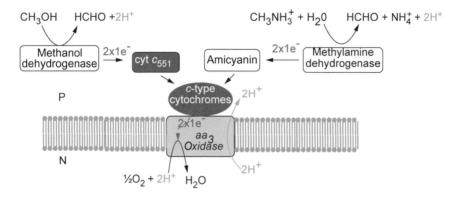

Figure 5.21   Schematic representation of periplasmic oxidation of methanol or methylamine in *P. denitrificans*.

5.14.1.2 Denitrification

The sequential reduction of NO_3^-, NO_2^-, NO, and N_2O is catalysed by anaerobically grown *P. denitrificans* in the process known as denitrification (hence the name of the organism). Five reductases carry out this process, receiving electrons from the underlying electron transport system used in aerobic respiration (Figures 5.20 and 5.22).

The *membrane-bound* NO_3^- *reductase* (often called Nar) receives electrons from UQH_2 towards the P-side of the membrane. $2H^+$ are released to the periplasm and $2e^-$ pass inwards across the cytoplasmic membrane via two *b*-type haems to the site of nitrate reduction on the N-surface—a Mo atom coordinated by sulfur atoms provided by two molecules of a cofactor known as MGD (Figure 5.22). The $E_{m,7}$ for the NO_3^-/NO_2^- couple is +420 mV. The inward movement of the electrons is equivalent to the transfer from cytoplasm to periplasm of two positive charges per $2e^-$. As in the case of mitochondria (Section 5.8), UQ was originally reduced at the N-face of the membrane; thus, the outward transfer of UQH_2 and the return to the N-phase of $2e^-$ is an example of a redox loop mechanism (Figure 5.8). The same enzyme occurs in *E. coli* and is discussed further in Section 5.14.2.1.

In common with many other organisms, *P. denitrificans* possesses a *periplasmic nitrate reductase* (often called Nap). As with the membrane-bound enzyme, nitrate is reduced at a Mo/MGD centre. The important bioenergetic distinction between the two types of nitrate reductase is that the periplasmic enzyme is not associated with proton translocation. The electrons pass from UQH_2 to Nap via a tetra-haem *c*-type cytochrome known as NapC. The structure shows that haem groups are located in a periplasmic globular domain, but this domain dips sufficiently into the bilayer to provide a binding site for ubiquinol, analogous to the monotopic ETF–ubiquinone oxidoreductase (Section 5.7.2). Explanations as to why two nitrate reductases can be present are complex and largely beyond the scope of this book. Suffice it to say that one role for Nap is

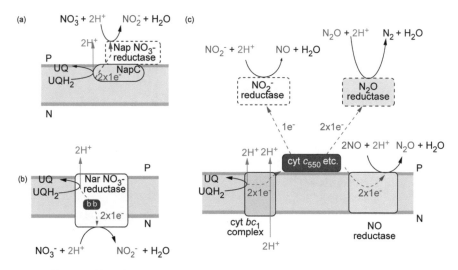

Figure 5.22   Electron transport pathways associated with denitrification in *P. denitrificans*.
There are two pathways for electron flow from UQH$_2$ to nitrate. (a) The Nap pathway results in electron and proton release to the periplasm and hence no generation of Δp. The tetra haem c-type cytochrome NapC protein catalyses oxidation of UQH$_2$ and transfer of electrons into the periplasm. (b) The second pathway uses the membrane-bound Nar reductase that has two b-type haems distributed across the membrane and a cytoplasmic-facing site at which there is a molybdenum centre containing a specialised pterin cofactor known as MGD. As explained in the text, this loop mechanism is associated with the same net positive charge translocation across the membrane as electron flow to the other three nitrogenous acceptors. The charge and proton translocation stoichiometry catalysed by the ubiquinol cytochrome c oxidoreductase is explained in Section 5.8. Note that the reduction of nitrate by the membrane-bound nitrate reductase requires that negatively charged nitrate enters the cell despite the $\Delta\psi$ being negative inside. A nitrate/nitrite antiporter (NarK) of the MFS family (see Chapter 8) is currently postulated to overcome this bioenergetic problem. (c) Oxidation of one UQH$_2$ by nitrite, nitric oxide, or nitrous oxide generates protonmotive force only as a result of the Q-cycle mechanism of the cytochrome bc_1 complex (Section 5.8). Note that 2e from UQH$_2$ can be delivered to any one of three electron acceptors; overall oxidation of three UQH$_2$ can reduce two nitrite ions (NO$_2^-$) to nitric oxide (NO), one molecule of NO to nitrous oxide (N$_2$O), and one molecule of N$_2$O to N$_2$.

to provide a pathway for loss from cells of excess reductant when they are growing on a carbon source (e.g., a fatty acid) that is more reduced than the average reduction state of the cell biomass.

Nitrite reductase is a soluble enzyme in the periplasm, containing both c- and d_1-type haem centres (hence usually called cytochrome cd_1) and can receive electrons from cyt bc_1 via either cyt c_{550} or a copper protein known as pseudoazurin. The active site is the d_1 haem, unique to this type of enzyme, contained in a propeller-shaped structure made up of eight blades of four stranded β-sheet. The unusual features (part saturation and the presence of carbonyl groups) of d_1-type haem enable the reduced state both to bind nitrite and to release nitric oxide, reaction steps that are strongly disfavoured by b-type haem. The cyt c domain interacts with an electron donor protein such as cyt c_{550} (Figure 5.22c). Note that these nitrate and nitrite reductases, and the E. coli enzymes discussed later (Section 5.13.2), are distinct from the widespread enzymes with the same names that are responsible for the assimilation of nitrogen in bacteria and plants and which are beyond the scope of this book. Remarkably, the cofactor in assimilatory nitrite reductase, sirohaem, is an intermediate on the pathway to d_1-type haem, suggestive of an evolutionary connection.

Nitric oxide reductase is an integral membrane protein containing both b- and c-type haems unexpectedly related to the cbb_3 oxidase but with the Cu_B replaced by an Fe. A critical difference is that the enzyme does not translocate charge across the membrane because both protons and electrons reach the active site from the periplasm. Electrons for NO reduction are supplied from cyt bc_1 via cyt c_{550}; thus, as in the case of nitrite reductase, the overall net outward charge transfer is $2q^+/2e^-$. The relationship between nitric oxide reductase and oxidases has led to the proposal that nitric oxides, formed by photochemical reactions between water and nitrogen, may have predated terrestrial oxygen as final electron acceptor.

Nitrous oxide reductase, a soluble periplasmic enzyme, contains the same Cu_A centre as in complex IV as well as a novel cluster of four copper atoms ('Cu_Z'), bridged by two sulfides, at the active site. Nitrous oxide binds to one side of the Cu_Z centre adjacent to the Cu_A centre in which a conformational change is induced in order to permit electron flow into the active site from c_{550}. It is notable that the $E_{m,7}$ of the N_2O/N_2 couple is even more positive than $\frac{1}{2}O_2/H_2O$, at $+1100\,mV$, although the concentration of N_2O *in vivo* may be so low that the actual E_h for the couple may be comparable to the $+800\,mV$ for the oxygen reaction.

Nitrite reductase, N_2O reductase, and NO reductase all serve as P-face electron sinks at the level of cyt c (Figure 5.22). Each is associated with $2q^+/2e^-$ stoichiometries for the spans from UQH_2, as indeed is Nar-type nitrate reductase, which unlike the other three reactions does not involve the bc_1 complex. Thus, despite the redox span from the UQ/UQH_2 couple to NO_2^-/NO_3^- being far smaller than to N_2/N_2O, the charge transfer is the same. One complication in the scheme of Figure 5.22 is that the role of cyt c_{550} in denitrification has been questioned by the finding that a mutant of P. denitrificans lacking cyt c_{550} is still able to denitrify. An explanation is that a copper protein known as pseudoazurin, similar to plastocyanin (Section 6.4.3), can substitute.

There is sometimes confusion regarding whether a soluble periplasmic enzyme such as N_2O reductase can 'participate' in the generation of Δp. It should be clear that although the activity of such enzymes per se does not contribute directly to Δp, their role as electron sinks is necessary for the electron transport chain to function. On the other hand, comparison of Figures 5.21 and 5.22 shows that transfer of electrons from the periplasmic methanol dehydrogenase to nitrous oxide reductase would not generate a Δp despite the large redox span. This illustrates the importance of considering not only redox spans but also the topology of electron flow in energy-transducing membranes.

5.14.2 *Escherichia coli*

Escherichia coli has two NADH dehydrogenases. One of these is proton translocating and very similar to that of *P. denitrificans* (Section 5.13.1) and mitochondrial complex I (Section 5.6). The second enzyme has a much simpler subunit composition and does not translocate protons.

Aerobically grown *E. coli* possesses no detectable *c*-type cytochromes. Two oxidases, cyt bo_3 (often just called *bo*) and cyt *bd*, can directly oxidise UQH_2 (Figure 5.23). Cyt bo_3 is a member of the haem–copper superfamily. Similarly to the *P. denitrificans* ba_3 oxidase, Cu_B is present and there is a UQH_2 binding site passing electrons singly to a cyt *b*; however, cyt a_3 is replaced by a cyt o_3 with a methyl residue replacing the formyl group of the *a*-type haem. Cyt bo_3 is a proton pump with stoichiometry $2H^+/2e^-$ and overall charge movement of $4q^+/2e^-$.

The cyt *bd* complex lacks Cu and does not show any sequence similarity with the haem–copper family. It has two *b*-type haems; b_{558} is the electron acceptor from UQH_2 and passes electrons to b_{595}, which forms a haem pair with the distinctive porphyrin ring of a *d*-type haem. Strictly speaking, this is a chlorin owing to saturation of one of the pyrrole rings, but it is markedly different from the d_1 haem in *P. denitrificans* nitrite reductase (Section 5.13.1). The *d* haem is the site of oxygen reduction. There is no evidence that cyt *bd* is a proton pump and thus the stoichiometry of charge translocation is

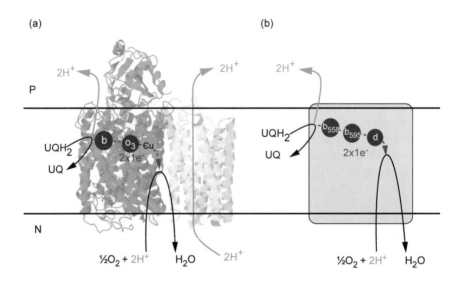

Figure 5.23   The *E. coli* aerobic electron transfer chain from ubiquinol to oxygen.
A crystal structure (a) is available for the bo_3 oxidase (PDB 1FT, Abramson *et al.*, 2000). The ubiquinol oxidation has been modelled at approximately the location shown. Note that the cytochrome is related to complex IV and essentially performs the same function. (b) Approximate cyt *bd*, which lacks a conformational proton pump and is not a member of the haem copper family of oxidases.

$2q^+/2e^-$, due purely to the inward movement of electrons meeting the outward flux of protons at the oxygen reduction site. All three haem groups are located towards the periplasmic side of the membrane. Cyt *bd* has a much higher affinity for oxygen than cyt bo_3 and is synthesised under low oxygen conditions; the lower stoichiometry of proton translocation may be the price that has to be paid to attain a high catalytic rate with no thermodynamic backpressure from Δp.

Clearly, *E. coli* has a truncated electron transport chain, in comparison with mitochondria and *P. denitrificans*, with lower $q^+/2e^-$ and $H^+/2e^-$ ratios. The H^+/ATP ratio for the ATP synthase is determined by the number of *c* subunits in the rotary motor (Section 7.5). The preferred, but still uncertain, value for *E. coli* is 10, meaning that 10 H^+ are required for one full rotation leading to the synthesis of 3 ATP. In contrast, animal mitochondria have 8 *c* subunits. The transfer of $2e^-$ through the *E. coli* proton-translocating NADH dehydrogenase and the bo_3 oxidase yields $8H^+/2e^-$ and will maximally yield 2.4 ATP, whereas $2e^-$ passing through the entire mitochondrial electron transport chain yields $10H^+/2e^-$ and hence up to 3.7 ATP can be generated in the matrix (although the additional $1H^+/ATP$ required for export (Section 9.5.1) decreases this to 2.7 for cytoplasmic ATP). At the other extreme, use of the *E. coli* non-translocating NADH dehydrogenase and cyt *bd* decreases the $H^+/2e^-$ to 2 and the maximal $ATP/2e^-$ to 0.6. The electron transport system of *E. coli* therefore illustrates that an organism may not always be seeking to maximise the stoichiometry of ATP production. Natural habitats may be rich in potential substrates, and the need to maximise ATP yield may not apply.

Figure 5.24 shows that the *E. coli* respiratory chain can receive electrons from many electron donors. The oxidation of D-lactate is interesting because the dehydrogenase is a peripheral monotopic protein that has to dip into the membrane sufficiently to make contact with ubiquinone or menaquinone. As in the case of the mitochondrial ETF/UQ

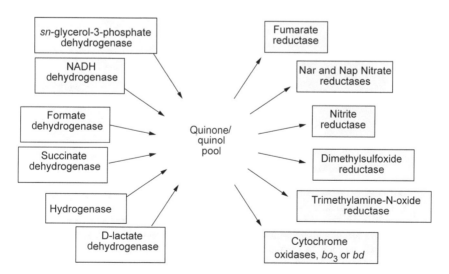

Figure 5.24   An overview of *E. coli* aerobic and anaerobic respiratory systems.
The components present depend on the growth conditions. Under anaerobic conditions, menaquinone replaces ubiquinone as the main quinone.

oxidoreductase (Section 5.7), transmembrane helices are not needed for the provision of a binding site for quinone.

5.14.2.1 Anaerobic metabolism

Escherichia coli is not restricted to aerobic growth, and a variety of anaerobic electron acceptors can be utilised. The expression of many of the required enzymes is dependent on the Fnr transcriptional activator. Under anaerobic conditions, the UQ/UQH_2 pool is replaced by menaquinone/menaquinol; the $E_{m,7}$ of the latter is approximately 130 mV more negative than ubiquinone (Table 3.2). In addition, the TCA cycle enzyme 2-oxoglutarate dehydrogenase ceases to function (in contrast to *P. denitrificans* and many other non-enteric bacteria). However, pyruvate can be converted to formate or fumarate. Under anaerobic conditions, formate can be oxidised by formate dehydrogenase, forming HCO_3^- (Figure 5.25), and the electrons transferred via the menaquinone pool to either fumarate reductase (forming succinate) or nitrate reductase (Figure 5.25).

Crystal structures of membrane-bound formate dehydrogenase, fumarate reductase and Nar nitrate reductase have been obtained (Figure 5.25). The formate binding site is in a globular domain exposed to the periplasm, which is connected to transmembrane

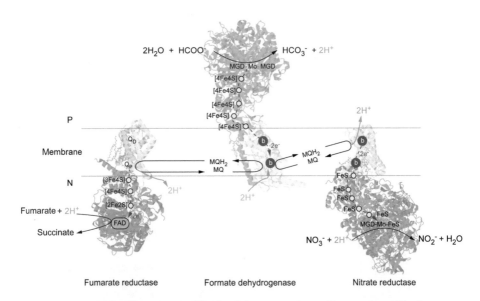

Figure 5.25 The structural basis of Δp generation as formate is oxidised by fumarate or nitrate in *E. coli*.
Formate oxidation and nitrate reduction each generate Δp principally as a result of transmembrane electron transfer by the two haems in each protein. They are connected by the movement of the uncharged hydrogen carrier menaquinone (MQ). This is a clear example of a redox loop mechanism. Reduction of fumarate does not generate Δp and so formate to fumarate can be regarded as a half loop. Structures: 1QG (Jormakka *et al.*, 2002; 1L0V (Iverson *et al.*, 2002) and 1Q16 (Bertero *et al.*, 2003) respectively.

helices between which two b-type haems are sandwiched—one towards the P-side and one towards the N-side. Protons are released to the periplasm, and electrons are transferred from the Mo/MGD centre in the active site via a wire of Fe–S centres to the haem at the N-side, where protons are taken up and MQH_2 is formed. Thus, the enzyme acts as a Δp generator (Figure 5.25). In contrast, fumarate reductase has its globular domain exposed to the N-side, and although related to succinate dehydrogenase, it lacks haem groups. Because quinol oxidation and proton release both occur at the N-side of the membrane, this reductase does not generate Δp. However, transfer of electrons from formate to nitrate via the Nar-type nitrate reductase constitutes a complete redox loop of the kind envisaged by Mitchell (Figure 5.25) – compare Figure 5.8. Note that Nar resembles formate dehydrogenase but with opposite membrane orientation.

Nitrate is reduced to NO_2^- by two reductases that are very similar to those described previously for *P. denitrificans* (the Nar-type reductase is shown in Figure 5.25). In contrast, however, NO_2^- is reduced to NH_4^+, rather than to NO, by a periplasmic nitrite reductase, containing five c-type haems, that receives electrons from the quinol pool via an Fe–S protein and NrfB (another c-type cytochrome also with five haems). Dimethylsulfoxide and trimethylamine-N-oxide (both occur in natural environments, the latter especially in fish) can also serve as terminal electron acceptors via one or more Mo-containing reductases, which are usually periplasmic (Figure 5.25).

5.14.3 Relationship of *P. denitrificans* and *E. coli* electron transport proteins to those in other bacteria

Many of the components of the electron transport systems of *P. denitrificans* and *E. coli* are also found in a range of other organisms in various combinations. Some general unifying themes include quinones and c-type cytochromes acting as mobile components to connect enzymes that handle different electron donors and acceptors and an appropriate spatial distribution of enzymes between the P- and N-sides of the membrane that will lead to the generation of Δp. Thus, quinone pools serve as a general 'crossroads' for electron transfer (Figure 5.24), compatible with their mobility within the bilayer, although we still do not fully understand how electrons distribute themselves between the different acceptors. The c-type cytochromes are far more varied and play a greater range of roles, often serving also as 'crossroads' for electron transfer, than in mitochondria; they are often water-soluble and almost invariably found in the periplasm of gram-negative bacteria. In some cases, the c-type cytochromes are replaced by copper proteins similar to plastocyanin (Section 6.4.3) and called azurins or pseudoazurins, or even periplasmic Fe–S proteins known as Hipips. In gram-positive bacteria, which do not have a periplasm, the c-type cytochromes appear to be more tightly associated with the cytoplasmic membrane, and the range of metabolic activities associated with periplasmic dehydrogenases and reductases is much more restricted. It is notable that in many organisms, periplasmic c-type cytochromes function at a junction point. NapC (Figure 5.22) is an example of a large class of such proteins that are involved in electron transfer into and out of the periplasm. Cyt bc_1 and aa_3 are also widely distributed. Indeed, the various oxidases of the haem–copper family are also widespread, one of which is the ba_3 type found in *T. thermophilus*, a protein that confusingly is distinct from the ba_3 oxidase of *P. denitrificans*. *Thermus thermophilus* also contains an example of yet another type of haem–copper oxidase, the caa_3 type. This is interesting because the cytochrome *c* domain is fused to one of the other subjects, and a structure has shown that the distance from the haem edge of cytochrome *c* to Cu_A is just 9Å (Lyons *et al.*, 2012). It is also clear in this case that the electron entry route to this fused cytochrome *c* must be different from the exit route to Cu_A. In the case of mitochondrial cytochrome *c*, it is still uncertain whether the electron entry and exit routes are the same (Section 5.5).

Nitric oxide reductase is widespread, even in organisms that cannot denitrify. The possession of this enzyme can allow potentially pathogenic bacteria to remove nitric oxide that is produced by certain eukaryotic cells as a toxic defence molecule against invading organisms. As with cytochrome oxidases, there are variants of nitric oxide reductases that can directly oxidise quinols, whereas others contain a Cu_A centre. Electron transfer components of *E. coli* are also found elsewhere. For example, a similar cyt *bd* oxidase with high affinity for oxygen terminates a ubiquinol oxidase system in *Azotobacter vinelandii* and *Klebsiella pneumoniae*. In these organisms, a role of this oxidase is to maintain low oxygen concentrations to protect an oxygen-sensitive nitrogenase enzyme, a function that cyt ccb_3 fulfills in rhizobial species

In the next section, we illustrate how some of the electron transfer components identified in *P. denitrificans* and *E. coli* appear in organisms that have very different physiologies than these two models. A caveat is that there are very large variations and diversity among these proteins because evolution has been able to shuffle genes around, as exemplified by oxidase enzymes with fused *c*-type cytochrome domains. We can only be illustrative and note that their electron transfer chains contain few similarities to the examples we discuss, for example, sulfate reducing bacteria which are rich in multi-haem *c*-type cytochromes. With the exception of methane synthesis, we do not discuss archaeal electron transfer systems.

5.14.4 Helicobacter pylori

Helicobacter pylori grows at very low oxygen concentrations and has attracted attention as a cause of gastric ulcers and gastric cancer. It is an example of an organism for which more knowledge of its electron transport system has been gained from the sequencing of its genome than from biochemical analyses. Thus, most of the respiratory chain components (Figure 5.26) can be identified from sequence similarities with known bacterial electron transport components. Several features are notable, some of which are in common with *P. denitrificans*, others with *E. coli*. First, and unusually for a bacterial respiratory chain, there is only one oxidase, the high-affinity cbb_3 type (Section 5.13.1). Second, like *E. coli*, the organism can use fumarate as an electron acceptor, while in common with *P. denitrificans*, it can reduce hydrogen peroxide to water. Third, the cyt bc_1 complex uses menaquinol as an electron donor. Finally, succinate dehydrogenase is absent, although how aerobic growth is possible without succinate dehydrogenase is beyond the scope of this book. The hydrogenase is critical for this organism; strains lacking this component of the respiratory chain are not pathogenic.

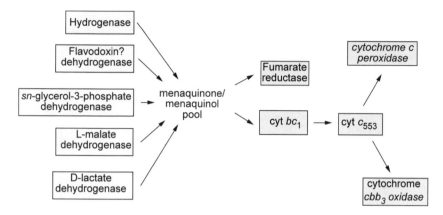

Figure 5.26   The electron transfer chain of *Helicobacter pylori* as deduced from the genome sequence.
The components were identified almost exclusively by assigning open reading frames in the genome sequence using sequence databases. Cytochrome c_{553} is related to a c-type cytochrome found in other bacteria, including *Campylobacter*. The other components shown in the figure have been introduced in the discussions of *P. denitrificans* and *E. coli*.

Not everything about the bioenergetics of *H. pylori* can be immediately deduced from the genome sequence. The critical NuoE and NuoF subunits of NADH dehydrogenase that provide the catalytic site for NADH oxidation are absent despite the presence of orthologs of other Nuo subunits. It is believed that they are replaced by distinct subunits that allow for the oxidation of a flavodoxin by the variant of complex I; reduction of the flavodoxin is associated with the oxidation of pyruvate and α-ketoglutatate in *H. pylori*. However, both a D-lactate dehydrogenase (similar to the enzyme in *E. coli*) and an L-malate dehydrogenase feed electrons to the menaquinone pool, again emphasising the role of quinone as a junction point in electron transport systems. It should be clear that the bioenergetic interpretation of the genome sequence relies on the knowledge of bacterial electron transport systems gained previously by biochemical studies on a limited number of model organisms.

5.14.5 *Nitrobacter*

If an organism grows on a substrate with a relatively positive redox potential, it can be faced with the problem of how to generate NADH or NADPH for biosynthetic reactions. The example of *Nitrobacter* illustrates this aspect of electron transport.

Nitrobacter grows by oxidising nitrite to nitrate ($E_{m,7} + 420\,mV$) by a nitrite oxidoreductase, transferring electrons via a c-type cytochrome to a cyt aa_3 oxidase and reducing oxygen to water ($E_{m,7} + 820\,mV$) (Figure 5.27). This immediately poses a problem because the $E_{m,7}$ for the cytochrome is $+270\,mV$, or $150\,mV$ more electronegative than for NO_3^-/NO_2^-. Although there is still controversy, it is likely that NO_2^- oxidation occurs on the N-face of the membrane, transferring electrons to cyt c on the P-face. This may

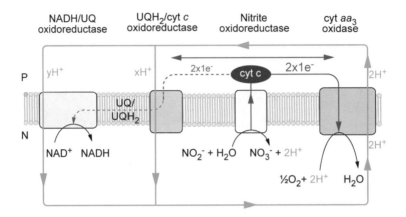

Figure 5.27   Protonmotive force generation and reversed electron transport in *Nitrobacter*.
As explained in the text, it is proposed that $\Delta\psi$ drives electrons energetically uphill from nitrite to the cytochrome c and that cytochrome aa_3 acts as proton pump. The dashed line indicates reversed electron flow, which will be of much smaller magnitude than the route of the majority of electrons flowing to oxygen. Not all details of this scheme have been fully substantiated, although the sequence of the cytochrome aa_3 shows the presence of all the amino acid residues implicated as important in proton pumping by this type of enzyme. The nitrite oxidase has sequence similarities with the membrane-bound type of nitrate reductase (Nar) (Figure 5.22).

seem strange because this charge movement *collapses* rather than generates Δp, but there is a thermodynamic reason. The 170 mV $\Delta\psi$, which is typical for bacterial cytoplasmic membranes, shifts the apparent midpoint potential for the NO_3^-/NO_2^- couple 170 mV more negative (i.e., it effectively becomes $420-170 = 250$ mV) relative to cyt c, greatly facilitating electron transfer. A similar utilisation of $\Delta\psi$ has been discussed in the context of the mitochondrial bc_1 complex (Section 5.8). Consistent with this model, protonophores inhibit electron transfer from nitrite. Because electrons start and finish on the N-face of the membrane, the only process generating a Δp is the proton pumping by cyt aa_3.

We have still not explained how the organism can reduce NAD^+ to NADH ($E_{m,7}$ -320 mV) for biosynthetic reactions. The genome sequence shows that it possesses both a bc_1 complex and an NADH dehydrogenase (complex I) that, as in other electron transfer chains, are expected to be reversible. The solution is that the Δp generated by a majority of electrons flowing through cyt aa_3 is used to drive a minority of electrons 'backwards,' with both these complexes creating reversed electron transfer leading to the reduction of NAD^+, as introduced in a mitochondrial context in Section 4.9.

The energetics of *Nitrobacter* illustrate the beautiful economy of the chemiosmotic mechanism. Δp drives the initial step of substrate oxidation and reversed electron transport as well as more conventional processes such as ATP synthesis and substrate transport.

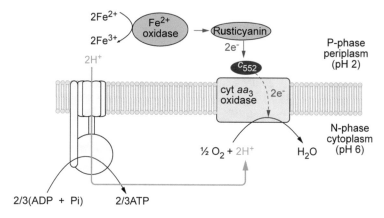

Figure 5.28   Electron transfer and ATP synthesis by *T. ferrooxidans*.
Only at external pH values of approximately 2 does the organism grow by
oxidising Fe(II) at the expense of oxygen. The low pH means that the Fe(II) is
more soluble and its noncatalysed oxidation to Fe(III) is much slower than at
pH 7. Furthermore, the free energy change for the overall oxidation of Fe(II)
by oxygen is much greater at pH 2 owing to the pH dependence of the redox
potential of the oxygen/water reaction. Nevertheless, the overall energy
available to the cell is still small. The oxidation of iron occurs in the
periplasm, and oxygen is reduced by a cytochrome aa_3 oxidase. It is thought
that as in all cytochrome oxidases, the protons required for reduction of
oxygen are taken from the cytoplasm, but in contrast to other cytochrome aa_3
molecules, there is reason to believe that there is no additional proton
pumping. Thus, the action of the protonmotive activity of the oxidase will
contribute to making the electric potential more negative inside and also to
consumption of cytoplasmic protons. In the absence of any electron transfer,
the protonmotive force is zero; the large pH gradient is balanced by a
membrane potential, positive inside the cells. The operation of cytochrome
oxidase scarcely changes the pH gradient but results in the membrane
potential inside becoming less positive by approximately 180 mV, thus giving
a net protonmotive force. This drives ATP synthesis (as shown) assuming that
$3H^+$ are needed for 1 ATP, and reversed electron transport (not shown).

5.14.6 *Thiobacillus ferrooxidans*

Further reading: Ferguson and Ingledew (2008)

Nitrobacter is by no means the only example of an organism in which reversed electron
transport is important. Another instance is *Thiobacillus ferrooxidans*, which oxidises
Fe^{2+} to Fe^{3+} ($E_{m,7} = +780$ mV). As with the oxidation of nitrite to nitrate, this reaction
cannot directly reduce NAD^+, and thus a small proportion of the electrons derived from
Fe^{2+} are transferred 'uphill' to NAD^+, whereas the remainder flow to oxygen with concomitant generation of Δp. The oxidation of Fe^{2+} occurs in the periplasm, and electrons
are transferred to an oxidase via a copper protein known as rusticyanin (Figure 5.28).
Thiobacillus ferrooxidans typically grows at an external pH of 2, which is important
for two reasons. First, the rate of uncatalysed oxidation of ferrous to ferric ion is slower

than at pH 7, and, second, the reduction of oxygen to water, being a reaction that consumes protons, has a more positive E_h at the lower pH.

In assessing the bioenergetics of this organism, it is important to keep in mind that the limits are set by the E_h values for the Fe^{2+}/Fe^{3+} and O_2/H_2O reactions at pH 2. With reasonable estimates of the actual concentrations or partial pressure of the substrates, we can calculate that at pH 2 the redox span is approximately 300 mV. This means that if transfer of one electron from Fe^{2+} to oxygen is associated with the movement of one charge across the membrane, then the maximum value of Δp could be 300 mV, whereas movement of two charges (Figure 5.28) would restrict the Δp to 150 mV.

The low pH outside the cells has important consequences for the relative contributions of $\Delta\psi$ and ΔpH to the total Δp. During steady-state respiration, the cytoplasmic pH is estimated to be approximately 6, giving a ΔpH of 4 units, equivalent to 240 mV. If the total Δp were to be 300 mV, and thus much larger than found in other systems, $\Delta\psi$ would be 60 mV, positive inside. On the other hand, if Δp were to be 150 mV, then the $\Delta\psi$ would have to be 90 mV, negative outside. Currently, there is some uncertainty about the size of Δp and whether or not the terminal oxidase, now known from the genome sequence to be cytochrome aa_3, is the first example of this class of enzyme that does not pump protons (Figure 5.27). Some key residues thought to be involved in proton pumping are absent from the sequence of the *T. ferrooxidans* oxidase. If it does turn out to have the 'standard' stoichiometry of proton pumping, then Δp will be restricted to 150 mV. The implications for the stoichiometry of ATP synthesis are explained in Figure 5.28.

A final point of confusion concerning the bioenergetics of *T. ferrooxidans* concerns the magnitude of Δp in the absence of respiration. The pH difference can still be 4 units, apparently equivalent to a Δp of 240 mV, but there is no net Δp under these conditions. The $\Delta\psi$ is approximately 240 mV, positive inside the cells, and arises from an inwardly directed diffusion potential of protons. The onset of respiration and thus of outward positive charge movement effectively lessens the magnitude of this potential. It is a common mistake to imagine that the large pH gradient can be regarded as some type of gratis and bonus contribution to Δp.

5.14.7 Electron transfer into and out of bacterial cells

Further reading: Richardson *et al.* (2012)

Some species of bacteria can reduce or oxidise extracellular solids. How are electrons transferred to or from the respiratory chain? In the case of *Shewanella* species, a molecular description of how electrons are transferred from the respiratory chain in the cytoplasmic membrane across the periplasm and outer membrane is emerging. A tetra-haem *c*-type cytochrome analogous to NapC (Figure 5.22) is reduced by ubiquinol and in turn transfers electrons to other periplasmic multi-haem *c*-type cytochromes. A deca-haem *c*-type cytochrome known as MtrA can be modelled on the basis of its similarity to NrfB (Section 5.14.2) to have an elongated structure with length 80Å, sufficient to span the outer membrane. It seems unlikely that such a protein would be stable by itself in the outer membrane, and it is notable that also needed for optimal electron transfer to extracellular solids is an outer membrane porin known as MtrB, which is modelled as a 28-strand transmembrane β-barrel with an interior pore diameter of 30–40Å. It is probable that the cytochrome MtrA is enclosed by the porin and provides the electron wire to conduct electrons from the periplasm to the outer surface of the cell. Other multi-haem *c*-type cytochromes are located at the exterior surface of the outer membrane, often containing 10 or 11 haems organised in a 'cross' pattern with some β-sheet flanking regions. Because the iron atoms in all these haems are hexacoordinate, with two histidine side chains as axial ligands, it remains to be determined how exactly these proteins transfer electrons received from MtrA to solid electron acceptors, such as Fe^{3+}, but for which specificity is low.

The covalent anchoring of haem groups to proteins via CXXCH motifs allows 'wires' of haems to extend over large distance within relatively small proteins; for example, the molecular weight of MtrA is 40kDa, and thus in effect one haem is enclosed by only 4kDa of polypeptide. The covalent attachment means that individual haem binding pockets are not needed and that elongated proteins are possible. Genome sequences predict the presence of such proteins in many species of bacteria, such as *Geobacter* and the *Desulfovibrio* group of sulfate-reducing bacteria in which the role may be to provide long-distance electron transfer along even extracellular wires such as pili.

5.14.8 The problem of generating reductant with a more negative redox potential than NAD⁺/NADH; reversed electron transfer or electron bifurcation

Further reading: Biegel *et al.* (2011), Buckel and Thauer (2013)

In many bacterial species, there can be a requirement for a more powerful reductant than NADH or NADPH. A good example is the nitrogenase enzyme system that catalyses reduction of nitrogen gas to ammonia. The reductants for this system are ferredoxins or flavodoxins. There are at least two sets of proteins associated with bacterial cytoplasmic membranes that can catalyse the reduction of ferredoxins or flavodoxins. The first is known as the Rnf system, which is a protonmotive force-driven system for driving electrons from NADH to ferredoxins; this is another example of reversed electron transfer. Originally described as a route for generating reductant for nitrogenase in one species of the photosynthetic bacterium, it has subsequently proved to be more widely implicated in a range of reactions in bacteria, but it turns out this is by no means always the route to provide reductant for nitrogen fixation. In other organisms such as the *Rhizobia*, an electron bifurcation mechanism operates in which electrons from NADH are first passed to the FixAB complex that is related to the mitochondrial ETF protein described previously. Oxidation of FixAB results in one electron being passed to ferredoxin, whereas the second passes to FixCX, which has similarity to the ETF–ubiquinone oxidoreductase. Thus, oxidation of two molecules of NADH will produce two molecules of reduced ferredoxin and one molecule of ubiquinol. What is the driving force for the endergonic transfer of electrons from NADH to ferredoxin? In this case, it is not the protonmotive force but, rather, the energy released by the oxidation of NADH by ubiquinol, a reaction that is substantially exergonic. This is effectively a disproportionation reaction. A further driving force may be the transfer of electrons from ubiquinol to oxygen, although at least part of the energy from this process will be otherwise conserved in proton translocation.

5.14.9 The bioenergetics of methane synthesis by bacteria

Further reading: Thauer *et al.* (2008)

Methanogenic bacteria are archaea that obtain energy from several types of reaction in which methane is an end product. It was only established in the late 1980s that this methane formation is associated with electron transport-driven H^+ or Na^+ translocation and that resultant ATP synthesis is by a chemiosmotic mechanism.

Two methanogenic organisms, *Methanosarcina barkeri* and *Methanosarcina mazei* strain *Göl*, provided important clues about the bioenergetics of methanogenesis; both gain energy for growth from the reduction of either CH_3OH or CO_2 by H_2:

$$CH_3OH + H_2 \rightarrow CH_4 + H_2O$$

or

$$CO_2 + 4H_2 \rightarrow CH_4 + 2H_2O$$

The organisms can also grow on methanol alone, but discussion of this will be reserved until the fundamental electron pathways for reduction of CH_3OH and CO_2 have been described. A striking feature of methanogenesis is that it involves a number of water-soluble molecules that are rarely found outside methanogens, including coenzyme M ($HSCH_2CH_2SO3^-$; CoM), coenzyme B (a molecule with a thiol group at the end of a chain of six CH_2 groups attached to a threonine phosphate; CoB) and F_{420}. The latter is a 5′-deazaflavin with an $E_{m,7}$ of $-370\,mV$ and is a structural and functional hybrid between nicotinamide and flavin coenzymes; it is a diffusible species in the cytoplasm of methanogenic bacteria.

There are a number of other unusual cofactors bound to the enzymes of methanogenesis (Figure 5.29): some have now been discovered in bacteria that oxidise methanol.

5.14.9.1 Reduction of $CH_3OH \rightarrow CH_4$ by H_2

Intact cells of *M. barkeri* can synthesise ATP as well as CH_4 in the presence of CH_3OH and H_2. ATP synthesis is chemiosmotic (rather than a result of substrate-level phosphorylation by a soluble enzyme system) by the following criteria:

(a) Protons are extruded.
(b) DCCD, presumed to be a specific inhibitor of an ATP synthase as in other organisms, inhibits ATP production, increases Δp and slows the rate of methane formation. Genome sequencing for a methanogen has shown that an ATP synthase of the type discussed in Chapter 7 is present, although certain distinctive features cause it to be classified as an A (archaeal)-type enzyme.
(c) Protonophores dissipate Δp but increase the rate of methane formation.

These observations parallel what would be observed in the analogous mitochondrial proton circuit (Chapter 4). An involvement of Na^+ can be eliminated because this ion is not needed for methanogenesis from CH_3OH plus H_2, although Na^+ is required for growth of methanogens and some reactions of methanogenesis (discussed later).

Further understanding required preparation of functional inside-out membrane vesicles from *M. mazei Gö*1. Addition of H_2 and CH_3SCoM to crude *M. mazeii Gö*1 vesicles resulted in ATP synthesis. (CH_3SCoM is formed in cells from CH_3OH and coenzyme M in a reaction catalysed by methanol: CoM methyltransferase (Figure 5.29, step 9), an enzyme that has at its active site a haem-type ring with a Co, to which the methyl group is transiently attached, instead of Fe at the centre (a corrinoid)). The methyl group of CH_3SCoM was converted to methane through reaction with a second thiol-containing compound known as *CoB*-SH, and present in the preparation of vesicles, to give a heterodisulfide:

$$CH_3SCoM + CoB\text{-SH} \rightarrow CH_4 + CoM\text{-S-S-}CoB$$

The enzyme catalysing this reaction is known as methyl coenzyme M reductase and was fortuitously present in the vesicle preparation. The crystal structure of this water-soluble enzyme reveals a deep channel, at the bottom of which is a cofactor known as F_{430}, which is a porphinoid; this is related to a haem group but contains an Ni atom rather than Fe at the centre. It is not known exactly how this unusual enzyme works, but the methyl group is widely believed to transfer from CH_3SCoM to the Ni ion. The *CoB*-SH also enters the channel, and its oxidation to give the heterodisulfide is linked to the cleavage of the Ni-methyl bond and the release of methane.

The previous reaction (step 7), catalysed by a water-soluble enzyme, cannot directly drive ATP synthesis; rather, it is the reduction of *CoM*-S-S-*CoB* back to the two separate thiol species that is catalysed by a membrane-bound enzyme system that translocates protons across the cytoplasmic membrane and thus makes Δp. The electron source for this reduction is hydrogen, from which electrons are transferred via a membrane-bound hydrogenase with similarities to complex I that pumps protons. In line with a chemiosmotic mechanism, the rate of step 8, Figure 5.29, catalysed by vesicles was accelerated by onset of ATP synthesis or addition of protonophores. The foregoing description applies to methanogens with cytochromes. In organisms without cytochromes, the reduction of *CoM*-S-S-*CoB* by hydrogen is associated with electron bifurcation whereby for every two molecules of hydrogen consumed, two molecules of a reduced ferredoxin, an uphill reaction, are made along with formation of *CoM* and *CoB*, a downhill reaction. There is no associated ion translocation. Oxidation of the reduced ferredoxin is then linked to generation of an ion electrochemical gradient, which in turns drives ATP synthesis. The ATP synthase in methanogens is an A-type enzyme and is probably driven by Na^+ ions. There is a sodium/proton exchanger in the cytoplasmic membranes of methanogens that permits the interconversion of proton and sodium electrochemical gradients, the details of which are not fully understood.

5.14.9.2 Reduction of $CO_2 \rightarrow CH_4$ by H_2

Growth of *M. barkeri* is also supported by the reduction of CO_2 by H_2. CO_2 is first taken up by covalent attachment to methanofuran (Figure 5.29). The first reduction step to a

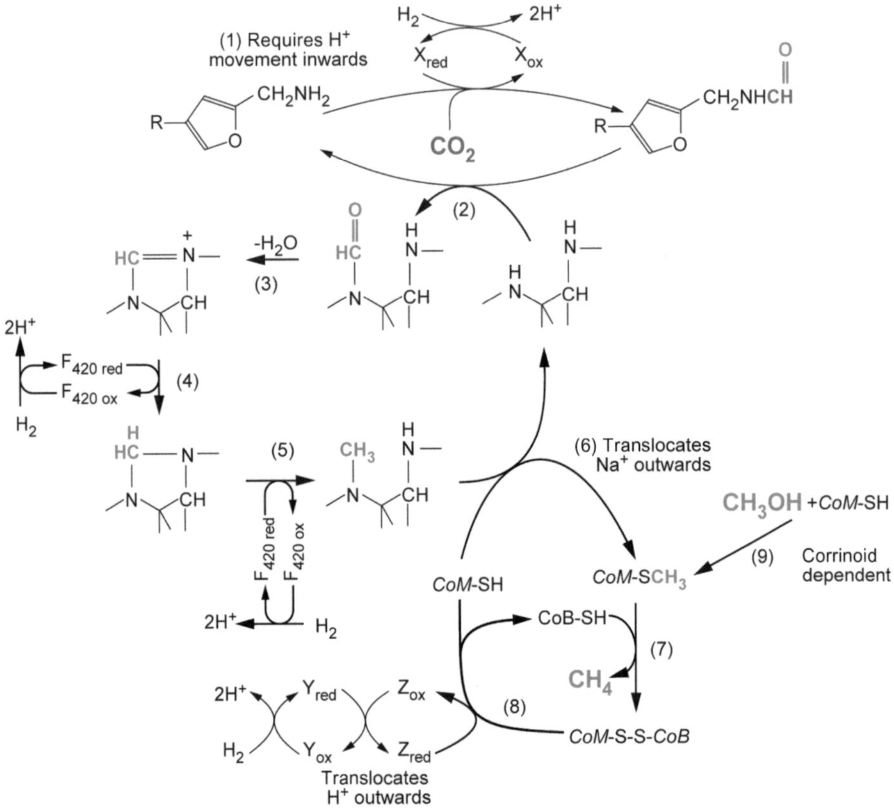

Figure 5.29   Sequence and energetics of reactions involved in methane formation from CH₃OH or CO₂ plus H₂ in methanogenic bacteria.
As explained in the text, reaction step 1, in the direction as written, is endergonic, whereas steps 6 and 8 are significantly exergonic. These are the three reactions in which at least some components are membrane-bound and in which, therefore, coupling to ion translocation across the cytoplasmic membrane is necessary. The reductive steps 4 and 5 are catalysed by water-soluble enzymes for which the reduced form of F420 is the electron donor. Re-reduction of F420 is catalysed by a water-soluble hydrogenase. Electrons for the reduction of *CoM-S-S-CoB* in step 8 are transferred from a membrane-bound hydrogenase, with some similarities to complex I, and in some methanogens *b*-type cytochromes schematically shown as Y and Z in the figure, to the reductase. This step is linked to proton translocation. Another membrane-bound hydrogenase, also with similarities to complex I, participates in step 1 to generate the reduced form of a specific ferredoxin, X from hydrogen, with the reaction being driven by inward movement of protons through the hydrogenase. As explained in the text, there is variation in the biochemistry of methanogens, and in those lacking cytochromes the energy available from reduction of *CoM-S-S-CoB* by hydrogen is not linked to proton translocation but to electron bifurcation whereby the downhill movement of two electrons to reduce the disulfide drives the uphill movement of two electrons to reduce two molecules of ferredoxin X.

formylated derivative requires a reduced ferredoxin, which is obtained via a membrane-bound hydrogenase that is driven in the uphill direction of hydrogen oxidation and ferredoxin reduction by proton translocation into the cell driven by the ion electrochemical gradient. This hydrogenase has some similarity to bacterial and mitochondrial NADH–ubiquinone oxidoreductases (Section 5.6); this formyl group is transferred to a pterin compound (Figure 5.29, step 2). After two further reductions in which electrons from H_2 are transferred, using water-soluble enzymes, via F_{420}, a methyl group is formed that is then transferred to coenzyme M (discussed previously) in a reaction that is linked to generation of a sodium ion electrochemical gradient (Figure 5.29). The CH_3SCoM then reacts as described previously (Section 5.14.9.1) to generate CH_4 and translocate protons. Overall, for each methane synthesised, it is believed that $4H^+$ and $2Na^+$ are moved out of the cell. $2H^+$ move in for each two reduced ferredoxins made, leaving four positive charges to drive the A-type ATP synthase. Thus, if $4q^+$ are needed per ATP, then one ATP will be made per methane. However, as discussed in Chapter 7, the number of ions moving through the ATP synthases per ATP made differs between enzymes and is not known for sure for methanogens. It could be either more or less that $4q^+/$ATP with consequent implication for the ATP/methane yield. As with growth on methanol plus hydrogen, a variation occurs for organisms that lack cytochromes. In this case, electron bifurcation gains occur at the reduction of the *CoM*-S-S-*CoB* species, and in this instance the reduced ferredoxin is used directly for the reductive trapping of carbon dioxide (Figure 5.29). In this case, the only ion translocation step is the outward movement of sodium ions coupled to methyl transfer to *CoM* (Figure 5.29). The yield of ATP is predicted to be smaller than in organisms with cytochromes because only a net $2q^+$ is moved out per methane made.

5.14.9.3 Growth by disproportionation of CH₃OH

Methanosarcina barkeri and *M. mazei Gö*1 can grow by disproportionation of methanol in the absence of H_2:

$$4CH_3OH \rightarrow 3CH_4 + CO_2 + 2H_2O$$

The stoichiometry of this reaction shows that one molecule of methanol is used to provide the reductant required for methane formation from the other three molecules of methanol. ^2H-labelling shows that three hydrogens in each of the three methane molecules are derived from the methyl groups in CH_3OH via CH_3SCoM. The electrons released in the oxidation of the fourth methane are, of course, those needed for the reduction of three molecules of CH_3SCoM to CH_4. The fourth molecule of CH_3SCoM is converted to CO_2 by the reverse of the reactions 1–6 shown in Figure 5.29, and the electrons released are used to drive the reductive reaction of methane formation (step 8).

5.14.9.4 Growth on acetate

In many natural environments, most methane is formed from acetate. This involves some sophisticated enzymology whereby the methyl group of acetyl-CoA, formed from acetate in an ATP-consuming reaction, is separated by carbon–carbon bond cleavage and transferred to the pterin compound of methanogenesis. The resulting methyl derivative is thereafter processed by steps 6–8 in Figure 5.29. The electron source for step 8 comes from taking the carbonyl group from acetyl-CoA and oxidising it to carbon dioxide. The electrons released by this oxidation are used to reduce protons to hydrogen using a membrane-bound hydrogenase that couples the reaction to proton translocation. The hydrogen so obtained is then used as the electron donor for reduction of mixed disulfide (proton translocating; step 8). Thus, there are two translocation steps involved in forming methane from acetate, but the net stoichiometry of ATP production will be low because one ATP is consumed for each acetate converted to acetyl-CoA in the first step. This is one of many examples in bacterial energetics in which the net yield of ATP is small.

5.14.9.5 The energetics of methanogenesis

The stoichiometries of ion movements, and haem of ATP synthesis, have been covered in the previous sections. The complexity of having both sodium and proton movement is perplexing, but it should be noted that methanogenic bacteria generally require Na^+ for growth. The proton and sodium electrochemical gradients appear to be interchangeable via an Na^+/H^+ exchanger.

Whereas $\Delta G^{o\prime}$ for the reduction of CO_2 by H_2 is $-131\,kJ\,mol^{-1}$ per mol CH_4 produced, the actual ΔG is likely to be closer to $-35\,kJ\,mol^{-1}$ for an organism without cytochromes. Because ΔG_p, the free energy for ATP synthesis, is likely to be approximately $+50\,kJ\,mol^{-1}$, this means that a theoretical maximum of approximately 0.6 ATP can be generated per mole of CH_4 formed from CO_2. This is a further example of the chemiosmotic mechanism allowing non-integral stoichiometries (Chapter 4). For organisms with cytochromes the available free energy is closer to $-70\,kJmol^{-1}$ because they require a much higher partial pressure of H_2 for growth, and this is consistent with the estimated stoichiometry of charge movements, which would suggest a stoichiometry of closer to 1 ATP for organisms with cytochromes.

5.14.10 *Propionigenium modestum*

Further reading: Dimroth *et al.* (2001)

We turn finally to an example of bacterial energy transduction that does not involve electron transport but that, in common with electron transport-dependent energy transduction, involves the cooperation of two ion pumps and so is appropriately discussed in this chapter. *Propionigenium modestum* is an anaerobic bacterium that ferments succinate to propionate by a short reaction sequence:

Succinate $\rightarrow$ succinyl CoA $\rightarrow$ methylmalonyl CoA $\rightarrow$ propionyl CoA $\rightarrow$ propionate

The decarboxylation of methylmalonyl CoA has a ΔG of approximately $-27\,kJ/mol$, close to that for the overall fermentation. The decarboxylase is a membrane-bound, biotin-dependent enzyme that pumps $2Na^+$ out of the cell for each CO_2 released. The Na^+ electrochemical gradient thereby set up could be up to 12–15 kJ/mol at equilibrium (because two ions are pumped) and is known to drive ATP synthesis through an Na^+-dependent ATP synthase that is discussed further in Chapter 7. Because a typical ΔG_p in a bacterial cell might be 45–50 kJ/mol, it is likely that three, or more likely four, Na^+ ions might be required per ATP synthesised for energetic reasons and thus 0.5 or 0.66 ATP made per succinate consumed. The energetics of this organism reinforces the significance of non-integral coupling stoichiometries in bacterial energetics; this organism could not by definition exist if one ATP had to be formed by a soluble enzyme system for each molecule of succinate fermented. This is not the only example of bacterial energy conservation being linked to a decarboxylation reaction. A further example is given in Chapter 8.

6 PHOTOSYNTHETIC GENERATORS OF PROTONMOTIVE FORCE

6.1 INTRODUCTION

A central feature of photosynthesis is the conversion of light energy into redox energy, meaning that photon capture causes a component to change its redox potential from being relatively electropositive to relatively electronegative. The electrons released from this component are utilised to generate a Δp, flowing either through a cyclic pathway back to re-reduce the original component or in a non-cyclic pathway to reduce another electron acceptor (ultimately $NADP^+$ in the case of photosynthesis catalysed by thylakoids in chloroplasts). In this latter case, a continual electron supply to the photon-sensitive component is required (obtained from H_2O in the case of the thylakoid).

The production of ATP by photosynthetic energy-transducing membranes involves a proton circuit that is closely analogous to that already described for mitochondria and respiratory bacteria. Thus, a Δp in the region of $200\,mV$ across a proton-impermeable membrane is used to drive a proton-translocating ATPase in the direction of ATP synthesis. In the case of photosynthetic bacteria, Δp may also drive other endergonic processes (see Figure I.1, Part 1), including reversed electron transport to generate NADH (discussed later). The ATPase (or ATP synthase) is very similar to, but distinct in some respects from, the mitochondrial enzyme (Chapter 7). The distinction between the respiratory and photosynthetic systems is in the nature of the primary generator of Δp, yet even here a number of familiar components recur, including cytochromes, quinones and Fe–S centres. Photosynthetic activity dependent on a retinal-containing protein (see Section 6.5) is distinctive: photon capture leads to a direct generation of Δp in the absence of electron transfer.

The two features unique to photosynthetic systems are the antennae or light-harvesting systems, responsible for the trapping of photons, and the reaction centres, to which the photon energy is directed. In the generalised process, denoted by the shorthand

$$P \xrightarrow{h\nu} P^* \xrightarrow{-e^-} P^+ \xrightarrow{+e^-} P$$

Bioenergetics. Doi: http://dx.doi.org/10.1016/B978-0-12-388425-1.00006-3

a pigment in the reaction centre (P) becomes electronically excited (P*) as a result of the absorption of a photon ($h\nu$). An electron can be released from this excited state (which is a relatively strong reductant, i.e., electron donor) at a potential that is up to 1 V more negative than the potential of donors to the reaction centre. The loss of this electron to a specific molecular component of the reaction centre produces P^+, the cationic form of the ground state of the pigment, P, that is a relatively strong oxidant and accepts an electron from a relatively positive potential molecule that acts as a donor to the reaction centre, regenerating P. These acceptors and donors are described later.

In the case of the representative purple photosynthetic bacterium *Rhodobacter sphaeroides*, the electron released from the reaction centre feeds into a bulk pool of ubiquinone/ubiquinol, from which it passes via a proton-translocating cytochrome bc_1 complex (Section 5.8) to a cyt c_2 (closely related to mitochondrial cyt c and *Paracoccus denitrificans* cyt c_{550}; Section 5.13.1). Cyt c_2 in turn acts as the donor of electrons to the P^+ species in the reaction centre, thus completing the cyclic electron flow (Figure 6.1).

Although a cyclic electron transfer pathway is also present in the thylakoids of chloroplasts (Section 6.4), the key difference from photosynthetic bacteria is the non-cyclic pathway, in which electrons are extracted from water, pass through a reaction centre, a proton-translocating electron transfer chain (which has similarities to the mitochondrial complex III), then through a second reaction centre before being ultimately donated to $NADP^+$, at a redox potential 1.1 V more negative than the $\frac{1}{2}O_2/H_2O$ couple (Section 6.4). The thylakoid membrane of the chloroplast thus not only accomplishes an 'uphill' electron transfer but also at the same time generates the Δp for ATP synthesis. The ATP and NADPH are typically used in the Calvin cycle, the dark reactions of photosynthesis in which CO_2 is fixed.

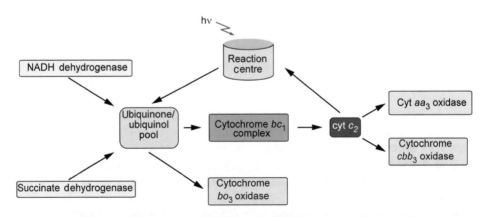

Figure 6.1   Simplified light-driven cyclic electron transport system and its relationship to respiratory electron transport in *R. sphaeroides*. Deletion of the gene for cytochrome c_2 does not prevent cyclic electron transport because an alternative c-type cytochrome (not shown) can act as substitute. Electron transport in the closely related organism *Rhodobacter capsulatus* is similar except that cytochrome aa_3 is absent.

6.2 THE LIGHT REACTION OF PHOTOSYNTHESIS IN *RHODOBACTER SPHAEROIDES* AND RELATED ORGANISMS

The heavily pigmented membranes of photosynthetic organisms act as antennae, absorbing light and funnelling the resultant energy to the reaction centres. In the case of *R. sphaeroides*, the photochemically active pigment has an absorption maximum at 870 nm and thus is known as P870. The equivalent energy of an 870-nm photon amounts to 1.42 eV (Section 3.5); thus, the energy transfer process is highly effective in the sense that 70% of the energy captured by the reaction centre is conserved in the resulting redox change of more than 1 V (comparing the potential of the P870$^+$/P870 redox couple with that of the P870$^+$/P870* redox couple) (Figure 6.2).

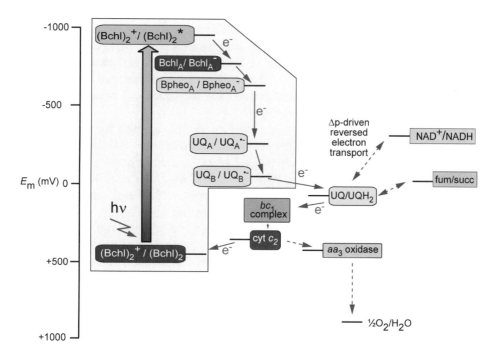

Figure 6.2   Pathways of electron transfer in *R. sphaeroides* in relation to the redox potentials of the components.
(Bchl)$_2$, Bchl$_A$, Bpheo$_A$, UQ$_A$ and UQ$_B$ are redox centres within the reaction centre (green box). The horizontal lines show their approximate midpoint potentials. Cyclic electron transport is completed by the bulk UQ pool, the bc_1 complex and cyt c_2. Alternative non-cyclic pathways are shown dashed. Note that photon absorption by (Bchl)$_2$ and generation of the excited (Bchl)$_2$* causes a negative shift in midpoint potential by more than 1 V.

6.2.1 Antennae

Further reading: Cogdell *et al.* (2006), Frank and Cogdell (2012)

The photochemical activity of reaction centres depends on the delivery of light at a specific wavelength (870 nm for the commonly studied reaction centre of *R. sphaeroides*). Energy of this wavelength can be obtained by direct absorption of incident light with this wavelength and also by transfer, through mechanisms described later, from components of the reaction centre that absorb at shorter wavelengths. However, even in bright sunlight, an individual pigment molecule will be hit by an incident photon only approximately once a second. Much higher rates of energy arrival at the reaction centre are required to match the turnover capacity of the centre, which is greater than 100/s. Some collecting process is evidently required.

Furthermore, most of the incident photons will have the wrong wavelengths to be efficiently absorbed by the pigments of the reaction centre. Effective light absorption over a wide range of wavelengths shorter than 870 nm is achieved by an assembly of polypeptides with attached pigment molecules. These polypeptides, which bind a variant of the terapyrrole chlorophyll known as bacteriochlorophyll (Bchl) and carotenoid pigments, are known as light-harvesting, or antenna, complexes and surround reaction centres. That such antennae are not strictly necessary for photosynthesis is established by the existence of bacterial mutants lacking them but which will nevertheless grow photosynthetically, albeit only in bright light.

Use of antennae, which provide a high surface area of pigments for absorption of light, to speed up the rate of photochemistry in the reaction centres is more effective in biosynthetic terms than inserting very many copies of the complicated reaction centre into the membrane. Thus, more than 99% of the Bchl molecules in an *R. sphaeroides* photosynthetic membrane are involved, together with carotenoid molecules, in absorbing light at shorter wavelengths than 870 nm and transferring it down an energy gradient to the lower energy absorption band at 870 nm. Light of wavelength greater than 870 nm, and therefore of lower energy, cannot be used by this type of reaction centre.

The transfer can occur by one of two mechanisms. The first, known as fluorescence resonance energy transfer (FRET), is intermolecular, depending on an overlap between the fluorescence emission spectrum of a donor molecule and the excitation (corresponding to the absorption) spectrum of an acceptor. Factors that affect the efficiency of such transfer include the relative orientation of donor and acceptor as well as the distance between the donor and acceptor (an inverse sixth power relationship). Energy transfer by this mechanism (which is *not* an emission followed by re-absorption of light) occurs over a mean distance of 2 nm in approximately 10^{-12} s.

At intermolecular separations of less than approximately 1.5 nm, direct interactions between molecular orbitals can occur, such that excitation energy is effectively shared between two molecules in a process known as delocalised exciton coupling and involving electron exchange. This process, the underlying principles of which are outside the scope of this book, occurs at faster rates than FRET and is the favoured mechanism at small intermolecular separations. We discuss FRET again in Chapter 9 with regard to the design of probes of cell function.

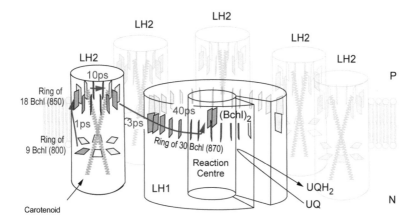

Figure 6.3   A representative organisation of the two light-harvesting complexes, LH1 and LH2, and the reaction centre in a purple non-sulfur photosynthetic bacterium.

Each LH2 cylinder represents an $\alpha_9\beta_9$ polypeptide unit; several LH2 complexes surround one LH1. LH1 in turn is an $\alpha_{15}\beta_{15}X$ assembly almost surrounding a reaction centre but forming an incomplete cylinder to allow access of UQ from the bulk phase. Alternative polypeptide chain stoichiometries are also found – see text. The coplanar arrangement of Bchls 850 (LH2) and 870 (LH1) facilitates excitation energy transfer to the reaction centre. An illustrative pathway and time course of excitation energy transfer is shown starting from a Bchl (800) molecule on LH2, transferring to a ring of Bchl (850) within the same LH2, and migrating to an LH1 and into the reaction centre special pair of chlorophylls $(Bchl)_2$. Other possibilities include initial capture by the carotenoids of LH2 and migration between several LH2 molecules before reaching LH1. There is variation in the detail of the organisation of LH1 and LH2 (see text).

In an organism such as *R. sphaeroides*, the light-harvesting or antennae Bchls are associated with two light-harvesting complexes known as LH1 and LH2. The former protein complex is closely associated with the reaction centre, whereas LH2 is located further away but sufficiently close to LH1 to permit energy transfer to it (Figure 6.3). LH1 and LH2 each have two types of polypeptide chains, known as α and β (which differ somewhat in the two complexes). X-ray diffraction analysis of LH2 from *Rhodopseudomonas acidophila* at 0.25 nm resolution showed that the polypeptides are arranged as an $\alpha_9\beta_9$ complex organised in two concentric circles with the nine β chains on the outside. Nine of the 27 Bchls (known as B800 because that is the wavelength of maximum absorbance) are intercalated between the β chains and lie more or less parallel to the plane of the membrane, at approximately 2.7 nm below the periplasmic surface of the protein (Figure 6.3). The radius of the outer ring of β subunits is 3.4 nm, leaving ample room for the 9 Bchl rings, which are approximately 2 nm from each other, to be included. In contrast, the inner ring of α subunits is only 1.8 nm in radius and thus there is insufficient room for Bchl molecules between the helices. Consequently, the other 18 Bchls (B850) lie parallel to the transmembrane helices and thus perpendicular to the plane of the membrane (Figure 6.3).

The difference in absorbance maxima arises from the different environments provided by the protein scaffold for the chemically identical Bchl molecules. Note that these Bchls also absorb at various wavelengths shorter than 800 or 850 nm and thus provide very effective light capture. The B850 chlorophylls are only 1 nm below the periplasmic surface of the protein and approximately 0.9 nm from each other. Thus, the majority of the Bchls in LH2 are located toward the periplasmic side of the membrane, which, as discussed later (Section 6.2.2), is the location of the Bchls that absorb light in the reaction centre to initiate photochemistry. Curiously, a very similar LH2 from a related bacterium *Rhodospirillum molischianum* has an $\alpha_8\beta_8$ structure with proportionately fewer Bchl molecules. The functional significance of the variation, if any, is not known. LH2 complexes have carotenoid molecules, nine for the $\alpha_9\beta_9$ structure, which extend from the periplasmic to the cytoplasmic side of the protein, making close (van der Waals) contacts with the Bchls.

LH1 comprises a similar set of α and β chains as LH2 and surrounds reaction centres, but it cannot form a closed cylinder because ubiquinone and ubiquinol must be able to move between the reaction centre and the bulk phase of the membrane bilayer (Section 6.2.2). Structural studies of LH1 from *R. palustris* have shown that the stoichiometry is $\alpha_{15}\beta_{15}X$, where X is a polypeptide that forms an α-helix and prevents a closed cylinder from forming, thus leaving a 'gap' for entry and exit of ubiquinone/ubiquinol (Figure 6.3). There are variations on this theme; in the much studied *R sphaeroides*, two reaction centres are surrounded by a dimer of $\alpha_{14}\beta_{14}$ LH1 molecules. The PufX protein in this organism promotes dimerisation and prevents the closure of the LH1 rings; cells lacking this protein cannot grow photosynthetically. The Bchl in LH1 has its maximum absorbance at approximately 870 nm, with the difference from LH2 again arising from the environment within the protein. Note that the longer wavelength maximum for LH1 permits downhill energy transfer from LH2 to LH1.

From the moment of absorption of light by a component in LH2, it takes approximately 100 ps for the excitation energy to reach the reaction centre via LH1 by FRET (Figure 6.3). In contrast, transfer within a light-harvesting complex between closely adjacent Bchl molecules, or from carotenoid pigments to Bchl, is always by delocalised exciton coupling. Because the excited state lifetime of carotenoids is too short to permit FRET, this restriction means that at least part of the carotenoid molecule must be very close (slightly further than the van der Waals distance) to a Bchl. Very rapid transfer of energy between pigments and onward to the reaction centre is essential if loss of energy by fluorescence or conversion to heat is to be avoided.

The LH2 structure suggests the following sequence of light-harvesting events (Figure 6.3). Note that light harvesting is concerned with *energy* and not *electron* transfer. Incident light shorter than 800 nm is absorbed by the carotenoids or Bchls of LH2. Within 100 fs, the carotenoid energy will be transferred to the B850 or B800 Bchls. Any energy absorbed by B800 will be transferred downhill within 1 ps to the B850 chlorophylls. Energy absorbed by the B850 molecules is effectively mobile on the femtosecond timescale within the LH2 ring; that is, it can be thought of as hopping around a storage ring. Such hopping can continue for up to 1 ns before the energy is dissipated, for instance as fluorescence. However, usually the energy is transferred after approximately 3 ps from LH2 to the Bchls of LH1, which are the same depth in the membrane. The circular nature of LH2 means that no particular defined orientation relative to LH1 will be needed because each B850 in LH2 is equivalent in a topological sense and thus energy transfer from any one of them is equally probable. It is estimated that the donor on LH2 and an LH1 acceptor Bchl come within 3 nm of each other for energy transfer by FRET.

Once absorbed by LH1, the energy can again migrate among the ring of Bchl molecules, this time B870, before—on the 30- to 40-ps timescale—migrating to Bchl in the reaction centre (Figure 6.3), particularly the 'special pair' $(Bchl)_2$ (Section 6.2.2), which is at the same depth in the membrane as the B870 ring (Figure 6.3). Because neither the reaction centre nor LH1 is circular, this transfer will not occur with equal probability from all parts of the LH1, but it is thought that many of the Bchls of the LH1 will be within approximately 4 nm of the special pair, thus again facilitating FRET.

The energy transfer process is very efficient; it is estimated that as much as 90% of the absorbed photons are delivered to the reaction centres of photosynthetic bacteria, aided by the ability of both Bchls and carotenoids of absorbing light at a range of wavelengths shorter than 800 nm.

6.2.2 The bacterial photosynthetic reaction centre

Further reading: Wraight (2004), Jones (2009), Gibasiewicz *et al.* (2009), Heathcote and Jones (2012)

The first two membrane proteins for which high-resolution structures were obtained by X-ray diffraction analysis were both bacterial photosynthetic reaction centres. Although that from *Rhodopseudomonas viridis* was the first and seminal structural determination, we mainly discuss what has been learned about photosynthesis from study of the *R. sphaeroides* centre because this has been studied much more extensively at the functional level.

Purified reaction centres from *R. sphaeroides* comprise three polypeptide chains—H, L and M—together with four molecules of Bchl; two molecules of bacteriopheophytin (Bpheo), which is a chlorophyll (Chl) derivative in which the Mg^{2+} is replaced by two protons; two molecules of UQ; and one molecule of non-haem iron (Figure 6.4). The reason why Mg^{2+} rather than another redox inactive metal is in the centre of Bchl and Chl is uncertain but may relate to the advantageous lifetime of the Mg–Chl/Bchl excited state and the optimal oxidation/reduction potentials of the Mg–Chl/Bchl species. In addition, Mg^{2+} may confer rigidity on the tetrapyrrole ring and give the possibility of binding to preferred amino acid side chains in the axial positions. However, there is evidence from some *in vivo* studies (Lin *et al.*, 2009) that Zn^{2+} can adequately replace Mg^{2+} without significant change in the oxidation/reduction potential of the P species, and it may be simply that greater bioavailability favours the latter in Bchl and Chl. The reason why Fe^{2+} would not be suitable for Bchl or Chl is discussed later.

Spectroscopic and biochemical studies on isolated reaction centres correlate in a very satisfying manner with the structure. First, we review the key findings from the functional studies (Figure 6.5).

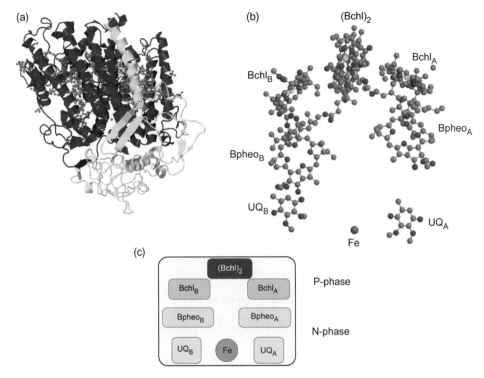

Figure 6.4   The three-dimensional structure of the _R. sphaeroides_ photosynthetic reaction centre.
(A) Subunit arrangement: H chain, green; L chain, blue; M chain, magenta.
(B) Location of ligands (side chains removed for clarity). (C) schematic arrangement of ligands.
Adapted from RCSB 1PCR.

6.2.2.1 P870 to Bpheo

Spectroscopic studies of reaction centres showed that illumination caused a loss of absorbance (bleaching) at 870 nm, consistent with the loss of an electron from a component absorbing at this wavelength. The component was termed P_{870}, and it was proposed that the absorption of a quantum led (within ~1 fs) to a transient excited state, P^*_{870}, in which an electron was raised to a high energy level. The electron was then lost (to an unknown component) to form the bleached (oxidised) product, P^+_{870}. Note that P_{870}, P^*_{870} and P^+_{870} are three states of the same chemical species. In the dark, P_{870} could be directly oxidised to P^+_{870} (without going through the excited state) by the nonphysiological electron acceptor ferricyanide ($E_{m,7}$ for $Fe(CN)_6^{3-}/Fe(CN)_6^{4-}$ + 420 mV), implying that the P^+_{870}/P_{870} redox couple has a distinctly _positive_ $E_{m,7}$ (approximately +500 mV). In contrast, estimates for the $E_{m,7}$ for P^+_{870}/P^*_{870} are more negative (by more than 1 V) relative to P^+_{870}/P_{870}. Thus, P^*_{870} can act as an electron donor, and P^+_{870} can act as an electron acceptor, driving the cyclic electron transport system (Figure 6.2).

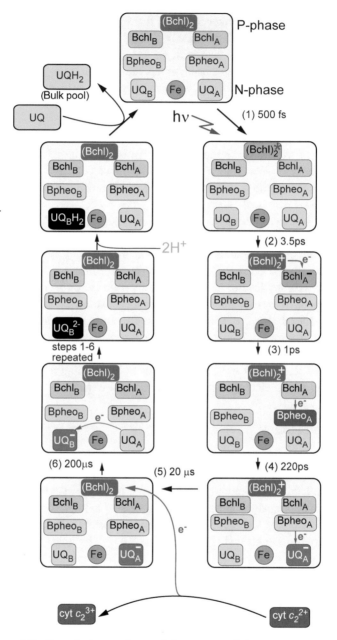

Figure 6.5   Two electron gating and time course of electron movement through the bacterial photosynthetic reaction centre from *R. sphaeroides*. The electron is shown as migrating down only one branch (A; see text) of the reaction centre. Transfer of the electron from (Bchl)$_2$ to Bpheo via the monomeric Bchl was at one time controversial but is now widely accepted.

Spectroscopic experiments further indicated that P_{870} was a noncovalent bacteriochlorophyll dimer $(Bchl)_2$. The oxidised state, P_{870}^+, had an electron spin resonance (ESR) spectrum with a linewidth consistent with an unpaired electron delocalised over both Bchl rings. Note that in contrast to a haem group, the electron is lost from the tetrapyrrole rings of the Bchl dimer. In chlorophyll, a redox-inactive metal is required, and of course unlike Fe^{2+}, Mg^{2+} cannot give up an electron. The crystal structure of the reaction centre is consistent with this model, with two closely juxtaposed Bchl molecules (Figure 6.4).

Rapid excitation of reaction centres with picosecond laser pulses, combined with rapid recording of visible absorption spectra, initially suggested that the immediate acceptor of the electron lost from P_{870}^+ was a Bpheo molecule. The transfer of an electron from P_{870}^* to Bpheo can be detected in less than 10 ps (Figure 6.5), and the resulting $(Bchl)_2^+ \ldots (Bpheo)^-$ biradical (often termed P^F) has a characteristic absorption spectrum. Studies of isolated Bpheo in nonpolar solvents suggest that E_m for Bpheo/Bpheo$^-$ in the reaction centre is approximately -550 mV, or more than 1 V more negative than for P_{870}^+/P_{870} in the unexcited state. This implies that E_m for P_{870}^+/P_{870}^* (Figure 6.2) must be even more negative to account for the electron transfer from P_{870}^* to Bpheo.

The two additional molecules of Bchl in the reaction centre were originally thought to be inactive and were termed the voyeur chlorophylls. However, subsequent to the elucidation of the crystal structure (Figure 6.4), which showed that they flanked $(Bchl)_2$, additional rapid spectroscopic measurements in conjunction with femtosecond flash excitation studies have indicated that one of the 'voyeur chlorophylls' is an intermediate in the passage of electrons from the special pair to Bpheo. Figure 6.5 shows a current view of the timescale of electron transfer in these initial stages.

6.2.2.2 Bpheo to UQ

The reaction centre has two bound UQ molecules, designated A and B. The biradical P_F (i.e., P_{870}^+.Bpheo$^-$) is highly unstable, and within 200 ps, the electron is transferred from Bpheo$^-$ to UQ_A, resulting in the formation of the free radical semiquinone anion, $UQ^{•-}$, which we previously discussed in the context of the bc_1 complex (Section 5.8). The effective $E_{m,7}$ of the $UQ^{•-}/UQ$ couple is approximately -180 mV. The electron is further transferred to the second bound quinone, UQ_B. The intervening Fe atom does not participate in this electron transfer. The timescale for these electron transfers is very rapid (Figure 6.4). We now have a UQ at A and a $UQ^{•-}$ at B. The latter must be stabilised by its binding site within the protein because in solution there is a strong tendency for this radical ion to disproportionate into $UQ + UQH_2$.

6.2.2.3 Transfer of the second electron and release of UQH_2

P_{870}^+ is reduced back to P_{870} by the thermodynamically favourable transfer of an electron from reduced-cytochrome c_2 external to the periplasmic side of the reaction centre ($E_{m,7}$ for the Fe^{3+}cyt c_2/Fe^{2+}cyt c_2 couple $+340$ mV) to P_{870}^+ ($E_{m,7}$ of the P_{870}^+/P_{870} couple $+ 500$ mV) (Figure 6.2). A second photon now causes a second electron to pass via the same route from P_{870}, via a transient $UQ^{•-}$ at A to the quinone binding site B. The ubiquinone at site A thus switches between the oxidised and anionic semiquinone forms and never becomes fully reduced (Figure 6.5). In contrast, the original UQ at B now becomes fully reduced, transiently forming UQ^{2-} before binding two protons to give UQH_2. This UQH_2 is then released to the bulk UQH_2/UQ pool. The two bound UQ molecules thus together act as a *two-electron gate*, transducing the one-electron photochemical event into a $2e^-$ transfer. The protonation of UQ_B plays an essential role in the generation of Δp, and cyclic electron transfer is completed by an external pathway from the bulk UQ back to cyt c_2, which is discussed in Section 6.3.

6.2.2.4 Structural correlations

The pathway of electron transfer deduced by spectroscopy correlates well with the structure of the reaction centre (Figure 6.4), with transfer of electrons from the $(Bchl)_2$ via Bpheo to UQ_A and then UQ_B. However, although there appear to be two branches down which the electron might flow, since the redox carriers are arranged in the reaction centre with near twofold symmetry, all the evidence, including photoreduction of only one (that on the A branch) of the two Bpheo molecules in crystals, points to only the right-hand 'A' branch (as depicted in Figure 6.4) being significantly active. The reason why the electrons flow more slowly down the left-hand 'B' branch is not known for certain. Distance can be excluded, but studies with reaction centres containing site-directed mutations suggest that the precise environments of the Bchl and Bpheo in the two branches are not identical, and that as a result the redox potentials of the $P^+(Bchl)^-$ state, and to a lesser extent the $P^+(Bpheo^-)$ state, are significantly more negative for the B branch, meaning that transfer of an electron from P_{850} through the B branch would be less energetically favourable. Note that the non-haem iron that lies between the two quinone sites has a primarily structural role and can be removed without unduly affecting the performance of the reaction centre.

The two quinone binding sites have distinct properties. Both sites must be able to stabilise $UQ^{\bullet-}$, but the B site must additionally allow entry of UQ from the bulk membrane phase, protonation of the fully reduced UQ^{2-} and release of UQH_2. A pathway can be discerned for UQ from the bulk phase to enter the B site 'head first.' Because the B site is not in direct contact with the aqueous phase, protein side groups may be responsible for transferring protons to the site. Site-directed mutagenesis of a glutamate residue (212) on the L-chain drastically attenuates the rate of protonation without affecting the rate of electron transfer to B, implicating the carboxylate side chain in proton transfer. Other adjacent acidic residues plus chains of bound water molecules seen in very high-resolution structures may also contribute. The properties of the B site must enormously retard the rate of dissociation of $UQ^{\bullet-}$ while promoting the dissociation of UQH_2 on a millisecond timescale.

X-ray diffraction data for purified reaction centres do not give information on the orientation of the complex in the intact membrane. However, cyt c_2 is located in the periplasm, in common with other bacterial c-type cytochromes, whereas the H subunit was only susceptible to proteolytic digestion and recognition by antibodies in inside-out membrane vesicles (i.e., chromatophores; see Chapter 1). Thus, the reaction centre is orientated with the special pair $(Bchl)_2$ towards the outside (periplasmic) surface of the bacterial cytoplasmic membrane, with the two UQ binding sites toward the cytoplasm.

The L- and M-chains each have five transmembrane α-helices, whereas the single α-helix of the H subunit also spans the membrane (Figure 6.4). All redox groups of the active A branch are bound to the L subunit, except that the Q_A site is provided by the M-chain. These α-helices contain predominantly hydrophobic amino acids and appear to provide a rigid scaffold for the redox groups. The importance of minimising relative molecular motion of these groups is illustrated by the finding that the rates of some of the electron transfer steps from the $(Bchl)_2$ to the Q_A site increase with decreasing temperature.

6.2.2.5 Charge movements

With the orientation shown in Figure 6.4, absorption of light will cause an inward (from periplasmic to cytoplasmic side) movement of negative charge from cyt c_2 to UQ_B, where the translocated electrons meet with protons coming from the cytoplasm. Thus, the net effect of both of these charge movements is to transfer negative charge into the cell (i.e., from P-phase to N-phase), contributing to the generation of a Δp (positive and relatively acidic outside). It is important to note that the reaction centre is not a proton pump as such but still contributes to the generation of a Δp. This key point is analysed again in the discussion of plant reaction centres in Section 6.4.

The inward movement of the electron through the reaction centre had already been detected using the carotenoid band shift (see Chapter 4) as an indicator of membrane potential. Using chromatophores, three phases of development of a membrane potential followed a short saturating flash of light. The first corresponded kinetically to the transfer of the electron from P_{870} to UQ_B; the second to the reduction of P_{870}^+ by cytochrome c_2; and the third, and the slowest, corresponded to the return of the electrons from UQH_2 to cyt c_2 via the cytochrome bc_1 complex and could be blocked by antimycin or myxothiazol (Chapter 5). Note that the transfer of electrons between the UQ A and B sites is parallel to the membrane and does not contribute to the establishment of a membrane potential.

The contribution of each electron transfer within the reaction centre to the development of $\Delta\psi$ is related to the distance moved by the electron perpendicular to the membrane and to the local dielectric constant. Thus, the movement of electrons from cyt c_2 to $(Bchl)_2$ makes less contribution than the electron transfer from Bpheo to Q_A through a more hydrophobic (low dielectric constant) environment. Finally, the uptake of the protons into the Q_B site contributes approximately 10% to the overall charge separation across the membrane.

Despite the satisfying correlation between structure and function, it should be appreciated that not everything is known about the functioning of the reaction centre. The necessity for the special pair as a central component of a reaction centre is not so clear now that it is known that both types of plant photosystem (Section 6.4) have somewhat different organisations of the pairs of chlorophylls that are equivalent to the $(Bchl)_2$ in bacteria. The origin of the close to twofold symmetry and the reasons why only one branch is photochemically active may relate to the evolution from a primitive reaction centre in which both branches are active, as indeed is thought to be the case for the photosystem I of green plants (Section 6.4.3).

An essential feature of a reaction centre is that it is not reversible. For example, the electron must not be allowed to return from $UQ^{\bullet-}$ at site A to reduce P_{870}^{+}, even though this is thermodynamically favourable. Reversal occurs 10^4 times more slowly than the forward reaction, accounting for the almost perfect quantum yield—that is, one photon results in the transfer of one electron to the quinone species at the Q_B site. However, the cost of this irreversibility is the large loss of redox potential (almost 1 V) as the electron passes from P_{870}^{*} to the quinone at the Q_B site. Additional factors that prevent reversed electron flow from components in the reaction centre to P_{870}^{+} include the large distance (~30Å) from the Q_A site (limiting the direct back reaction to less than $10\,s^{-1}$), the thermodynamically and kinetically unfavourable reversal of the reaction $UQ^{\bullet-}$ to Bpheo (even though the 10-Å separation of the two centres is consistent with rapid electron transfer) and, counterintuitively, the fact that the rate of electron transfer *declines* dramatically when the driving force exceeds a certain value. This is the so-called *inverted region* that is found when a plot of rate against driving force between two electron transfer centres is made on a theoretical basis (Section 5.4), and it can explain why $Bchl_A^{-}$ to $(BChl)_2^{+}$ does not occur at a significant rate.

6.2.3 The *R. viridis* reaction centre

Further reading: Deisenhofer and Michel (1989), Deisenhofer *et al.* (1995)

The *R. viridis* reaction centre differs in one major respect from that in *R. sphaeroides* by having an additional polypeptide subunit that contains four *c*-type haems (Figure 6.6). The haem nearest the special pair of Bchls (designated P_{960} in this organism because the environment and thus absorption maximum are different than in *R. sphaeroides*) is the immediate donor to the reaction centre. The electron is thus transferred over a distance of approximately 20Å (Section 5.4) from this haem ($E_{m,7} = +370\,mV$). The other haems have redox potentials of $+10$, $+300$ and $-60\,mV$, listed in order of increasing distance from the special pair. It is not known which of the haems accepts electrons from the cyt c_2, but as explained in Section 5.4, it is likely that the electron passes via the lower potential haems on a route that starts with a periplasmic cyt c donating an electron to the haem furthest from the reaction centre. There is no satisfactory explanation as to why this tetrahaem *c*-type cytochrome is dispensable (i.e. there is no gene coding for it) in *R. sphaeroides* and certain other organisms (e.g., *Rhodobacter capsulatus*). The other difference in *R. viridis* is that the Q_A site is occupied by menaquinone rather than UQ. The Q_B site was found to be empty in some crystals of the reaction centre, which is important support for the ability of UQ or UQH_2 at this site to equilibrate with the bulk quinone/quinol pool in the bilayer.

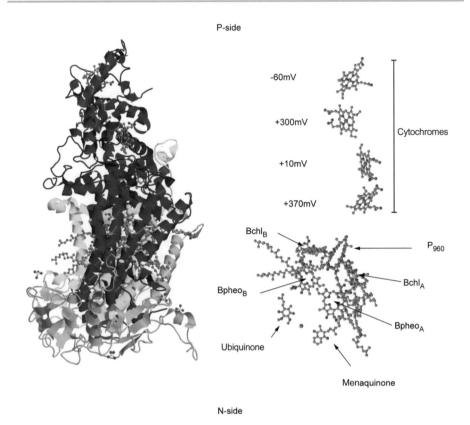

P-side

-60mV

+300mV

+10mV

+370mV

Cytochromes

Bchl$_B$

P$_{960}$

Bpheo$_B$

Bchl$_A$

Bpheo$_A$

Ubiquinone

Menaquinone

N-side

Figure 6.6 Subunit and ligand structure of *R. viridis* reaction centre.
Note the additional cytochrome polypeptide carrying four *c*-type haems
(designated by their midpoint potentials) and the presence of menaquinone
rather than UQ in the 'Q$_A$' binding site. See Section 5.4.1 for a discussion of
the midpoint potentials of the haems.
Source: RCSB 1PRC.

6.3 THE GENERATION BY LIGHT OR RESPIRATION OF ΔP IN PHOTOSYNTHETIC BACTERIA

We have already seen that absorption of light causes the movement of negative charge
into the cell and that the optical changes in carotenoids can be used to follow this charge
separation (Section 6.2.2). Three phases of development of the carotenoid shift are
observed following exposure of chromatophores to very short saturating flashes of light.
The slowest is blocked by inhibitors of the cyt bc_1 complex (Section 6.2.2) and corresponds to the movement of charge across the membrane through this complex. Such
carotenoid measurements, together with measurements of light-dependent proton uptake
by chromatophores, provided important evidence in favour of a Q-cycle mechanism for
the complex, similar to that described for the cytochrome bc_1 complex in Chapter 5.

Because electrons are retained within the cyclic pathway while protons are taken up
and released, only the latter need be considered when calculating the overall charge

movements per cycle. For each electron cycled, the reaction centre takes one proton from the cytoplasm while the bc_1 complex takes up one proton from the cytoplasm but releases two protons to the periplasm. Thus, two protons are translocated for each electron handled (or $4H^+/2e$, equivalent to $4q^+/2e$).

Generation of the NAD(P)H required for biosynthetic reactions often requires Δp to drive reversed electron transport (from an electron donor) in addition to ATP synthesis. For example, an organism growing on H_2 and CO_2 requires NADPH for CO_2 fixation. Electrons from H_2 are first fed, via a hydrogenase, into the cyclic electron transport system at the level of ubiquinone (at a redox potential close to $0\,mV$) and driven by reversed electron transfer through a rotenone-sensitive NADH dehydrogenase (an analog of complex I; Section 5.6) to give NADH ($E_{m,7}$ of the NADH/NAD$^+$ couple~$-320\,mV$). A subsequent transhydrogenase reaction (Section 5.13) leads to formation of NADPH. On the other hand, if the organism is growing on malate, malate dehydrogenase is able to generate NADH without involving the reaction centre. Thus, the extent to which electrons are fed into the cyclic electron transport system is dependent on the substrate.

Sulfide and succinate are further examples of substrates that feed in electrons at ubiquinone, and there are also electron donors in some organisms that donate to the c-type cytochromes. It is crucial that the cyclic electron transport system does not become over-reduced; if every component were to be reduced, then cyclic electron transport could not occur because there would be no available electron acceptor from P*. The mechanism whereby overreduction is avoided is not fully understood, but it is clear that for each electron entering the cyclic electron transport system, one must be delivered to oxidants, for example, to NAD$^+$ by Δp-dependent reversed electron transfer from ubiquinol (see Chapter 5).

Some photosynthetic bacteria have nitrogenase, which catalyses the reductive fixation of nitrogen to ammonia (Section 5.13.5). The reaction requires a reductant (a flavodoxin) with a more negative potential than NAD(P)H. For a photosynthetic bacterium growing on malate, there is increasing evidence that the cytoplasmic membrane contains a novel component, Rnf, that utilises Δp to drive electrons uphill from ubiquinol or NADH to generate the required reductant. In some organisms, an electron bifurcation system (Section 5.13.8) may play this role.

In common with many other photosynthetic bacteria, *R. sphaeroides* can grow aerobically in the dark. Oxygen represses the synthesis of bacteriochlorophyll and carotenoids, and so the reaction centre is absent. However, the b and c cytochromes are retained, and three terminal oxidases can be induced, including, in the case of *R. sphaeroides*, a cytochrome aa_3 oxidase (Figure 6.1) very similar to the mitochondrial complex IV (Section 5.9). By using some constitutive cytochromes, the bacterium can therefore switch very economically between anaerobic photosynthetic growth and aerobic growth in the dark by assembling respiratory and photosynthetic chains with common components (Figure 6.1).

Cytochrome c_2 has long been regarded as an essential component in the cyclic electron transport system. However, gene deletion experiments with both *R. sphaeroides* and the related *R. capsulatus* have shown that the cytochrome is dispensable and that an alternative c-type cytochrome can substitute. In some organisms, the bc_1 complex is absent and substituted by another complex, whereas the cyt c_2 can be replaced by a periplasmic iron–sulfur protein known as HiPiP (high potential iron–sulfur protein), the name reflecting an unusually positive oxidation/reduction potential for an Fe–S protein. Degeneracy among electron transfer proteins is common in bacteria. Organisms such as *R. sphaeroides* can also use certain anaerobic electron acceptors, including at least some of the oxides of nitrogen. Their electron transport chains are thus even more versatile than Figure 6.1 suggests because they can also incorporate some of the components shown for *P. denitrificans* in Section 5.15.1. Use of exogenous electron acceptors can provide an anaerobic route for ensuring that the cyclic electron transfer system does not become over-reduced.

6.3.1 Photosynthesis in green sulfur bacteria and heliobacteria

Further reading: Hohmann-Marriott and Blankenship (2011)

Whereas the *R. sphaeroides* reaction centre is closely related to the oxygen-evolving photosystem II in green plants (Section 6.4.2), the green plant photosystem I (Section 6.4.3) has a close relative in green sulfur bacteria and heliobacteria, based on sequencing and cofactor content, although structures are not yet available. The light-harvesting apparatus in green sulfur bacteria comprises a self-assembled system of BChls, carotenoids, quinones and lipids, with only a limited amount of protein for binding Bchl molecules. This 'chlorosome' system is very effective at low light intensities. There is one important difference from plants: in the bacterial protein, the two major subunits of the reaction centre are identical. We will return to the significance of this later (Section 6.4.3).

Possession of a photosystem I-type reaction centre has important bioenergetic consequences. As Figure 6.7 shows, the reaction centre does not directly reduce a bulk phase quinone; instead, electrons are transferred to iron–sulfur proteins with a more negative redox potential, known generically as ferredoxin. Ferredoxin can reduce NAD^+ without the need for the Δp-driven reversed electron flow required by *R. sphaeroides*.

Electrons from substrates such as sulfide have a variety of entry points, via the quinone pool or cyt *c*, and then either directly into the reaction centre or through cyt bc_1 or an equivalent complex. Δp is generated by the electron flow through the reaction centre from the periplasmic to the cytoplasmic side of the membrane, in parallel with the cyt bc_1 complex, if involved. Rather than reducing NAD^+, an alternative fate for the reduced ferredoxin could be as a reductant for menaquinone or even the cyt bc_1 complex. This would permit cyclic electron transfer, but the extent to which this occurs is unclear. Speculatively, the cyclic process might be analogous to that operating in thylakoids (Section 6.4.5).

6.4 LIGHT-CAPTURE AND ELECTRON TRANSFER PATHWAYS IN GREEN PLANTS, ALGAE AND CYANOBACTERIA

Just as mitochondria have evolved from bacteria incorporated into a nonrespiring eukaryotic cell ('endosymbiosis'), green plant chloroplasts in eukaryotic photosynthetic organisms are derived from endosymbiotic ancestors similar to present-day cyanobacteria.

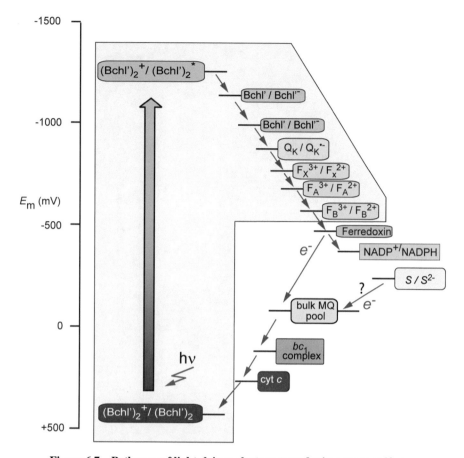

Figure 6.7   Pathways of light-driven electron transfer in a green sulfur bacterium in relation to the redox potentials of the components.
Bchl' designates that the bacteriochlorophyll in these organisms may be modified relative to that in purple bacteria. By analogy with photosystem 1 (Figure 6.14), Q_K is phylloquinone and F_x, F_A and F_B are iron–sulfur centres; together with the Bchl' molecules, these make up the cofactors of the reaction centre. The entry point into the electron transport system of electrons derived from sulfide oxidation is uncertain. MQ, menaquinone.

Photosynthetic electron transfer in eukaryotic thylakoid membranes can be either cyclic or non-cyclic, the latter resulting in a stoichiometric oxidation of H_2O and reduction of $NADP^+$ and requiring two independent reaction centres to act in series to drive electrons from $H_2O/\frac{1}{2}O_2$ to $NADP^+/NADPH$, a redox span of more than 1 V (Figure 6.8). To date, the only atomic-resolution structures obtained for these reaction centres have come from thermophilic cyanobacteria.

The presence of two reaction centres in thylakoid membranes was discovered by a classical observation known as the 'red drop.' Illumination at 400–680 nm resulted in oxygen evolution, but longer wavelength light was ineffective. There was thus a component that required light of no greater than 680 nm, either by direct photon absorption or by energy transfer from shorter wavelengths via antennae. What was more striking was

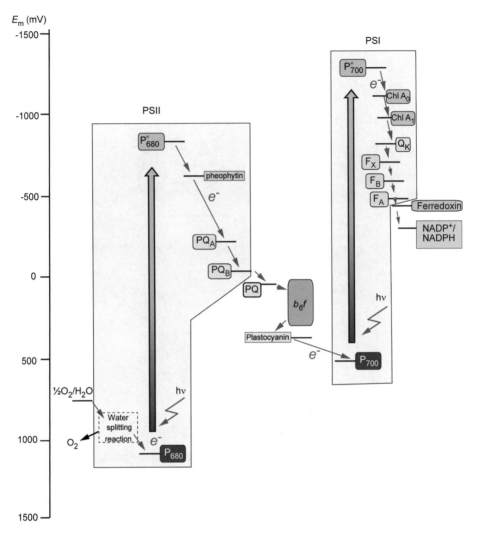

Figure 6.8   Non-cyclic electron transport in green plants, algae and cyanobacteria arranged by redox potential.
Electrons from water pass through photosystem II (PSII), the $b_6 f$ complex and photosystem I (PSI) before reducing NADP$^+$. The horizontal bars show the approximate midpoint potentials in PSII—P_{680}^+/P_{680}, the excited $P_{680}^+/P*_{680}$, pheophytin, $PQ_A/PQ_A^{\bullet -}$ and $PQ_B/PQ_B^{\bullet -}$, the bulk PQ/PQH_2 pool (PQ is plastoquinone) and plastocyanin (Cu^{2+}/Cu^+)—and in PSI— $P_{700}^+/P_{700}, P_{700}^+/P*_{700}, Chl\,A_0/Chl\,A_0^-, Chl\,A_1/Chl\,A_1^-$, the phylloquinone Q_k, three Fe–S centres (Fe^{3+}/Fe^{2+}), and ferredoxin (Fe^{3+}/Fe^{2+}). The water-splitting reaction is described in Figure 6.12.

that oxygen evolution produced by low-intensity light at 400–680 nm could be increased by simultaneous illumination at 700 nm. The interpretation was that two photosystems were involved, one of which had an absorbance centred at 700 nm and could be supplied either by 700-nm light or by energy transfer from antennae-absorbing light below

680 nm. The other photosystem, with an absorbance maximum at 680 nm, could be excited by wavelengths up to that value but not by the lower energy 700-nm light.

The two photosystems are connected in series by other electron transfer components (Figure 6.8). The system that requires 680 nm light for excitation is photosystem II (PSII), which abstracts electrons from water and raises them to a sufficiently negative potential to reduce plastoquinone (PQ) to plastoquinol (PQH_2); the latter are respectively very similar to UQ and UQH_2 in structure and function (Figure 5.6). PQH_2 supplies electrons to the cyt $b_6 f$ complex (the '6' is redundant but persists in the literature), which is functionally very similar to the cytochrome bc_1 complexes of respiratory chains. Whereas the mitochondrial bc_1 complex is an electron donor to cytochrome c, the $b_6 f$ complex passes electrons to plastocyanin, which is a peripheral copper protein located at the luminal side of the thylakoid membrane. Photosystem I (PSI), which is excited by 700 nm light, takes electrons from plastocyanin ($E_{m,7}$ +370 mV) and generates reductant sufficiently powerful (Figure 6.8) to reduce ferredoxin ($E_{m,7}$ −530 mV) in a one-electron reaction. Most of the reduced ferredoxin (Fd) reduces $NADP^+$ to NADPH via Fd–$NADP^+$ oxidoreductase. Two molecules of reduced Fd are re-oxidised for each NADPH produced.

Figure 6.8 shows that the redox potentials of the oxidation–reduction couples, as electrons are driven from water to NADPH, follow the shape of a distorted letter N. This scheme has become known as the Z-scheme because it was once presented in a format displaced by 90° from the present convention in which redox potential is shown on the vertical axis.

We now progress along the Z-scheme, starting with the light-harvesting antenna complex II (LH II) feeding energy into PSII.

6.4.1 Light-harvesting complex II

Further reading: Liu *et al.* (2004), Daum *et al.* (2010)

As in bacterial photosynthesis, light harvesting, or antennae, complexes are required in the thylakoid membrane even though the two photosystems, especially PSI, themselves contain much antennae chlorophyll (Chl). Distinct polypeptide complexes, LHC I and LHC II (not to be confused with LH1 and LH2 of bacterial membranes), are normally associated with PSI and PSII, respectively, although LHC II can migrate to PSI upon phosphorylation of a threonine residue near the stromal N-terminus of the polypeptide to aid balance between the two photosystems (Section 6.4.7).

LHC II is a homotrimer that contains approximately half the total Chl found in a green plant (Figure 6.9). Each polypeptide has a molecular weight of approximately 25 kDa and is non-covalently associated with eight molecules of chlorophyll *a* (Chl*a*), six molecules of chlorophyll *b* (Chl*b*) and four carotenoid molecules. The high-resolution structure shows that each monomer forms three transmembrane α-helices, with one of these being tilted significantly away from an angle of 90° to the membrane plane and two being significantly longer (~35Å) than typical transmembrane helices (Figure 6.9). These two helices, together with two carotenoid molecules, cross over in the centre of the membrane. Between them, the three helices provide a scaffold for the binding of the

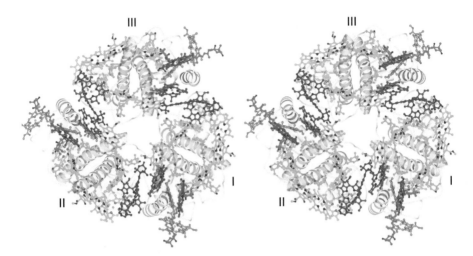

Figure 6.9   The structure of thylakoid light-harvesting complex II.
Stereo view of trimeric LHC II from spinach looking at the membrane from
the stromal (N-side) of the membrane. Monomers are labelled I–III, with each
having three transmembrane helices. Side chains and lipids are omitted.
Green, Chl*a*: blue, chl*b*; yellow, orange and magenta, carotenoids. Note that
many of the chlorophylls are oriented almost perpendicular to the membrane
surface. Phytyl side chains have been omitted. Two molecules of carotenoid
(yellow) in the centre of each monomer act as an internal cross brace, linking
loops of polypeptide on opposite surfaces of the membrane.
Adapted from Liu et al. (2004) with permission. PDB 1RWT.

chlorophyll molecules that are distributed as two layers—one with five Chl*a* and three
Chl*b*, closer to the stromal (N) side, and the other (three Chl*a* and three Chl*b*) towards
the luminal side (P). In the centre of the complex, Chl*a* molecules are in close contact
with Chl*b*, allowing rapid energy transfer, whereas the nearby carotenoids are available
to quench any toxic singlet oxygen that might be formed. Much of the energy transfer
within LHC II will be by FRET (Section 6.2.1), although some chlorophylls may be sufficiently close for delocalised exciton coupling (Section 6.2.1). An important role of LHC II
is to allow the nonradiative dissipation of excessive energy as heat by non-photochemical
quenching, one of the photoprotective strategies that have evolved in plants.

LHC II is found on the periphery of PSII close to a number of 'minor' monomeric
light-harvesting complexes, some of which are actual components of PSII and exist in
a roughly 1:1 stoichiometric ratio with the photosystem. Because LHC II is present in
a vast molar excess, only a fraction can be directly associated with PSII. Nevertheless,
'super-complexes' of PSII and LHC II have been described. Based on the X-ray structure of the D1 and D2 subunits of the cyanobacterial PSII and the spinach LHC II structure, it has been possible to combine these with cryoelectron microscopy to deduce the
locations of LHC II and several minor chlorophyll binding proteins (Figure 6.10). This
super-complex can account for approximately 75 Chl*a* and 15 Chl*b* molecules per PSII
reaction centre.

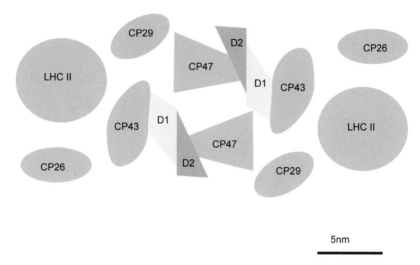

**Figure 6.10   Schematic view of the arrangement of the spinach LHCII–
PSII supercomplex derived from cryo-EM and single particle analysis.**
D1 and D2 are the major subunits of PSII and are shown in relation to 'minor'
light-harvesting complexes CP26, −29, −43, and −47 and LHII.
Adapted from Nield and Barber (2006).

6.4.2 Photosystem II

Further reading: Umena *et al.* (2011)

The green plant photosystems have proved more difficult to purify and characterise than
the single reaction centre of the purple bacteria. One difficulty is that the CP43 and CP47
chlorophyll binding proteins (Figure 6.10) are part of the complex, giving approximately
36 Chl and 11 β-carotene molecules per PSII core. To date, no atomic-resolution structure
is available for the eukaryotic PSII. Instead, crystal structures have been obtained from
thermophilic cyanobacteria at increasing resolution, culminating in a 1.9-Å resolution
structure from *Thermosynechococcus vulcanus* (Figure 6.11). PSII has many similarities
with the purple bacterial photosynthetic reaction centre, including an approximate twofold
symmetry relationship within the molecule, with the two major polypeptides D1 and D2
each having five transmembrane helices (analogous to the L and M subunits of the bacterial system) providing binding sites for a 'voyeur' Chl, pheophytin, and plastoquinone at
A and B sites (Figure 6.11). It is presumed that only one branch is photochemically active.

6.4.2.1 The oxygen-evolving complex

Further reading: Umena *et al.* (2011), Kawakami *et al.* (2011)

The ability of PSII to extract electrons from water, the 'water-splitting reaction,' is one
of the most intriguing reactions in bioenergetics. In air, the E_h for the $O_2/2H_2O$ couple is

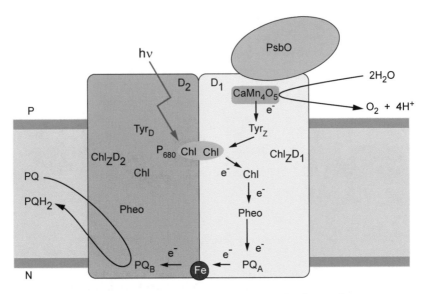

Figure 6.11   Organisation of core polypeptides and cofactors in photosystem II.
The diagram is based on the 3.8-Å crystal structure of the PSII from the thermophilic cyanobacterium *Synechococcus elongatus*. By analogy with the bacterial reaction centre, it is thought that there is only one active branch from the P_{680} centre to plastoquinone at Q_A. The roles of monomeric Chl and pheophytin (Pheo) are deduced from the structure and by analogy with the bacterial reaction centre. Only the core D1 and D2 subunits and the PsbO subunit, which appears to cap the oxygen-evolving centre, are shown. *Based on Zouni* et al. *(2001).*

+810 mV (Section 5.9); therefore, to abstract electrons from water requires a redox centre that is even more electropositive than this value. Indeed, the midpoint potential of the P_{680}^+/P_{680} couple in the ground state is estimated to be +1.3 V.

The water-splitting centre consists of a Mn_4CaO_5 cluster that looks like a distorted chair, with three of the Mn atoms, calcium and four oxygen atoms forming an approximate cube (cubane) and the fourth Mn and final oxygen forming the seatback (Figure 6.12). The metals in the cube are bound to amino acid side chains (aspartate, glutamate and histidine), mainly from the D1 polypeptide but with one glutamate and possibly an arginine from CP43. The octahedral coordination spheres of the three Mn in the cube are saturated with six ligands each, whereas the Ca and the fourth Mn(Mn) each have two water molecules in the crystal structure, making their respective coordination numbers 7 and 6 (Figure 6.12). Because these coordination shells are full, it is likely that the oxygen atoms of two of these water molecules are those destined to become oxygen.

Because Ca is redox inactive, the oxidation of a water must take place on an Mn atom. Although not fully established, it seems likely that an oxy species is formed from one of the waters on the fourth Mn atom and that the oxygen–oxygen bond formation is facilitated by calcium polarising the second water so that it can attack the oxygen bound to the Mn. If this is correct, it provides a relatively rare example of calcium polarising

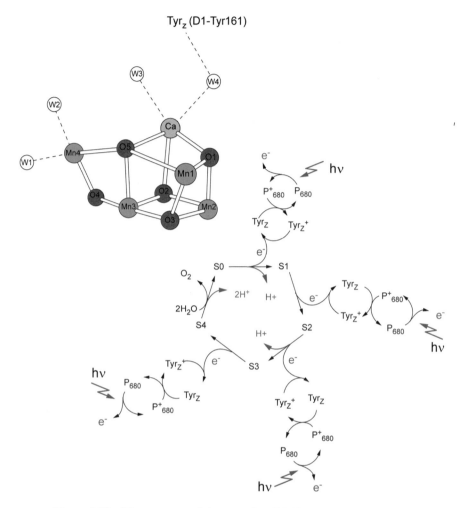

Figure 6.12   The oxygen evolving complex (OEC) and its oxidation states during the the water-splitting reaction of photosystem II.
The OEC is believed to access five sequential states, S0–S4, during a cycle of water binding, absorption of four quanta and release of oxygen. The structure of the OEC Mn_4CaO_5 cluster is shown at the top left. Note the associated water molecules (W1–W4) and the proximity of the redox active Tyr_z. Four quanta are required to abstract $4e^-$ from two H_2O. The S0 state has three Mn^{3+} plus one Mn^{4+} and the S1 state has two Mn^{3+} and two Mn^{4+}. In the simplest models, two further consecutive oxidations occur to give four Mn^{4+} in the S3 state, although it is possible that one of the electrons lost going from S1 to S3 is taken from an oxo ligand to a Mn atom. Modelling indicates that in S4, Mn4 exists as an MnO oxy species in which the metal is formally in the +5 oxidation state. Attack by a polarised water molecule bound to the Ca atom could then result in the formation of oxygen. Note that contrary to intuition, S1 and not S0, which is the least oxidised species, is the major populated species in the dark; this arises because in the dark an electron is lost from S0, probably to Tyr_z. The magenta electrons are transferring at a relatively very positive redox potential, whereas the red electrons released from P_{680} after excitation are transferring at a relatively very negative potential.

a substrate at an active site, a process that is more often carried out by Zn. The abstraction of $4e^-$ from two molecules of water to yield O_2 releases $4H^+$ into the lumen of the thylakoid.

6.4.2.2 The electron transfer pathway through PSII

It is assumed that the P_{680}^+ species formed in PSII is associated with the two adjacent Chls (Figure 6.10), but it is now clear that these cannot be considered as a special pair in the same way as in the purple bacterial system because they are further apart and the positive charge appears to be largely associated with one of these two Chls. The significance of this difference is not clear.

Electrons from the oxygen-evolving complex (OEC) are not transferred to P_{680} directly but travel via a specific tyrosine residue side chain (Tyr$_z$, D1-Tyr161). In the crystal structure, this side chain on the D1 polypeptide is 7Å at its closest approach to the Mn_4CaO_5 cluster, within hydrogen bonding distance to one of the waters bound to Ca, and approximately 12Å from the photochemically active chlorophyll (P_{680}). These distances are sufficiently short to allow rapid electron transfer. Loss of an electron to P_{680}^+ from this tyrosine generates a neutral radical because a proton is also lost, probably to a neighbouring histidine. The tyrosine residue in turn regains an electron from the OEC. There is a tyrosine residue in the equivalent position on polypeptide D2, but although it can form a radical, its distance from the OEC cluster is thought to preclude any role for it in electron transfer.

The organisation of the cofactors shown by the crystal structure is clearly consistent with the pathway of electron flow from P^* to quinone being very similar to that in the purple bacterial reaction centre (Figure 6.4). The electron paramagnetic resonance spectrum of bound $UQ^{\bullet-}$ in the latter has similarities with that of $PQ^{\bullet-}$ bound to PSII. The Q_b site is believed to be the site of action of the inhibitor DCMU, and the similarity to the bacterial reaction centre is strengthened by the finding of a mutant of the latter which is sensitive to DCMU.

There is one molecule of a b-type cytochrome, cyt b_{559}, in the PSII structure (not shown in Figure 6.11 but sandwiched between two single transmembrane helical subunits with the haem nearer the N- than the P-side) that is not found in the purple bacterial reaction centre. Its function is believed to be as an electron donor to any unwanted, relatively long-lived and potentially damaging oxidised species in the reaction centre, such as a carotenoid or a chlorophyll including P_{680}^+. There is evidence for oxidation of b_{559} under these conditions, and a carotenoid is sufficiently close to allow electron transfer from the haem on the millisecond timescale. Such a role for b_{559} clearly requires a route for electron transfer to b_{559}. Despite the distance (more than 14Å), there is evidence that electron transfer between the Q_a or Q_b sites and b_{559} can occur. Thus, cyt b_{559} provides a path for electron flow from PQH_2 back to the P^+. Such a short-circuit is what normally the purple bacterial reaction centre and PSII are designed to avoid because it would dissipate the energy captured from light as heat. However, it may be needed under conditions of dryness and/or high light and temperature when electron flow from OEC to P^+ might be inadequate to reduce the latter sufficiently rapidly. P^+ is such a highly oxidising species (E_m +1.3V) that it would cause damage to the components of the thylakoid membrane if it were to persist for a significant period. Indeed, D1 is damaged during normal conditions of illumination, with the result that it is one of the fastest turning over polypeptides known in biology. Nevertheless, return via b_{559} of an electron to P$^+$ from the Q_a or Q_b sites must be carefully controlled so that it does not compete with the normal operation of the reaction centre. Finally, there is also (not shown in Figure 6.11) a c-type cytochrome in PSII from *T. vulcanus*; its role is unknown, and because it is not present in higher plant PSII, we assume that it cannot have a key role in energy transduction by PSII.

6.4.3 Cytochrome b_6f and plastocyanin

Further reading: Darrouzet *et al.* (2004), Baniulis *et al.* (2008), Cramer *et al.* (2011), Cruz-Gallardo *et al.* (2012)

The final electron acceptor from PSII is PQ, which is reduced to PQH$_2$. PQH$_2$ in turn is the electron donor to the cyt b_6f complex, which operates a Q-cycle functionally similar to mitochondrial complex III (Section 5.8), with two b-type haems (differing in $E_{m,7}$, similar to their counterparts in the cyt bc_1 complexes)—a Rieske-type high-potential Fe–S protein and a c-type cytochrome, which is, however, referred to as cyt f (cyt f is a c-type cytochrome because the haem is attached to the protein through the characteristic C-X-X-C-H motif). The evolutionary origin of cyt f is currently unknown (f refers to *frons*, which is Latin for 'foliage' or 'leaf').

However, there are notable structural differences between bc_1 and b_6f. First, the transmembrane subunit that binds the two b-type haems has four transmembrane helices rather than eight as in the bc_1 complex; there are an additional three, rather than four, helices, but these are on a separate subunit in the b_6f complex. The haems, one towards the P-side and the other on the N-side, are sandwiched between conserved histidine residues similar to the organisation in bc_1, two on each of the two helices. However, the b_6f complex has an extra haem on the same subunit, which is covalently attached by a single thioether bond; hence, it is regarded as a variant c-type cytochrome. This third haem is located towards the N- or stromal side of the membrane and is often called c_i. We return to its possible function later. Whereas the globular domain of mitochondrial cyt c_1 is similar to the α-helical structure of cyt c, cyt f is a mainly β-sheet protein with the N atom of the amino terminus of the polypeptide acting, uniquely, as one of the axial ligands to the haem iron. The b_6f complex is insensitive to both antimycin and myxothiazol. 2,5-Dibromo-3-methyl-6-isopropylbenzoquinone (DBMIB) does inhibit, acting at the Q_p site (i.e., on the lumenal side of the membrane), and is thus equivalent to the locus of action of myxothiazol on the cyt bc_1 complex (Section 5.8.1). The electron acceptor for cyt b_6f is plastocyanin. Note that the E_{m7} span from PQH$_2$ to plastocyanin is very similar to UQH$_2$ to cytochrome c in respiratory systems.

The redox centre in plastocyanin is a Cu ion that undergoes 1e$^-$ oxidation–reduction reactions between its +1 and +2 oxidation states. The environment of the Cu within the protein is such that its $E_{m,7}$ is +370 mV, very different from what it would be in aqueous solution. This, together with the characteristic ESR and blue absorbance spectra in the Cu^{2+} state, is diagnostic of what is called a type I Cu centre. These are also found in azurin and pseudoazurin—common components of bacterial electron transfer chains (Section 5.13). The Cu in plastocyanin has a highly distorted tetrahedral coordination geometry with ligands from the sulfurs of cysteine and methionine, as well as two histidine side chains. This geometry is imposed on the Cu by the polypeptide chain and thus overcomes the tendency of Cu to prefer different geometries in its two oxidation states and thereby tunes the Cu to its biological function.

The nonphysiological electron acceptor ferricyanide (hexacyanoferrate) allows the Hill reaction, a light-dependent oxygen evolution in the absence of $NADP^+$, to be observed. The $E_{m,7}$ of the $Fe(CN)_6^{3-}/Fe(CN)_6^{4-}$ couple, $+420$ mV, is sufficiently positive to accept electrons from reduced plastocyanin. However, because plastocyanin is on the luminal side of the ferricyanide-impermeable membrane, ferricyanide can only accept electrons from a donor on the stromal side (N-side of the thylakoid membrane) of the PSI complex. Thus, oxygen evolution with ferricyanide as acceptor requires the operation of both photosystems. Plastocyanin donates electrons to PSI.

6.4.4 Photosystem I

Further reading: Amunts *et al.* (2007, 2010), Busch and Hippler (2011)

The understanding of PSI has been considerably enhanced by the acquisition of a high-resolution (2.5Å) crystal structure for the protein isolated from the thermophilic cyanobacterium *Synechococcus elongatus*. A slightly lower resolution (3.4Å) structure of the plant PSI has been obtained from pea (Figure 6.13). One difference is that PSI is monomeric in plants and green algae, whereas the cyanobacterial PSI is crystalised as a trimer. Both the cyanobacterial and plant PSI contain at least 90 antenna Chl molecules and 22 carotenoids associated with the core, in addition to the six chlorophylls, two phylloquinones and three Fe–S centres that make up the reaction centre. Furthermore, the plant PSI has a 'belt' of four additional light-harvesting complexes, named Lhc1–Lhc4, that contribute a further 61 Chl molecules per photosystem (Figure 6.13A). Finally, under high light conditions, LHC II, normally associated with PSII, can become phosphorylated and translocate to PSI. This 'state transition' is discussed in Section 6.4.7.

With the exception of the light-harvesting complexes, the overall structure and mechanism is well conserved between cyanobacteria and green plants, with at least 12 subunits (PsaA–PsaL in the green plant nomenclature). PsaA and PsaB are the largest subunits; they are highly homologous and have been formed by gene duplication. As in PSII and the purple bacterial reaction centre, there is a twofold axis with each set of cofactors duplicated on each side of the molecule—that is, bound to PsaA or PsaB. However, in contrast to the other systems, it appears that both branches are active, although one of them probably operates at a faster rate than the other. Note that an important difference between PSI and either the bacterial reaction centre or PSII is that single electron delivery is required at the N-side of the membrane; there is no requirement for one side to deliver two electrons sequentially to distinct quinone binding sites. In common with PSII, two closely neighbouring Chl molecules provide the centre at which the P^+ species is assumed to form, but they are not organised in the same juxtaposed special pair arrangement as seen for the bacterial reaction centre. Furthermore,

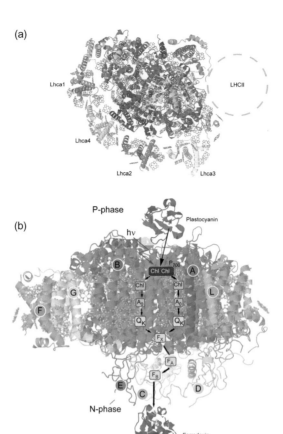

(a)

Lhca1

LHCII

Lhca4

Lhca2   Lhca3

P-phase

(b)

Plastocyanin

hν

P$_{700}$

B   Chl Chl   A

G   L

Chl   Chl

F

A$_0$ A$_0$

Q$_K$ Q$_K$

F$_X$

F$_A$

F$_B$

E   C   D

N-phase

Ferredoxin

Figure 6.13   Structural organisation of photosystem I.
The model is based on the crystal structure of the protein from the pea [PDB
1QZV]. (A) View from the stroma of a monomer of PSI with associated light-
harvesting proteins (Lhca1–Lhca4, each of which is similar to LHC II with
three transmembrane helices) and possible location of phosphorylated LHCII
after translocation from PSII. Note that PS1 is a trimer. (B) Side view of a PSI
monomer without light-harvesting proteins, showing the approximate
pathway of electron transport (for key, see legend to Figure 6.8). There are
12 polypeptides, PsaA–PsaL, of which 8 are visible in this view. PsaA and
PsaB have significant sequence similarity; each forms 11 transmembrane
α-helices and binds a total of 79 chlorophyll molecules. Another 8
transmembrane α-helical subunits bind a further 11 Chls. P$_{700}$ is a chlorophyll
dimer, but the two molecules are not chemically identical; that on PsaA is an
epimer of the normal chlorophyll a. Loss of an electron from P$_{700}$ results in a
cation radical being formed on the chlorophyll bound to the PsbB subunit. It
is currently thought that the electron can migrate down both sides of the
reaction centre but at unequal rates. It is not known whether the electron
migrates to the A$_0$ chlorophylls via the chlorophylls that are adjacent to the
P$_{700}$. P$_{700}^+$ is reduced by plastocyanin [PDB 1IUZ], which docks onto two short
helices, one provided by each of PsaA and PsaB, which run along the surface
of the membrane and thus connect two transmembrane α-helices. Electrons
from Chl A$_0$ are further transported via a phylloquinone at Q$_K$ and a chain of
3 Fe–S centre to ferredoxin [PDB 1FER]. Approximate distances are as
follows: P to Chl, 12 Å; Chl to A$_0$, 8 Å; A$_0$ to Qk, 9 Å; Qk to Fx, 14 Å; Fx to
Fa, 15 Å; Fa to Fb, 12 Å.

the unpaired electron of the P^+ species is associated with only one of the two Chl molecules and thus functionally the two chlorophylls, which are not chemically identical, may not act as a unit in the manner of the special pair that operates in the bacterial reaction centre. Currently, the advantage of a functional pair of Chls in the bacterial reaction centre is not clear, given that the feature is agreed to be absent in PSII and its occurrence in PSI is debated. The electron lost from the Chls on the PsbB subunit (Figure 6.13), to give the P^+ species, passes to the A_0 Chls on either side of the twofold axis. How it gets there is not certain, but the Chls that lie between the 'pair' of Chls at the P_{700} centre and the two A_0 Chls seem likely to be involved, just as the analogous molecules are in the bacterial reaction centre.

Kinetic studies have shown at least one of the A_0 Chls to be reached by an electron after approximately 10 ps. After approximately a further 100 ps, the electron has reached a Q_K (also called A1) centre, which is a phylloquinone (also known as vitamin K_1) (Figure 6.13) with a very low redox potential attributed to its hydrophobic environment. From there, the electron migrates to the cluster of three iron–sulfur centres that are located at the N-side of the complex. These in turn can reduce the water-soluble ferredoxin iron, which has an $E_{m,7}$ of -530 mV and so is extremely electronegative. The latter is the electron donor for several reactions, including ferredoxin–NADP$^+$ oxidoreductase (FNR)-catalysed reduction of NADP$^+$, mainly needed for the Calvin cycle, and reduction of nitrite to ammonia, for assimilation of nitrogen. The carotenoid band shift (Section 6.3), which was first detected in chloroplasts, indicates that the electron is transferred across the membrane in less than 20 ns. The way in which biology is able to tune the properties of Chls in proteins is exemplified by the finding that the two A_0 Chls have sulfur axial ligands from methionines; in pure chemical systems, sulfur is not a ligand for S, and this unusual ligation may be responsible for a very low redox potential for A_0/A_0^-.

At the lumenal side of the PSI, the P^+ is reduced by plastocyanin, whose copper centre is thought to dock within 14 Å, thus permitting direct electron transfer on an adequate timescale. The redox potential of the P^+ species in PSI is estimated as $+450$ mV and is thus appropriate to accept electrons from plastocyanin (E_h approximately $+250$ mV).

As discussed previously (Section 6.3.1), green sulfur bacteria have a reaction centre with considerable resemblance to that of PS1, with the intriguing difference that PsaA and PsbB are replaced by two copies of a single core subunit. The bacterial analog is a true homodimer that can be expected to have a twofold axis. This strongly suggests that electron transfer may occur at equal rates down both sides of the bacterial reaction centre. This supports the use of both branches in PSI. An intriguing issue concerning photosynthesis is exactly how the reaction centres evolved and when the two classes (I and II) of reaction centre diverged from one another.

6.4.5 Δp generation by the Z-scheme

Carotenoids orientated perpendicular to the plane of the membrane respond to fluctuations in membrane potential with small but extremely rapid spectral changes. Observation of such carotenoid shifts indicates that both PSII and PSI translocate charge across the membrane. Evidence for the orientation of PSII comes from the observation that the protons liberated in the cleavage of H_2O are initially released into the lumen, indicating that oxidation of water occurs on the P-side of the membrane. Also, a radical anion form of plastoquinone bound to the reaction centre must be located close to the N-side of the membrane because it can be made accessible to impermeant electron acceptors such as ferricyanide after brief trypsin treatment. Ferredoxin and ferredoxin–NADP$^+$ reductase are accessible to added antibodies, whereas plastocyanin is not. These observations all suggest that PSI is oriented across the membrane as shown in Figure 6.13.

For each water molecule split, the OEC generates 2H$^+$ in the lumen, with the 2e$^-$ from the oxidation being ultimately used to generate NADPH in the stroma. In addition, the b_6f complex releases 4H$^+$ into the lumen. Note that two of these stromal protons are recruited for the reduction of PQ to PQH$_2$ at the PQ$_B$ site of PSII (Figure 6.8). Overall, 6H$^+$ are released into the P-phase, 4H$^+$ are removed from the N-phase and two electrons are translocated from the P-side to the N-side of the membrane. Thus, six positive charges reach the P-phase for each 2e$^-$ passing from H_2O to NADP$^+$. Note that any protons taken up in the stroma when NADP$^+$ is reduced to NADPH are released again as the NADPH is reoxidised by the Calvin cycle and can be ignored (Figure 6.14). It is important to grasp that although neither PSI nor PSII are proton pumps, each is contributing to the generation of protonmotive force by moving electrons across the membrane in the appropriate direction.

The H$^+$/ATP stoichiometry of the thylakoid ATP synthase is a matter of uncertainty (Chapter 7), but its value should be considered in the context that the Calvin cycle uses 1.5 ATP per NADPH. Thus, if $n = 4$, there is an exact matching of the synthesis of 1 NADPH

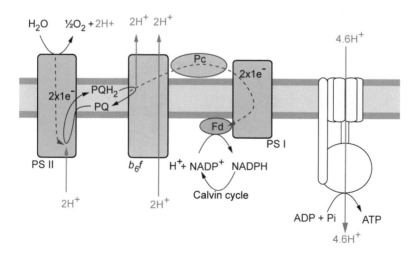

Figure 6.14   Stoichiometry of charge translocation and generation of Δp associated with electron transfer from water to NADP$^+$.
The OEC generates 2H$^+$ in the lumen for each H_2O split. The 2e$^-$ from the reaction plus 2H$^+$ from the stroma reduce PQ to PQH$_2$. The b_6f complex, using a Q-cycle mechanism, delivers these 2H$^+$ to the lumen and pumps a further 2H$^+$ across the membrane. The 2e$^-$ finally reduce NADP$^+$ to NADPH in the stroma. Thus, six charges are moved across the membrane per H_2O split. The H$^+$/ATP ratio is shown here as 4.6 (see text and Chapter 7).

and 1.5 ATP for each two electrons transferred from water to ferredoxin. However, as discussed in Chapter 7, there are reasons to suppose that $n = 4.6$, in which case the yield of ATP would not be sufficient to match the requirements of the Calvin cycle.

6.4.6 Cyclic electron transport

Further reading: Iwai *et al.* (2010)

The main fate of the NADPH produced by non-cyclic electron flow is for the Calvin cycle, which fixes CO_2 in an overall process that requires 3 ATP for every 2 NADPH. However, as discussed previously, there may be a shortfall of ATP, assuming of course that the Calvin cycle is the dominant consumer of NADPH and ATP and that these molecules can only be supplied to the chloroplast stroma by the thylakoid system. The situation is more extreme in C4 plants, in which 5 or 6 ATP molecules are consumed per CO_2 fixed. One mechanism to make good this shortfall is cyclic electron transport, which occurs when electrons are able to return from ferredoxin to the b_6f complex (Figure 6.15). Cyclic electron transport can occur, and ATP synthesis can be observed, when thylakoids are illuminated with 700 nm light so that only PSI is active. Furthermore, there are cells in which only PSI appears to be active and cyclic electron transport would seem obligatory. These include heterocysts of cyanobacteria, which fix nitrogen and thus require an anaerobic environment, and the bundle-sheath cells

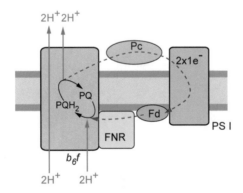

Figure 6.15   Physiological cyclic electron transport in thylakoids—a plausible scheme.
FNR's established function is to transfer electrons from the reduced ferredoxin, produced by photosystem I, to $NADP^+$. As the figure shows, there is now evidence that FNR can dock onto the cytochrome *bf* complex and thus act as electron donor. Exactly how the electrons enter the *bf* complex and how there is net proton translocation across the membrane is not known (see text). Note that a complete circuit of an electron alone would not generate any protonmotive force, and so it is presumed that elements of the Q-cycle mechanism must be involved. It is possible that PQH_2 is formed at the Q_n site and then diffuses to the Q_p site, whereupon electron would flow toward plastocyanin and the second through the transmembrane haems to the Q_p site, where PQH_2 could be again formed, with the second electron coming from FNR; this scheme would be protonmotive with overall $q^+/2e = 4$.

of some C_4 plants. However, measurement of quantum efficiencies of PSI and PSII in green plants at ambient CO_2 suggests that cyclic phosphorylation is in general a minor contributor, but it may be important during an induction phase when leaves are illuminated. Although the idea of cyclic electron flow has been accepted for some time, and appears indispensable for C4 plants, the pathway has been difficult to identify. There is now evidence that ferredoxin–NADP$^+$ oxidoreductase, in addition to its 'normal' role transferring electrons from reduced ferredoxin to NADP$^+$, can under some conditions dock with, and return electrons to, the b_6f complex. The exact route of electron transfer has not been fully elucidated, including the question as to whether the c_i haem of the b_6f complex is involved. There are also other possible routes of cycle electron flow that are outside the scope of this book. Cyclic electron flow in the thylakoid has a close relative in the cyclic electron transfer system of green sulfur bacteria (Section 6.3.1). In these bacteria, it is not known how ferredoxin transfers electrons to the electron transport system; possibly, it will prove to be analogous to the scheme suggested for thylakoids. Figure 6.16 shows that assuming a Q-cycle mechanism for the b_6f complex, each turn of the cycle would result in four protons, and also 4 charges, translocated per 2e$^-$ per cycle. The latter stoichiometry is the same as for cyclic electron transport in bacteria.

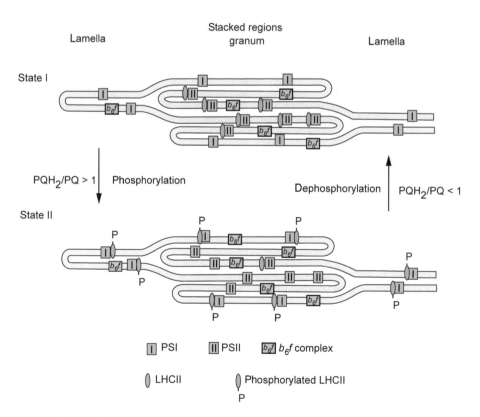

Figure 6.16   Phosphorylation of light-harvesting complexes may affect their distribution between stacked (rich in PSII) and unstacked (rich in PSI) regions of the thylakoid membrane.

6.4.7 **Photosynthetic state transitions**

Further reading: Haldrup *et al.* (2001)

What regulates the relative activities of the two photosystems and thus of cyclic and non-cyclic electron transport? Some clue may come from the arrangement of the photosystems in the thylakoid membrane. PSII can be found in the stacked regions of the thylakoids, whereas PSI, which has to deal with the large ferredoxin substrate, is restricted to the unstacked regions (Figure 6.16). Because the light-harvesting complexes, carotenoids and chlorophylls, transfer energy to the photosystems by resonance energy transfer, the effectiveness of which decreases as the sixth power of the distance, the separation between light-harvesting complexes and photosystems will be critical. LHC II can be phosphorylated on a threonine residue exposed to the stroma, and this is thought to cause it to be excluded from the stacked regions, thus decreasing the energy transferred to PSII. The activity of the kinase responsible for the phosphorylation increases as the ratio PQH_2/PQ increases (implying that PSII is becoming more active than PSI). Thus, the light-harvesting complexes are directed towards PSI in the nonstacked regions, restoring the balance. Subunits PsaH, PsaL and PsaO form the docking site for phosphorylated LHC II.

   The condition to which chloroplasts revert in the dark, when LHC II is predominantly associated with PSII, is often termed state I, whereas state II refers to the situation in which at least some of the LHC II has migrated to PSI, the interconversions being termed 'state transitions.' The terminology is not to be confused with the description of mitochondrial respiratory states (Chapter 4). State I was originally defined as the condition in which PSI was overexcited (in which case the thylakoid showed a relatively high fluorescence) and state II the condition in which PSII received excess excitation and fluorescence was relatively low. State transitions are thus induced by absorption of excess excitational energy by one of the two photosystems, with the changes occurring reversibly over several minutes. State II favours cyclic electron flow around PSI, and because cyclic electron transport generates ATP, but not NADPH, additional ATP requirements may control the transition to state II.

6.4.8 Δ*p* and Δ**pH**

The steady-state Δp in thylakoids is, as discussed previously (Section 4.2.4), present almost exclusively as a ΔpH due to the permeability of the thylakoid membrane to Mg^{2+} and Cl^-. One important consequence of this is that electron transport can be uncoupled from ATP synthesis by ammonium ions or other weak bases that enter as neutral species, increasing the internal pH as they protonate. Additional Cl^- uptake occurs in response to further proton pumping, with the result that a massive accumulation of NH_4Cl occurs and the thylakoids burst. In the steady state, ΔpH can exceed 3 pH units, estimated from the accumulation of radiolabelled amines or the quenching of 9-aminoacridine fluorescence. The transient $\Delta\psi$ decays too rapidly to be measured by radiolabelled anion distribution, but it can be followed from the decay of the carotenoid shift following single and flash-activated turnover of the photosystems. The timescale of the electron transfer reactions under these conditions is much shorter than for ion movements. The chloroplast ATP synthase is essentially the sole consumer of Δp in the thylakoid, and it is important that it does not wastefully hydrolyse ATP in the dark (Section 7.6).

6.5 BACTERIORHODOPSIN, HALORHODOPSIN AND PROTEORHODOPSIN

Further reading: Lanyi (2004), Hirai *et al.* (2009), Lanyi (2012)

Bacteriorhodopsin (BR) was first discovered in the archaebacterium *Halobacterium salinarum*, where it acts as an ancillary generator of Δp by capturing energy from light and using it to pump protons outward across the cytoplasmic membrane, supplementing the Δp generated by respiration under conditions of limiting oxygen.

We are closer to a molecular description for the proton pumping mechanism for BR than for any other pump, but as we shall see, active transport is a subtle process, depending on a carefully orchestrated series of sequential events within the protein. Unfortunately, active transport in this case, and perhaps in all transporters, is not explained at the molecular level by a simple two- or three-state model of the system. In fact, as we shall see, it is not just the amino acid side chains but also water molecules that make a significant contribution to the mechanism.

There are several reasons why BR has revealed more about an active transport mechanism than any other protein. First, the events that follow the absorption of light have been characterised in some detail by spectroscopy, particularly fast recordings of the time dependence of the visible absorption spectrum of the retinal. This has enabled the formulation of a photocycle involving at least seven species (Figure 6.17c). Second, high-resolution structures have been obtained for the protein not only in its dark state but also in states that are believed to correspond to discrete intermediates within the photocycle. Third, expression systems have allowed extensive study of molecules carrying specific mutations.

6.5.1 The bacteriorhodopsin photocycle: structure and function

Further reading: Lanyi (2004), Morgan *et al.* (2008)

BR is a protein with seven transmembrane α-helices (A–G) connected by short loops such that little of the protein protrudes from a membrane bilayer (Figures 6.17a and 6.17b). Retinal is covalently attached via its aldehyde group of the side chain of a lysine, forming a Schiff's base, on the G helix. The retinal molecule is oriented roughly transversely relative to the membrane.

In the dark bR state, the nitrogen atom of the Schiff's base (Figure 6.17d) is protonated (this was established by resonance Raman spectroscopy), and the carbon–carbon double bonds of the retinal are all in the *trans* geometry. The *extracytoplasmic* channel between this nitrogen atom and the aqueous phase on the P-side of the protein contains water molecules and the side chains of several key amino acid residues. The opposite

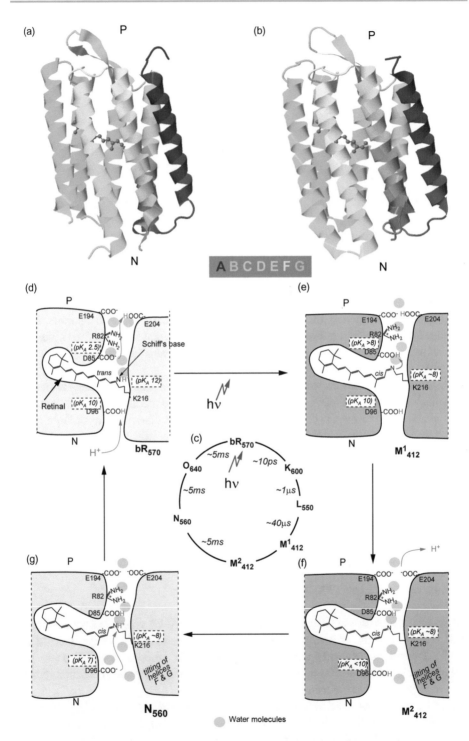

Figure 6.17   Representation of some of the conformational changes in bacteriorhodopsin, together with proton movements in the photocycle. *Continued →*

intracytoplasmic channel appears to be both narrower and lacking any well-defined water molecules. A so-called π bulge distorts helix G; this means that instead of the normal hydrogen-bonding pattern within the α-helix, two peptide bond carbonyls are hydrogen-bonded to water molecules.

Information for individual states of the photocycle has come from X-ray crystallographic or electron diffraction analysis of illuminated mutant proteins that stall at different stages, combined with spectral analysis of the protonation states and pK_a of key residues. The assumption in this type of approach is that the mutation has not generated an irrelevant structure. Such mutagenesis has strongly implicated two aspartate residues, D85 in the extracytoplasmic channel and D96 in the intracytoplasmic channel, as key participants in the pumping of protons. Fourier transform infrared spectroscopy has indicated that in the dark state, their respective pK_a values are approximately 2.5 and 10, respectively. This means that the side chain carboxyl of D85 is deprotonated, whereas that of D96 is protonated. The pK_a of the Schiff's base is 12 or 13; this is unusually high and may be due to directly interacting water molecules and the influence of the negative charge on the nearby aspartate D85.

Illumination of the dark bR state causes the following events to occur in strict sequence (Figure 6.17c):

1. Photon absorption by the retinal causes it to isomerise from all-*trans* to the 13/14 *cis* isomer within approximately 10 ps, generating a transient K_{600} state. In a general sense, the photocycle and associated proton pumping is driven by thermal relaxations of the retinal back from this *cis* state to the original all-*trans* conformation. These relaxations will involve transfer of energy from the retinal to the polypeptide chain, thus driving protein and water conformational movements. As a first step, transfer of energy from the twisted *cis*-retinal to the helices initiates changes in the environments of the Schiff's base and aspartate D85, together with other structural changes, occurring within 1 μs, to form the L state.

Figure 6.17 *Continued*
Seven-transmembrane structure in dark bR$_{570}$ (a) and illuminated M$_{412}$ (b) conformations. Note the change in retinal conformation from *trans* to *cis*. The link between helices E and F is not shown in panel (a) [PDB 1C3W and PDB 1DZE]. The chain is 'rainbow' colour coded from the N-terminal (blue) to the C-terminal (orange), and the transmembrane helices are named A–G. (c–g) Bacteriorhodopsin photocycle. The colour coding corresponds approximately to the peak absorption of the retinal. For simplicity, the J state, a precursor to K that forms on a 550-fs timescale, is not shown. It is also likely that there are other substates within the M intermediate. M1/412 and M2/412 are sometimes called the early and late M states, respectively. The numbers associated with each state are the absorption maxima for each species; note that these differ slightly from one investigator to another. The diagram should not be taken to mean that a single retinal conformation is retained throughout the cycle. Some of the approximate pKa values are shown for the key functional groups involved in proton translocation. The proton translocation mechanism is not simple, and indeed other side chains apart from those shown play some role in the process.

2. After a further $40\,\mu s$, the M^1 state (Figure 6.17e), with the Schiff's base deprotonated, is formed irreversibly with a large decrease in free energy. Release of a proton to the medium from the 'proton-release group' (comprising glutamates E194 and E204 plus additional amino acids and associated water molecules) at the mouth of the extracytoplasmic channel now occurs. The transition is characterised by a pK_a decrease for the Schiff's base, which may be associated with a redistribution of water molecules, and an increased pK_a for aspartate D85. This results in proton migration from the Schiff's base to aspartate D85. At the same time, relaxation of the now deprotonated Schiff's base allows movement of helices to widen the intracytoplasmic channel, in turn allowing water molecules to enter between aspartate D96 and the Schiff's base. This is the M^2 state (Figure 6.17f).

3. In the next step, there is a decrease in the pK_a of D96 and an increase in the pK_a of the Schiff's base, leading to transfer of a proton between the two groups, which are separated by approximately $10\,\text{Å}$ (Figure 6.17). Such a large separation requires a pathway because whereas an electron can migrate from one centre to another on a millisecond timescale over distances of up to $14\,\text{Å}$ (Chapter 5), a proton can only move unaided 1 or $2\,\text{Å}$ in a comparable time. This forms the N state (Figure 6.17g).

4. The Schiff's base isomerises back to the all-*trans* conformation, and an increase in the pK_a of aspartate D96 allows it to be reprotonated from the cytoplasm, forming the O state.

5. The dark bR state is regained by transfer of a proton from D85 to the proton-release group. This means that the now unprotonated D85 can regain its favourable electrostatic interactions with the Schiff's base nitrogen atom.

6.5.2 Proteorhodopsin and halorhodopsin

Further reading: Venter *et al.* (2004)

BR has two close relatives, proteorhodospsin and halorhodopsin. For many years, it was believed that a protein of the BR-type only occurs in the archaeal group of organisms. However, the bulk sequences of DNA collected from the Sargasso Sea by Craig Venter and colleagues showed that BR-like proteins are widely distributed among eubacteria. Subsequently, they have been identified elsewhere, including in organisms isolated from permafrost as well as in some marine eukaryotes, where it is believed to have arrived by lateral gene transfer. This type of retinal-containing proton pump is called proteorhodopsin. When a carotenoid molecule is also bound, as a light-harvesting pigment, the name xanthorhodopsin has been adopted. The structure of xanthorhodopsin shows that although it is broadly similar to that of BR, there are significant differences. Thus, there is no direct equivalent to the set of water molecules associated with D85 or the R92 movement that is linked to the participation of E194 and E204 in releasing protons to the P-side of the membrane, and the structure is much more open at the P-side of the membrane. The equivalent of R92 is more than $10\,\text{Å}$ further away from a single glutamate at the surface of the protein than is arginine R92 from the E194/E204 surface in

BR. This and other features make it unlikely that the movement of an R-side chain has any mechanistic role in xanthorhodopsin.

As discussed previously, D85 in BR has a rather low pK_a of 2.5, making it a good proton acceptor from the Schiff's base. In xanthorhodopsin, the corresponding aspartate has a pK_a of approximately 7, and this can be attributed to the carboxylate side chain being strongly hydrogen-bonded to a nearby histidine that is conserved in proteorhodopsins and xanthorhodopsins. This feature can help explain why the non-archael proton pumps only function at alkaline pH, at which the aspartate–histidine counterion complex could be active. However, this complex is believed to retain the proton acquired from the Schiff's base for longer in xanthorhodopsins. Thus, in this class of molecule, reprotonation of the Schiff's base by protonation from a glutamate, or even in some cases a lysine (recall the recently proposed role of lysines in the proton translocation activity of complex 1; Chapter 5), acting equivalently to D96 in bacteriorhodopsin is the next step, followed by capture of a proton from the N-side of the protein by the glutamate or lysine. The final step in the proton translocation process is thought to be direct loss of a proton from the aspartate–histidine complex to the P-side of the membrane. A further difference is that throughout the photocycle, the environment of the glutamate on the N-side is less hydrophobic than that of the D96 in BR. Thus, overall, although there are many similarities between BR and its counterparts in eubacteria, it would have been incorrect to assume that the mechanisms would be the same; this illustrates the common observation that biology has been able to evolve many variations on a theme and that it can be a mistake to extrapolate unthinkingly from one protein to a closely related one.

Halorhodopsin acts as an inwardly directed chloride pump. The high-resolution structure of this protein has shown that it is remarkably similar to BR. There are important differences, especially the absence of residues equivalent to the carboxylate side chains of D85 and D96 and that the Schiff's base nitrogen remains protonated throughout the photocycle. The crystal structure of halorhodospsin in the dark state shows that a chloride ion is bound where the carboxylate side chain of Asp85 is located in BR. Thus, it is proposed that transport is initiated by a light-driven conformational change in the retinal that results in movement of the Schiff's base nitrogen toward the cytoplasmic side of the membrane and drags the chloride with it so that the chloride is moved to the cytoplasmic side of the protein. As with BR, no obvious route exists for ion movement between the Schiff's base nitrogen and the cytoplasm in the structure of the unilluminated protein. Therefore, it is expected that structures of photocycle intermediates, obtained by studying mutant proteins or low-temperature species, will in due course show an opening up of the channel to the cytoplasmic side. It may also be possible to track the pathway of the chloride down this route, which cannot be done for the proton. Thus, insight into the mechanism of BR may also ensue from the studies of halorhodopsin, especially when it is recalled that it is possible that some steps in the BR reaction may be hydroxyl rather than proton migration.

In particular, comparison of halorhodopsin with BR suggests that the deprotonation of the Schiff's base in the latter may reflect movement of hydroxide toward the nitrogen atom rather than of the proton away from this atom of the Schiff's base. Indeed, because it is very difficult to distinguish movement of protons in one direction from that of hydroxides in the opposite, it cannot be rigorously excluded that BR is a hydroxide pump. It is usual to postulate proton movement rather than that of hydroxide because the former has much faster mechanisms for moving through proteins. However, although we can be fairly sure that protons flow through the ATP synthase (Chapter 7) and are taken up and released by the cytochrome bc_1 complex (Chapter 5), there are many cases in which we cannot be so certain.

7

ATP SYNTHASES AND BACTERIAL FLAGELLA ROTARY MOTORS

7.1 INTRODUCTION

Further reading: Muench *et al.* (2011), Walker (2013)

We emphasised in Chapters 1 and 3 that the protein complex we term the ATP synthase can be thought of as a reversible ATP-hydrolysing proton pump (a 'proton-translocating ATPase') whose direction depends on the thermodynamic balance between Δp and ΔG_p (the Gibbs free energy for ATP synthesis). Under most conditions and for most energy-conserving membranes, the balance is in favour of ATP synthesis; however, in Part 3, we discuss conditions in which the ATP synthase in intact mammalian cells reverses, hydrolysing cytoplasmic ATP and generating a Δp (Section 10.8.2); indeed, this is the normal mode for some organisms, such as fermenting bacteria, or trypanosomes in their bloodstream form, that utilise ATP generated from glycolysis to maintain their Δp required for solute transport and other processes.

7.1.1 F$_1$ and F$_o$

Mitochondria, thylakoids, and almost all eubacteria have F-type ATP synthases, also termed F$_o$F$_1$-ATPases, comprising two interlocked subcomplexes, F$_1$ and F$_o$. F$_o$ is an integral membrane complex, but F$_1$ can be released, for example by 8 M urea, as a water-soluble protein capable of catalysing ATP hydrolysis, but not ATP synthesis, in an oligomycin-insensitive manner. Classic electron micrographs of negatively stained inner mitochondrial membranes (IMMs) showed that 10-nm diameter 'knobs' are released from inverted submitochondrial particles (see Figure 1.4) by such procedures. The residual F$_1$-depleted membranes are leaky to protons and cannot maintain a significant respiration-dependent Δp, but a low proton conductance and significant Δp can be restored by addition of oligomycin, showing that F$_o$ (the 'o' stands for oligomycin) can conduct protons across the membrane and indicating that the block of this pathway by oligomycin explains the ability of the inhibitor to prevent both ATP synthesis and hydrolysis. ATP is always hydrolysed or synthesised on the side of the membrane from which F$_1$ projects (the N-phase), whereas during ATP synthesis protons cross

Bioenergetics. Doi: http://dx.doi.org/10.1016/B978-0-12-388425-1.00007-5

from the opposite P-phase (see Figure 1.1). Thus, F_1 faces the mitochondrial matrix, the bacterial cytoplasm and the chloroplast stroma.

High-resolution crystal structures have been obtained for parts of F-type ATP synthases from mitochondria, several bacteria and thylakoid membranes, although crucial components, particularly in F_o, are at the time of writing unresolved. The most extraordinary conclusion, which we develop in this chapter, is that F_o acts as a proton-driven turbine, spinning the eccentric (in the sense of not having a central axis about which it is symmetric) γ subunit that drives conformational changes in the three α,β subunit pairs, but principally in the β subunits, causing them sequentially to bind ADP + P_i, to form tightly bound ATP, and then to release the bound ATP to the N-phase as a result of lowering the affinity for ATP.

7.2 MOLECULAR STRUCTURE

The 370-kDa *Escherichia coli* and mitochondrial F_oF_1-ATP synthases have very similar overall structures, although there is some difference in both nomenclature and exact subunit composition (Figure 7.1). F_1 has three α and three β globular hydrophilic subunits alternating rather like segments of an orange and surrounding the central rotary γ subunit. A further subunit (termed ε for *E. coli* and δ for mitochondria) is involved in the attachment of the γ subunit to F_o, whereas the confusingly named mitochondrial ε subunit has no counterpart in *E. coli*. Both ATP synthases require a means, the stator, to prevent the $\alpha_3\beta_3$ assembly from rotating. In *E. coli*, this 'peripheral stalk' comprises b_2 and δ subunits, the corresponding subunits in mitochondria include the b and OSCP, the oligomycin-sensitivity conferring factor, which is *not* the oligomycin binding site but is required for the mitochondrial F_oF_1 to be sensitive to oligomycin (this will be explained later).

The b_2 subunit (b in mitochondria) bridges F_1 and F_o. F_o has two additional subunits, a and c; the latter is present in multiple copies, creating a largely hydrophobic 'c ring' with 8 members for mammalian mitochondria, 10 for yeast mitochondria and perhaps *E. coli* and 14 for the plant thylakoid F_o. Note, however, that some of these stoichiometries are controversial. F-ATP synthases from other sources have very similar, but not completely identical, subunit structures, although those from mitochondria also possess extra subunits (f, A6L, d, e and g), the functions of which are not all clear, although some are components of the stator. Note that the inhibitor protein for the mitochondrial enzyme (Section 7.7) has no direct counterpart in the bacterial enzyme. The occurrence of differences between ATP synthases of different organisms explains why inhibition by the classic inhibitor oligomycin is essentially restricted to the mitochondrial enzyme and why a candidate reagent against *Mycobacterium tuberculosis* binds to F_o of that organism but not to the animal mitochondrial enzyme. On the other hand, DCCD, which binds to a critical residue in F_o (see later), is a universal inhibitor.

7.2.1 A rotary mechanism

Further reading: Spetzler *et al.* (2009), Adachi *et al.* (2012)

With the exception of bacterial flagellar motors (which we discuss later in this chapter), until relatively recently it had always been assumed that nature avoided rotary

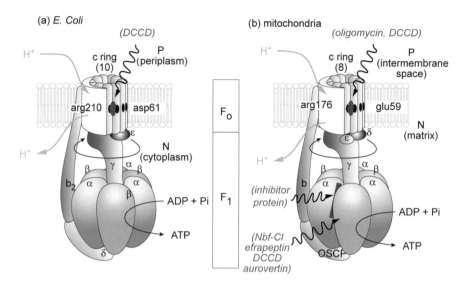

Figure 7.1   Schematic diagrams of the overall structure of F_oF_1-ATP synthases from (a) *E. coli* and (b) mitochondria showing the similarities, equivalencies and differences.

The complexes are shown working in the direction of ATP synthesis. The rotation of the γ chain is shown as anticlockwise when viewed from the N-phase. Diagrams are based on the X-ray diffraction data available for the $\alpha_3\beta_3\gamma\delta\epsilon$ $c_{8/10}$ subunit assemblies of the yeast and bovine heart mitochondrial enzymes. Note that: (1) OSCP in the mitochondrial enzyme is equivalent to the δ subunit in *E. coli*; (2) the mitochondrial ε subunit has no equivalent in bacteria; and (3) for the mitochondrial enzyme, the a and c subunits are sometimes called subunits 6 and 9, respectively. The locations of binding of principal inhibitors of the ATP synthase are also shown. Nbf and DCCD (at relatively high concentrations) bind to one of the three β subunits, aurovertin to two of the three β chains and efrapeptin binds in the central cavity between the three α and three β subunits into which both the C- and the N-terminal ends of the coiled coil γ subunit penetrate. Oligomycin binds to the surface of the c ring and H-bonds to the glutamate side chain essential for proton translocation (Symersky *et al.*, 2012a). DCCD at relatively very low concentrations inhibits by binding to an essential carboxylate group in a c subunit. Only one β subunit is represented as catalysing ATP synthesis, but as discussed in the text, all three sequentially play this role. The F_o 'turbine' comprises a ring of c subunits (8 for mammalian mitochondria, 10 for yeast mitochondria and perhaps *E. coli* and 14 for chloroplasts). This means that between 8 and $14\,H^+$ are required for one rotation of the c ring and hence the synthesis of three ATP molecules. A large c ring decreases the ATP/H^+ stoichiometry but also lowers the Δp required (think about gears on a bicycle). Whereas the b subunit is thought to be dimeric in *E. coli*, it is argued to be monomeric in mitochondria. The inhibitor protein is a dimer that has a coiled coil structure which is believed to be able to bind simultaneously to two F_1 molecules on the same membrane.

mechanisms. However, hypotheses for a cyclic alternating site mechanism for ATP synthesis date back more than 30 years (Cross, 1981) and received strong support from the first high-resolution crystal structures. In turn, the structure suggested that rotation of the γ chain within the $\alpha_3\beta_3$ assembly might be at the heart of the mechanism. But how could one obtain direct confirmation that rotation occurs?

In the first experiment, a bacterial $\alpha_3\beta_3\gamma$ complex was modified to remove all the naturally occurring cysteine residues and one inserted in the γ subunit at a position distant from the $\alpha_3\beta_3$ core. In addition, polyhistidine tags were added to the β chains, enabling the protein to be anchored to a microscope slide that was coated with nickel ions that bind strongly to polyhistidine (Figure 7.2). The single introduced cysteine was used to attach, via a streptavidin linker, a long actin filament (much bigger than the enzyme) to which a fluorescent label had been added. Having assembled this modified ATPase, it was viewed under a fluorescence microscope and ATP-Mg^{2+} was added. The actin filament was observed to rotate in one direction only with 120° steps (Figure 7.2). Later studies using attached gold beads with less viscous drag, or gold bars, combined with high-speed imaging (or, in the latter case, observations with polarised light) showed 90° and 30° substeps, each taking only a fraction of a millisecond and, at saturating ATP, allowing rotation at up to 130 revolutions per second. Rotation in the opposite direction, and observation of concomitant ATP synthesis, was achieved by again binding the $\alpha_3\beta_3$ part (the stator in this type of experiment) of the F$_1$ molecule to the glass, attaching a magnetic bead to the γ subunit, and exploiting magnetic tweezers to rotate it in the direction (i.e., opposite to that for hydrolysis) for ATP synthesis.

Because F$_1$ was removed from the membrane, these experiments could only be used to follow events in F$_1$ and gave no information about events in F$_o$. In the most recent type of experiment (Ernst et al., 2012), E. coli F$_1$F$_o$-ATP synthase was engineered with enhanced green fluorescent protein (EGFP) attached to the b$_2$ subunit of the peripheral stalk, and two fluorescent probes were attached respectively to a c ring subunit and to the γ subunit (Figure 7.2c). The synthase was reconstituted into liposomes, and ATP synthesis was initiated by a K$^+$/valinomycin diffusion potential (Section 2.5).This allowed both γ subunit and c ring rotation to be monitored by Förster resonance energy transfer (FRET), as the distance between the EGFP and the FRET partners oscillated.

Rotation of the c subunit assembly could also be deduced from an experiment with the E. coli enzyme in which the ε subunit was chemically cross-linked to both γ and c subunits. This modified enzyme was still able to make ATP, implying that the cross-linked c–ε–γ unit can rotate.

7.3 F$_1$

Further reading: Walker (2013)

The central γ subunit has two interacting α-helices with N- and C-termini both located within the $\alpha_3\beta_3$ assembly (Figure 7.3). The remainder of the γ subunit extends approximately 30Å towards the membrane and constitutes much of the central stalk seen in early micrographs. The 'foot' (not shown in Figure 7.3) at the membrane end of the γ subunit is important for the interaction of F$_1$ with F$_o$.

The α and β chains have broadly similar structural features, which include an N-terminal domain with six β-strands, a central nucleotide binding domain comprising both α-helices and β-sheets, and a C-terminal domain containing six (β subunit) or seven (α subunit) α-helices (Figure 7.3). A hydrophobic sequence in each α and β

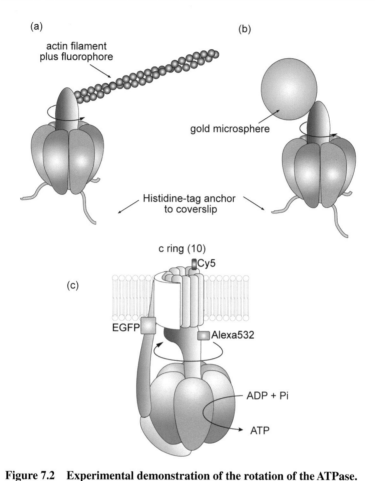

Figure 7.2 Experimental demonstration of the rotation of the ATPase.
The anticlockwise rotation of γ driven by ATP hydrolysis refers to looking
down from the membrane towards the F_1 molecule. (a and b) The enzyme was
immobilised on a microscope slide coated with Ni via his tags attached to the
N-termini of the three β-chains by recombinant DNA methodology. (a) A
fluorescently labelled actin molecule actin filament (1–4 μm in length)
carrying a fluorophore was attached via a streptavidin linker that was bound
via a covalent bond to the thiol of a cysteine introduced at the end of the
γ-chain most distant from the $\alpha_3\beta_3$ assembly. (b) A gold bead, with a diameter
of 40 nm, was linked via bovine serum albumin and streptavidin via the same
engineered cysteine of the γ-chain as in panel (a). The gold bead, being
smaller than the actin filament, permitted higher speed rotation owing to less
viscous drag. A maximum rotation of 130 revolutions per second would
correlate with an ATP hydrolysis rate of 390 per second, similar to the
turnover number (k_{cat}) for the enzyme. The use of the gold bead allowed the
steps in the rotation to be split into 90° and 30° substeps. (c) The *E. coli* ATP
synthase was engineered with the EGFP fluorescent protein at the base of the
b_2 subunit and two FRET partners (Section 7.2.1) attached to the γ-subunit
and a c-ring subunit. As these rotated in the ATP synthesis direction, the
distance from the EGFP varied, affecting the extent of resonance energy
transfer.

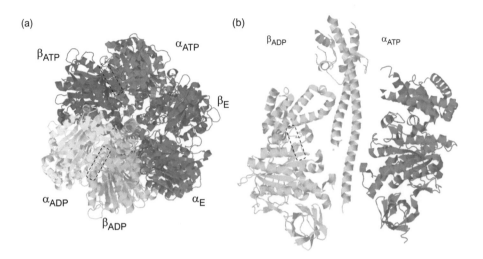

Figure 7.3   The organisation of the α, β and γ chains in the F₁ part of the beef heart ATP synthase.

(a) View from the matrix, showing alternating α and β subunits together with the α-helix of the central rotating γ subunit. The approximate positions of the nucleotide binding sites at the interface between α and β subunits are indicated by the dotted boxes, with ADP + P$_i$ bound between the α$_{ADP}$ and β$_{ADP}$ subunits and ATP between the α$_{ATP}$ and β$_{ATP}$ subunits. Note the open conformation of the 'empty' site between α$_E$ and β$_E$. As the γ subunit is driven anticlockwise (during ATP synthesis), each αβ subunit pair changes its conformation cyclically. Thus, 120° of γ subunit rotation causes the magenta α$_{ATP}$/β$_{ATP}$ pair to open, forming an α$_E$/β$_E$ pair and releasing ATP. After a further 120° of γ subunit rotation, the magenta pair adopts the α$_{ADP}$/β$_{ADP}$ conformation with bound ADP + P$_i$. (b–d) Transverse views of opposing pairs of subunits: Note that while the conformation of the α subunits is essentially identical, the β subunits undergo considerable conformational changes. *Continued* →

subunit contributes to a hydrophobic 'sleeve' that can be thought of as a 'bearing' guiding the spinning γ subunit.

The sequence motif GXXXXGK(T/S) (the Walker A motif), commonly found in proteins binding ATP, is found in a P-loop ('phosphate-binding loop') region on both α and β subunits, and it contributes to a nucleotide binding site. ADP-Mg or ATP-Mg binding to the α subunits is permanent and nonexchangeable, and it will not be considered further. However, the key to the mechanism of ATP synthesis is the change in conformation of the β subunits' nucleotide binding sites as the subunit is distorted by the γ subunit, which rotates anticlockwise viewed from the N-phase, during ATP synthesis (Figure 7.3).

Figure 7.3 shows a snapshot of F₁ with the γ subunit frozen in a particular rotational orientation. Viewed from the matrix (Figure 7.3a), the β subunit nucleotide binding sites are located on the clockwise edge of the subunit, adjacent to the neighbouring α subunit. Regarding the nomenclature of the subunits, TP (for ATP or ATP analogue bound), DP (for ADP bound), and E (for empty) refer to the catalytic mechanism, which will be explained later. Figures 7.3b–7.3d show cross-sectional views (with the matrix at the top) of the same structure. Note that the α subunits retain essentially the same conformation throughout,

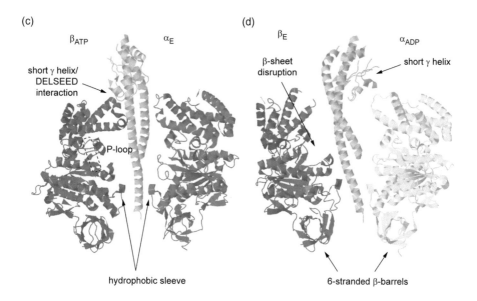

Figure 7.3 *Continued*
In the β_{ATP} to β_E conformational change, interaction between the conserved
'DELSEED' region (see text) and a short (residues 73–90) α-helix of the γ
subunit breaks, and there is disruption of β-sheet structure adjacent to the P
loop region, which interacts with the terminal phosphates on ATP. Two out of
six β-barrels that hold the α and β subunits together are shown. The
hydrophobic sleeve may allow rotation of γ within the central cavity in the
core of the $\alpha_3\beta_3$ assembly. Ground state structures from Abrahams *et al.* (1994)
and Bowler *et al.* (2007), PDB 2JDI. Non-catalytic sites are omitted here. The
structure corresponds to panels (d) and (e) in Figure 7.4.

whereas the β subunits display different conformations depending on the orientation of the
γ subunit. After a further 120° rotation of γ, the magenta α_{TP} and β_{TP} subunits in Figure 7.3
become α_E and β_E, the blue α_E and β_E subunits become α_{DP} and β_{DP}, and so on.

How does rotation of the γ subunit transmit conformational changes to the β subunits? The clearest interaction is between a short stretch of α-helix (18 amino acids)
projecting from the γ subunit (Figure 7.3) and a short stretch of the β subunit known as
the DELSEED sequence (the letters represent the one-letter amino acid code), which is
conserved between ATP synthases from different sources. This interaction is seen in the
β_{ATP} conformation (Figure 7.3c) and may drive the necessary conformational changes in
β. However, the protein can tolerate significant mutation (e.g., substitution of the acidic
residues) in this region without major deleterious effects, but certain deletions adjacent
to the DELSEED loop result in an enzyme that exerts half of the normal torque, thus
implicating an important role for this part of the β chains (Usukura *et al.*, 2012). Three
residues, a His–Gly–Gly sequence close to the P-loop region, have distinct dihedral
angles in the different β subunits and may form an important hinge for permitting conformational transitions between the different β subunits. Consistent with this interpretation, mutation of all three of these residues is very inhibitory.

Note that sufficient conformational energy must be transferred to the β subunit for ATP to be released from its tight binding site. Also remember that F_1, and indeed the entire complex, is reversible; that is, ATP hydrolysis drives the rotation of the γ subunit as in Figure 7.2, and this causes the rotation of the entire γ subunit/c ring assembly in the intact F_1F_o.

7.3.1 The binding change mechanism

After discussing the conformational changes at the molecular level, we now need to move down to the atomic level to understand the actual mechanism of ATP synthesis and hydrolysis by F_1. The β subunits exhibit cooperativity; in other words, the three separate active sites on the three β subunits cannot function independently. For example, the reagent Nbf-Cl reacts specifically with the hydroxyl group of one tyrosine residue of one β subunit adjacent to the P-loop with the Walker A motif, and this completely inactivates the enzyme. Similarly, DCCD (which also reacts with F_o) totally inhibits F_1 by binding to a single glutamate of one β subunit, which has no counterpart on the α chain. This glutamate is implicated in polarising water for attack on the γ phosphate of ATP to facilitate the ATP hydrolysis reaction. In the ATP synthesis direction, it presumably facilitates the loss of water from phosphate as the bond between P_i and ADP is formed to give ATP. A catalytic role for the β chains was supported by the use of chemically reactive derivatives of ATP or ADP that bind covalently on exposure to UV light. These were found selectively bound to β chains after separation of the subunits.

A surprising feature of the mechanism of ATP synthesis is that energy is not required to synthesise enzyme-bound ATP from bound ADP and P_i. This does not contradict the laws of thermodynamics because the ATP is very tightly bound and energy is required to change the conformation of this binding site allowing the nucleotide to be released. The evidence for this came from experiments that showed that the hydrolysis of bound ATP to bound ADP + P_i by F_1 or by submitochondrial particles in the absence of a proton-motive force was to some extent reversible. In one such experiment, the hydrolysis by isolated F_1 of ATP labelled on its γ oxygen with ^{18}O was studied. Hydrolysis, of course, involves the addition of water, and the result of a simple hydrolysis is that one of the oxygen atoms in the P_i that is produced will be normal ^{16}O from H_2O. All four oxygen atoms on P_i are equivalent, and if the reaction is reversible while the ADP and P_i are still bound at the site, then, if the P_i can reorientate at the site, a random oxygen atom from P_i will be lost to water during resynthesis and then replaced by ^{16}O during subsequent rehydrolysis. Eventually, the P_i will be released and the deficit of ^{18}O can be analysed. The experiment was performed using extremely low concentrations of labelled ATP (~10^{-8}M) both to increase sensitivity and to confirm the extraordinarily tight binding of the substrate ATP. The content of ^{16}O label in P_i was more than the one per ATP expected from an irreversible hydrolysis reaction, indicating that exchange had occurred within a catalytic site. This conclusion could be reinforced by similar experiments in which excess loss of ^{18}O originally in P_i was observed in ATP synthesised during steady-state oxidative phosphorylation.

We now have to convince the reader that this is not contrary to the laws of thermodynamics. An equilibrium constant is equal to the ratio of the forward and reverse rate constants. Because the K' for ATP hydrolysis in free solution under physiologically relevant conditions is approximately 10^5M (Section 3.2.1), the reverse reaction should be undetectably slow. Tight binding to F_1 therefore appears to alter the equilibrium constant for

ATP hydrolysis to make the rate of the reverse reaction significant. How can this occur without contradicting the first law of thermodynamics, since we appear to be making ATP with no energy input? The answer is that we are not making *free* ATP but, rather, ATP tightly bound to F_1—that is, at the bottom of an energy well. Confirmation of this model has come from measurements of the dissociation constant for the ATP from F_1 in the absence of Δp; a value of approximately 10^{-12} M was obtained.

Consider the ΔG values associated with the following steps:

$$ADP_{free} \rightleftharpoons ADP_{bound} \quad \Delta G_1$$

$$P_{i\ free} \rightleftharpoons P_{i\ bound} \quad \Delta G_2$$

$$ADP_{bound} + P_{i\ bound} \rightleftharpoons ATP_{bound} \quad \Delta G_3$$

$$ATP_{bound} \rightleftharpoons ATP_{free} \quad \Delta G_4$$

The overall reaction is the sum of these steps:

$$ADP_{free} + P_{i\ free} \rightleftharpoons ATP_{free} \quad \Sigma \Delta G = +40\ kJ\,mol^{-1*}$$

*A typical value for the mitochondrial matrix

What is being observed in the ^{18}O exchange reaction is not the overall reaction but, rather, reaction 3, which occurs with a ΔG close to zero. Nearly all the input of $+40\,kJ\,mol^{-1}$ is required for the final step, the removal of very tightly bound ATP from the catalytic site. It is the conformational change driven by the protonmotive force that releases the bound ATP.

With very low concentrations of ATP ($<10^{-10}$ M), labelled with ^{32}P on the γ-phosphate, hydrolysis proceeds very slowly. If, however, a higher concentration of cold ATP is subsequently added in a chase experiment, the rate of hydrolysis of the already bound ATP-$\gamma^{32}P$ considerably increases. It appears that the higher concentration of ATP occupies one or more lower affinity ATP binding sites on different β subunits of the enzyme and that this causes a conformational change allowing release of the products at an accelerated rate. Thus, there are site-to-site interactions, mediated through subunit interfaces. This negative cooperativity of binding, but positive cooperativity of catalysis, explains why the ^{18}O exchange experiment discussed previously must be performed at very low ATP concentrations: at higher concentrations of ATP, the ADP formed would be released rather than remaining, to allow the reverse reaction (and thus $^{18}O/^{16}O$ exchanges) to occur.

Assuming the molecular basis of ATP synthesis is the reverse of the F_1 hydrolysis discussed previously, it follows that the major ΔG changes are associated with the binding of ADP and P_i and/or the release of ATP, and that these are the steps that must in some way be coupled to Δp through a conformational change. A crucial experiment using submitochondrial particles involved loading the very high-affinity ATP binding site on F_1 in the absence of a Δp and then initiating respiration. The generated Δp caused a release of this tightly bound ATP, clearly demonstrating that Δp decreases the binding affinity of ATP. The change in binding affinity has to be dramatic: from a value of approximately 10^{-12} M in the absence of Δp to a sufficiently loose binding such that ATP can dissociate in the presence of the normal N-phase concentration of the nucleotide. Experiments with a bacterial vesicle system and submitochondrial particles have suggested that during ATP synthesis, the binding of ATP to catalytic site has a dissociation constant in the range of 10^{-6} to 10^{-5} M, a change of 10^6- to 10^7-fold induced by the protonmotive force. This has been deduced in part from the observation that ATP is a very weak product inhibitor of ATP synthesis. The Δp-induced change in affinity of ATP also, via rotation of γ, explains why the ATP analogue, AMP-PNP, is a powerful inhibitor of ATP hydrolysis in the absence of Δp but an extremely weak inhibitor of ATP synthesis.

The combination of the cooperative properties and the evidence for bound ATP formation from ADP and P_i at an active site led to the formulation of the 'binding change' mechanism in which each β subunit would be in a different conformation at any given instant, the three conformations reflecting different affinities for ATP and ADP plus P_i. Thus, the three catalytic sites for nucleotide exist in O (open), L (loose) and T (tight) conformations, and the Δp-induced conformational change causes a T-site with bound ATP to become an O-site, and thus release its bound ATP, while at the same time causing a second site to change from an L-site, with loosely bound ADP and P_i, to a T-site, where the substrates are tightly bound, allowing bound ATP to be formed (Figure 7.4). Thus, each of these catalytic sites has at any instant a distinct conformation, but all the sites pass sequentially through the same set of at least three conformations.

7.3.2 Conformational changes at the catalytic site during ATP hydrolysis

Further reading: Menz *et al.* (2001), Rees *et al.* (2012)

The F_1 β subunit structures discussed to date represent 'ground states' after completion of conformational changes associated with hydrolysis of ATP. To obtain some insight into intermediate states, and hence the catalytic mechanism at the active site, a structure has been obtained in the presence of an Mg^{2+} chelator (the 'F_1-PH' structure), which enables an active site to be detected that, during ATP hydrolysis, has lost P_i and Mg^{2+} but still retains ADP. A second structure in the presence of Mg.ADP and AlF_4^- (termed F_1-TS because ADP.AlF$_4$ is regarded as a transition state analogue for ATP hydrolysis) is consistent with a post-hydrolysis but pre-release conformation; note that AlF_4^- substitutes for P_i but is not released. Putting these conformations together with the ground-state structures enables a plausible cycle to be devised (Figure 7.5). In the direction of ATP hydrolysis, shown in Figure 7.5 with black arrows, and starting with the empty ground state (β_E-GS), binding of Mg.ATP induces a conformational change that drives a 120° clockwise (viewed from the matrix) rotation of the γ subunit generating the next ground state, β_{TP}-GS. Because this is probably the step with the greatest free energy change, it follows that it would provide the majority of the torque for the rotation.

A major feature of β_{TP}-GS is the formation of a hydrophobic pocket with the adenosine of ATP sandwiched between the aromatic rings of Tyr345 (Y345) and Phe424 (F424). The Mg^{2+} (green dot in Figure 7.5 but hidden in the β_{TP}-GS conformation) is coordinated by the nucleotide's β- and γ-phosphates, three hydrating water molecules (red dots) and Trp163 (T163). In addition, one arginine residue from the adjacent α subunit (αR373) brings its positive charge close to the γ-phosphate of ATP. This extremely stable binding is consistent with a dissociation constant of approximately 10^{-12} M. A further 120° clockwise rotation of the γ subunit driven by the E–TP transition of the

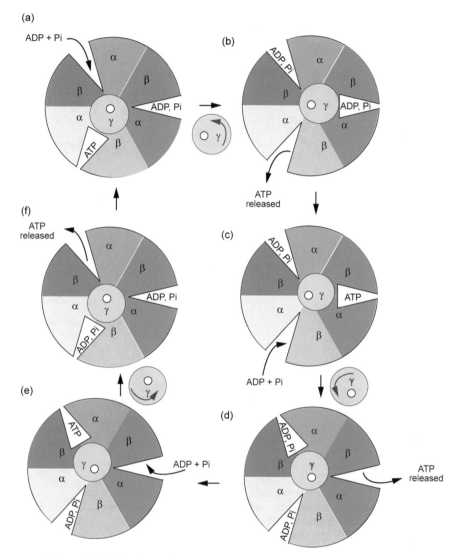

Figure 7.4   The three-site alternating binding site mechanism for ATP synthase.

This diagram shows the consequences of the anticlockwise rotation of the γ subunit, looking from below F_1 and towards the bilayer during ATP synthesis. The colour coding corresponds to Figure 7.3. Starting from the top left (a), the scheme shows binding of ADP and P_i to an empty site (β_E) while ADP and P_i are already bound at a second site. Tightly bound ATP is at the third site, β_{ATP}. Rotation of γ causes conformational changes in β_{ATP}, releasing ATP and generating the β_E conformation. Concomitantly, the original β_E becomes β_{ADP} and the initial β_{ADP} becomes β_{ATP}. Thus, in panel (c), the structure has returned to the original (top left) conformation, except that three conformations have migrated around the ring of α and β subunits. Repeat of these steps releases a further two ATP molecules and returns the structural arrangement to the original (top left).

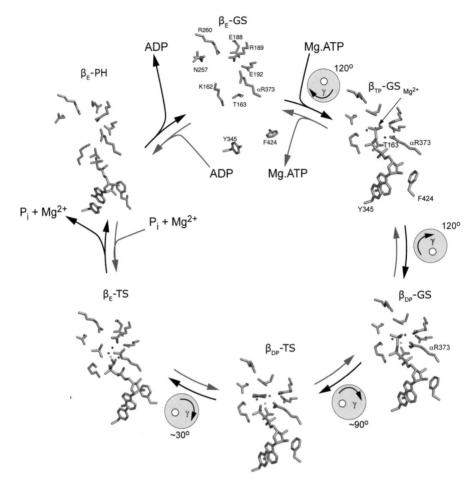

Figure 7.5   Conformational changes deduced from different crystal structures in one catalytic site of bovine F_1-ATP synthase during ATP hydrolysis and, by extrapolation, ATP synthesis.
The cycle for ATP hydrolysis is shown with black arrows and that for ATP synthesis with blue arrows. The sequence is taken from Rees *et al.* (2012). The cycles start at the top with the β_E-GS conformation and proceed clockwise (hydrolysis) or anticlockwise (synthesis). The probable rotation steps of the γ subunit during hydrolysis are illustrated. Mg^{2+} is shown in green (note that it is obscured in β_{TP}-GS), waters in red and the nucleotide in grey. The hydrolysis cycle is described in the text.
Adapted from Rees et al. (2012), with permission.

adjacent β subunit gives the β_{DP} ground-state conformation. The major changes here are that αR373 and the ordered water molecules move still closer to the γ-phosphate, allowing nucleophilic attack and hydrolysis of the β–γ bond.

The $Mg.ADP\text{-}AlF_4^-$ (F_1-TS) structure has nucleotides bound to all three catalytic sites, and it may represent an intermediate conformation consistent with a partial rotation of the γ subunit, driven by the E–TP transition of the third β subunit. The post-hydrolysis, pre-release conformation of the β_{DP}-TS subunit is similar to the β_{DP}-GS.

Completion of the γ subunit rotation generates the β_E-TS conformation. Unlike the β_E-GS structure, the β_E-TS site still retains ADP and the P_i analogue AlF_4^-. The second β_E state, β_E-PH, allows release of P_i. The coordination of the Mg^{2+} is greatly weakened from β_E-TS to β_E-PH, allowing its release. Finally, the C3 α-helix (residues 418–426 in the beta subunit) swings out, allowing release of the weakly bound ADP and regenerating the β_E ground state. This is a notable mechanism because as water attacks the γ-phosphate, the initial leaving group at the active site would be the Mg.ADP complex, which could itself—as in many enzymes—using ATP, dissociate from the active site. Conformational changes leading to release first of Mg^{2+} and then of ADP are seen in some other energy-transducing ATPases or GTPases. By extrapolation in the direction of ATP synthesis, the initial binding would be of ADP, followed by Mg^{2+}, and these discrete steps would be important in the overall energetics of ATP synthesis.

7.4 THE PERIPHERAL STALK OR STATOR

The peripheral stalk (Figure 7.1) plays at least a dual role in anchoring the $\alpha_3\beta_3$ assembly to a subunit of F_o in the membrane and preventing the $\alpha_3\beta_3$ assembly from rotating. The mitochondrial stalk is made up of a long kinked α-helix from the b subunit, augmented by helical stretches from the minor F_6 and d subunits, and the oligomycin sensitivity-conferring protein OSCP, whose C-terminal domain associates with the C-terminal of the b subunit and F_6. The OSCP N-terminal is non-covalently attached to one of the α subunits and in effect sits in a dimple at the top of the $\alpha_3\beta_3$ assembly. There is believed to be some flexibility within OSCP that could permit slight torsion to accommodate the mismatch between the rotational steps in F_o (36° in the case of a 10 c subunit structure) and F_1 (multiples of 30°; Figure 7.5), although flexibility in the γ subunit is believed to be more significant (discussed later). The b subunit is probably largely a stiff structure, but it is possible that it plays a part as an elastic element in storing energy from the rotation of the c assembly until it is transduced to rotation of γ. The structures of the long helix and OSCP fit well into the overall envelope established by separate low-resolution studies of the enzyme by cryoelectron microscopy. We can now explain why OSCP is not the site of oligomycin binding (which is actually in F_o, some distance from OSCP; Figure 7.2). The absence of the peripheral stalk, as a result of loss of OSCP, will presumably allow ATP hydrolysis to drive rotation of the $\alpha_3\beta_3$ unit about the γ subunit even if the latter is immobilised by oligomycin bound to the c ring in F_o. ATP synthesis is impossible in the absence of OSCP, irrespective of the presence of oligomycin, because the integrity of the stator is destroyed. The bacterial peripheral stalk contains the δ subunit, analogous to the OSCP, although the b subunit is dimeric.

7.5 F_o

The analogy is frequently made between the F_oF_1-ATP synthase and a hydroelectric power station, with F_1 corresponding to the generator and F_o to the water turbine, linked by the γ subunit shaft (Figure 7.1). F_o comprises the two transmembrane helices of subunit b

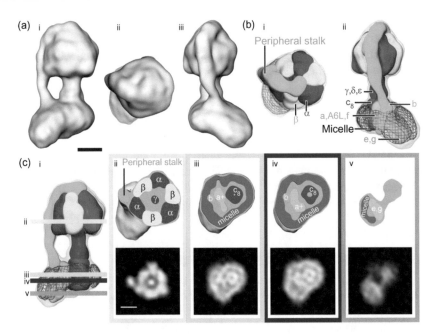

Figure 7.6 Electron cryomicroscopy of bovine F_oF_1-ATP synthase at 18-Å resolution.
(a) Three surface views. F_1 is at the top of the image in i and iii. A view from the matrix is shown in ii. (b) Views after segmenting individual polypeptides. The intact surface is shown as a transparent grey surface. The detergent micelle required for stability is also shown. (c) Slices through the structure. Lower images show actual electron micrograph images, whereas upper images show slices through the segmented reconstruction. (ii) In F_1, the α and β subunits surround the γ subunit, with the peripheral stalk running along an α/β interface. (iii) Near the middle of the membrane region, subunit a contacts the c_8 ring. (iv) Further toward the intermembrane space, the a subunit is separate from the c_8 ring. (v) At the intermembrane space side of the membrane region, the e and g subunits can be seen running along the detergent. Scale bars = 50Å.
Reproduced from Baker et al. (2012).

anchoring the peripheral stalk, an a subunit, multiple copies of identical c subunits and e and g subunits. In contrast to F_1, high-resolution structural information is incomplete; although crystal structures are available for assemblies of c subunits, the full F_oF_1 assembly has to date only been visualised at lower resolution (~18Å) by electron cryomicroscopy (Figure 7.6). However, this resolution is sufficient to distinguish the gross structure, with a ring of c subunits, an adjacent a subunit making only limited contact with the c ring and the membrane domain of the b subunit. Higher resolution for an A-type synthase has been observed (Section 7.8) and supports the conclusions drawn from the 18-Å map.

7.5.1 The c ring

Isolated rings of c subunits have been crystallised and possess two transmembrane helices, arranged as a hairpin. The *in situ* ring is arranged with each N-terminal helix on the inside, the C-terminal on the outside, and with the loop between the two helices on the N-side of the membrane. There is evidence that the interior of the rings is filled with phospholipid. The number of c subunits in the ring defines the 'gearing' of the F_1F_o-ATP synthase—that is, the number of protons required for a 360° rotation of the γ subunit and hence the synthesis of 3 ATP molecules (Section 7.6). Careful counting of electron densities indicates that all animals possess 8 subunits; yeast and other lower eukaryotes have 10, as perhaps does *E. coli*; thylakoids have 14; and between 11 and 15 have been observed in other organisms. Although electron microscopy (or eventually, it is hoped, a high-resolution crystal structure) of intact F_oF_1 is the least ambiguous method for determining stoichiometry, it is also possible to visualise self-assembled rings of puri-fied c subunits; however, the assumption is that these will always correspond to the *in situ* rings. In the case of yeast, bulky conserved residues seem to preclude a smaller ring than 10. Apart from a limited contact with the a subunit, the c ring is surrounded with phospholipid; thus, there is no 'corset' limiting ring stoichiometry. Instead, it seems to be determined by the subunit structure, despite the high sequence similarity between c rings from different sources. The important presence of a critical small alanine side chain in a key position in the mitochondrial enzyme is a determinant of ring size, whereas in bacte-rial enzymes, deviations from a conserved stretch of glycines (GxGxGxGxG) have been shown to allow stoichiometry to vary from less than 11 to greater than 14. An interest-ing aside here is the problem of alkaliphilic bacteria, in which the protonmotive force is very small (the pH gradient is alkaline outside and the membrane potential does not fully compensate). An unusually high number of c subunits might explain how these organ-isms function.

7.5.2 c ring rotation

Further reading: von Ballmoos *et al.* (2004), Symersky *et al.* (2012a, b)

A key to the mechanism of c ring torque generation came from early studies with the covalent inhibitor dicyclodihexylcarbodiimide (DCCD). DCCD at extremely low con-centrations will specifically label c subunits in mitochondrial, thylakoid, or bacterial membranes, reacting with the carboxylate side chain of a conserved aspartate (Asp61 in *E. coli*) or glutamate (Glu59 in mitochondria and Glu61 in thylakoids) in the subunit's outer (C-terminal) helix. As can be seen from the space-filling model in Figure 7.7, these are the only hydrophilic residues in a uniformly hydrophobic cylinder. Incorporation of approximately one DCCD molecule per c ring is sufficient to completely inhibit activ-ity, strongly suggesting that the assembly of c subunits works cooperatively and not independently.

Whereas protons are ubiquitous, sodium ions can be easily detected in spe-cific locations by X-ray crystallography. The Na^+-dependent F_1F_o-ATP synthase of *Propionigenium modestum* (Section 5.13.10) appears to function by a closely analo-gous mechanism to the proton-dependent enzyme. The reaction of the c subunit aspar-tate with DCCD is blocked by the presence of Na^+, suggesting that the translocated ion comes into close proximity to the carboxylate side chain. A crystal structure of a c ring from a sodium ATP synthase shows that the sodium coordination sphere, provided by the interface of two c subunits, is full; one ligand is the equivalent of Glu59.

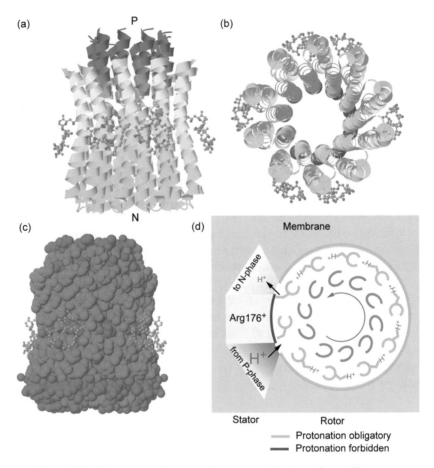

Figure 7.7   Structure and proposed rotary mechanism of yeast F_o.
Structure of a 10-membered c ring from *S. cerevisiae* showing bound oligomycin
molecules: (a) plane of membrane; (b) viewed from the N-face. 'Rainbow' view
(N-terminals shown in blue and C-terminals in red). (c) Corresponding space-
filling view; hydrophobic surfaces are shown in red and hydrophilic surfaces in
blue. Note the central ring of hydrophilic but presumed protonated Glu59 side
chains; from PDB 4F4S (Symersky *et al.*, 2012a). (d) Proposed rotary mechanism
viewed from the matrix; adapted from Symersky *et al.* (2012b). The stator subunit
'a' contains two half-channels for protons, connecting the rotor to the P- and
N-phases, respectively, and a critical arginine residue 176. Each c subunit Glu59
(green) on the C-terminal outer helix can be reversibly protonated. (i) The
positively charged Arg176 side chain in proximity to this negatively charged
Glu59 lowers the pK_a of the latter, meaning that it requires an effective pH value
perhaps as low as 2 so as to match the low pK_a value (the glutamate side chain
'normally' has a pK_a of around 4) for protons delivered via the P-phase half-
channel for it to be protonated. (ii) Protonation releases the attraction between
Arg176 and Glu59, causing the c ring to rotate one step and the Arg176 to change
conformation as shown. The pK_a of Glu59 increases probably by 3 or 4 units
due to the removal of the arginine, but it cannot deprotonate when in contact with
the hydrophobic core of the membrane because it is in a 'closed' conformation.
(iii) When protonated Glu59 reaches the N-phase half-channel, it releases its
proton, which is discharged to the N-phase.

Trialkyl tin compounds are also potent inhibitors of F-type ATP synthases. Studies of a sodium-translocating enzyme, analogous to that in *P. modestum*, have shown that this type of inhibitor binds to the a subunit and that this binding is again attenuated by Na^+, implying that the a subunit provides at least part of a pathway for Na^+, or by analogy H^+, across the membrane. The exact binding site for oligomycin is also known (Figure 7.7) and in yeast involves bridging between two c subunits and hydrogen bonding with the crucial Glu59 residue.

Based on the critical role of the Glu59 (or Asp61 residue in the case of *E. coli*), four properties are required for proton re-entry to drive c ring rotation and hence ATP synthesis. First, there must be a pathway from the P-phase that can deliver protons to this residue; second, there must be a mechanism by which the free energy of Δp is transduced into torque in the c-ring. Third, there must be an additional pathway to accept protons from the ring and deliver them to the N-phase. Finally, there must be some mechanism to confer directionality on the rotation; however, the process must be fully reversible to account for proton pumping during ATP hydrolysis.

Although the lack of high-resolution structures is a limitation, current evidence favours the presence of two distinct proton half-channels within the a subunit: one (the P half-channel) delivering protons to the c ring acidic amino acid and the second (the N half-channel) receiving protons from c subunits further round the ring and delivering them to the N-phase (Figure 7.7). At this stage, we need to clarify how the ATP synthase can be driven by either the $\Delta\psi$ component (which dominates in mitochondria and most bacteria) or ΔpH (for the thylakoid complex). The critical c ring acidic amino acids are located roughly in the midpoint of the membrane. It is therefore reasonable to assume that the protons experience a 50% drop in $\Delta\psi$ on reaching this point via the P half-channel. This energy is not lost but instead can be thought of as leading to an increase in local proton concentration (i.e., decreased pH), although there is no evidence for an aqueous vestibule in which protons can accumulate. If we assume no bulk ΔpH, a membrane potential of 180 mV, and a P-phase pH of 7, then the local pH at the site of injection into the c ring after the protons have dropped through 90 mV could be 7–90/60 = 5.5 (where the '60' term is the approximate factor that allows millivolts to be converted to ΔpH at 30°). Conversely, at the start—again approximately in the middle of the membrane—of the N half-channel, protons will be depleted as they are driven to the N-phase by the residual 90 mV of $\Delta\psi$. This could raise the formal local pH there to 7 + 90/60 = 8.5. Thus, a Δp of 180 mV could be seen by the F_o rotor as a 3 pH unit gradient, regardless of whether bulk electrochemical gradient was in the form of a membrane potential or a pH gradient or any intermediate combination. Suggestions that ΔpH was not equivalent to $\Delta\psi$ have been found to relate to the interaction of F_o with F_1 and not to F_o proper.

7.5.3 Mechanisms of torque generation

Further reading: Feniouk *et al.* (2004), Junge *et al.* (2009), Ishmukhametov *et al.* (2010), Toyabe *et al.* (2011), Saroussi *et al.* (2012)

In the direction of ATP hydrolysis, the torque for rotation of the γ subunit/c ring assembly is generated in F_1. Amazingly, it has been calculated that this transduction takes place at virtually 100% thermodynamic efficiency in the sense of power out divided by power in. For ATP synthesis, in contrast, it is F_o that must generate torque. Remember that the ATP synthase under physiological conditions operates rather close to equilibrium. In other words, referring to Figure 7.7, the anticlockwise (viewed from the matrix) torque generated by proton movement from P-side to N-side in the forward direction of ATP synthesis is continuously opposed by the clockwise torque from F_1 trying to hydrolyse ATP. Which process wins depends purely on the thermodynamic balance between Δp and ΔG_p, taking account of the 'gearing'—the number of c subunits in the ring. This is discussed later.

Models for F_o torque generation focus on the interaction between the critical Asp or Glu side chains in the c ring and the positive charge on an essential arginine residue in the a subunit (Arg210 in *E. coli* and Arg176 in mitochondria). There is a general consensus that in the direction of ATP synthesis, protons are delivered by the P-channel adjacent to the critical Glu59 (mitochondria) or Asp61 (*E. coli*) of the c subunit whose carboxylate anion (e.g., Glu59 or Asp61) forms a salt bridge to the positive charged arginine on the a subunit. The arrival of the proton compensates for the loss of electrostatic interaction between the positive arginine and negative glutamate (aspartate) and allows the ring to rotate by one subunit per proton bound; the glutamate remains protonated as the c ring continues to rotate in the hydrophobic core of the membrane. This incremental rotation means that a protonated c subunit will move towards the arginine residue, and loss of a proton will be compensated by the formation of a salt bridge between carboxylate and arginine. Thus, when the protonated amino acid meets the N half-channel, the proton is unloaded, and repeating this sequence of events *n* times, where *n* is the number of c subunits, results in the c subunit completing a circuit (Figure 7.7). This is the basis of a 'Brownian ratchet' mechanism, in which random Brownian motion is proposed to allow a step rotation, with the intervention of another residue on the a subunit (a 'ratchet' or 'leash') to prevent random reversal, although of course the direction must reverse during ATP hydrolysis.

In this simple form, this mechanism seems to be a version of 'Maxwell's demon,' and it is not immediately apparent how this develops sufficient torque to oppose the counter-torque from F_1. Maxwell's demon, by the way, sat astride a tiny hole between two compartments, only letting gas molecules pass in one direction, building up a pressure difference with no apparent energy input (for those who want the original description, see Maxwell's *Theory of Heat*, published in 1871, p. 308). In Figure 5.9, we reviewed the generic properties required of a redox-driven proton pump and concluded that it is necessary to coordinate redox, conformational and proton binding affinity changes. In the case of F_o, the key requirement is to transduce the energy available from the effective nearly 1000-fold difference in proton concentration between the input and output sites of the c ring into torque with high efficiency.

We have seen that in the case of bacteriorhodopsin (Section 6.5), conformational changes in proteins can be driven by changes in pK and vice versa. It is perhaps simplest to discuss F_o when, together with F_1, it is in its ATP-hydrolysing, proton-pumping mode, generating and maintaining a 3-unit pH gradient (or equivalent $\Delta\psi$) between the N- and P-phases. F_o is now spinning clockwise viewed from the matrix. Glu59 binds a proton from the N-channel, even though the local pH can be 8.5. This demands a comparably high pK for the glutamate carboxyl. After rotating through almost a full revolution, the same carboxyl must unload its proton to the P-channel, where the local pH may be 5.5. Consistent with this, one model (Cherepanov *et al.*, 1999) estimated a pK

of 6.4 at the P-channel injection site and 8.3 at the N-channel. The critical arginine residue (210 in *E. coli* and 176 in yeast) may lower the pK for the Asp or Glu carboxylate side chain at the P-channel, favouring an electrostatic interaction with the deprotonated state, which would make protonation more difficult.

It has to be asked if the c ring should be thought of as a rigid cylinder even though the interior must be filled with phospholipds. There is evidence that it undergoes multiple modes of elastic deformation that are likely to be involved in torque generation (Saroussi *et al.*, 2012), although crystal structures suggest a relatively rigid tightly packed assembly of c subunits. There is a high-resolution structure of c ring from an H^+ translocating ATP synthase, and it can be seen that the critical carboxylate is held between two subunits, where it can be deduced to be stabilised in the protonated form (Pogoryelov *et al.*, 2009).

The rotation of the c ring does not depend on the presence of F_1, something that can be deduced by the flow of protons through the membranes of submitochondrial particles from which F_1 has been stripped (Section 7.1.1). Thus, the rotation of the c subunit assembly does not depend on its interaction with γ or ε. It has also been shown with experiments with such stripped membranes that the rate of proton movement by a c ring is a nonlinear function of the Δp, which may account, at least in part, for why the rate of ATP synthesis by intact membranes declines sharply upon even moderate attenuation of Δp.

For those ATP synthases that are driven by a sodium electrochemical gradient, a very similar description for the generation of torque can be given as described previously for protons. Notably, there is a crystal structure for a c ring that shows how the critical carboxylate residue helps occlude sodium ions (directly visible unlike protons by X-ray crystallography) between neighbouring c subunits and therefore from contact with the hydrocarbon bilayer.

7.6 THE STRUCTURAL BASIS FOR H$^+$/ATP STOICHIOMETRY

Further reading: Ferguson (2000), Watt *et al.* (2010), Pogoryelov *et al.* (2012)

One complete rotation of the c ring causes a complete rotation of the $\alpha_3\beta_3$ unit and thus the formation of three molecules of ATP. The models we have discussed suppose that a complete 360° rotation of an n subunit c ring will require n protons. Thus, the mammalian F_o with 8 subunits will have an H$^+$/ATP ratio of 8/3 (i.e., 2.67); for yeast mitochondria ($n = 10$), the ratio will be 3.33; and for the 14 subunits of the thylakoid enzyme (deduced from atomic force microscopy studies on isolated assemblies of c subunits), the ratio will be 4.67. Assuming the validity of these assumptions, this approach leads to H$^+$/ATP ratios more accurate than those estimated from kinetic or thermodynamic approaches. Note that these ratios are not integers. This means that there must be some torsional flexibility between F_o and F_1 to allow both components to 'click' into their individual rotatory steps. The peripheral stalk is contemplated to accomplish this (see Section 7.8), but some elastic flexibility in the γ subunit is more probable.

The c ring size can be thought of as analogous to the gears on the back wheel of a bicycle: moving the chain to a larger ring lowers the effort (decreases the Δp required to maintain a given ATP/ADP ratio) but increases the amount of pedalling (increases the proton current required for a given rate of ATP synthesis). The small ring (high gearing) of the mammalian mitochondrion may have evolved because the adenine nucleotide exchange and phosphate transport across the IMM use an additional proton per ATP (Section 9.5.1), raising the H^+/ATP ratio for synthesis and export to the cytoplasm from 8/3 to 11/3, or 3.67. Nevertheless, if thylakoid membranes have the same Δp as for mitochondrial membranes, then from Section 3.6.2, the maximal ΔG of ATP synthesis will be higher for thylakoids. Currently, this variability in H^+/ATP has not been fully explained, but it is important to note that the lower the H^+/ATP, the higher the P/O for mitochondria but the lower the maximum ΔG of ATP synthesis.

7.7 INHIBITOR PROTEINS

Further reading: Bason *et al.* (2011), Faccenda and Campanella (2012)

Mitochondria contain an ATPase inhibitory factor 1 (IF_1) protein of approximately 10 kDa that is capable of inhibiting ATP synthase selectively in the direction of ATP hydrolysis. This directionality occurs because its interaction is pH dependent, binding under conditions of lowered matrix pH, normally associated with a decreased Δp. However, it has also been suggested that whereas the direction of rotation of γ in the ATP hydrolysis direction favours binding of IF_1, rotation in the (opposite) ATP synthesis direction favours dissociation of this inhibitor protein. This explains why *in vitro* experiments have shown inhibition of ATP hydrolysis but not of ATP synthesis under similar conditions of pH and ionic strength. A monomeric fragment of IF_1 binds in a cleft between α_{DP} and β_{DP} subunits, but it is thought that IF_1 initially binds in the more open α_E/β_E cleft, with the complex jamming at the DP conformation. The complete protein is active as a homodimer, and it is (controversially) also implicated in dimerization of the complex in the mitochondrial membrane (Section 10.2). The physiological roles of the protein are discussed in Section 10.8.2.

The regulation in bacteria is less well understood. In the case of *Paracoccus denitrificans* and many other organisms in which respiration, and hence oxidative phosphorylation, is obligatory, ATP hydrolysis and an ATP–Pi exchange reaction are both very feeble in the absence and sometimes even in the presence of a Δp. There is evidence that these bacteria possess a distinct inhibitor protein. However, for the *E. coli* enzyme, there is evidence that the ε subunit might confer this unidirectional behaviour by acting like a ratchet. It appears to take up two very different conformations. In one of these, the C-terminal domain extends towards the F_1 part and allows the enzyme only to synthesise ATP. An interaction of this domain with one αβ pair might be broken only by rotation of the $\alpha_3\beta_3$ assembly in the direction associated with ATP synthesis; this would be a ratchet-like action.

In thylakoids, an essential requirement is to avoid ATP hydrolysis in the dark. In the light, activation of the ATP synthase follows exposure of a disulfide bridge in the γ subunit and its reduction by a thioredoxin. This bridge is near the interface with F_o and can be modelled into the mitochondrial enzyme structure. The thioredoxin is reduced in turn by ferredoxin and the activity of photosystem I. In the dark, ATP synthase relaxes back into an inactive state because reduced ferredoxin and thioredoxin are no longer formed and the disulfide bridge, formed between cysteines six residues apart in the sequence, regenerates under the more oxidising conditions. Indeed, it has been possible to induce this thiol sensitivity into an enzyme by incorporating the two cysteine residues by recombinant DNA methodology.

7.8 PROTON TRANSLOCATION BY A-TYPE ATPASES, V-TYPE ATPASES AND PYROPHOSPHATASES

Further reading: Kellosalo *et al.* (2012), Lin *et al.* (2012), Wilson and Rubinstein (2012)

The archaea do not have the F-type ATP synthase, but instead they possess a related enzyme, known as the A type, which is even more complex than the F type. For example, there are two peripheral stalks. This type of enzyme is closely related to the V-type ATPase, which eukaryotes have as proton pumps in vacuoles and intracellular vesicles where the function is to generate pH gradients. It is clear, although the evidence is less extensive than for the F type, that A- and V-type ATPases/synthases function via a rotary mechanism.

There are a small number of examples of eubacteria having an ATP synthase that is not of the F type but is similar to the A-type ATP synthase found in archaea (but which has confusingly often been called a V type when it occurs in eubacteria); it presumably arrived in those organisms via a gene transfer process from archaea. Study of this type of enzyme, particularly that from *Thermus thermophilus*, has given important information that is relevant to understanding the F type, especially with respect to how the a and c subunits of any type of ATP synthase interact. A 9.7-Å cryoelectron microscopy analysis of this enzyme has shown that its equivalent of the c subunit, called L, forms a ring of 12 subunits, with each L subunit having two helices, just as for c but in fact unusual for A types, in which L usually has 4 helices. The equivalent of the a subunit, I, has 8 transmembrane helices that can be seen as forming two clusters each with four helices. These two clusters are in contact with two adjacent L subunits within one ring, with one contact being slightly farther towards the P-side than the other. The other 10L subunits in the ring do not make protein–protein interactions with I and thus are directly in contact with the hydrocarbon phase of the bilayer. It is possible that the two separate clusters of four helices within I are sufficient to provide two 'pores' for protons, one to take protons from the P-phase to the critical arginine-carboxylate salt bridge (there is a conserved and essential arginine in I as in a of the F-type enzymes) and the other to take protons from a protonated carboxylate on L, after rotation, to the N-side. Four helices could conceivably provide an ion pathway through the membrane (but whether the subunit in the F-type enzyme, with only five predicted transmembrane helices in total, could do this remains to be seen). The structural information for the *T. thermophilus* enzyme rules out any model in which protons move across the membrane at an interface between the L ring (or c ring in the F type) and the I subunit (a in the F type).

The *T. thermophilus* enzyme has provided a structure of the stator stalk, which in this case is a heterodimer of the E and G subunits. These two polypeptides form an unusual right-handed coiled coil that is thought to be able to undergo bending and twisting motions that would allow a radial wobbling of the A_3B_3 head group (confusingly, it is the A and not the B subunits that bear catalytic sites in the A-type enzymes) as the catalytic cycle is followed while on the other hand retain the stiffness necessary to oppose the rotation of the central stalk. These observations may be relevant for understanding the function of the stator in the F-type ATP synthases.

The cytoplasmic membranes of some bacterial species, such as *Rhodospirillum rubrum*, contain a proton-translocating pyrophosphatase. The same enzyme is found in vacuolar membranes in some eukaryote cells. It is not clear why either *R. rubrum* or these other membranes also contain a conventional proton-translocating ATPase, but in the case of *R. rubrum*, it seems that illuminated cells can drive the synthesis of ATP and pyrophosphate simultaneously. A structure of a proton-translocating pyrophosphatase and of a sodium-translocating pyrophosphatase from the thermophilic bacterium *Thermotoga maritima* shows that the protein is a homodimer, with each subunit having 16 transmembrane helices. The catalytic site for pyrophosphate is located within a cavity among these helices at the cytoplasmic side of the membrane. From the structure, it can be deduced that the proton or sodium translocation pathway is formed by six core transmembrane helices. Proton or sodium pumping would be initiated by pyrophosphate hydrolysis with protons transported through an Arg, Asp, Lys, Glu pathway (provided by helices 5, 6 and 16) and two bound waters. Conformational changes would sequentially render the protein open to first the cytoplasmic side, then open to neither side, and finally the other side of the membrane (i.e., the periplasm if it were a bacterial enzyme). Reversal of these steps would give Δp-driven pyrophosphate synthesis. Therefore, phosphoanhydride bond formation is achieved in this case by a mechanism similar to bacteriorhodopsin (Section 6.5) and other transport proteins (Chapter 8). It is speculated that pyrophosphate synthesis occurred before the evolution of the rotary ATP synthase.

7.9 BACTERIAL FLAGELLAE

Further reading: Berry and Sowa (2012), Sowa and Berry (2012)

Many, but not all, species of bacteria are motile, usually as a consequence of the rotation of one or more helical flagella that extend from the surface of the cell (Figure 7.8). The basal body of a flagellum is embedded in the cytoplasmic membrane; traverses the periplasm, the peptidoglycan, and the outer membrane; and connects to a filament that extends into the external phase. Movement of the filament, which propels the cell at several body lengths per second, is driven by a 'motor' embedded in the cytoplasmic membrane. The motor rotates at up to 3000 rpm about an axis perpendicular to the plane of the membrane and is usually driven by the protonmotive force, although a Na^+ electrochemical gradient is used in some organisms in which sodium circuits are generally important, such as *Vibrio* species.

The bacterial flagellum was the first rotary electric motor identified in nature, and the first model for generation of rotation was successfully adapted for the F_oF_1-ATP

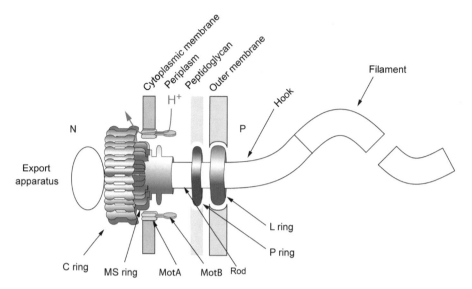

Figure 7.8   Side view of a proton-driven bacterial flagellar motor.
MotA and MotB form stator complexes in an A_4B_2 stoichiometry. The
stoichiometry of the C ring may vary from 32 to 36. The MS (an old term)
ring comprises multiple copies of FliF. The rotor spans the cytoplasmic
membrane, the peptidoglycan cell wall and the outer membrane.
Adapted from Berry and Sowa (2012).

synthase (Chapter 7). Ironically, this model was later shown not to apply to the flagellum. The part of the structure where rotation is generated is known as the C ring (not to be confused with the c ring in ATP synthases). The C ring comprises a ring of approximately 26 copies of the FliG protein, which is at the centre of generating rotation, along with a similarly large number of copies of FliM and FliN. Each molecule of FliG is folded into three domains with a hinged loop, allowing the formation of the ring and conformational changes that allow the motor to switch from counterclockwise to clockwise rotation (discussed later).

In contrast to the c ring of the ATP synthase (discussed previously), the passage of protons or, in certain cases, sodium ions through the flagellum does not involve the binding of these ions to the ring. This means that whereas, for example, 10 protons binding to a c-10 assembly in the ATP synthase will cause a 360° rotation, it should not be expected that 26 protons will drive a 360° rotation of the C ring; indeed, current estimates indicate that for a tightly coupled motor with approximately 26 steps per rotation, in the order of 300 protons are required to drive a full 360° rotation. There is good evidence that in the flagellum, the protons pass down their electrochemical gradient through the MotA and MotB proteins, located in the cytoplasmic membrane, which are effectively both the stators and the torque generators of the motor (it may well turn out that the a subunit of the F_oF_1-ATP synthase functions similarly if it is shown that it provides the passage for ions across the membrane).

Each stator has four copies of MotA that surround two copies of MotB. Approximately 11 such assemblies surround the C ring. MotB Asp32 in *E. coli* (and its equivalent in other organisms) is the only conserved charged residue in MotA or MotB that is essential for function. It is postulated that each stator (Figure 7.8) contains two ion channels, each containing one MotB Asp32 residue. Loss by mutation of proline residues P173 and P222 in *E. coli* MotA severely impairs motor function. Thus, a putative mechanism for the motor is that proton flux coordinates conformational changes in MotA via MotB-Asp32, and that these conformational changes involve motions about the MotA proline residues that lead to an alternating interaction with FliG that generates torque. It is possible that the interactions involve formation and breakage of salt bridges (or other equivalent electrostatic interactions between an aspartate on FliG and an arginine on MotA and between a glutamate on MotA and an arginine on FliG). These interactions will be repeated 11 times around the FliG ring as it interacts with the 11 MotA/B assemblies. As indicated by Figure 7.8, a cytoplasmic loop region on MotA interacts with a part of FliG, which lies immediately below it. This again is in large contrast with ATP synthase, in which the stator–motor interaction point lies in the middle of the membrane. For a single flagellar motor, there will be 11 sets of interactions between sets of MotA/B proteins and FliG polypeptides, but these need not be synchronised. Together, they account for the progressive, but stepping, rotation of FliG.

An important difference between the ATP synthase and the flagellum motor is that whereas in the former, protons moving down their electrochemical gradient, as they pass through the protein, will always drive the same direction of c ring rotation, in the flagellum the direction can be either clockwise or anticlockwise depending on whether the organism is swimming smoothly or undertaking a tumble and change in direction. It is difficult to explain this bidirectionality of rotation because any model in which passage of protons through the stator, and resulting in electrostatic interactions to drive rotation of FliG, would be expected to drive a single direction of rotation. The mechanism underlying the switch in direction is not fully understood but must be related to conformational changes ensuing from the binding to FliM and FliN of the phosphorylated form of CheY, a protein that is central to relaying from sensor proteins to the flagellar the information that indicates a need to switch direction of rotation. In turn, this can influence the hinged loops on FliG. Switching direction of rotation happens very rapidly, and current models suggest a conformational spread mechanism in which there is a switch in direction of rotation when a critical number of rotor proteins have changed conformation.

8 TRANSPORTERS
Structure and Mechanism

8.1 INTRODUCTION

Mitochondria and bacteria continuously exchange metabolites and end products with the cell cytoplasm or external environment. In almost all cases, transport proteins are required for the entry and exit of these molecules. At the same time, the membranes maintain a high Δp for ATP synthesis. Because most metabolites are charged and/or weak acids, it follows that their distribution will be affected by $\Delta\psi$ or ΔpH (Chapter 3). In practice, transport mechanisms are not only designed to operate under the constraints of a high $\Delta\psi$ and/or ΔpH gradient but also may exploit these gradients to drive the accumulation of substrates or the expulsion of products across the membrane. The following are some of the more common strategies:

(a) Proton symport with a neutral species leading to an accumulation driven by the full Δp; for example, the *lac* permease of *E. coli* (Section 8.3.2.1).
(b) Electroneutral proton symport, or equivalent hydroxide antiport, leading to an accumulation driven by ΔpH alone; for example, the mitochondrial P_i^-/H^+ symporter (Sections 8.2.2 and 9.4.1.3).
(c) Electrical uniport of a cation driven by $\Delta\psi$; for example mitochondrial Ca^{2+} accumulation (Sections 8.2.5 and 9.5).
(d) Electroneutral or electrogenic exchange of two metabolites; for example, the mitochondrial ATP^{4-}/ADP^{3-} antiporter (Sections 8.2 and 9.5.1). Such exchange is a common device to allow the entry of a polyanionic species, which would be effectively excluded from the mitochondrial matrix due to the high inside negative membrane potential if transported by a uniport mechanism. Other examples include carriers for $citrate^{3-}$ or $malate^{2-}$ (Section 9.5.3).
(e) Electroneutral antiport of an ion or metabolite with protons; for example, the mitochondrial Na^+/H^+ antiporter, which expels any excess Na^+ that leaks into the matrix in response to the membrane potential.

In general, mitochondria and bacteria have very distinctive transport proteins, and these will be considered separately.

Bioenergetics. Doi: http://dx.doi.org/10.1016/B978-0-12-388425-1.00008-7

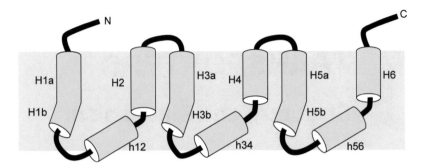

**Figure 8.1   Two-dimensional view of a representative member of the
SLC25A mitochondrial transport protein family.**
The six transmembrane helices are numbered from the N-terminal. Note the
kink in the odd-numbered helices due to the presence of a proline residue.

8.2 THE PRINCIPAL MITOCHONDRIAL TRANSPORT PROTEIN FAMILY

Further reading: Robinson *et al.* (2008), Kunji (2012)

The integration of metabolism between the matrix and cytoplasm requires continuous
two-way traffic across the inner mitochondrial membrane. Although the membrane has
a variety of different transport systems, most are catalysed by a single group of proteins
comprising, in man, the SLC25A family (Section 9.5.3). The nomenclature is somewhat
loose, and the proteins are referred to as transporters, translocators, transport proteins,
or carriers. Sequence analyses, alongside biochemical and structural studies, indicate
that they share the following structural features (Figure 8.1): (1) a single polypeptide
chain in the size range 30–40 kDa; (2) three homologous amino acid repeats within each
protein, including PX[D or E]XX[R or K] and [Y or F][D or E]XX[K or R] consensus motifs, which are respectively present at the matrix and cytoplasmic sides of the
proteins in each repeat; (3) six transmembrane helices within which the odd-numbered
helices 1, 3 and 5 are related to each other, as are helices 2, 4 and 6; and (4) three short
α-helices running parallel to the membrane on the matrix side of the protein.

Although the mechanism of transport has not been fully worked out, it is very probable that the SLC25A family function as monomers with a single substrate binding site
in the large central water-filled cavity formed by the transmembrane α-helices. Access
to this binding site from either side of the membrane during the transport cycle is regulated by two gates formed by two or three salt bridges between charged residues of the
consensus motifs on either side of the cavity. At any one time, one gate is very probably
open and one closed, and the alternate disruption and formation of these salt bridge networks allows net transport across the membrane without ever allowing an open channel,
which could dissipate ion gradients across the membrane. There must also be a transient
occluded state with both gates closed (discussed later).

We know most about the adenine nucleotide transporter (ANT), which exchanges
ADP^{3-} for ATP^{4-} (a rare example of these molecules being handled without bound
Mg^{2+}), and so this transport protein will serve as the paradigm. It is an anion exchanger,

and it should again be noted that entry of ADP^{3-} via a uniport step would be thermodynamically prohibitive owing to the large $\Delta\psi$, negative on the matrix side, a point first recognised by Mitchell and addressed in his fourth postulate (Section 1.3), which envisaged anion exchangers for this reason. However, the long-standing assumption that transporters in this family are all anion exchangers now appears to be invalid, and we shall see that sequence variations can alter specificity and also explain activities other than antiport.

8.2.1 The adenine nucleotide translocator

Further reading: Pebay-Peyroula *et al.* (2003)

An important development for understanding membrane transport in general came with the identification of two very specific and tight binding inhibitors, carboxyatractylate and bongkrekic acid. The former binds from the intermembrane space (IMS) side (sometimes called the cytoplasmic side, also the P-side) of the membrane and the latter from the matrix (the N-side). These two molecules cannot bind simultaneously, and either ADP or ATP is necessary to induce the conformations required for each inhibitor to bind. Although both inhibitors block ADP/ATP exchange, carboxyatractylate facilitates the damaging permeability transition (Section 9.4), whereas bongkrekate tends to protect the mitochondrion (and is sometimes erroneously marketed as an apoptosis inhibitor; Section 10.7). Thus, the ANT can adopt at least two conformations, one in which the substrate binding site is open to the IMS and can be 'clamped' by the bound carboxyatractylate and another in which the substrate binding site is open to the matrix and can be clamped by bound bongkrekic acid. These observations suggest a transport cycle in which a single substrate binding site is alternately accessible to either side of the membrane. This is likely to be a core common mechanism for all transporters, irrespective of their structure and detailed mechanistic variations. Furthermore, the fact that these two inhibitors, which structurally have little obvious similarity to ATP and ADP, bind to the protein much more tightly than the actual substrates can be used to understand the energetics of transport. The substrates must not bind as tightly to the transport protein as the inhibitors because the binding of a substrate at one side needs to induce conformational changes leading to release on the other side of the membrane, whereas inhibitors must prevent conformational changes and trap the transport protein in an aborted state.

Initially, there were indications that the stoichiometry of inhibitor binding was one per two polypeptide chains, leading to the suggestion that the adenine nucleotide translocator functioned as a dimer with the substrate binding site located between the two monomers. However, a projection map in the membrane and the crystal structure in detergent demonstrated that the transport protein has a central translocation pathway and is structurally monomeric; one molecule of carboxyatractylate thus binds per monomer (Figure 8.2). The assumption is that the other members of the transporter family also function as monomers.

The crystal structure of the bovine heart ANT is shown in Figure 8.2. The common features (mentioned previously) of this class of protein are clear. Sequence analysis and mutagenesis studies support the idea that the substrate binding site is in the middle of the membrane, as is the case with other transporters. Carboxyatractylate is bound in the crystal structure shown in Figure 8.2, meaning that the binding site is open to

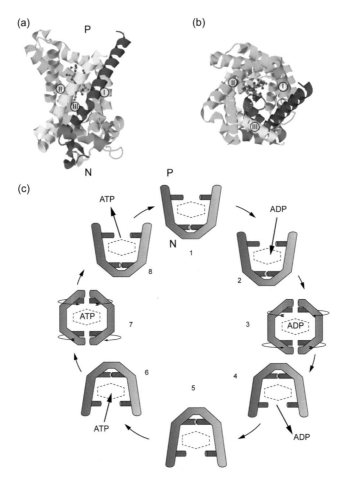

Figure 8.2   The crystal structure and proposed transport mechanism for the adenine nucleotide translocator.
(a) Side view and (b) view from the IMS of the bovine heart protein complexed with carboxyatractylate. Roman numerals represent locations of residues believed to interact with bound adenine nucleotide. Structure 1OKC (Pebay-Peyroula *et al.*, 2003). (c) Two-dimensional representation of a model for the transport mechanism. The hexagon represents the nucleotide binding site, and the red and the blue cylinders represent respectively negatively and positively charged residues making salt bridges. 1, Empty C state (i.e., open to the IMS or P-side); 2, ADP binds from IMS; 3, α-helices twist, passing through a transition state to produce (4) the M-conformation (i.e., open to the matrix or N-side), allowing ADP to leave; 5, empty M-state; 6, ATP binds; 7, α-helices twist back, passing through the transition state to restore (8) the C state, allowing ATP to leave.
Panel C after Kunji (2012).

the intermembrane space—that is, it is in the C (for cytoplasmic) conformation. The matrix side is closed by three salt bridges (only one is shown in Figure 8.2) between positively and negatively charged residues of the threefold repeated PX[D or E]XX[R or K] motif—for example, between E of helix 1 and R of helix 3.

Unfortunately, there is currently no crystal structure for the conformation that is open to the mitochondrial matrix (M-state)—that is, stabilised by bongkrekic acid—so this conformation is a matter of conjecture. One proposal is that the protein undergoes a slight twist following substrate binding, and that this is sufficient to disrupt the salt bridge gate at the matrix side and reform a closed gate at the IMS-side, for example, between K of helix 2 and D of helix 4, or between the positively and negatively charged residues of the [Y or F][D or E]XX[K or R] motifs in the three repeats. The ADP can now be released to the matrix. Binding of an ATP from the matrix reverses the process. This is shown schematically in Figure 8.2. Based on analogy with other transporters, it is probable that there will be an occluded state in which the salt bridge networks are substantially, if not fully, formed at both sides of the membrane such that transiently the bound substrate does not have access to either side of the membrane (Figure 8.2, Nos. 3 and 7).

It is important to note that the ANT will not catalyse uniport (Section 2.2.3); for example, having imported ADP, it cannot reorientate without a bound nucleotide. A possible reason is that there is a low probability that either of the salt bridge networks disrupts spontaneously in the absence of substrate. This supports the idea that the binding energy from the substrate–protein interactions is required to drive the necessary conformational changes in carriers with potentially strong salt bridges.

8.2.2 The phosphate carrier

The concept of anion exchange across the inner mitochondrial membrane has become firmly entrenched and thus phosphate transport into mitochondrial matrix is usually shown as antiport of $H_2PO_4^-$ against OH^-, which is experimentally impossible to distinguish from an $H^+:H_2PO_4^-$ symport. Uniport of the negatively charged phosphate is very unlikely because the matrix side concentration would be restricted to approximately $10\,\mu M$, one-thousandth of the cytoplasmic side concentration due to the membrane potential of approximately $180\,mV$ negative on the matrix side. The phosphate carrier has a hydrophobic residue in place of a charged residue in one of the salt bridge networks discussed previously for ANT. As in the case of the glutamate carrier (discussed later), this will lower the interaction energy of the networks considerably and increase the chance that the protein could interconvert between conformations without any substrate bound; that is, the phosphate transporter may catalyse electroneutral $H^+:H_2PO_4^-$ symport during the transition from the IMS-side facing to matrix-facing and then switch back to the IMS-facing conformation with no ligand bound.

There is another reason for considering $H^+:H_2PO_4^-$ symport rather than the 'traditional' phosphate/hydroxide antiport. The residues in locations corresponding to the contact points for the nucleotides in ANT include a well-positioned and conserved glutamate residue, which could bind a proton. Note that such a negatively charged residue is not needed to bind either phosphate or hydroxide. On the other hand, the presence of two conserved arginines is appropriate for phosphate binding. Generally, the nature of one of these contact points seems to determine the class (antiporter, symporter, or uniporter), and the other determines the type of substrate. These contact points are relatively close to the salt bridge networks. Thus, using sequence information, there is a basis for modelling binding sites for the quite diverse range of substrates that can be handled by this class of protein. Note that it is not meaningful to try to distinguish experimentally between an H_3PO_4 uniport and an $H^+:H_2PO_4^-$ symport.

8.2.3 Other transporters

Further reading: Kunji (2012)

There is evidence for a glutamate uniporter. A similar analysis to that described for the adenine nucleotide and phosphate transporters when applied to this protein shows that no salt bridges are possible on the IMS side because two of the consensus motifs have lost their charged residues. Thus, although glutamate entry from the IMS would be needed to trigger breakage of the salt bridges on the matrix side, once the glutamate is released, the protein may reorientate spontaneously without any bound substrate and with no necessity to break IMS-face salt bridges.

It is becoming increasingly possible to understand how key, but limited, sequence changes can endow a large range of substrate specificities (implying quite wide variation in substrate binding site size) and antiport or symport/uniport onto one basic structural framework. Interestingly, analysis of the acylcarnitine/carnitine exchangers shows that the predicted binding pocket is large enough to accommodate the acyl chain and that there will be a network of three salt bridges on either side of the membrane, consistent with an 'ANT-like' antiport activity. Further support for deductions made from sequences is that the different substrate specificities of the two isoforms of ornithine carriers, ORC1 and ORC2, which share 87% identical amino acids, were essentially swapped by exchanging a single residue predicted to be a contact point; this is arginine in ORC1 and glutamine in ORC2.

Uncoupling protein 1 (UCP1, SLC25A7) behaves functionally as a regulated proton conductance pathway (Section 12.4) and perhaps surprisingly turns out to be a member of the SLC25A family of mitochondrial transporters. Under physiological conditions, it binds a purine nucleotide with relatively low specificity, but unlike ANT, the nucleotide is not transported. Arginine residues at the contact points may be involved in the nucleotide binding, which is highly pH sensitive, and binding may be controlled by the protonation state of a nearby glutamate residue. An aspartate residue may be involved in the proton transport. Both the matrix and the cytoplasmic salt bridge networks appear complete, which would seem to indicate an antiport mechanism; indeed, UCP1 is most closely related to the keto acid exchangers. However, there is no evidence for keto acid translocation, and it is possible that proton translocation can be uncoupled from substrate transport such that UCP1 can provide a uniport pathway for protons from the IMS to the matrix side of the membrane.

8.2.4 Transport of pyruvate into mitochondria

Further reading: Herzig *et al.* (2012), Bricker *et al.* (2012)

The transporter that permits entry of pyruvate from the cytoplasm into the mitochondrial matrix has been difficult to identify, although it has long been assumed to be a member of the extensive mitochondrial transporter family. However, that expectation was confounded by the discovery of a novel class of inner mitochondrial membrane proteins, now termed the mitochondrial pyruvate carrier (MPC) family. There are two or three of these proteins in eukaryotes, and each is predicted to form three transmembrane α-helices. Because it seems unlikely that three helices could provide a transport pathway, it is speculated that a functional transporter must contain at least two, and possibly more, subunits organised as a multimers of homo- or heterodimers. These proteins were only discovered as a result of advanced multidisciplinary studies of yeast and *Drosophila* systems. To date, the detailed bioenergetics of these proteins are not characterised, but there is evidence that pyruvate entry into mitochondria is accompanied by one or more protons (or by equivalent exit of hydroxide anions); recall that pyruvate entry via a uniport would be opposed by $\Delta\psi$.

8.2.5 The mitochondrial Ca^{2+} uniporter and other cation transporters

Further reading: De *et al.* (2011), Baughman *et al.* (2011), Wei *et al.* (2012), Palty and Sekler (2012), Liao *et al.* (2012)

Mitochondria from many sources possess a Ca^{2+} uniporter allowing calcium to be transported into the matrix. This is a component of a complex Ca^{2+} regulatory pathway that is described in Section 9.4. Although this process has long been known to be inhibited by ruthenium red, it proved very difficult to identify the protein responsible for this activity. Finally in 2011, using a multidisciplinary approach, a 40-kDa protein named MCU was identified. Notably, MCU is absent from yeasts (in which mitochondrial calcium transport is absent). Each molecule of MCU contains two predicted transmembrane helices with globular regions that are external to the bilayer. At the time of writing, there was uncertainty regarding whether the N- and C-termini face the matrix or the IMS. Because two helices can hardly by themselves provide a path for Ca^{2+}, it is assumed that MCU functions as a multimer. A specific mutation of a single serine in MCU abolished the sensitivity to ruthenium red, which is convincing evidence that MCU is at least part of the uniporter. A second protein, MCU1, is associated with MCU and has EF hands, a structural feature characteristic of binding calcium. It is thought that MCU and MCU1 act together to give the functional Ca^{2+} uniporter.

Mitochondria from many sources possess Na^+/Ca^{2+} exchange activity. A protein responsible, termed NCLX, was identified in 2011. It is related to plasma membrane exchangers and was originally thought to be located in the plasma membrane. There is no structure for the mitochondrial protein, but a crystal structure is available for an analogue (NCX_Mj) from the archaea *Methanococcus jannaschii* (Figure 8.3). NCX_Mj comprises 10 transmembrane helices, formed from two structural repeats of 5 transmembrane helices with opposing topology. Five-transmembrane helical inverted repeat proteins are a vast structural class of transmembrane transport proteins, but NCX_Mj has an entirely new structural fold with the same principal elements of structural symmetry. NCX proteins have conserved α repeats in transmembrane helices 2 and 3 (α_1) and 7 and 8 (α_2), which are critical for transport function. The two α repeats are packaged in the core of the protein, forming an ion-binding pocket with four cation binding sites that are arranged symmetrically in a diamond pattern. Two high-affinity Na^+ sites are located closest to the extracellular and intracellular surfaces. A twofold rotational axis, centred on the plane connecting a low-affinity Na^+ site and a high-affinity Ca^{2+} site, coincides with the 5-transmembrane helical inverted topology of the protein. This arrangement is consistent with a three Na^+ to one Ca^{2+} stoichiometry.

(a)   (b)   P

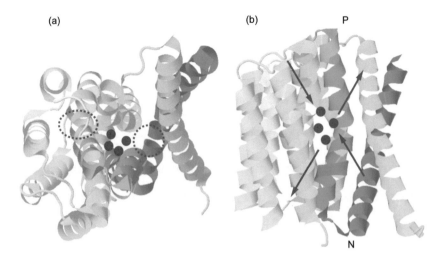

N

Figure 8.3   The 3Na$^+$/Ca^{2+} exchanger from *Methanococcus jannaschii* (NCX_Mj).
(a) View from extracellular side of the membrane. Na$^+$ ions shown in red; Ca^{2+} ions shown in purple; red oval, approximate position of external Na$^+$ channel; purple circle, approximate position of external Ca^{2+} channel. (b) Side view. Arrows show approximate entry and exit pathways for Na$^+$ and Ca^{2+} during Na$^+$ entry and Ca^{2+} extrusion. *Data from PDB 3V5U (Liao et al., 2012).*

8.3 BACTERIAL TRANSPORT

Bacteria survive in environments that are far more variable, and usually more hostile, than anything experienced by a mitochondrion or chloroplast. As a consequence, they have developed a variety of mechanisms for the transport of metabolites, such as amino acids and sugars, from the external medium where such molecules occur at very low concentrations to the cytoplasm where the concentrations must be considerably higher to sustain metabolism. The origins of the chemiosmotic theory lay in Mitchell's desire to explain such 'active transport' in bacteria, and there is now extensive evidence that many such transport processes are directly driven by Δp. However, there are also active transport processes that are powered by an Na$^+$ electrochemical gradient or by the direct hydrolysis of ATP, whereas a fourth class relies on phophoenolpyruvate as immediate energy source. Finally, there are examples of anion exchange reactions, such as hexosephosphate/phosphate antiports.

8.3.1 Proton symport and antiport systems

Further reading: Henderson (2012)

One of the triumphs of Mitchell's ideas was the proposal that symporters exist to cotransport one or more protons with an uncharged molecule across a membrane. Thus, the movement of protons down their electrochemical gradient would provide the driving force for the transport of the uncharged molecule up its concentration gradient. In some cases, it is now known that sodium is the ion moved down its electrochemical gradient, but the principle is the same. Before we describe examples of the structural basis of symporter and antiporter activity, we review some basic principles. Proton symporters drive the uptake of substrates, whereas proton antiporters expel metabolites.

Symporter proteins contain multiple transmembrane α-helices that create a pathway within the protein for the substrate and proton (or sodium) to move across the membrane. At some stage in its cycle of operation, an uptake symporter must exist with binding sites for its substrates and the proton empty and facing the P-side of the membrane. Although the affinity of these sites must be sufficient to bind the proton and substrate from the external phase, excessively tight binding would simply stabilise this conformation and there would then be no transport, as is the situation when carboxyatractylate binds to the mitochondrial ANT (Section 8.2.1). Instead, a cycle of conformational changes similar to that described for the ANT (Figure 8.2) must occur, but with the difference that the inward (P- to N-phase) part of the cycle requires binding of both proton and substrate, whereas the binding sites are empty for the return to the P-facing conformation (as is the case for the mitochondrial glutamate carrier; Section 8.2.3).

We previously discussed the concept that a change in binding affinity can drive a conformational change (and vice versa) in the context of proton pumps (see Figure 5.9) and the ATP synthase (see Figure 7.4). The symporter substrate binding site must have a high affinity when facing the P-phase (where the concentration of substrate may be very low), but after the conformational change making the binding site accessible to the interior N-phase, this must change to a sufficiently low affinity to release the substrate. As in the case of the ATP synthase, a tight-to-loose binding change requires an energy input, and for a proton (or Na^+) -coupled symporter the energy is supplied by a loose-to-tight binding conformational change of the cation (Figure 8.4a). As in the case of the F_o-ATP synthase proton turbine, if we assume that the proton (or Na^+) binding site is towards the middle of the membrane, then the entire Δp can be regarded as presented as a ΔpH, (ΔpNa) with protons (sodium ions) being driven into the loose binding site in the P-facing conformation and extracted from the tight binding site in the N-conformation by the $\Delta \psi$. The implication is that the empty P-conformation is more stable than the empty N-conformation. What the symporter must not do is to move inwards carrying only the proton or outwards carrying just the substrate because these would uncouple.

In the case of an antiporter, the scheme has to be modified (Figure 8.4b). Proton binding from the P-phase induces a P- to N-conformational change without additional substrate binding. The low-to-high affinity change in proton binding introduces stored conformational energy in the protein that is used to drive the return to the P-conformation and the change from high to low affinity binding of the substrate, leading to its release to the medium. Structural similarities between symporters and antiporters (discussed later) suggest that this is the broad mechanistic principle and that an alternative in which both substrates are simultaneously bound and effectively pass each other as they move through the protein is an unlikely mechanism.

We now discuss some proteins for which this description can be related to structural changes in the symporter protein.

8.3.2 Members of the major facilitator superfamily proteins

Major facilitator superfamily (MFS) proteins are defined by usually having 12 transmembrane helices (occasionally 14), with the N-terminal helices 1–6 related to the

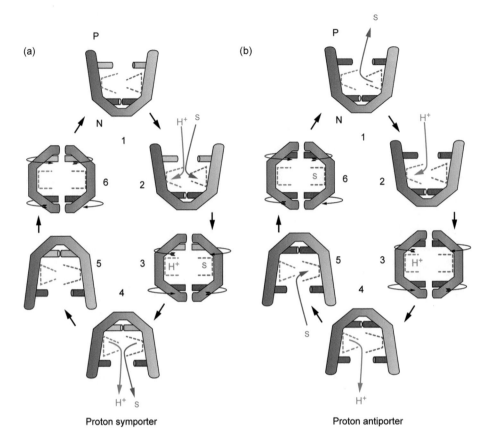

Figure 8.4   Schematic models of the catalytic cycles for a proton symporter (a) and proton antiporter (b).
Red and blue cylinders represent respectively positive and negative charge residues of salt bridges. The P-face salt bridges for the symporter are shown to be relatively weak. Tight and loose binding sites for protons and substrate are shown. The transition from the P- to N-side conformation is believed to involve a twisting of the α-helices as shown.
Two-dimensional models based on Kunji (2012).

C-terminal helices 7–12 (Figure 8.5). A consensus sequence motif (DRXGRR) is found between helices 2 and 3 and sometimes between helices 8 and 9. The driving force is usually the proton electrochemical gradient, but there are examples in which the sodium or phosphate gradients are involved.

8.3.2.1 The lactose (galactoside)/H+ symporter

Further reading: Abramson *et al.* (2003), Guan and Kaback (2006)

Arguably the best known, and certainly the most studied, MFS protein is the lactose (galactoside)/H+ symporter (LacY or Lac permease). Early experiments with *Escherichia coli* cells, and with the purified protein incorporated into the bilayer membrane of

(a)

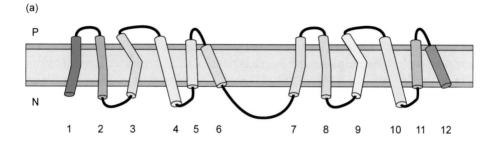

(b)

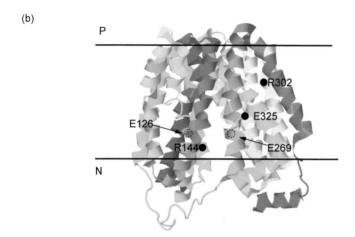

Figure 8.5   The LacY lactose transporter.
(a) Two-dimensional representation. Note that helices 7–12 are homologous
to helices 1–6. (b) Crystal structure showing position of key amino acid
residues.
PDB 1PV6 (Abramson et al., 2003).

phospholipid vesicles, established that the uptake of lactose was accompanied by the entry
of $1H^+$ and hence 1 positive charge, meaning that the uptake is driven by the full Δp. The
protein functioned as a monomer, with a turnover number of approximately $20\,s^{-1}$.

The first crystal structure of LacY showed a protein with 12 transmembrane helices with a pseudo two-fold symmetry between the N- and C-terminal six helix bundles (Figure 8.5a). A nontransportable substrate analogue, TDMG, bound in a position
approximately corresponding to the middle of the bilayer, but at the bottom of a cavity that led to the N-phase. In this conformation, corresponding to 4 in Figure 8.4a, the
transmembrane helices are packed in such a way that this substrate binding site is sealed
off from the periplasm. An essential arginine residue, R144, forms a bifurcated interaction with the substrate, and in the vicinity are other residues deduced from site-directed
mutagenesis experiments to be important for either sugar/substrate binding or proton
binding and translocation, including R302 (helix IX) and E 325 (helix X).

A second structure was obtained, still with the substrate binding site facing towards
the N-phase but without ligand bound (Figure 8.5b). Interestingly, a salt bridge between
R144 (helix V) and E126 (helix IV) had now formed, and certain other structural

changes suggested that a proton bound to E126 when TMDG was bound may have been released. As explained previously, understanding of the mechanism of LacY requires that we know more about the organisation of the binding site when it has two other orientations—facing the periplasm and the transitional, occluded conformation. However, if the salt bridge between R144 and E126 is preserved in the outward-facing conformation, then we can deduce that binding of lactose from that side is likely to disrupt this salt bridge. LacY has been subjected to a large number of noncrystallographic experiments designed among other things to probe proximities between side chains. Some of the residues deduced to be neighbours by this approach are far apart in the crystal structure; this discrepancy is indicative of the occurrence of at least one more conformation, presumably outward facing in LacY.

8.3.2.2 The fucose:proton symporter

Further reading: Dang *et al.* (2010)

Although there has been no success in obtaining an outward-facing binding site structure for LacY, a second MFS protein, FucP (a fucose-proton symporter), has given such a structure. In broad outline, the organisation of the 12 helices is similar to that in LacY. By comparing the FucP and LacY structures, one can see how, in terms of overall shape, they can interconvert through the rotation as a rigid body of a few helices around an axis roughly perpendicular to the membrane (Figure 8.6). Thus, a comparison of the two structures gives a very good idea of how the alternating access mechanism occurs. However, it is important to note that there are not great similarities in the details of the proteins, and key residues implicated in proton translocation in LacY do not line up with counterparts in FucP. Thus, although there is probably a common mechanism of overall structural transitions, the fine details of the mechanism appear to be specific to each individual protein. An interesting fact about FucP is that unlike LacY, it is one of a few bacterial sugar transporters that have sequence similarity with the glucose transporter in liver cell membranes, a protein that catalyses glucose movement down its concentration but without any symport (or antiport of an ion). Evidently, only subtle changes are needed to switch from symport to facilitated diffusion. In fact, there is now a crystal structure of a close relative of the glucose family of transporters, the xylose/proton symporter XylE of *E. coli*.

Regarding both LacY and FucP, crucial questions are how protons move through the protein and how energy is extracted from this movement to cause the conformational and binding affinity changes necessary to pump the substrate into the cell. Protons are not seen in X-ray structures; thus, protonation states of amino acid groups have to be deduced from considerations of their environments or the results of site-directed mutagenesis studies. By comparing the P-conformation of FucP with the N-conformation of the LacY protein, it seems likely that quite dramatic rigid body rotations can occur between the N-terminal and C-terminal domains, driven by the translocation of the proton within the protein from an Asp (D46) to a Glu (E135), causing the disruption of a salt bridge between the latter and a tyrosine side chain (Y365). Note that this model of alternating salt bridges is similar to that previously discussed for ANT; however, the likely key residues for proton movement in LacY and FucP are not conserved and so no generalisation

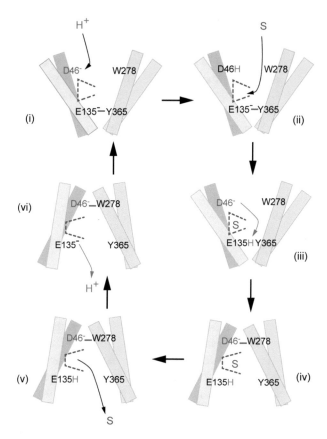

Figure 8.6 Proposed transport mechanism for the fucose:proton symporter FucP.
After Dang et al. (2010).

can be made. The advantages of studying sodium symporters include the ability to 'see' bound Na^+ and to detect sodium by radioisotope methods (discussed later). An interesting consideration is whether there are many binding sites for protons, leading from one side of the membrane to the other, in this class of protein. It appears that fewer such sites are needed than in bacteriorhodopsin because the opening up of the transporters allows protons direct access through water to binding sites in the centre of the membrane.

An example of an MFS antiporter is the glycerol-3-phosphate/phosphate exchanger (GlpT) from *E. coli*. No substrates are bound, but strikingly the structure closely resembles LacY, again with a binding site facing inwards. The similarity of this structure to those of LacY and FucP is fully consistent with an alternating access model in which transfer of glycerol-3-phosphate into the cell would be followed by export of phosphate.

8.3.2.3 EmrD, a putative multidrug efflux pump

EmrD is a multidrug-resistance transporter from *E. coli* implicated in the efflux of amphipathic compounds from the cytoplasm. Unlike the published LacY and GlpT structures, which are both in an N-side-facing conformation, that for EmrD probably represents an intermediate state. The periplasmic loops in EmrD are more embedded in the cell membrane, and the central loop linking helices 6 and 7 is considerably shorter. A molecular twofold axis relates the N- and C-terminal halves of EmrD and supports the notion that the MFS arose from recurrent gene duplication of an ancestral six-helix domain. The two halves of EmrD, however, are less symmetric than those of LacY or GlpT. The most notable difference between the structure of EmrD and the structures of LacY and GlpT is in the internal cavity. Whereas LacY and GlpT have hydrophilic interiors, the internal cavity of EmrD comprises mostly hydrophobic residues, consistent with its function of transporting amphipathic compounds.

8.3.2.4 The proton-dependent oligopeptide transporter symporter family

Further reading: Newstead *et al.* (2011); Solcan *et al.* (2012)

This type of MFS protein is found in some bacterial cytoplasmic membranes but also in various mammalian cell membranes; the latter examples are of medical interest because they are able to transport a wide spectrum of orally administered drugs. A structure of a bacterial version, $PepT_{So}$ (from *Shewanella oneidensis*), shows that the general features of the helices 1–12 are organised as in other MFS proteins of known structure. A surrogate substrate can be detected at the same relative position in the protein as the lactose analogue TDMG in the original LacY structure. Interestingly, the $PepT_{So}$ structure is in an intermediate occluded state (similar to 3 in Figure 8.4a), with closed P-phase and N-phase gates. The N-terminal domain is clearly involved in both proton binding and regulating peptide specificity. Structural comparison with a related N-facing conformation structure indicates that the C-terminal domain is more dynamic, with helices H10 and H11 facilitating opening of the central peptide binding site to the interior of the cell during transport. This view of the fundamental mechanism of an MFS protein thus differs somewhat from that of the model proposed for FucP, which involves extensive rigid body rotation between the N-terminal and C-terminal domains. However, the pathways and energetics of the proton pathways in the MFS family are far from understood, and there does not appear to be a conserved set of amino acids providing for this in the family. On the other hand, models are in agreement about the importance of salt bridges on both sides of the transporter and within the structure (Figure 8.6).

Similarly to FucP, a model for proton-dependent oligopeptide transporters can be formulated that employs symmetrical salt bridge interactions at both extracellular and intracellular sides of the peptide binding site that alternatively form and break during transport. In the outward-facing P-conformation, the central binding site should be accessible to the extracellular side of the membrane to receive peptides, resembling the crystal structure of FucP. Conserved side chains involved in proton binding should be accessible because proton binding is predicted to occur before recognition of peptide. Following proton binding, entry of either a di- or a tripeptide into the central cavity should facilitate closure of the extracellular gate to occlude the peptide within the binding site. Closure is facilitated by the formation of conserved salt bridge interactions between conserved arginine and glutamate residues. The transitional occluded state is likely to require substantial interhelical movements within the C-domain, which should in turn weaken salt bridges forming part of the intracellular gate. Comparison of occluded and N-facing conformations (obtained for POT protein from *Streptococcus thermophilus*) suggests that this occurs as a result of localised hinge-like bending in helices H10 and H11 at the conserved pivot points formed by Gly407 and Trp427.

8.3.2.5 The bioenergetics of bacterial symporters

An alternating gate model does not in itself explain the vectorial nature of the transport and its coupling to Δp. As in the case of the ATP synthase, changes in binding affinity, both for the proton and for the substrate, play a central role. We have summarised this in Figure 8.6 using the FucP protein as an example. Protonation of D46 towards the P-face of the protein allows the substrate access to a binding site whose affinity is sufficiently high to bind the trace concentrations of the substrate that may be present in the environment. Once the substrate has bound, the proton on D46 has access to the protonatable E135. However, this requires breaking of the salt bridge between E135 and Y365. Because of the stability of the bridge, a very high proton concentration is required to protonate E135 and break the bond (i.e., E135 will have a very low pK). Since E135 is close to the N-face of the membrane, on passing through the structure to E135 a proton will be driven by $\Delta\psi$ and, in response to a $\Delta\psi$ of 180 mV, might be accumulated 1000-fold relative to the external phase, sufficient for this unfavourable protonation. Breaking the salt bridge initiates the conformational changes that make the substrate binding site accessible to the N-phase and greatly lowers the affinity of the substrate binding site, allowing it to dissociate and enter the cell cytoplasm. A weaker salt bridge can now form between D46 and W278 at the P-face of the protein. At the same time, E135, now no longer part of a salt bridge, resumes a more normal pK; that is, the thermodynamic energy from Δp has been transferred to the protein. Once the E135 proton dissociates, the stronger N-phase salt bridge reforms, breaking the P-face bridge.

We have focused on the symport of a neutral species with one proton; however, the components of Δp can in principle drive the uptake of any type of nutrient molecule. In the case of a positively charged molecule, such as lysine at neutral pH, a possible mechanism would be a uniport driven by $\Delta\psi$ and the logarithm of the equilibrium accumulation ratio would be a function of $\Delta\psi$. If a proton were to be cotransported with lysine, then both ΔpH and $\Delta\psi$ would contribute to the driving force and the equilibrium distribution would be a function of ($\Delta pH + 2\Delta\psi$). A monoanionic species in symport with one proton would be driven only by ΔpH. However, because Δp in most species of bacteria is usually predominantly present as $\Delta\psi$, such a symport is improbable. On the other hand, symport of an anion with $2H^+$ would allow an equilibrium to be attained that would be proportional to ($2\Delta pH + \Delta\psi$). These should be viewed as illustrations because there is no *a priori* reason why other proton stoichiometries should be excluded.

A variant on the 12-transmembrane helical transporter, as exemplified by those discussed previously, has been discovered because a group of proton symporters possess two additional subunits—a 4-transmembrane α-helix subunit and (for gram-negative bacteria) a water-soluble periplasmic protein. The latter is often called an extracytoplasmic solute-binding receptor and belongs to a class of proteins more widely found in transporters where ATP hydrolysis directly powers transport (Section 8.3.5).

In fact, it had been assumed that such receptors were not found in association with proton-driven symporters. The class of such symporters is known as TRAP (*tri*partite *A*TP-independent *p*eriplasmic) transporters; one of the best characterised is that for C4-dicarboxylates (e.g., malate) in *Rhodobacter capsulatus*.

8.3.3 Sodium symport and antiport systems

Protons were once thought to be the only cations to move in symport with sugars and amino acids into *E. coli*. It is now known that melibiose and proline transport both occur in symport with Na^+, driven by the Na^+ electrochemical gradient $\Delta\tilde{\mu}_{Na^+}$ (note that the term 'sodium motive force' is sometimes used). Unlike *Propionigenium modestum* (Section 5.15.7), there is no primary Na^+ pump. Instead, $\Delta\tilde{\mu}_{Na^+}$ is maintained by an electroneutral Na^+/H^+ exchanger in the cytoplasmic membrane that equalises the Na^+ and H^+ concentration gradients. $\Delta\psi$ is, of course, a delocalised parameter, and so the $\Delta\psi$ generated by proton pumping will be a component of $\Delta\tilde{\mu}_{Na^+}$. There is a structure of the *E.coli* Na^+/H^+ antiporter, NhaA, which suggests that there is a single ion binding site, in the middle of the membrane and provided by two pairs of short helices connected by extended chains, that is alternately accessible to the two sides of the membrane and able to bind Na^+ or H^+ directly from the two aqueous phases without pathways provided by other side chains for these ions.

There are many instances in which the role of an Na^+ circuit has been established for various species of bacteria. The INDY proteins catalyse dicarboxylate transport in symport with two Na^+ ions. A structure shows that this type of protein is a dimer with each protomer having 11 transmembrane helices and providing two binding sites for Na^+, one of which is occupied and provides five oxygen ligands for the ion (similar to Mhp1; Section 8.3.3.1); as in NhaA, there is no evidence for a chain of Na^+ sites through the protein. In some cases, such as active transport into alkaliphilic bacteria, a sodium circuit has been rationalised on the basis that Δp is too small to drive active transport. An electroneutral Na^+/H^+ exchanger is not sufficient because the Na^+ gradient out/in would be reversed to the same extent as the proton gradient. Instead, alkaliphiles possess electrogenic $nNa^+/(n + x)H^+$ exchangers, where x represents the excess of protons over sodium ions transported (Morino *et al.*, 2010). It is not clear why *E. coli* should use both proton and sodium symports, nor is it appreciated why methane synthesis (Section 5.14.9) seemingly involves both Na^+ and H^+ translocation.

8.3.3.1 The five-helix inverted repeat LeuT family

Further reading: Weyand *et al.* (2008), Shimamura *et al.* (2010)

The bacterium *Microbacterium liquefaciens* has a sodium–hydantoin symporter, Mhp1. Hydantoin is a potential precursor of tryptophan or phenylalanine. Structures for this protein have been obtained for an outwardly facing conformation with sodium, but not hydantoin, bound, an occluded state with both substrates bound, and an inwardly facing form with no bound substrates (Figure 8.7). The structure of this protein shows that transmembrane helices (TMs) 1–5 and 6–10 are inverted relative to each other but otherwise have

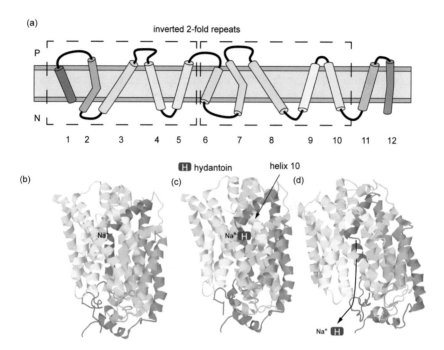

Figure 8.7   The LeuT family of Na⁺-coupled transporters, exemplified by Mhp1 of *M. liquefaciens*.
(a) Arrangement of helices. Note that helices 6–10 are an inverted repeat of helices 1–5. (b–d) Crystal structures of three conformations of the sodium: hydantoin symporter Mhp. (b) Unloaded conformation with binding site accessible from the P-phase. PDB 2JLN from Weyand *et al.* (2008). (c) Substrate-bound occluded conformation. PDB 2JLO (Weyand *et al.* 2008). (d) N-phase accessible conformation. PDB 2 × 79 (Shimamura *et al.*, 2010).

similar conformations (the 11th and 12th helices can be ignored for this discussion). Thus, a loop connecting helix 2 to helix 3 is on the N-side, whereas the corresponding loop between helices 7 and 8 is on the P-side. The N terminal end of helix 3 is at the inside, whereas the N-terminal end of helix 8 is at the outside; both helices make a similar angle relative to the vertical. Overall, helices 1–5 are related to helices 6–10 by a pseudo-twofold axis. The protein can approximately be regarded as having two parts, a four-helix bundle comprising TM1 and TM2 together with their symmetry-related counterparts TM6 and TM7. TM3 and TM4, along with their counterparts, TM8 and TM9, pack together such that from a certain angle the crossing over the four helices gives an appearance of the keyboard hash (#) symbol. The inverted repeat pattern is thus not the same as in MFS proteins.

The structure of the outwardly facing (P-side of membrane) conformation of Mhp1 (Figure 8.7b) shows a relatively large cavity with sodium but without hydantoin bound. The sodium is five coordinated by either main chain carbonyls or hydroxyl side chains provided by helices 1 and 8. The packing of the helices provides a strong barrier or gate between this outwardly directed site and the N-side of the membrane. Binding of hydantoin results in a reorientation of helix 10 such that it presents a relatively thin barrier between the bound substrates and the aqueous P-phase (Figure 8.7c); this is called the occluded state, which has been trapped by the particular crystallisation conditions employed. However, this cannot be a stable state, and the interaction between the carbon substrate and the side chains drives further conformational changes that open up a route from the occluded substrates to the N-side of the membrane. A third structure of the protein has been obtained in which both sodium and hydantoin are absent (Figure 8.7d), but this route is clearly seen while TM 10 remains in place to block access to the P-side. The conformational change that achieves this reorientation of the binding sites can be described as a rigid body rotation by 30° of the hash domain relative to the bundle (in other words, there are no rearrangements within the hash or bundles). In the N-facing conformation, the rigid body rotation has destroyed the Na binding site as helices 1 and 8 have moved apart. This rigid body rotation must subsequently spontaneously reverse, with neither hydantoin nor sodium bound, to restore the outwardly facing conformation, which presumably—possibly enhanced by the binding of sodium—is the most stable conformation.

It is very important that the protein can only reorientate its binding sites for hydantoin and sodium when either both or neither are bound. If, for example, it were able to reorientate from inward facing to outward facing with hydantoin bound, then there would be no net uptake of hydroxytanin but simply a net import of sodium. Transport proteins must have rules for reorientation. It appears that there is just one binding site for sodium, which is intact when the protein is in the outward-facing conformation but is destroyed when the protein reorientates. There does not appear to be a chain of sodium binding sites passing through the protein—a point that was raised previously when considering proton symporters.

Mhp1 is a representative of a family of proteins, but with little sequence similarity, that have the same overall fold and presumably fundamentally the same mechanism. Another example is LeuT, which has given its name to the family; it catalyses sodium/leucine symport. This is of interest because it is a homologue of the mammalian serotonin neurotransmitter transporter. There are many members of what is effectively a superfamily of five-helix inverted repeat proteins in the mammalian plasma membrane.

8.3.4 Δp-driven transport across the bacterial outer membrane

8.3.4.1 The TonB system

Further reading: Noinaj et al. (2010)

The outer membranes of gram-negative bacteria contain large receptor molecules that participate in the active transport of iron–siderophore complexes and vitamin B_{12}. Remarkably, the energy for this active transport is provided by the Δp across the adjacent cytoplasmic membrane. It is believed that the outer membrane receptor interacts at its periplasmic side with a cytoplasmic membrane protein complex made up of three proteins, TonB, ExbB and ExbD, at least one of which presumably undergoes a conformational change in response to protons flowing through it from the periplasm and thus back into the cytoplasm. The crystal structure of one outer membrane receptor, FepA, shows that at its periplasmic surface there is a globular domain that folds into the major β-barrel structural feature of the protein and thus blocks access from the interior of the barrel into the periplasm. The globular domain also provides two loops for potential ligand binding and interaction with the TonB/ExbB/ExbD complex. Thus, Δp-dependent conformational change in the latter complex could in turn drive a conformational change in the globular domain of FepA and thus permit ligand entry into the periplasm. This would thus achieve the coupling of Δp across the cytoplasmic membrane to a transport process across the outer membrane.

8.3.4.2 The Acr B multidrug effluxer from E. coli

Further reading: Murakami *et al.* (2006)

The natural function of the Acr B protein is the proton motive force-dependent pumping of bile salts out of the *E. coli* cell. Acr B will also transport a variety of hydrophobic drug molecules out of the cells. The crystal structure of Acr B shows that it has a trimeric organisation. Each monomer, a very long polypeptide with more than 1000 residues, can be regarded as having three domains. The transmembrane domain has 12 α-helices with, as for many other transporters, an approximately symmetric relationship between the 6 at the N-terminus and the 6 at the C-terminus. Helices 6 and 7 are connected by a short helix that runs along the surface of the protein on the cytoplasmic side.

A relatively extended connection between these helices is also found in other 12-transmembrane helical transporters. At the periplasmic side of the membrane, each of the three monomers contributes 3 helices to a single funnel-shaped pore domain that extends approximately 40Å from the membrane surface, beyond which there is a further 30-Å extension that is the site of docking to an outer membrane protein known as Tol C. The latter protein has a long helical region that extends for approximately 100Å into the periplasm. Thus, it is possible for a molecule to be transported across the cytoplasmic membrane and carried across the periplasm (a distance of 170Å and therefore important for the packing of periplasmic electron transfer proteins; Chapter 5) before being moved across the outer membrane via the β-barrel domain of Tol C. The connection between Acr B and Tol C is believed to be enhanced by a third protein, Acr A, for which a structure is not available.

The trimeric organisation of Acr B provides single pore and Tol C docking domains. Each of the three sets of 12 transmembrane helices delivers a substrate into a very large central cavity of approximately 5000Å^3 in volume. From there, a substrate must pass through a narrower pore before moving into a wider funnel region that is provided by the Tol C docking domain. Each of the three Arc B subunits has been found to be in a different conformation at any one time, with only one subunit having a substrate bound which is located at the periplasmic side of the membrane. From this observation it has been deduced that each the three subunits undergo sequential conformational change which in effect is a functional rotation mechanism. This would lead to a peristaltic mechanism whereby the substrates were effectively 'squeezed' into the single funnel-shaped pore domain. In the structures of both unliganded and liganded Acr B, a pore helix from each monomer makes a direct coil–coil interaction in a central position between the three subunits, thus closing the pore. It is possible that the proton flux through the 12 α-helical domain also generates a conformational charge that opens the pore. There appears to be a flexibility in the action of Acr B in that there is access from the periplasm or periplasmic surface of the membrane for substrates to the central cavity via what are known as vestibules.

For those molecules that are transported out of the cytoplasm, one would like to know how this is coupled to the protonmotive force; in other words, what is the basis of a drug–proton antiporter? A clue here is that lysine 940 of helix 10 and both aspartate 407 and aspartate 408 of helix 4 are essential for activity on the basis of site-directed mutagenesis experiments. These residues might tentatively be suggested to participate in reversible ion pair formation (analogous perhaps to the glutamate–arginine interaction between subunits a and c of the ATP synthase; Chapter 5) as part of a proton translocation mechanism. If there are to be general principles concerning the mechanism of substrate/proton symporters/antiporters, then implication of these particular helices correlates well with ideas concerning proton translocation by the lactose/proton symporter. An important point is that the space between the three sets of 12 transmembrane helices must be filled by phospholipids to prevent proton leakage across the membrane.

One of the questions concerning multidrug effluxers is how a variety of substrates (sometimes called allocrites) can be handled. The second structural analysis of Acr B included details of how four different drugs—rhodamine, ethidium, dequalinium and ciprofloxacin—are bound. In each case, three drug molecules are bound in the central cavity. It is clear that four conserved phenylalanines (i.e., 12 in total) are important for this binding. The ethidium and rhodamine binding sites overlap. The positive charge on quinolinium is accommodated by an interaction with the side chain of aspartate 99, which is also a conserved residue. Thus, the central cavity contains features provision of hydrophobic, aromatic stacking and van der Waals interactions that can accommodate these diverse molecules, each of which uses a slightly different subset of Acr B residues for binding. There are interactions between some of the bound ligands, a feature that also contributes to the binding. Certainly, all four compounds seem to induce a similar conformational charge. The four drugs observed in the crystal structure may enter the central cavity via the vestibules. In the structures of both unliganded and liganded Acr B, a pore helix from each monomer makes a direct coil–coil interaction in a central position between the three subunits, thus closing the pore. It is possible that proton flux through the 12 α-helical domain generates a conformational charge that opens the pore.

8.3.5 Transport driven directly by ATP hydrolysis

Although transport driven by an electrochemical gradient across cytoplasmic membranes of bacteria is a very widespread mode of transport, there nevertheless exist a range of processes that are directly driven by ATP hydrolysis. We discuss here the two principal classes of ATP-powered transporters, both of which are also commonly found in eukaryotic plasma, endoplasmic and cytoplasmic membranes.

8.3.5.1 ABC-type transporters

Further reading: Ward *et al*. (2007), Callaghan *et al*. (2012)

A major group of transport systems in gram-negative bacteria are the ATP-binding cassette (ABC) proteins. They are defined because they have regions of sequence, including the Gly-X-X-X-X-Gly-Lys-Walker A motif seen in F_1F_o ATP synthases (Section 7.6.2.2) and the Walker B motif, as well as a signature motif Leu-Ser-Gly-Gly-Gln. Many are involved in the import of nutrients including P_i, SO_4^{2-}, ribose, maltose and histidine, for example, in *E. coli* and *Salmonella typhimurium*. The fundamental organisation of these proteins is that they usually have 12 transmembrane helices, generally with two separate subunits providing 6 helices each, together with two identical globular polypeptides that are bound to the cytoplasmic side of the transmembrane helices; each of these contains an ATP binding site that is called the nucleotide binding domain (NBD). These import systems also have a periplasmic binding protein that first captures the substrate for the transport system. The periplasmic binding proteins have a high affinity for their substrates; their specificity is known from a series of high-resolution X-ray diffraction structures to be conferred by a set of hydrogen bonds that interact with the substrate. If the contents of the periplasm are released, for example, by exposing the cells to an osmotic shock, transport is greatly inhibited.

Crystallographic studies of the maltose import proteins have identified discrete steps in the reaction. Transport commences by maltose binding to the open conformation of the periplasmic binding protein (MalE), which then adopts a closed conformation (Figure 8.8a,i). The MalE.maltose complex then docks onto an outward-facing organisation of the 12 transmembrane helices (MalF and MalG) (Figure 8.8.a,ii). This docking event is conformationally relayed to the NBD domains on the cytoplasmic side of the

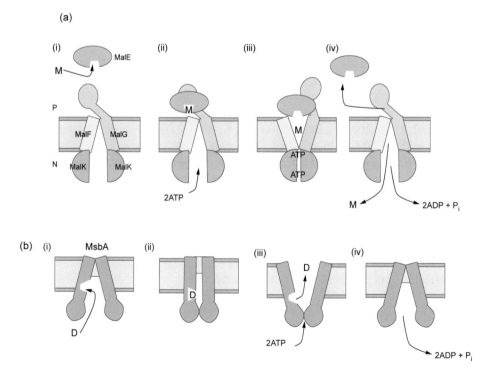

Figure 8.8   Models of ABC symporters and antiporters.
(a) Maltose import cycle. M, maltose. (b) The multidrug effluxer MsbA. D, drug.

membrane. Interactions with substrate-loaded maltose binding protein in the periplasm induce a partial closure of the MalK dimer in the cytoplasm. ATP binding to this conformation then promotes progression to the outward-facing state (Figure 8.8a,iii). ATP hydrolysis causes rearrangement of the transmembrane helices such that the maltose is released from the periplasmic binding protein and moves to a binding site approximately halfway across the membrane. Subsequent conformational changes allow it to exit into the cytoplasm, presumably via an interface region between the transmembrane subunits (MalF and MalG) and the globular pair of MalK subunits (Figure 8.8a,iv). The binding of maltose to the maltose binding protein is extremely strong; thus, the overall hydrolysis of ATP must play a role in causing the conformational change that releases the maltose from MalE into the transmembrane pathway. It is believed that the stoichiometry of the reaction is 2ATP hydrolysed per molecule of maltose transported. This stoichiometry correlates with very high accumulation ratios, up to 10^5, that are achieved by these transport systems. Indeed, if a dianion such as sulfate or phosphate at pH values above 7 were to be taken up by a chemiosmotic mechanism, then up to four protons would need to move in symport at a cytoplasmic membrane potential of 180 mV to achieve accumulation ratios of this order.

There may well be variation in the detail of the mechanism of ABC importers. For example, the BtuCDF protein, which imports vitamin B_{12}, has a cytoplasmic gate that is closed in a conformation in which the substrate is fully occluded in the middle of the membrane (Figure 8.9). This protein also shows a variation in structure in that apart from the periplasmic binding protein BtuF, only two polypeptides make up the rest of the molecule.

Analogous ATP-dependent transport systems have been identified in gram-positive organisms that do not have a periplasm. In these instances, the equivalent to the periplasmic binding protein appears to be anchored to the cytoplasmic membrane. By analogy, it seems likely that in gram-negative organisms, the binding protein functions to deliver the bound substrate to the membrane-spanning components of the system.

Other ABC transporters in bacteria are efflux systems. One of the best known is the multidrug effluxer MsbA (Figure 8.8b). This type of ABC protein does not have the periplasmic binding protein and can comprise just one polypeptide chain that nevertheless provides two ATP hydrolysis sites and 12 transmembrane helices, as do MalFGK acting together. The exporter ABC proteins are thought to be in their lowest energy state when in the inward (cytoplasm)-facing conformation. The binding energy released when ATP binds causes the two NBDs to dimerise, and this in turn promotes the transmembrane helical domain to adopt a conformation facing the periplasm. Subsequent hydrolysis of ATP and release of the substrate (sometimes called an allocrite in these systems) allows relaxation to the inward-facing conformation.

Interest in the bacterial ATP and periplasmic binding protein-dependent transport systems has been significantly enhanced by the finding that there are considerable similarities with many mammalian proteins. Prominent among these is the multidrug resistance (MDR) protein, which is an ATP-dependent system of low specificity that exports drugs from cells and thus causes problems in certain drug treatments. It is related to MsbA but is often called the P-glycoprotein because it is glycosylated on the external face. Another bacterial MDR occurs in a *Streptomyces* species that produces antibiotics to which it is resistant, presumably as a result of expelling the molecules from the cell via this system.

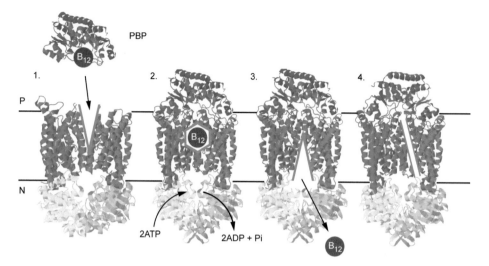

Figure 8.9   Structure and transport mechanism for the B_{12} transporter BtuCD-F of *E. coli*.
(1) The periplasmic binding protein (PBP) delivers B_{12} to the transporter in the P-accessible conformation (PDB 17JV, Locher *et al.*, 2002), generating (2) the occluded state (PDB 2QI9, Korkhov *et al.*, 2012). No crystal structure is currently available for the N-accessible conformation created by ATP hydrolysis (3). Finally, a closed, ATP-free conformation is attained (4) (PDB 4DBL, Korkhov *et al.*, 2012).

A second gene product with resemblance to the bacterial ABC transporters is called the CTFR (*cystic fibrosis transmembrane regulator*) protein; a common mutation leads to a mistargeting of this protein with the consequence that cystic fibrosis develops. The comparison with the bacterial proteins initially suggested to some investigators that CFTR would be part of an ATP-dependent transport system. It is now accepted to be a chloride channel. The ATP binding site is present to permit a regulatory role, analogous to that exerted by GTP and GDP binding to G proteins. The channel is opened maximally as a result of both phosphorylation of the protein and occupancy of two ABC sites by ATP. Slow hydrolysis of the latter (analogous to GTPase activity of G proteins) promotes closing of the channel. The example of CFTR serves as a warning that all proteins identified bioinformatically as ABC proteins may not be transporters.

8.3.5.2 P-type ATPases

Further reading Arnou and Nissen (2012)

P-type ATPases are so named because their reaction mechanism proceeds via a phosphorylated protein intermediate. Detailed structural analysis of the calcium ATPase of the sarcoplasmic reticulum has shown that the single polypeptide has three globular domains and 10 important transmembrane helices. Binding of ATP, transfer of the terminal phosphate to a conserved aspartate group, relay of conformational changes in the globular domains to the transmembrane helices, hydrolysis of the aspartyl phosphate and a further series of conformational changes summarise the transport cycle for Ca^{2+}. It is important to note that although this protein is usually described as moving only Ca^{2+} alone, it is now generally accepted that this protein catalyses exchange of protons for calcium, analogously to the Na/K ATPase, another member of this family. Movement of only Ca^{2+} would, for the capacitance reasons discussed in Chapter 3, result in impossibly high membrane potentials.

P-type ATPases are important in transporting metals across bacterial cytoplasmic membranes, such as the *Legionella pneumophila* CopA Cu^+-ATPase, for which an X-ray structure is available. Interestingly, the structure also provides a framework to analyse missense mutations in the human Cu P-type ATPases that are associated with Menkes' and Wilson's diseases. The mechanism of the Ca-ATPase forms the basis for understanding how the Cu ATPase functions. A critical sequence change, the introduction of cysteine in the transmembrane metal binding site, essentially causes the switch from Ca to Cu specificity in this part of the protein. CopA occurs in *E. coli*, in which another P-type ATPase is Kdp, which is induced when the external K^+ is very low. Three gene products—KdpA, KdpB and KdpC—constitute a K^+-dependent ATPase.

8.3.6 Other transporters

Bacterial cytoplasmic membranes harbour many transport proteins originally recognised in a mammalian context. In many cases, the physiological role of these proteins is not clear in their bacterial contexts. Nevertheless, whatever their physiological roles, many of these proteins have proved amenable to crystallisation and thus attract a great deal of attention. We describe some of these here, along with other important transporters in bacteria not yet described in this chapter:

8.3.6.1 Relatives of channels in higher cells

A surprising development from the 1990s onwards has been the identification of proteins in bacterial cytoplasmic membranes that are highly homologous to ion channels in eukaryotes. These include the KcsA potassium channel from the soil bacteria *Streptomyces lividans*, which is activated by changes in pH, and the voltage-gated sodium channel from *Arcobacter butzleri*. Crystal structures of these two proteins have been of enormous value for the understanding of sodium and potassium channels in animal cells. However, their roles in the bacteria in which they occur are not fully understood. Presumably, they catalyse a controlled uniport of sodium or potassium into the bacteria, driven by the $\Delta\Psi$ component of Δp.

The chloride channel/transporter in *E. coli* was first assumed to function as a channel, analogous to animal chloride channels. However, the *E. coli* protein turned out to be a dimer with functional monomers that did not enclose a pore but comprised 18 helical elements, many of which do not fully span the membrane. Subsequently, it was discovered that the protein is a fast transporter, exchanging two chlorides for one proton at a turnover of up to 2000 per second; in other words, each turnover is envisaged as moving three net charges. The role of this protein is thought to be in acid resistance. Transient survival in the gut at pH 2 requires some way to avoid the intracellular pH declining to 2.

Uptake of arginine (1+) followed by cytoplasmic decarboxylation and proton uptake to give agmatine (2+), which in turn is exported in exchange for the arginine (1+) by a protein related to LeuT, is crucial to acid resistance. However, this would result in positive charge accruing on the P-side of the membrane to give potentials well in excess of the normal range. This assumes that at pH 5 respiration is not generating a Δp; if there is a respiration-dependent Δp, then export of agmatine in exchange for arginine would be even more problematic. However, for every three turnovers of the arginine/agmatine exchanger, one turnover of the chloride channel would move two chlorides out of the cell and one proton in. In other words, it would electrically compensate for the arginine/agmatine activity (equivalent to the export of 3 negative charges), with the intracellular chloride originating from the passive diffusion of HCl into the cells. This mechanism does imply that for every three intracellular protons removed by agmatine formation, one is replaced by the action of the chloride transporter. In general, the chloride transporter would be a very effective Δp-dependent system for expelling intracellular chloride because both the normal $\Delta\Psi$ and ΔpH would favour entry of protons and exit of chloride. It is notable that a variant of the chloride transporter in *E. coli* acts to expel fluoride from the cytoplasm, albeit at a stoichiometry of $1H^+$ entering per F^- exiting.

An intriguing aspect of the chloride transporter is that some forms in eukaryotic cells truly function as channels; only slight sequence changes are needed to switch from a transporter mode to a channel mode.

8.3.6.2 Glycerol, NirC and FocC type

Further reading: Lü et al. (2013)

The protein that facilitates the passage of glycerol across the *E. coli* membrane is not an active transporter and does not comprise 12 transmembrane α-helices but instead is a member of the aquaporin family, which has 10 such helices.

Formate and nitrite have to enter or exit from a variety of bacterial species. This occurs via proteins known as FocA and NirC, respectively. The high-resolution crystal structures of these proteins, as well as for a bacterial hydrosulfide channel, show that these have an unprecedented, pentameric architecture with structural similarity to aquaporins and glyceroporins. These are thought to function as uniporters, although it may still emerge that they are proton symporters because the $\Delta\Psi$ would oppose the entry of nitrite and formate into cells.

8.3.6.3 Magnesium and zinc transport

Further reading: Lu and Fu (2012)

Uptake of Mg^{2+} and Zn^{2+} across the bacterial cytoplasmic membrane is vital. The CorA family of magnesium transporters is the primary Mg^{2+} uptake system of most prokaryotes and a functional homologue of the eukaryotic mitochondrial magnesium transporter. The transporter is a funnel-shaped homopentamer with two transmembrane helices per monomer. The channel is formed by an inner group of five helices and putatively gated by bulky hydrophobic residues. The large cytoplasmic domain forms a funnel whose wide mouth points into the cell and whose walls are formed by five long helices that are extensions of the transmembrane helices. The cytoplasmic neck of the pore is surrounded, on the outside of the funnel, by a ring of highly conserved positively charged residues. Two negatively charged helices in the cytoplasmic domain extend back towards the membrane on the outside of the funnel and abut the ring of positive charge. CorA is believed to be a uniporter.

YiiP is a structurally distinct membrane transporter that catalyses exchange of Zn^{2+} for protons across the inner membrane of *E. coli*. Mammalian homologues of YiiP play critical roles in zinc homeostasis and cell signalling. YiiP is a homodimer held together in a parallel orientation through four Zn^{2+} ions at the interface of the cytoplasmic domains, whereas the two transmembrane domains swing out to yield a Y-shaped structure. In each protomer, the cytoplasmic domain adopts a metallochaperone-like protein fold; the transmembrane domain features a bundle of six transmembrane helices and a tetrahedral Zn^{2+} binding site located in a cavity that is open to both the membrane outer leaflet and the periplasm.

8.3.7 Transport driven by anion exchange

There is evidence, especially for the gram-positive organism *Streptococcus lactis*, that transport can occur by anion exchange systems analogous, at least in principle, to those found in the inner mitochondrial membrane (Section 8.2). The system identified is concerned with the uptake of glucose-6-phosphate (G6P) into the cell. In one mode, two $G6P^-$ anions move into the cell in exchange for the export of one $G6P^{2-}$. This apparently curious exchange is therefore electroneutral and is equivalent to the net entry of one $G6P^{2-}$ and $2H^+$ and so is thermodynamically equivalent to a $2H^+$:$G6P^{2-}$ symport with ΔpH acting as the sole driving force. The exchange system can also operate by moving two $H_2PO_4^-$ anions out and one $G6P^{2-}$ in. The exchange is electroneutral, and the advantage to the cell is that growth on G6P provides too little carbon and too much phosphorus, and therefore extrusion of excess phosphate in exchange for G6P is favourable. As currently understood, these exchange systems do not involve electrogenic movement of molecules across membranes.

8.3.8 Transport driven by phosphoryl transfer from phosphoenolpyruvate

The fourth general class of transport mechanism is the phosphotransferase system (PTS), which catalyses the transport of several hexose and hexitol sugars, such as glucose and mannitol, in many different bacterial genera, including *E. coli* and *Staphylococcus aureus*. A distinctive feature of this system (Figure 8.10) is that an integral membrane polypeptide, specific for a particular sugar and generally called enzyme II, binds the sugar from the external surface and phosphorylates it to give, for example, G6P. The phosphorylated sugar is released into the cytoplasm. This has often been called a group translocation mechanism because substrates and products have to

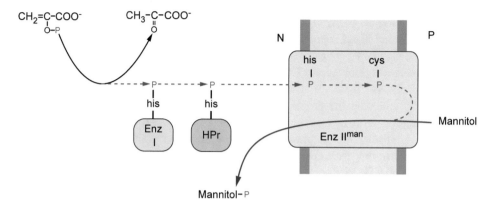

Figure 8.10   Schematic representation of the phosphotransferase system (PTS) for mannitol in *E. coli*.
In this case, the mannitol-specific enzyme II^{man} accepts phosphate onto a histidine residue from which it moves to a cysteine residue (a thus far unique example of a phosphocysteine catalytic intermediate in biology) before transfer to the incoming mannitol. As mentioned in the text, in other PTS transport systems (e.g., for glucose in *E. coli*), a separate enzyme III catalyses transfer of phosphate from HPr to the enzyme II^{glu}.
Sequence analysis of the enzyme II^{man} shows that a cytoplasmic domain is in effect a fused type III enzyme. The N-terminal region contains the mannitol-binding domain. The enzyme II molecules (excluding where appropriate fused enzyme III domains) all have in the order of 650 amino acids, between 350 and 380 of which could fold to form transmembrane α-helices. There may be extensive regions of relatively hydrophilic polypeptide that extend from the bilayer. Such proteins must presumably fold to give a hydrophilic central channel in which phosphorylation occurs.

approach and leave the catalytic site of enzyme II along defined pathways (i.e., glucose should not have access to this site from the cytoplasm).

The ultimate source of the phosphate group for the phosphorylation is cytoplasmic phosphoenolpyruvate (PEP), which first autophosphorylates the N-1 nitrogen of a histidine residue in a soluble protein known as enzyme I. The phosphoryl group is subsequently transferred to the histidine phosphocarrier protein (HPr). HPr then transfers its phosphoryl group to the A domain of the sugar-specific enzymes II, which are divided into four structurally distinct families corresponding to which sugar is handled. All enzymes II have similar minimal organisations comprising A and B cytoplasmic domains and a membrane-bound sugar transporter comprising the C domain. In some instances, the domains are expressed as a contiguous protein, whereas in other cases they are expressed as separate proteins. From enzyme IIA, the phosphoryl group is transferred to IIB and finally on to the incoming sugar molecule bound to the transmembrane IIC domain. Despite the similar domain organisation of the enzymes II, the A and B cytoplasmic domains from the different branches of the PTS bear no sequence similarity to one another. However, structural analyses of all components, apart from the transmembrane domain C, have shown how binary protein interactions and phosphoryl transfer between histidine residues can occur, including the basis for how these interactions can occur between structurally diverse partners.

Under physiological conditions, the ΔG for PEP hydrolysis is more negative than for ATP hydrolysis and is sufficient for phospho transfer to histidine. The driving force for transport is considerable; it has been calculated that at equal concentrations of pyruvate and PEP, the equilibrium intracellular concentration of a phosphorylated sugar would reach approximately $100\,M$ at $10^{-6}\,M$ external sugar. Such equilibration is apparently not achieved, indicating the requirement for tight control of these transport systems. Another energetic facet is that the bacterial cell expends one PEP molecule for the acquisition of an intracellular phosphorylated sugar. Other active transport processes (e.g., energy-consuming proton symports) are followed by intracellular phosphorylation at the expense of ATP. The conversion of PEP to pyruvate in the pyruvate kinase reaction yields only one molecule of ATP, in a reaction with a large negative Gibbs energy change under cellular conditions, and thus it is energetically advantageous to harness more fully the energy associated with PEP hydrolysis to drive both the transport and the phosphorylation events.

8.3.9 Generation of Δp by transport

Although transport processes are usually consumers of Δp, some species of bacteria have evolved strategies for coupling the export of end products of metabolism to the generation of Δp. For example, in fermenting bacteria, there is evidence that under some conditions the monovalent lactate anion leaves the cells together (i.e., in symport) with more than one proton, thus generating Δp. The driving force for this is the movement of lactate from a high concentration in the cell to a lower external concentration.

A different type of exchange system that generates Δp has been described. In the anaerobe *Oxalobacter formigenes*, oxalate is taken up as a dianion. Once in the cell, it is decarboxylated:

$$(COO)_2^{2-} + H^+ \rightleftharpoons HCOO^- + CO_2$$

Formate as a monoanion exits from the cell in exchange for the oxalate, with the overall effect that the exchange is responsible for generating a membrane potential, positive outside (Figure 8.11). There is also a tendency for a ΔpH to develop, alkaline inside, owing to the consumption of a proton during the decarboxylation reaction. This mechanism requires that the CO_2 leaves the cell in the gaseous form. If it were to be hydrated to HCO_3^- and exported as such from the cell, no $\Delta \psi$ would be generated. In line with this requirement, it would be expected that carbonic anhydrase activity is very low or absent in these cells.

The mucosal pathogen *Ureaplasma urealyticum* provides a most unusual final example. This organism contains very high levels of urease and thus generates large amounts of ammonium cations in the cytoplasm as a result of the hydrolysis of urea. The cytoplasmic membrane contains transporters that catalyse efflux of ammonium from the cell. These transporters are known as the Amt proteins, which are present in this organism. There is debate as to whether Amt moves NH_3 or NH_4^+. Clearly, only the latter substrate would generate $\Delta \psi$, positive outside, which could provide a driving force for the F_oF_1-ATP synthase that is known to be present from the complete genome sequence of the organism. This scheme supposes that the uncharged urea molecule enters the cell via

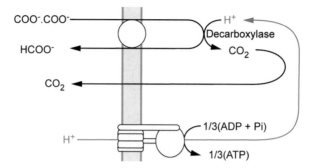

Figure 8.11   Generation of Δp by anion exchange in *Oxalobacter formigenes*.
The cycle of influx, decarboxylation and efflux effectively constitutes a proton pump with a stoichiometry of $1H^+$ per turnover. If the H^+/ATP ratio for the ATP synthase is 3, then the maximum stoichiometry of ATP synthesis would be 1 for each three oxalate molecules decarboxylated.

facilitated diffusion and that the other product of urease action, carbon dioxide, leaves the cell as the uncharged gas. It is also apposite here to remind the reader of the decarboxylation system that is linked to sodium transport (Chapter 5).

8.3.10 Transport of macromolecules across the bacterial cytoplasmic membrane

Further reading: Park and Rapoport (2012), Palmer and Berks (2012), Rollauer *et al.* (2012)

Bacteria synthesise all of their proteins in the cytoplasm, but the destination of some of the proteins may be the periplasm, the outer membrane, or even the external medium. In these cases, the newly synthesised polypeptide must be transported across the cytoplasmic membrane. Such transport appears in general to require the participation of the *sec* gene products, the retention by the polypeptides of a relatively unfolded state (the role of binding to SecB), ATP hydrolysis by the SecA protein, and finally a leader or signal sequence of approximately 20–25 amino acids at the N-terminus of the polypeptide. In some cases, the roles of SecA and SecB may be bypassed by the nascent protein being 'pushed' into the Sec translocon by being coupled to protein synthesis with a ribosome being effectively docked on to the Sec proteins. The signal sequence is not conserved between proteins, but some features are always found, including the presence of several positively charged residues at the N-terminus and a hydrophobic sequence in the middle. A high-resolution structure of this Sec system is available.

  The main connection with the subject matter of this book is that there is evidence that Δp is involved, as well as ATP hydrolysis by SecA, in driving the translocation of nascent polypeptides across the cytoplasmic membrane. The membrane must remain sealed as a polypeptide is in transit because otherwise unrestricted proton flow through the Sec system would dissipate Δp. The structure of the Sec protein complex suggests a mechanism whereby the route for polypeptide movement is closed by hydrophobic interactions until it is opened by the entry of a polypeptide chain to be transported. Coupling of polypeptide transport to Δp is still a matter of debate. Presumably, one of the components of the Sec system must act as a proton–polypeptide antiport system. In this context, protein movement within the Sec apparatus could be the basis of the dependence on Δp. In the previous edition, we suggested that an exaggerated 'breathing' movement of a channel-type protein in the membrane may be responsible for squeezing the polypeptide across the membrane. It has subsequently been suggested that the proton translocating activity resides in the SecD and SecF proteins, which constitute a set of 12 transmembrane helices that sit side by side with the main part of the Sec complex that provides the actual transit pathway across the membrane. In addition to their 12 helices, these two polypeptides, or in some organisms a single fused polypeptide, make up six periplasmic regions, two of which—P1 and P4—constitute separate domains. It is hypothesised that P1 moves in response to proton passage through the SecDF pair of polypeptides. Such movement could allow P1 to catch an emerging polypeptide as it emerges from the Sec tunnel. Subsequently, P1 would move around before returning to capture a subsequent section of the polypeptide as it emerges from the tunnel. Of course, this description does not explain how the protonmotive force is involved. A clue is that some resemblance can be discerned between the SecDF proteins and the Acr B protein (discussed previously) that moves drugs out of bacteria by a peristaltic-type mechanism in a Δp-dependent manner. There are some conserved charged residues that line up between the two proteins and that have been previously demonstrated to be important for protonmotive force utilisation by Acr B. Variants of the corresponding residues in SecDF lead to loss of activity, thus reinforcing the connection between the two apparently unconnected protein systems. Although plausible, this scheme needs to be confirmed by further work.

It used to be considered axiomatic that proteins were translocated across membranes in an unfolded state. However, it has been discovered that certain proteins, especially those containing complex redox cofactors such as Fe–S centres, found for example in the Riekse centres of bacterial bc_1 complexes and thylakoid b_6f complex, and molybdenum-containing enzymes (e.g., periplasmic nitrate reductase (Nap); Chapter 5), are transported to the bacterial periplasm or thylakoid lumen as folded proteins with such cofactors already bound. The transport system is known as Tat (twin arginine transporter) in recognition of the presence of two sequential arginine residues in the targeting sequences of proteins destined to be transported by this system. Before its molecular characterisation, this Tat system, comprising TatABC proteins, was called the ΔpH transporter and known to import some proteins into the lumen of thylakoids. As the name suggests, this thylakoid system is powered by ΔpH, which, as you may recall (Chapters 3 and 4), is the dominant component of Δp in thylakoid membranes. It is now known that the Tat system generally depends on Δp. How the movement of protons back into the cell through the Tat complex drives movement of a folded polypeptide in the opposite direction is unknown. A recent crystal structure of TatC shows that a highly conserved glutamate is located on a transmembrane helix and is at a position in the middle of the bilayer. The role of carboxylate side chains in proton translocation has been discussed previously (e.g., bacteriorhodopsin in Chapter 6), and thus it is tempting to implicate this residue in the transduction of Δp into macromolecular movement, especially because conversion of this residue to alanine severely attenuates the activity of the Tat system. However, there is no obvious route for protons to access this glutamate and an alternative role for this glutamate could be in an intersubunit interaction. It is likely that this glutamate must form an ion pair; if TatC participates in proton movement it could be that the process occurs in conjunction with another subunit. A squeezing or peristaltic mechanism as envisaged for Sec seems unlikely; somehow, the TatABC has to rearrange such that the pathway, which has to be of variable size, for the protein is blocked in the absence of the protein substrate, to avoid dissipation of Δp, and then alternately faces opposite sides of the membrane to effect transport. It is possible that the role of Δp is not in driving transport per se but, rather, in organising the structural arrangement of the Tat proteins. The exact role(s) of Δp in the Tat system will only be clarified by additional research.

The import of proteins into mitochondria poses related problems and is addressed in Chapter 9.

INTRODUCTION TO PART 3

In Part 1, we described the basic principles of bioenergetics, focusing on the proton circuit; Part 2 detailed the molecular mechanisms of the pumps and carriers that drive the circuit and in Part 3 we apply this information to the mitochondrion in the cell. The growth of what can be termed 'mitochondrial physiology' has been explosive since the third edition was published in 2002. Indeed, there are currently few aspects of human physiology and disease that are not associated with some evidence or hypothesis concerning a mitochondrial involvement. If only 10% of these withstand the test of time, this will still mean that an understanding of mitochondrial function and dysfunction in the cell, tissue and body is of crucial importance.

We start by reviewing the key methodologies required to extend the study of mitochondrial bioenergetics from the isolated organelle to the intact cell and beyond. We follow this with a chapter that we term 'mitochondrial cell biology,' in which we review the structure, biogenesis and turnover of mitochondria in the cell. This is followed by a chapter reviewing the multiple signalling pathways that match mitochondrial function and capacity to demand. In the final chapter, we bring together all this information to review our current understanding of the bioenergetics of specific tissues, their dysfunction in disease states, and the still largely mysterious process of aging.

The task we are faced with in Part 3 is to condense the vast literature from the past decade (a single representative topic, apoptosis, has produced almost 15,000 mitochondria-related publications since the third edition) into a few paragraphs without losing the key principles. Our approach has been to restrict ourselves wherever possible to mammalian systems and to focus on bioenergetic aspects by simplifying the protein chemistry, giving key reviews for further reading. We have tended to cite the most recent authoritative reviews, with the result that the discoverers of key mechanisms and pathways are frequently not acknowledged. Should any of these scientists read this section, we hope they will understand that this is the only way we can condense 100,000 mitochondria-related papers during the past decade into this book.

9 CELLULAR BIOENERGETICS

9.1 INTRODUCTION

Traditionally, it was usual to disrupt a tissue and isolate the mitochondria in order to investigate the role of the organelle in the tissue's bioenergetics. In recent years, however, a battery of techniques have appeared that allow the researcher to investigate the bioenergetics of mitochondria *in situ* within the cell with a precision approaching that for the isolated organelle. 'Cellular bioenergetics' has advantages relative to classical studies using isolated mitochondria, but it also has limitations. The undisturbed cytoplasmic environment and retained interactions with the other components of the cell can greatly increase the physiological relevance of the investigation. On the other hand, cells are more complex with multiple compartments, and the restricted permeability of the plasma membrane means that many substrates and reagents that work with isolated mitochondria cannot be used with intact cells. Also, although mitochondria can be isolated from animals of any age, it is frequently difficult or impossible to culture primary cells such as neurons from adult animals, while cell lines derived from tumour cells may diverge in function from cells in the native tissue. Finally, it is important to expose the cells to appropriate media, particularly with regard to the choice of substrates, and to remember the importance of growth factors contained in any added serum. It should not be assumed that all that cells require is a high concentration of glucose; indeed, glucose transport is highly regulated in many cells.

Bioenergeticists are frequently asked by colleagues to help them investigate mitochondrial function in their cell preparations, and so a few definitions need to be clarified. Although the primary role of mitochondria is to supply ATP to the cell, these organelles are involved in many additional processes, including the generation and removal of reactive oxygen species, ion transport, and metabolic interconversions. A failure of any one of these processes can be termed 'mitochondrial dysfunction'; however, caution must be exercised before concluding that an altered cell bioenergetic behaviour is the cause, rather than a downstream effect, of the primary dysfunction. The focus of this chapter is the methodology for the investigation of *in situ* mitochondrial bioenergetic function and dysfunction.

Bioenergetics. Doi: http://dx.doi.org/10.1016/B978-0-12-388425-1.00009-9

Chapter 4 emphasised the central role of the proton circuit in defining the bioenergetics of isolated mitochondria and bacteria, and it is of equal relevance in the investigation of the bioenergetics of intact eukaryotic cells. Oxygen electrodes can be used to quantify the proton current, with the proviso that non-mitochondrial oxygen uptake (which can be considerable in cells such as neutrophils with active oxygenases) must be allowed for. This is typically done by determining the residual respiration after inhibition of mitochondrial electron transport. Quantitative determination of Δp in intact cells is complex, prone to errors, and rarely attempted; however, considerable information can frequently be obtained by semiquantitative determinations of mitochondrial membrane potential, $\Delta\psi_m$, (typically monitoring changes in potential from an assumed starting point) with (or usually without) parallel determinations of ΔpH. Fluorescence techniques are usually employed, and although these are not without disadvantages (discussed later), they do have the advantage that fluorescence microscopy allows determinations to be performed at single-cell and even single-mitochondrial levels of resolution. Unfortunately, these levels of resolution are not currently available for respiratory studies, but the combination of proton current determinations on cell populations with parallel single-cell monitoring of $\Delta\psi_m$ is a powerful approach to monitor *in situ* mitochondrial function and dysfunction. Finally, multiple wavelength fluorescence microscopy allows $\Delta\psi_m$ to be monitored in parallel with additional parameters such as $\Delta\psi_p$ (plasma membrane potential), $[Ca^{2+}]_c$ or $[Ca^{2+}]_m$ (cytoplasmic and matrix free Ca^{2+} concentrations, respectively), and levels of reactive oxygen species.

9.2 THE CYTOPLASMIC ENVIRONMENT

The cytoplasm of a typical mammalian cell contains approximately 120 mM K^+ and 10 mM Na^+, contrasting with approximately 4 mM K^+ and 120 mM Na^+ in the extracellular fluid. This imbalance is maintained by the action of the Na^+/K^+-ATPase, expelling $3Na^+$ in exchange for $2K^+$ for each ATP hydrolysed. Although this sodium pump is potentially very active, its consumption of ATP under most conditions is rather modest, being limited by the slow resting inward Na^+ flux through ion channels and the Na^+ symporters and antiporters moving ions and metabolites across the membrane. These latter vary from cell to cell, but examples of ion antiporters include the Na^+/H^+ exchanger at the plasma membrane involved in cytoplasmic pH regulation and a $3Na^+/Ca^{2+}$ exchanger that contributes to Ca^{2+} extrusion.

In most cells, the constitutive K^+ conductance of the plasma membrane is high and in excess over the Na^+ conductance. As a consequence, the resting $\Delta\psi_p$ more closely approximates to the Nernst potential (Eq. 3.41) for K^+ rather than for Na^+, which is why the extracellular medium has a positive potential relative to the cell cytoplasm.

In this book, we use a convention for plasma membrane potentials ($\Delta\psi_p$) such that a positive sign means that the external medium is positive with respect to the cytoplasm. This allows us to be consistent across the plasma and mitochondrial membranes. Note, however, that this is the opposite of the convention used by electrophysiologists for $\Delta\psi_p$.

In neurons and muscle cells, whose plasma membranes are 'excitable' by virtue of their ability to fire action potentials, the constitutive K^+ conductance is even higher, and these cells have a high $\Delta\psi_p$ that approaches the K^+ Nernst potential. More exactly, in cells in which K^+, Na^+ and Cl^- are the dominant permeable species, $\Delta\psi_p$ is a function of the ion gradients of these species weighted by their respective permeabilities, and it is calculated from the Goldman equation (Eq. 3.42).

The membrane potentials across eukaryotic plasma membranes, where relatively slow transport processes often enable ion gradients to be sustained for several hours in the absence of ion pumping, are largely defined by these transmembrane ion gradients and ion permeability pathways. This is in contrast to the mitochondrion, with its enormous surface-to-volume ratio and rapid movement of counter-ions, where the membrane potential is controlled by the electrogenic proton pumps.

The free cytoplasmic Ca^{2+} concentration, $[Ca^{2+}]_c$, is only about 0.1 µM under resting conditions, contrasting with approximately 1 mM in plasma. The enormous Ca^{2+} concentration gradient across the plasma membrane is maintained by two transport processes whose relative importance depends on the cell type. At least four isoforms of the plasma membrane Ca^{2+}-ATPase are expressed in different cell types. Each expels 1 Ca^{2+} in exchange for $2H^+$ per ATP hydrolysed (Brini and Carafoli, 2009), whereas a plasma membrane $3Na^+/Ca^{2+}$ exchanger utilises the Na^+ electrochemical potential maintained by the Na^+/K^+-ATPase to expel Ca^{2+} in parallel, although it can reverse and allow Ca^{2+} entry when Na^+ gradients are collapsed. Thermodynamic calculations indicate that the Ca^{2+}-ATPase should be capable of lowering $[Ca^{2+}]_c$ to less than 1 nM, indicating that $[Ca^{2+}]_c$ is defined under resting conditions not by approach to a thermodynamic equilibrium but, rather, by a steady-state cycling between the Ca^{2+}-ATPase and one or more independent influx pathways. As discussed later (Section 9.4), steady-state Ca^{2+} cycling also occurs across the mitochondrial inner membrane and provides a means of rapidly restoring basal conditions (at the expense of a small but continuous drain on the proton circuit).

In addition to these transport processes, a plethora of ion channels and ionotropic ('ion-moving') as opposed to G protein-coupled ('metabotropic') receptors affect the fluxes of Na^+, K^+, or Ca^{2+} across the plasma membrane. A description of these is beyond the scope of this book, but those of most relevance to cellular bioenergetics include voltage-activated Na^+ and K^+ channels that control plasma membrane potentials, ATP-inhibited K_{ATP} channels that respond to low-energy conditions in cardiac muscle and play a central role in the triggering of insulin secretion from pancreatic β cells, voltage-activated Ca^{2+} channels that modulate metabolism and control neurotransmitter and hormone exocytosis, and neurotransmitter receptors that initiate or inhibit action potentials (Figure 9.1).

9.3 MITOCHONDRIAL MONOVALENT ION TRANSPORT

Further reading: Brierley *et al.* (1994), Garlid and Halestrap (2012)

Mitochondria cannot survive in a cytoplasm containing high millimolar concentrations of Na^+ and K^+ without mechanisms to expel any cations that slowly leak across the inner mitochondrial membrane (IMM) by electrical uniport driven by the 150-mV $\Delta\psi_m$.

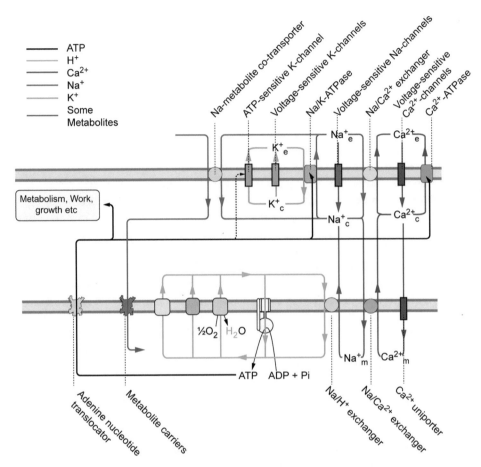

Figure 9.1   Simplified schematic of interrelations between mitochondrial and plasma membrane transport processes in a typical cell.
The primary proton circuit (light blue) drives the cycling of Ca^{2+} (magenta) across the IMM with the intermediacy of an Na^+ circuit (red). The ATP synthase generates ATP for use by metabolism, etc. and to drive the Na^+/K^+-ATPase at the plasma membrane, cycling Na^+ and K^+ (green) across the plasma membrane. K^+ cycling is completed via a range of K^+ channels, some of which are activated by a decreased $\Delta\psi_p$, whereas others are inhibited by ATP. Na^+ cycling can drive Na^+–metabolite co-transporters and Na^+/Ca^{2+} exchange. Plasma membranes also contain Ca^{2+}-ATPases to expel the cation. Note that only some metabolites are accumulated across the plasma membrane by Na^+ co-transport and that this varies from cell to cell. Endoplasmic reticulum transport and several additional processes are omitted for clarity.

As originally demonstrated by Mitchell using passive swelling experiments of the type detailed in Chapter 2, mitochondria show both Na^+/H^+ and K^+/H^+ exchange activity in the inner membrane, with the former being more active. Because mitochondria operate with a ΔpH of approximately -0.5, equivalent to a threefold gradient of H^+ out/in, these exchangers would be capable at equilibrium of lowering the matrix K^+ or

Na^+ to approximately 30% of the concentration in the cytoplasm. A dynamic balance between monovalent cation uptake and efflux is important for maintaining the matrix volume. Note that although this cycling utilises the proton circuit, the contribution to the inwardly directed proton current is usually very small.

The K^+/H^+ exchanger in yeast mitochondria, encoded in part by the LETM1/0LO27 gene, is well characterised, and its deletion, as would be predicted, leads to massive swelling of the *in situ* mitochondria. In most tissues apart from liver, the mitochondrial Na^+/H^+ exchanger is essential to couple the slow Ca^{2+} cycling across the inner membrane to the proton circuit (see Figure 9.3). Its molecular identity is controversial, in part because of difficulties in distinguishing between a plasma membrane, endosomal or mitochondrial location, and some initial reports have been contested.

The plasma membranes of many cells contain K^+ uniport channels that are activated by modest reductions in cytoplasmic ATP/ADP ratio (Figure 9.1). Such an ATP-sensitive K channel (K_{ATP}) plays an important role in controlling insulin secretion from pancreatic β cells (Section 12.5). The channel complex includes an inwardly rectifying K^+ (K_{IR}) channel and a sulfonylurea receptor (SUR). A number of reagents that affect the activity of the plasma membrane K_{ATP} have actions on isolated mitochondria, and this has led to the proposal that the mitochondrial inner membrane may also possess a K_{ATP} channel (mK_{ATP}) that could control K^+ entry into the matrix and hence control matrix volume in concert with the K^+/H^+ exchanger. By analogy with the plasma membrane channel, a drop in cytoplasmic ATP would activate the channel, leading to an osmotic expansion of the matrix that could control the activity of matrix NAD^+-linked dehydrogenases, whose ability to transfer NADH to complex I can be viscosity-limited when the matrix is condensed.

The history of the putative mitochondrial K^+ channel has been fraught with controversy. A simple re-targeting of the plasma membrane subunits to the mitochondrion is not supported because their sequences do not contain mitochondrial targeting sequences and are not found in mitochondrial proteomic databases. No convincing antibody data exist, and many reagents proposed to affect mK_{ATP} activity have off-target effects on mitochondrial bioenergetics. Knockout studies are not possible because of the failure to unequivocally identify a protein or gene and due to confounding effects of the plasma membrane K_{ATP} channel. Thus, although a splice variant of SUR2 has been identified in mitochondria, and complex II has been implicated in the control of mK_{ATP} activity, the channel itself remains elusive. Despite the controversy over the channel, a broad literature exists on the cardioprotective actions of putative mK_{ATP} activators in reperfusion injury (Section 12.3.2).

9.4 MITOCHONDRIAL CALCIUM TRANSPORT

Further reading: Nicholls and Chalmers (2004), Drago *et al.* (2011); see also *Cell Calcium* (2008), Vol. 44 (1)

Three Ca^{2+} transporting membranes bound the cytoplasm: the inner mitochondrial membrane, the endoplasmic reticulum membrane and the plasma membrane. The level at which $[Ca^{2+}]_c$ is maintained is a dynamic consequence of the net transport across all three membranes. In most cells, steady-state Ca^{2+} cycling across the plasma membrane maintains resting $[Ca^{2+}]_c$ at approximately $100\,nM$. Under physiological conditions, $[Ca^{2+}]_c$ can be raised by Ca^{2+} entry through plasma membrane voltage-activated Ca^{2+} channels or by Ca^{2+} release from endoplasmic (or sarcoplasmic) reticulum by inositol trisphosphate (or ryanodine) receptors. Mitochondria respond to Ca^{2+} elevations from either source, but endogenous cytoplasmic Ca^{2+} buffer proteins greatly slow the propagation of Ca^{2+} waves across the cell, so the physical location of the mitochondria is important in determining their response. For example, mitochondria located just below the plasma membrane preferentially accumulate Ca^{2+} entering across the plasma membrane, whereas a close association between mitochondria and endoplasmic reticulum facilitates signalling between the organelles (Section 10.3.3).

It is essential to distinguish between free and total Ca^{2+} concentrations. In the matrix, these may differ under extreme conditions by a factor of more than 100,000. Thus, under conditions of high matrix Ca^{2+} loading, as can occur in neurons (Section 12.6.2), the total matrix Ca^{2+} concentration can exceed $1\,M$, in the form of a calcium phosphate complex, whereas $[Ca^{2+}]_m$ may be a few micromolar. Total Ca^{2+} is most relevant to studies of cytoplasmic Ca^{2+} buffering and to Ca^{2+} overload under pathological conditions, whereas free Ca^{2+} is most relevant to metabolic studies.

A variety of optical techniques are available to monitor $[Ca^{2+}]_m$, although none of the techniques are without problems (Figure 9.2). Variants of the photoprotein aequorin can be targeted to the matrix and can detect $[Ca^{2+}]_m$ transients. Fluorescent Ca^{2+} indicators loaded into cells as the membrane-permeant acetoxymethyl (AM) esters are most commonly used to monitor $[Ca^{2+}]_c$ following their hydrolysis by nonselective cytoplasmic esterases to the impermeant, Ca^{2+}-sensing free carboxylic acid. Conditions can be tuned to optimise their accumulation into the matrix, sometimes after the cytoplasmic signal has been quenched with Mn^{2+}. Alternatively, a Ca^{2+} indicator can be used that is positively charged in its permeant esterified AM form (e.g., rhod-2) and would therefore be predicted to accumulate rapidly into the matrix before its hydrolysis to the active probe.

The most recent approaches involve targeting to the matrix of green fluorescent protein variants that have been engineered to respond to free Ca^{2+} concentrations in the sub-micromolar range. *Chameleons* are fluorescence resonance energy transfer-based indicators with a calmodulin motif linking cyan- and yellow-fluorescent proteins. Binding of Ca^{2+} induces a conformational change in the calmodulin that alters the separation between the CFP (donor) and YFP (acceptor) and hence the YFP emission when the CFP is excited. *Ratiometric pericam* is a fusion between calmodulin, its target protein M13, and modified YFP (Figure 9.2). $[Ca^{2+}]_m$ can be monitored by ratioing the emission after excitation at 488 and $415\,nm$. *Camgaroos* are fusions between YFP and calmodulin.

Remember that none of the previously discussed techniques detect the quantity of total matrix Ca^{2+}. Indeed, once a calcium phosphate complex is formed in the matrix (Figure 9.3), little change in $[Ca^{2+}]_m$ occurs even if the total Ca^{2+} load were to increase 50-fold. A qualitative technique to estimate total matrix Ca^{2+} is to rapidly release any Ca^{2+} into the cell cytoplasm with a protonophore and to determine the rise in $[Ca^{2+}]_c$

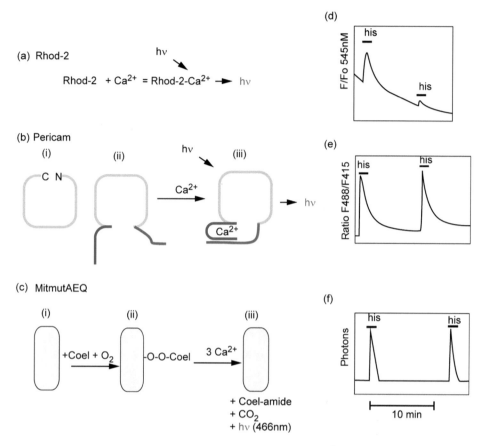

Figure 9.2   Techniques for monitoring matrix free Ca^{2+} concentrations.
(a) Rhod-2 is positively charged in the membrane-permeant acetoxymethyl
ester form and can accumulate in the matrix before hydrolysis to the active
rhod-2. The Ca^{2+}-bound form is fluorescent. (b) Mitochondrially targeted
pericams are engineered from yellow fluorescent protein (YFP) by fusing the
N- and C-termini and breaking the peptide bond at Tyr145 to insert
calmodulin (blue) and myosin light-chain kinase (red). (c) A mitochondrially
targeted apoaequorin variant, mitmutAEQ, forms a complex with added
coelenterazine (Coel) to form aequorin. When this binds Ca^{2+}, the complex is
disrupted, regenerating apoaequorin, liberating coelenteramide (Coel-amide),
and emitting photons. (d–f) The response of the three probes to the rise in
matrix free Ca^{2+} in HeLa cells exposed to histamine (his) to release Ca^{2+}
from endoplasmic reticulum, allowing it to be accumulated by the
mitochondria. Note that rhod-2 fails to respond to the second stimulus, and
each technique reports a different kinetic.
Adapted from Fonteriz et al. (2010).

with a low-affinity cytoplasmic indicator. Electron probe microanalysis (Stanika *et al.*,
2010) provides a quantitative measure of total Ca^{2+} accumulated in mitochondria and
cytoplasm, but it has the limitation that samples must be fixed and processed, although
other ions, such as Na^+ and K^+, can be quantified in parallel.

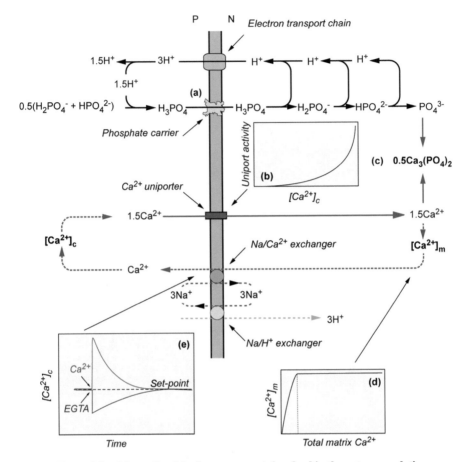

Figure 9.3   Schematic of the ion movements involved in the net accumulation (solid arrows) and steady-state cycling (dashed arrows) of Ca^{2+}.
(a) The phosphate carrier transports $H_2PO_4^-$ with H^+. Because three proton dissociations are required to form PO_4^{3-}, the concentration of this species is inversely proportional to the cube of the proton concentration in the matrix. (b) The uniport activity increases as the 2.5 power of cytoplasmic free Ca^{2+} concentration $[Ca^{2+}]_c$ (see insert). (c) The tricalcium phosphate complex forms when its ion activity product is exceeded. Because the concentration of PO_4^{3-} increases with pH, the solubility of Ca^{2+} decreases and is approximately $2\,\mu M$ when matrix pH is approximately 7.7 and external total phosphate is approximately $5\,mM$. (d) The matrix free Ca^{2+} concentration, $[Ca^{2+}]_m$, varies with total matrix Ca^{2+} until approximately $10\,nmol/mg$ is accumulated and the tricalcium phosphate complex starts to form. In this initial region, matrix Ca^{2+} can regulate tricarboxylic acid enzymes. Once the complex forms, $[Ca^{2+}]_m$ is invariant with matrix Ca^{2+} load, and the cytoplasmic Ca^{2+} buffering mode is seen (see insert). (e) The Na^+/Ca^{2+} exchanger is controlled by $[Ca^{2+}]_m$; when the matrix is in cytoplasmic buffering mode ($>10\,nmol\ Ca^{2+}/mg$ accumulated), mitochondria seek to accumulate (or release) matrix Ca^{2+} to restore a set point at which the kinetics of uptake via the uniporter exactly balance efflux via the Na^+/Ca^{2+} exchanger.
Adapted from Nicholls and Chalmers (2004).

9.4.1 Mitochondrial Ca^{2+} buffering

Further reading: Nicholls (2005b)

Although the extraordinary capacity of isolated mitochondria from vertebrate sources to accumulate Ca^{2+} has been known for more than 50 years, early studies paid little attention to its possible physiological significance, and it is still common to find the misleading statement that mitochondrial Ca^{2+} uptake is a 'low-affinity' process. A clue to the physiological function is to determine the level to which isolated mitochondria can reduce the free Ca^{2+} concentration in the medium, $[Ca^{2+}]_e$. Depending on the source of the mitochondria and the concentration of external Mg^{2+}, mitochondria can lower $[Ca^{2+}]_e$ to 0.5–2 µM. Interestingly, these values can be almost independent of the total load of accumulated Ca^{2+} over a range from 10 to 500 nmol Ca^{2+} mg protein^{-1}. Isolated mitochondria can thus act as almost perfect buffers of external free Ca^{2+}, $[Ca^{2+}]_e$. Note that Ca^{2+} chelators do not actually buffer the cation in the same way as the mitochondrion. At a given chelator concentration, a doubling of the total Ca^{2+} concentration at least doubles the free concentration, even though this may be a small fraction of the total cation. The first thing to notice about mitochondrial Ca^{2+} buffering is that the level at which the mitochondria attempt to buffer the Ca^{2+} in their environment is somewhat higher than typical values reported for $[Ca^{2+}]_c$ in cells that lie in the range 0.1–0.5 µM. This suggests that mitochondria will not accumulate Ca^{2+} in resting cells but may set an upper limit to $[Ca^{2+}]_c$ by reversibly accumulating the cation above this level. As discussed later (Section 12.6.2), there is considerable evidence for this role in neurons.

Mammalian mitochondria cycle Ca^{2+} between a Ca^{2+} uniport uptake pathway and a Ca^{2+}/3Na$^+$ (or possibly Ca^{2+}/2Na$^+$) antiport efflux pathway. In concert with the Na$^+$/H$^+$ exchanger mentioned previously the Ca^{2+} and Na$^+$ cycling are driven by net proton re-entry into the matrix (Figure 9.3). The cycling is slow (2–10 nmol Ca^{2+} min^{-1} mg mitochondrial protein^{-1}) but contributes to the endogenous proton leak discussed in Section 4.5.

9.4.1.1 The Ca^{2+} uniporter

Further reading: Griffiths (2009), Drago *et al.* (2011)

The activity of the liver mitochondrial Ca^{2+} uniporter increases as the 2.5th power of the free Ca^{2+} in the medium (Zoccarato and Nicholls, 1982). When $[Ca^{2+}]_e$ is sufficiently high, the conductance of the uniporter can utilise the total respiratory capacity of most mammalian mitochondria (except heart) for the accumulation of the cation. Note that yeast mitochondria do not transport Ca^{2+}. Although the hexavalent glycoprotein stain ruthenium red and its derivative Ru360 are effective inhibitors of the uniporter in isolated mitochondria, permeability problems limit their use with intact cells. In fact, there is controversy regarding whether they enter at all. As little as 1 pmol mg^{-1} of ruthenium red can inhibit the uniporter of isolated mitochondria, indicating that its abundance in the mitochondrial inner membrane is very low, and this helps to explain why the search for the responsible protein(s) took 40 years. Suggestions that novel uncoupling proteins were components of the channel have not been accepted, but a combination of comparative physiology (e.g., the absence of the uniporter from yeast mitochondria), proteomics, and evolutionary genomics highlighted two gene products, termed MICU1 and MCU (Section 8.2.5). Because MICU1 did not span the membrane, it was unlikely to be the uniporter channel. However, a search for proteins functionally related to MICU1 converged on a protein with two predicted transmembrane helices that forms oligomers in the inner membrane and has been named MCU, which stands for mitochondrial Ca^{2+} uniporter. In a key experiment, mice were treated with a siRNA duplex specific for MCU and targeted to the liver, producing an 84% mRNA knockdown. Subsequently isolated liver mitochondria showed unchanged respiratory control and $\Delta\psi_m$ responses to ADP but had an almost complete failure of Ca^{2+} uptake.

9.4.1.2 Mitochondrial Ca^{2+} cycling

Further reading: Crompton and Heid (1978), Jiang *et al.* (2009), Palty and Sekler (2012)

At equilibrium, the Ca^{2+} uniporter acting in isolation would be capable of developing a concentration gradient of the cation across the inner membrane approaching 10^5 (10-fold for each 30 mV of $\Delta\psi$; Eq. 3.39). The presence of the $Ca^{2+}/3Na^+$ antiport prevents this by providing an independent efflux pathway so that in the steady state there is a continuous cycling between uptake and efflux pathways (Figure 9.3). There is evidence that a transporter termed NCLX is the member of the $Ca^{2+}/3Na^+$ antiporter family that is located on the inner mitochondrial membrane.

$Ca^{2+}/3Na^+$ exchange is found in mitochondria from most tissues, including heart, brain and brown adipose tissue. The structure of a bacterial $Ca^{2+}/3Na^+$ exchanger was discussed in Section 8.2.5. Liver mitochondria have an Na^+-independent $Ca^{2+}/2H^+$ antiporter. Somewhat confusingly, LETM1, mentioned previously as a mitochondrial K^+/H^+ exchanger, has also been proposed to catalyse this $Ca^{2+}/2H^+$ exchange. Either exchange mechanism drives the efflux of Ca^{2+} from the matrix, the Na^+-coupled pathway requiring the additional participation of the Na^+/H^+ exchanger introduced previously (Figure 9.3). The existence of the independent efflux pathway can be demonstrated in isolated mitochondria most simply following the selective inhibition of the uptake pathway by ruthenium red once steady-state conditions have been obtained. Because the inhibitor does not affect the efflux pathway, a net efflux of Ca^{2+} from the matrix occurs. The exchanger can be inhibited in isolated mitochondria by CGP37157, but in intact cells this can also affect plasma membrane transporters.

9.4.1.3 Net Ca^{2+} uptake into the matrix

Further reading: Nicholls (2005b)

The steady-state cycling of Ca^{2+} and Na^+ across the inner membrane driven by the proton circuit (Figure 9.3) seems symmetrical and does not in itself explain the kinetics of Ca^{2+} uptake and release by the organelles. To understand this, it is necessary to review the distinctive kinetics of the Ca^{2+} uptake and efflux pathways and the way in which Ca^{2+} is reversibly stored in the matrix.

The ability of mitochondria to accumulate Ca^{2+} from media containing greater than $0.5\,\mu M$ $[Ca^{2+}]$ is truly spectacular. Under appropriate conditions, in excess of $1\,\mu mol$ of Ca^{2+} per milligram of mitochondrial protein can be sequestered within the matrix with no deterioration of bioenergetic function, equivalent to a concentration of total Ca^{2+} approaching $1\,M$. Such accumulation requires the presence of P_i, which is taken up in parallel (Figure 9.3). Ca^{2+} entry lowers $\Delta\psi$, allowing the net expulsion of more protons by the respiratory chain. If this were the only process, Ca^{2+} accumulation would soon stop as $\Delta\psi_m$ is converted to ΔpH (Figure 4.9). However, in the presence of external P_i, the increasing ΔpH drives the phosphate into the matrix via the phosphate transporter (Sections 8.2.2 and 9.5.1). Note that a H_3PO_4 uniport is energetically equivalent to an exchange of $H_2PO_4^-$ for OH^-. Phosphate transport serves two functions in this context: (1) It neutralises the increase in internal pH (Figure 4.9); and (2) the transported P_i combines with the accumulated Ca^{2+} to form a reversible calcium phosphate complex that is able to instantly dissociate when $\Delta\psi_m$ is collapsed, allowing Ca^{2+} to be released via the uniporter and P_i via the phosphate transporter. Nevertheless, the complex is osmotically inactive, preventing matrix swelling as ion accumulation proceeds.

Accumulation will continue until the mitochondrion succeeds in lowering $[Ca^{2+}]_e$ to a level at which the rate of uptake and efflux balance. When the mitochondrial matrix contains greater than $10\,nmol$ Ca^{2+} mg protein^{-1} in the presence of physiological concentrations of P_i, the efflux pathway becomes independent of the matrix Ca^{2+} content because the free Ca^{2+} in the matrix is essentially buffered by the formation of this calcium phosphate complex. The value of $[Ca^{2+}]_e$ at which this kinetic balance occurs has been termed the *set point* and varies from 0.5 to $2\,\mu M$ depending on incubation conditions. Isolated mitochondria in the presence of P_i seek to lower $[Ca^{2+}]_e$ to the set point and thus appear to be capable of acting as effective 'buffers' of $[Ca^{2+}]_c$ in the cell. This predicts that the mitochondrion in the cell could serve as a temporary store of Ca^{2+} under conditions of elevated local Ca^{2+}. Over a range from approximately 10 to $500\,nmol$ Ca^{2+} mg^{-1}, the set point is virtually independent of the matrix Ca^{2+} load. To understand why, it is necessary to discuss how Ca^{2+} is stored within the matrix.

9.4.1.4 Matrix free Ca^{2+} concentrations

Further reading: Chalmers and Nicholls (2003), Denton (2009)

When $1-10\,mM$ P_i (a physiological range) is present in the incubation medium for isolated mitochondria, the accumulation of the first $10\,nmol\,mg^{-1}$ Ca^{2+} causes the free matrix Ca^{2+} concentration, $[Ca^{2+}]_m$, to rise to approximately $2\,\mu M$ and thereafter to remain virtually constant as the total matrix load is increased from 10 to at least $500\,nmol\,mg^{-1}$. This implies that a Ca–P_i complex (or gel) forms in the matrix when $[Ca^{2+}]_m$ is only $2\,\mu M$. However, this differs dramatically from the test-tube, in which millimolar concentrations of Ca^{2+} and P_i can coexist (and do in most physiological cell incubation media). What is special about the mitochondrial matrix to promote complex formation? The answer seems to lie in the alkaline pH of the matrix under these conditions. P_i probably enters the matrix on the phosphate transporter as a $P_i^-:H^+$ symport (Section 8.2.2), equivalent to the phosphate entering as the fully protonated H_3PO_4. However, the form that interacts with Ca^{2+} in the matrix to form the complex is the trianion PO_4^{3-}, and so the phosphate has to undergo three sequential deprotonations (Figure 9.3). It was shown in Section 4.4.2 that a monovalent weak acid such as acetate accumulates as a function of ΔpH. However, a trivalent acid such as phosphate accumulates its trianion as the third power of the pH gradient, and the concentration of the trianion can easily be 100 times higher than that in the medium. The $Ca-P_i$ complex forms when the solubility product, $[Ca^{2+}]^3 \times [PO_4^{3-}]^2$, is exceeded. In an alkaline matrix, this can occur at $1-5\,\mu M$ $[Ca^{2+}]_m$, whereas in a medium close to neutrality, free Ca^{2+} can be in excess of $1\,mM$. The dramatic efflux of Ca^{2+} from the matrix when a protonophore is added is likely a consequence of the acidification of the matrix and consequent dissociation of the complex into free Ca^{2+} and P_i.

The activity of the inner membrane $3Na^+/Ca^{2+}$ exchanger, NCLX, seems to be controlled by $[Ca^{2+}]_m$, which explains why it is constant over a wide range of total matrix Ca^{2+} loads. In contrast, the uniporter activity increases as the 2.5th power of $[Ca^{2+}]_e$, as long as the proton current is not limiting. These distinctive kinetics are sufficient to explain the Ca^{2+} transport behaviour of isolated mitochondria (Figure 9.3). Above the set point, mitochondria accumulate Ca^{2+} because the uniporter is more active than the $3Na^+/Ca^{2+}$ exchanger. Below the set point, the reverse holds and mitochondria slowly release any accumulated Ca^{2+}.

What happens when $[Ca^{2+}]_e$ varies but does not reach the set point so that the $Ca-P_i$ complex is not formed? In this range, $[Ca^{2+}]_m$ varies with $[Ca^{2+}]_e$, and the activities of three enzymes can be regulated by the free matrix Ca^{2+} concentrations. When $[Ca^{2+}]_m$ increases:

(a) pyruvate dehydrogenase phosphatase is activated, removing P_i from the inactive, phosphorylated form of the pyruvate dehydrogenase complex, thus allowing the V_{max} of the complex to increase;

(b) the K_m for isocitrate of NAD^+-linked isocitrate dehydrogenase is decreased, allowing a given flux through the citric acid cycle to be achieved at a decreased substrate concentration; and

(c) the substrate affinity of 2-oxoglutarate dehydrogenase is increased.

In addition to the three enzymes activated by increased $[Ca^{2+}]_m$, FAD-linked α-glycerol phosphate dehydrogenase on the outer face of the inner mitochondrial membrane, an $ATP-Mg/P_i$ exchanger and two aspartate/glutamate carriers (Section 9.5.2) are activated by physiological increases in cytoplasmic Ca^{2+}. Overall, all these effects contribute to an activation of the tricarboxylic acid cycle. Because, in most cells, substrate delivery can exert significant control over the rate of electron transport (Section 4.8; see also Figure 9.7), this could lead to a mitochondrial hyperpolarisation, limiting the drop in cytoplasmic ATP/ADP ratio associated with increased rates of ATP utilisation when a cell is responding to an external energy-demanding signal linked to an increase in cytoplasmic and matrix calcium concentrations.

Note that there is no contradiction between the two modes of mitochondrial Ca^{2+} handling (regulation and buffering) because there is a smooth transition from the first to the second as the Ca^{2+} load is increased. The relative physiological importance of these two modes depends on the tissue: regulation may be more important in muscle, whereas buffering can be of key importance in firing neurons.

9.4.1.5 The permeability transition

Further reading: Halestrap (2009)

Although the ability of mitochondria to accumulate Ca^{2+} can be enormous, it is not infinite. Rather than uptake eventually simply ceasing, the IMM can suffer a catastrophic loss of integrity—the opening of the mitochondrial permeability transition pore (mPTP). The simplest way to observe the mPTP with isolated mitochondria is to follow the decrease in light scattering of the suspension as the mitochondrial matrix swells and the outer membrane ruptures. By determining the minimum molecular weight of solutes that continue to provide osmotic support, it was possible to determine that a pore had appeared in the inner membrane that was nonselectively permeable to solutes up to approximately 1.5 kDa. Naturally, the presence of such a pore is not compatible with the retention of Δp, and in intact cells the ATP synthase reverses, depleting cytoplasmic ATP.

The transition is facilitated by the presence of P_i and by oxidative stress, particularly matrix NAD(P)H oxidation, whereas low pH and high Δp protect. Thus, the conversion of acetoacetate to β-hydroxybutyrate, which oxidises NADH in liver mitochondria, or the addition of the nonphysiological t-butyl hydroperoxide, which oxidises GSH, each facilitate the transition. However, although the MPT is observed at high matrix Ca^{2+} loads, its induction does not appear to correlate directly with $[Ca^{2+}]_m$ because it is potentiated by elevated phosphate concentrations that should lower $[Ca^{2+}]_m$ by increased formation of the calcium phosphate complex.

The onset of the mPTP can be delayed by cyclosporine A (CsA), which is an inhibitor of the Ca^{2+}-dependent phosphatase calcineurin and is employed as an immunosuppressant. In the present context, CsA binds to cyclophilin D (CyP-D) localised in the mitochondrial matrix. CyP-D has the ability to catalyse the isomerisation of proline residues in proteins and can thus have a profound effect on protein structure. It is difficult to conceive of a normal physiological role for mPTP, particularly because mice lacking CyP-D show little phenotype, but its role under pathological conditions is firmly established in cardiac ischaemia–reperfusion injury (Section 12.3.2) and in equivalent pathologies in the brain (Section 12.6.2).

A number of observations implicate a role for the adenine nucleotide translocator (ANT) in the mPTP. First, the Ca^{2+} loading required to induce the transition is decreased by atractylate, which stabilises the leaky 'C' (cytoplasmic) conformation of the translocator, but increased by the 'M' (matrix) conformation-stabilising bongkrekate (Section 8.2.1). Second, the mPTP is most readily observed with isolated mitochondria incubated in the absence of adenine nucleotides; indeed, in the presence of physiological concentrations of ADP or ATP, high total Ca^{2+} loads are often required to induce the transition. It has been proposed that ANT and CyP-D can form a complex that is prevented by CsA. Mouse liver mitochondria lacking both ANT1 and ANT2 still have a CsA-sensitive permeability transition, although it must be remembered that the mitochondria would still retain ANT4 (otherwise it would be a dead mouse). Evidence has also been advanced for a role of the phosphate carrier, but knockout studies appear to eliminate a role for the outer membrane voltage-dependent anion channel. Although the mPTP releases the pro-apoptotic cyt c (Section 10.7.2) as a result of matrix swelling and outer membrane rupture (Section 2.5), the apoptotic pathway will proceed to completion only if the mPTP were to reclose sufficiently rapidly to prevent ATP depletion by the now reversed ATP synthase.

9.5 METABOLITE COMMUNICATION BETWEEN MATRIX AND CYTOPLASM

Further reading: Arco and Satrustegui (2005), Palmieri and Pierri (2010)

Although much of the research on mitochondrial metabolite carriers (MCs) has been carried out with yeast, we restrict our discussion to mammalian mitochondria. Structural aspects were reviewed in Chapter 8, and here we discuss their roles in controlling bidirectional metabolite traffic between matrix and cytoplasm. The human genome encodes 53 six-transmembrane MCs. With the exception of the ANT and uncoupling protein 1 in brown fat (UCP1), the carriers are present at very low abundance in the IMM. However, the catalytic activity of a number of the major carriers is sufficient to monitor their activity with techniques such as ammonium swelling (Section 2.5.1), isotope exchange, or oxygen electrode experiments designed so that transport has a high control over the rate of electron transport. Many MCs have isoforms that differ between tissues. For those carriers that have only been detected by genomic analysis, gene expression and reconstitution into liposomes have been used to investigate their kinetics and substrate specificities. Despite this, a number of MCs currently remain orphans with no known function; conversely, some established carrier-dependent transport pathways, including glutamine and glutathione, remain to be unequivocally characterised at the molecular level. A number of synonyms exist in the literature: mitochondrial transporters, six-transmembrane transporters, translocators and anion exchangers all refer to MCs. We use the term carrier, although they are not physically carriers in the sense that valinomycin is a carrier (Section 2.3.3), but follow the convention that refers to the adenine nucleotide carrier as a translocator, ANT. The human MC family has been given the nomenclature SLC25A followed by a number.

MCs have been divided into subfamilies based on their substrate specificities and phylogenetic relationships: carriers for nucleotides and related metabolites; for phosphate and other cofactors (sometimes considered a separate subfamily), citrate, glutamate, and oxo-dicarboxylic acids; and for dicarboxylic acids and uncoupling proteins (Table 9.1). The transport stoichiometries of the MCs have important consequences for their function. In particular, if the transport involves net charge transfer, then the equilibrium distribution of the metabolite will be affected by $\Delta\psi_m$, whereas proton transfer means that ΔpH will play a role.

Table 9.1 Major members of the mammalian mitochondrial carrier family[a]

Carrier	Substrates		Equilibrium gradient (in-out) $\Delta\psi_m = 150\,mV$ and $\Delta pH = -0.5$
Di-/tricarboxylates, oxo acids			
Citrate	(Citrate or isocitrate)+H^+/malate	AP	$[Citrate]_{in}/[citrate]_{out} = 30$
Dicarboxylates	(Malate or succinate)+H^+/P_i	AP	$[Malate]_{in}/[malate]_{out} = 10$
Oxoglutarate	Oxoglutarate/malate	AP	
Nucleotides			
ADP/ATP	ADP^{3-}/ATP^{4-}	AP	$([ATP]/[ADP])_{in}/([ATP]/[ADP])_{out} = 0.003$
ATP-Mg/P_i			
Amino acids			
Glutamate	H^+:Glu^-	SP	
Glutamate/aspartate	(Glu^{2-}+H^+)/Asp^{2-}	AP	$(Glu/Asp)_{in}/(Glu/Asp)_{out} = 1000$
Ornithine	H^+/ornithine$^+$	AP	
Carnitine	Carnitine/acylcarnitine	AP	
Others			
UCP1	H^+ (or OH^-)	UP	
UCP2,3	Controversial	?	
Phosphate	H^+:$H_2PO_4^-$	SP	$[H_2PO_4^-]_{in}/[H_2PO_4^-]_{out} = 3$

AP, antiport; SP symport; UP, uniport.
[a]For more details, see Arco and Satrustegui (2005) and Palmieri and Pierri (2010).

We now review the physiological role of the major transporters. The number of MCs expressed in the inner mitochondrial membrane varies from tissue to tissue. All mitochondria possess the adenine nucleotide and phosphate transporters, which are responsible for the uptake of ADP + P_i and the release of ATP to the cytoplasm. Virtually all mitochondria oxidise pyruvate and so possess the pyruvate carrier, although this has a distinct structure and is not a member of the MC family (Section 8.2.4). However, there is a tissue-specific expression of the other carriers that correlates with the range of metabolic pathways present in the cell. Thus, the liver, with its plethora of metabolic pathways, has mitochondrial transport pathways for most of the citric acid cycle intermediates, for a number of amino acids, and for carnitine and its fatty acyl ester. The more specialised metabolic role played by mitochondria in the heart is reflected in the more restricted variety of carriers, whereas the mitochondria of a highly specialised tissue such as brown fat can only transport the metabolites acylcarnitine, succinate, and pyruvate.

Although the uncoupling proteins share a common structure with the MCs, their role is to control thermogenesis and (controversially) oxidative stress, and this is discussed separately in Sections 9.12 and 12.4.

9.5.1 Adenine nucleotide and phosphate transport

Further reading: Satrustegui *et al.* (2007), Klingenberg (2008)

The molecular mechanism of the adenine nucleotide carrier (or translocator) was reviewed in Section 8.2.1. The four human isoforms, ANT1–4 (SLC25A4, 5, 6, 31), each catalyse the 1:1 electrogenic exchange of ADP^{3-} for ATP^{4-} across the inner membrane. Although ANT transports ADP and ATP symmetrically when there is no membrane potential, under normal respiring conditions uptake of ADP and efflux of ATP are preferred, corresponding to the usual physiological direction of the exchange. The reason for this asymmetry lies in the relative charges on the two nucleotides. ATP is transported as ATP^{4-}, whereas ADP is transported as ADP^{3-} (Figure 9.4). The resulting charge imbalance means that the equilibrium of the exchange is displaced 10-fold for each 60 mV of membrane potential.

ANT does not transport AMP, adenosine, or other purine nucleotides such as GTP. The total pool size of adenine nucleotides in the matrix (i.e., ATP + ADP+AMP) does not change as a result of translocator activity because the uptake of a cytoplasmic nucleotide is automatically compensated by the efflux of a nucleotide from the matrix. However, a second transporter, the ATP-Mg/P_i carrier, can change pool sizes (discussed later).

As discussed in Section 8.2.1, ANT is, at the time of writing, the only mitochondrial carrier to have a full crystal structure. A number of specific inhibitors exist. Atractyloside (also called atractylate), a glucoside isolated from the Mediterranean thistle *Atractylis gummifera*, is a competitive inhibitor of adenine nucleotide binding and transport. The closely related carboxyatractylate binds more firmly (K_d 10^{-8} M) and cannot be displaced by adenine nucleotides. Bongkrekic acid (bongkrekate) is produced by *Pseudomonas cocovenenans* and derives its name from its discovery as a toxin in contaminated samples of the coconut food product bongkrek. It is an uncompetitive inhibitor of the translocator. Atractylate locks the carrier in the C-conformation, in which the nucleotide binding site is accessible from the cytoplasm, whereas bongkrekate stabilises the opposite M-state where the site accesses the matrix.

Four isoforms of ANT have been detected in mammals with 80–90% sequence homology. ANT1 is the predominant form in heart, brain and muscle; ANT2 is the predominant form in proliferating tissues such as liver; ANT3 is expressed ubiquitously at low levels; and ANT4 is expressed at low levels in liver, testis and brain. The possibility that the different isoforms play subtly different roles, particularly in relation to apoptosis, is currently being investigated. ANT has been implicated in the mitochondrial permeability transition pore (Section 9.4.1.5) because atractylate facilitates, and bongkrekate protects against, pore formation under conditions of Ca^{2+} overload and oxidative stress.

Because ANTs catalyse the strict 1:1 exchange of adenine nucleotides across the IMM, they do not provide a mechanism for altering the total adenine nucleotide content of the matrix during, for example, cell growth and division. The ATP-Mg/P_i carrier (SLC25A23, 24, 25) fills this role, catalysing an electroneutral exchange of ATP. Mg^{2-} for HPO_4^{2-} (note that ANTs, in contrast, exchange the free uncomplexed nucleotide). Because P_i is accumulated as a function of ΔpH, its exchange for Mg.ATP will mean that at equilibrium the Mg.ATP gradient (in/out) across the IMM will reflect the

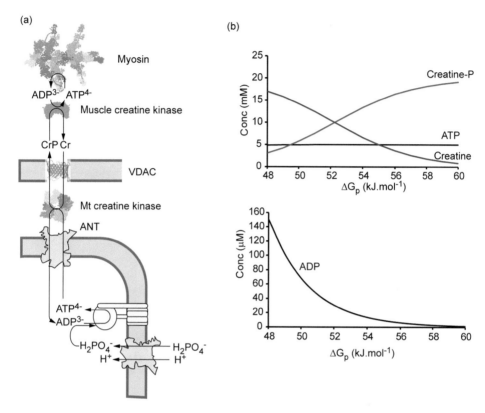

Figure 9.4   ATP export and the creatine/creatine phosphate pathway.
(a) The import of ADP and P_i together with the export of ATP utilises Δp and is driven by the entry of one proton. In highly active tissues (muscle and brain), a mitochondrial creatine kinase (mtCK) in the IMS generates creatine phosphate (CrP), whereas muscle CK located on actomysin filaments regenerates ATP. The Cr/CrP pathway helps to relieve a diffusion limitation for the low ADP concentration in the cytoplasm to return to the ANT and also provides a short-term buffer system if ATP demand temporarily exceeds supply. (b) Changes in cytoplasmic ATP, ADP, Cr and CrP as the free energy for ATP hydrolysis, ΔG_p, is varied from 48 to 60 kJ mol^{-1}. Note that the free ADP concentration falls to low micromolar levels, that ATP is essentially invariant and that the Cr/CrP ratio is poised around unity.

H^+ gradient (out/in). An interesting consequence is that oligomycin, by lowering matrix ATP, will favour Mg.ATP uptake, whereas carboxyatractylate, preventing ATP/ADP exchange, will raise matrix ATP and facilitate Mg.ATP efflux. However, the activity of the carrier in the mitochondrion is considerably lower than that of ANT.

The ATP-Mg/P_i carrier belongs to a subgroup of MCs termed Ca^{2+}-dependent mitochondrial carriers (CaMCs), which possess a Ca^{2+} binding N-terminal in the intermembrane space (IMS). In addition to three genes for ATP-Mg/P_i carriers (and numerous splice variants), the subgroup includes isoforms of the aspartate/glutamate carrier. The ATP-Mg/P_i carrier has an absolute requirement for Ca^{2+} in the IMS and is activated by physiologically elevated cytoplasmic Ca^{2+}.

The mitochondrial phosphate carrier (PiC; SLC25A3) catalyses the electroneutral transport of $H_2PO_4^-$. Although this is commonly referred to as an $H_2PO_4^-/OH^-$ antiport, current evidence favours an $H_2PO_4^-:H^+$ symport mechanism (Section 8.2.2); note, however, that the two are experimentally indistinguishable. The carrier is inhibited by mercurial reagents such as p-mercuribenzoate and mersalyl, and also by N-ethylmaleimide, although none of the inhibitors are completely specific. PiC is ubiquitous and extremely active. Because of the proton symport, the distribution of P_i across the membrane is influenced by ΔpH. Two isoforms of the phosphate transporter (A and B) exist in mammals, encoded by the same gene but generated by alternative splicing of two exons. Isoform A, found in heart and skeletal muscle, has a higher catalytic activity and affinity for phosphate than the more generally expressed B isoform. The phosphate carrier plays a key role in all mitochondria. Continuous uptake of P_i is required during oxidative phosphorylation, and matrix P_i exchanges with malate or succinate on the dicarboxylate carrier. In addition, uptake of P_i occurs in parallel with mitochondrial Ca^{2+} accumulation (Section 9.4.1.3).

Continuous mitochondrial ATP synthesis and export requires the coordinate activity of the ATP synthase, the ANT and the phosphate carrier (Figure 9.4). The combined effect of the phosphate carrier and ANT is to cause the influx of one additional proton per ATP synthesised. The thermodynamic consequences of this are considerable: based on an eight-membered c ring in the F_o subcomplex of the ATP synthase (Section 7.6), synthesis of 1 mol of ATP in the matrix requires 2.67 mol of H^+ (i.e., 8/3), meaning that 3.67 H^+ are required for synthesis plus export. This means that isolated mitochondria can maintain a theoretical ΔG of up to 64 kJ mol^{-1} for the extramitochondrial ATP pool, in contrast to a value of less than 50 kJ mol^{-1} for submitochondrial particles.

9.5.1.1 The creatine/creatine phosphate pathway

Further reading: Balaban (2009a)

When we discuss mitochondrial structure in detail (Chapter 10), we will see that ATP synthesis occurs deep within the cristae, and that communication with the IMS occurs via narrow cristae junctions. Once exported to the IMS, ATP must diffuse across the outer mitochondrial membrane (OMM) and through the cytoplasm to its site of action. Any or all of these events could potentially impose a kinetic limitation in rapidly metabolising tissues such as muscle and brain. More seriously, ADP, which is present in the cytoplasm at a much lower concentration, must at the same time make the return journey. In part, the problem is decreased by the cell architecture minimising the diffusion path of the adenine nucleotides. For example, in muscle the interfibrillar mitochondria are packed in close contact with the muscle fibres. However, the major mechanism in tissues with a high peak energy demand such as brain and muscle is to utilise the creatine/creatine-phosphate (Cr/CrP) pathway (Figure 9.4).

The respiration of isolated muscle mitochondria in state 4 can be increased to state 3_{ADP} levels by the addition of creatine. This is because mitochondria from brain and skeletal and cardiac muscle contain a bound form of creatine kinase (MtCK), regenerating ADP and forming CrP. The enzyme is located in the IMS, and in cardiac mitochondria it is equimolar to (and closely associated with) ANT. Under physiological conditions, the apparent equilibrium constant for the CK reaction

$$ADP + CrP \rightleftharpoons ATP + Cr \qquad [9.1]$$

is approximately 180. The presence of mtCK in the IMS, together with additional iso-forms of CK associated with myofibrils, the sarcolemmal Ca^{2+}-ATPase and the Na^+/K^+-ATPase (Figure 9.4), allows the Cr/CrP system to act as the bidirectional 'energy shuttle' in these energy-demanding tissues. The concentration of the Cr/CrP pool in muscle can exceed 20 mM, contrasting with a cytoplasmic adenine nucleotide pool of approximately 5 mM. The equilibrium constant means that the Cr/CrP pool is poised close to equilibrium over the physiologically relevant range of Gibbs free energies for ATP synthesis (Figure 9.4). In contrast, the concentration of free ADP in the cytoplasm is calculated to decrease to approximately 1 μM at a ΔG_p of 60 kJ mol^{-1}, and this could pose kinetic limitations from the requirement of this vanishingly low ADP concentration to diffuse from the myofibrils back through the outer membrane and into the cristae. In addition, the large CrP pool can act as a short-term buffer to minimise a decrease in Δp if ATP demand exceeds supply.

9.5.2 Electron import from the cytoplasm

NADH, which is produced in the cytoplasm of mammalian cells, for example by glycol-ysis, does not have direct access to complex I, whose NADH binding site is located on the inner face of the inner membrane. The inner membrane is essentially impermeable to NADH, and two strategies are employed to transfer the electrons to the respiratory chain (Figure 9.5).

The human genome encodes two aspartate/glutamate carriers (AGCs) termed aralar (AGC1; high in brain, skeletal muscle and heart) and citrin (AGC2; highest in heart and liver). They are, like the ATP-Mg/P_i carrier, examples of CaMCs. They exchange aspar-tate^{2-} for glutamate^{2-} plus a proton and are thus essentially unidirectional in energised mitochondria, expelling aspartate and accumulating glutamate.

AGC2 allows aspartate transported from the mitochondrion to participate in the urea cycle. Together with the malate/2-oxoglutarate carrier, AGC1 contributes to the *malate–aspartate shuttle* (MAS) (Figure 9.5), allowing the oxidation of cytoplasmic NADH by the respiratory chain. The thermodynamic problem posed by the fact that the cytoplas-mic NAD$^+$/NADH couple is considerably more oxidised (i.e., less reducing) than the equivalent matrix couple is overcome by the electrical imbalance of the AGC discussed previously. However, this apparently 'uphill' transport of electrons from the cytoplasmic to the matrix NAD pool occurs at the cost of the proton co-transported with glutamate. The Ca^{2+} activation of the AGCs, and hence the MAS, is thought to be important for the α-adrenergic stimulation of gluconeogenesis in hepatocytes. Thus the yield of ATP from cytoplasmic NADH is 90% of that from matrix NADH, assuming that the respiratory chain translocates 10 protons per 2e flowing from NADH to oxygen.

The *s,n-glycerophosphate shuttle* provides an alternative means for the oxidation of cytoplasmic NADH (Figure 9.5). This makes use of the two *s,n-glycerophosphate* dehy-drogenases present in some cells—a cytoplasmic NADH-coupled enzyme reducing

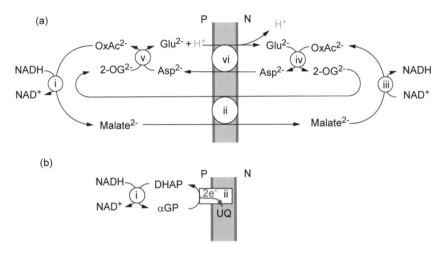

Figure 9.5 The oxidation of cytoplasmic NADH by mitochondria.
The malate /aspartate shuttle (MAS): (i) Cytoplasmic NADH is oxidised by
cytoplasmic malate dehydrogenase; (ii) malate enters the matrix in exchange
for 2-oxoglutarate; (iii) malate is reoxidised in the matrix by malate
dehydrogenase, generating matrix NADH; (iv) matrix oxaloacetate
transaminates with glutamate to form aspartate and 2-oxoglutarate (which
exchanges out of the matrix); (v) 2-oxoglutarate transaminates in the
cytoplasm with transported aspartate to regenerate cytoplasmic oxaloacetate
and to give cytoplasmic glutamate; and (vi) cytoplasmic glutamate re-enters
the matrix by proton symport in exchange with aspartate. (b) The *s,n*-
glycerophosphate shuttle for the oxidation of cytoplasmic NADH: (i)
cytoplasmic *s,n*-glycerophosphate dehydrogenase (NAD⁺-linked) and (ii)
inner membrane *sn*-glycerophosphate dehydrogenase (flavoprotein-linked).

dihydroxyacetone phosphate to *s,n*-glycerophosphate and an enzyme on the outer face
of the IMM reoxidising this and feeding electrons directly to the UQ pool. In this case,
the directionality is induced by feeding electrons to the quinone pool at a potential close
to 0 mV (Section 5.3). However, the resulting translocation of 6 protons per 2e means
that the ATP yield is less than when the MAS system is used.

The relative importance of the two shuttles varies with the tissue. The MAS is dominant in liver, heart, pancreatic β cells and brain so that inhibition of the two aspartate
amino transferases with aminoxyacetate results in a profound inhibition of pyruvate
delivery to the mitochondrion as the substrate is diverted to lactate to reoxidise glycolytic NADH. The *s,n*-glycerophosphate shuttle is highly active in brown adipose tissue
(Section 9.12.2).

9.5.3 Additional metabolite carriers

Further reading: Arco and Satrustegui (2005), Palmieri and Pierri (2010)

As shown in Figure 9.5, the MAS does not result in a net increase in matrix glutamate. For glutamate dehydrogenase to be supplied with substrate, a separate carrier is

required. This is the *glutamate carrier* (SLC25A18, 22), which catalyses the electroneutral symport of glutamate$^-$ and a proton (note that this is formally equivalent to a glutamate$^-$/OH$^-$ antiport). Structural aspects are discussed in Section 8.2.3. Two human isoforms have been identified: the lower affinity GC1 is highly expressed in pancreas and liver, and both forms are expressed in brain.

The *dicarboxylate carrier* (SLC25A10) allows the net export of citric acid cycle intermediates from the matrix for gluconeogenesis. The carrier catalyses the electroneutral exchange of malate^{2-} or succinate^{2-} for HPO$_4^{2-}$. The carrier is most abundant in liver and kidney, consistent with its role in gluconeogenesis, but it is also present in heart and brain mitochondria.

The *2-oxoglutarate carrier* (SLC25A11) exchanges the oxo-dicarboxylic acid for malate as an electroneutral exchange of dianionic species. Together with the electrogenic glutamate–aspartate transporter, the 2-oxoglutarate transporter is a component of the malate–aspartate shuttle.

The *tricarboxylate carrier* (SLC25A1) exchanges citrate^{3-} or isocitrate^{3-} for malate^{2-} and is important for the export of the tricarboxylic acids for fatty acid synthesis. The carrier catalyses an electroneutral exchange because it co-transports a proton together with citrate. The carrier is inhibited by 1,2,3-benzyltricarboxylate.

Fatty acids are activated to acyl-CoA on the OMM and converted to acylcarnitines via carnitine *N*-acyltransferase I. Acylcarnitines of various chain lengths enter via the *acylcarnitine/carnitine carrier* (SLC25A20, 29) in exchange for carnitine liberated by the matrix located carnitine *N*-acyltransferase II, which regenerates acyl-CoA for β oxidation and liberates the carnitine required for the exchanger. The carrier is not specific for the acyl chain length and will transport acetylcarnitine and short- and long-chain acylcarnitines.

The *ornithine/citrulline carriers* (SLC25A15, 2) catalyse the electroneutral exchange of cytoplasmic ornithine for matrix citrulline $+1H^+$ and function to connect the matrix and cytoplasmic reactions of the urea cycle in liver and kidney. The *deoxynucleotide carrier* (SLC25A19) transports deoxynucleoside diphosphates into the matrix. In addition to being involved in the import of nucleotides for DNA synthesis, the carrier transports toxic antiviral nucleoside reverse transcriptase inhibitors such as 3-azido-3'-deoxythymidine (AZT) and may be involved in the mitochondrial toxicity associated with these compounds.

Because pyruvate is a monocarboxylic acid, it has been argued that it would cross bilayer regions without the need for a carrier, following the precedent of acetate moving as acetic acid. However, α-cyanocinnamate and related inhibitors potently inhibit pyruvate transport into the mitochondrion, indicating that a carrier protein is involved. Surprisingly for such a major transport process, its molecular identity has only recently been established (Bricker *et al.*, 2012) (see Section 8.2.4).

9.5.4 Metabolite equilibria across the inner mitochondrial membrane

Table 9.1 shows the way in which a number of metabolites would distribute at thermodynamic equilibrium across the inner membrane of mitochondria maintaining a typical Δp. We have already discussed the ANT and the aspartate/glutamate carrier, but even pyruvate distribution is affected by ΔpH. Because ΔpH can be abolished or even reversed by addition of protonophores, the ability of excess uncoupler to inhibit cellular respiration could well be substrate starvation as the matrix acidifies, expelling substrates from the matrix. However, the presence of a carrier does not necessarily mean that the transported metabolite is a significant energy-yielding substrate for intact cells. For example, although succinate is an excellent substrate for isolated mitochondria, entering via the dicarboxylate carrier and generating malate that can leave via the same carrier, the plasma membrane is impermeable to succinate and no major pathways generate cytoplasmic succinate.

9.6 QUANTIFYING THE MITOCHONDRIAL PROTON CURRENT IN INTACT CELLS

Further reading: Brand and Nicholls (2011)

Despite the increased complexity of an intact cell, respiration remains the most powerful method to investigate the bioenergetics of *in situ* mitochondria. After subtracting the contribution from non-mitochondrial oxygen uptake (by determining residual respiration after adding electron transport inhibitors at the end of the experiment), respiration is directly proportional to proton current (for a given substrate) as before. Quantification, however, is more complex if two different preparations are being compared: should the rate be normalised to cell number, total protein, mitochondrial content, or some other parameter? Thus, it can be difficult to interpret changes in basal respiration between two cell populations, particularly if they are growing and mitochondria are proliferating.

There are technical challenges to measuring cell respiration. The classical Clark-type oxygen electrode and chamber (see Figure 4.5) is only useful for large quantities of cells that can be obtained in suspension (e.g., blood cells or acutely isolated hepatocytes), in which case the chamber can additionally be equipped with a TPP$^+$ electrode (see Figure 4.7) for the parallel determination of $\Delta\psi_m$ (note, however, that as we discuss in Section 9.7.1, most determinations of $\Delta\psi_m$ in intact cells are influenced in some way by the plasma membrane potential, $\Delta\psi_p$). In the majority of situations, the cells to be investigated are growing as monolayers on coverslips and do not respond well to trypsinisation and stirring, and are additionally often available in very limited amounts, requiring alternative approaches to the classical Clark electrode and chamber (see Figure 4.5).

A 'cell respirometer' (Figure 9.6) has been used to monitor oxygen uptake by a cell monolayer on a coverslip within a closed perfusion chamber, with micro flow-through oxygen electrodes placed upstream and downstream to allow continuous monitoring of the cellular respiration rate, while simultaneously monitoring fluorescence parameters. Several alternative techniques have been developed for multi-well plates with built-in fluorescent oxygen sensors, or by reversibly enclosing a very small volume (as little as $7\,\mu l$) above the cells, measuring oxygen uptake in that volume for a short period before allowing the bulk medium (~1 ml) to re-equilibrate before repeating the measurement (the Seahorse extracellular flux analyser).

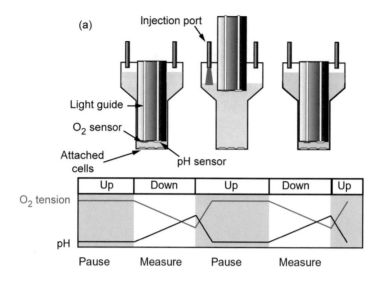

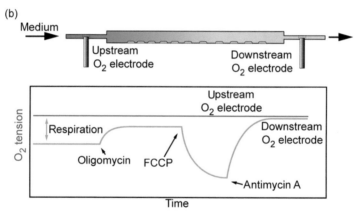

Figure 9.6 Techniques for monitoring cell respiration.
(a) Seahorse extracellular flux analyser: a piston equipped with solid-state sensors for O_2 and pH entraps a small volume of medium over the cells; the rate of O_2 uptake is determined. The piston is then raised, allowing O_2 to re-equilibrate and additions to be made before the cycle is repeated. (b) Cell respirometer: cells are cultured on a circular coverslip that is assembled in a thin closed-perfusion chamber. Medium is slowly superfused, and respiration is determined by the difference between upstream and downstream O_2 tension measured with micro flow-through O_2 electrodes (Jekabsons and Nicholls, 2004).

These technologies have allowed the concept of 'cell respiratory control' (CRC), analogous to mitochondrial respiratory control (Section 4.3), to be developed. The basic Chance and Williams respiratory states for isolated mitochondria (Section 4.3) are not readily applicable to intact cells. The basal respiration of the cell corresponds neither to state 4_o (zero ATP synthesis) nor to state 3_{ADP} (unlimited ADP availability), nor even to an intermediate condition ('state 3½'), because it is difficult to arrange conditions such that substrate delivery does not exert some control over the rate (in contrast to isolated mitochondria, which can be exposed to saturating substrate).

The information that can be obtained from CRC experiments can be illustrated with a study using synaptosomes (Figure 9.7). These are isolated resealed nerve terminals, obtained by homogenising brain tissue followed by gradient centrifugation to separate them from mitochondria, and can be considered as mini cells for bioenergetic studies. Because this approach provides such detailed information of cellular bioenergetics, we describe each stage in detail.

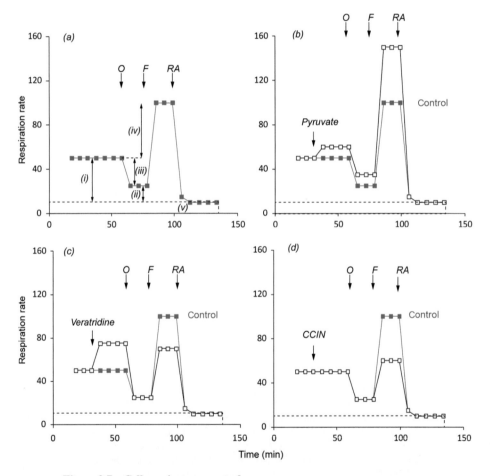

Figure 9.7   Cell respiratory control.
Seahorse respiratory traces of isolated nerve terminals (synaptosomes) in the presence of glucose. (a) Basic cell respiratory control experiment. Additions: O, oligomycin; F, FCCP; RA, rotenone plus antimycin A. Rates: (i) basal, (ii) oligomycin-insensitive, (iii) oligomycin-sensitive, (iv) spare respiratory capacity, and (v) non-mitochondrial respiration. (b) Response to addition of pyruvate; (c) addition of an Na^+ channel opener veratridine; and (d) addition of α-cyanocinnamate (CCIN), a partial inhibitor of mitochondrial pyruvate transport.
Adapted from Choi et al. (2009).

Non-mitochondrial respiration (Figure 9.7a,v): Most cells have some non-mitochondrial oxygen uptake caused by cytoplasmic oxidases and related enzymes. This is quantified at the termination of a CRC experiment following complete inhibition of the electron transport chain, here by adding rotenone plus antimycin A. It is assumed that this rate is constant during the experiment and is subtracted from all the other rates.

Basal respiration (Figure 9.7a,i): Basal respiration is primarily controlled by proton re-entry through the ATP synthase, with contributions from the endogenous proton leak and substrate oxidation. This means that any condition that changes the cell's ATP demand will be reflected in the respiration rate. An example of this is shown in Figure 9.7c; addition of the neurotoxin veratridine locks the voltage-activated Na^+ channels in an open state and results in rapid ATP hydrolysis as the plasma membrane Na^+ pump accelerates to extrude the Na^+ flooding into the cytoplasm, initiating an increased respiration.

ATP turnover (Figure 9.7a,iii): An approximate estimate of the rate of ATP turnover in the basal state can be calculated from the decrease in respiration following addition of oligomycin to inhibit the ATP synthase. Because this results in a slight mitochondrial hyperpolarisation, and the endogenous proton leak is voltage-dependent (Section 4.5.1), this approach tends to underestimate the proportion of the basal respiration devoted to the ATP synthase and to exaggerate the proton leak under basal conditions. The basic CRC experiment assumes that glycolysis has sufficient capacity to meet the cell's total ATP demand in the absence of oxidative phosphorylation. Although this is generally the case, a resulting ATP deficit could lead to a failure of glycolysis or fatty acid activation (each with its ATP-requiring steps). In such cases, the CRC experiment can be redesigned with separate parallel additions of oligomycin and FCCP. The lower maximal respiration seen in the veratridine-exposed synaptosomes in Figure 9.7c can probably be ascribed to ATP depletion by the activated Na^+ pump in the presence of oligomycin.

Proton leak (Figure 9.7a,ii): In the presence of oligomycin, the entire proton current re-enters through the endogenous proton leak, plus any uncoupling protein activity. Note that if rapid transport or cycling of metabolites, or ions such as Ca^{2+}, are occurring at the same time, these will also contribute to proton re-entry. By analogy with isolated mitochondria, this can be termed state 4_o. In this state, respiration is usually strongly controlled by the proton leak, with minor control by substrate oxidation. The fraction of the basal respiration used for ATP synthesis is sometimes termed the *coupling ratio*. A preparation of cells that shows an unexpectedly large oligomycin-insensitive respiration may be damaged. However, a modest increase could reflect either a change in leakiness or an increase in $\Delta\psi_m$, emphasising that although cell respiration is information-rich, a complete analysis requires parallel determination of changes in potential.

Maximal respiration: Ideally, we would like to determine the maximal state 3_{ADP} respiration of mitochondria in the intact cell. Unfortunately, this is usually not possible due to the impermeability of the plasma membrane to ADP. Respiration can be stimulated by increasing cytoplasmic ATP turnover, as is done in Figure 9.7c with Na^+ channel activators such as veratridine. However, these agents only act in excitable cells with voltage-activated Na^+ channels, and there is no guarantee that

Na$^+$ cycling at the plasma membrane can utilise the full ATP-generating capacity of the mitochondria. Instead, protonophores are used to relieve downstream control of respiration so that all control is now exerted upstream of the proton circuit, divided between substrate delivery and electron transport. This is not entirely satisfactory because, like all ionophores, protonophores such as FCCP show no selectivity between membranes and so produce a plethora of side effects (discussed in Section 9.6.1). It is most important to titrate the minimum concentration of protonophore required to attain maximal respiration. In this condition, the mitochondria still retain a substantial Δp (see Figure 4.12). Excess protonophore usually causes an inhibition of cell respiration, even in the presence of oligomycin, which would prevent ATP synthase reversal and cytoplasmic ATP hydrolysis. Reversal of ΔpH across the inner membrane by excess protonophore will tend to exclude mitochondrial substrates such as pyruvate from contributing to the decline in respiration. Substrate availability exerts a strong control over maximal respiration in most cells, even in the presence of saturating glucose. This is shown in Figure 9.7b, in which the most notable effect of the supplementation of glucose with pyruvate as additional substrate is to substantially increase the maximal respiration. Note in this trace that basal and oligomycin-insensitive respiration rates are also increased; this is a consequence of pyruvate hyperpolarising the mitochondria and increasing the leak proton current, rather than an 'uncoupling' action. Figure 9.7d illustrates an important principle: partial inhibition of a step will affect the respiration rate only if the experiment is designed so that the step exerts significant control over respiration. In this experiment, α-cyano-4-hydroxycinnamate partially inhibits the mitochondrial pyruvate carrier; however, no effect is seen on basal or oligomycin-insensitive respiration because proton re-entry, rather than pyruvate transport, has most control over respiration under these conditions. Only when control is moved upstream of the proton circuit by addition of protonophore does the limitation in the maximal rate of the carrier become apparent.

Spare respiratory capacity (Figure 9.7a,iv): One of the most important questions to ask when investigating mitochondrial function (or dysfunction) is whether the cell possesses sufficient spare respiratory capacity to cope with the maximal energy demand to which it would ever be exposed. Because it is not usually possible to experimentally increase ATP demand sufficiently, the difference between the basal and maximal protonophore-stimulated rates provides an indication of this reserve capacity. We discuss this again in the context of mitochondrial dysfunction in Chapter 12.

9.6.1 Ionophores and cells

Some of the complications in the use of protonophores in intact cells have already been mentioned; however, the point is so important that it bears repeating. The basic problem is that ionophores show no membrane selectivity so that a protonophore added, for example, to a neuronal cell culture will additionally acidify the cytoplasm, releasing Ca^{2+} from mitochondria and neurotransmitters from synaptic vesicles. Other ionophores have additional complications. Perhaps the most notorious are the electroneutral

$Ca^{2+}/2H^+$ antiport ionophores typified by ionomycin. These have been used in more than 20,000 cell biology studies, with the expectation that they will specifically raise $[Ca^{2+}]_c$. However, at the IMM, they create a dissipative Ca^{2+} cycle in concert with the mitochondrial Ca^{2+} uniporter, which, under the conditions of elevated $[Ca^{2+}]_c$ created by the ionophore in the plasma membrane, can be as potent as a protonophore. This seems to be forgotten in 90% of studies in the literature using these ionophores. The neglect of bioenergetic fundamentals extends to other agents such as the ANT inhibitor bongkrekic acid (Section 9.5.1), which is often marketed as a 'permeability transition inhibitor' while ignoring its primary effect in inhibiting ATP export to the cytoplasm.

9.7 MITOCHONDRIAL PROTONMOTIVE FORCE IN INTACT CELLS

Thus far, we have focused on the 'ammeter' of the proton circuit. We now turn our attention to the voltmeter—the monitoring of the components of the protonmotive force. Although techniques are becoming available for the optical estimation of the ΔpH component of Δp, absolute determinations of $\Delta \psi_m$ using membrane-permeant cations are exceedingly complex, rely on multiple assumptions of mitochondrial volume and activity coefficients in cytoplasm and matrix and usually require the plasma membrane potential, $\Delta \psi_p$, to be taken into account. The large majority of studies are either semi-quantitative, estimating changes in potential from an assumed starting value, or purely qualitative, determining whether mitochondria are polarised or depolarised and essentially functioning as a live–dead assay for the organelle.

Although the $\Delta \psi_m$ component of Δp is always dominant, it is important not to equate the two, particularly when monitoring small changes, because a change of 0.1 unit in ΔpH will alter Δp by 6 mV. The use of nigericin to minimise ΔpH in isolated mitochondria studies (Section 4.4.3) can be hazardous when applied to intact cells due to its effects at the plasma membrane. With these provisos, we start with approaches to monitori $\Delta \psi_m$ in intact cells.

9.7.1 Mitochondrial membrane potential

Further reading: Duchen *et al.* (2003), Nicholls (2012)

All techniques seek to determine the equilibrium distribution of a membrane-permeant cation between the cytoplasm and matrix, using this gradient to estimate $\Delta \psi_m$ by application of the Nernst equation (Eq. 3.39). The term *equilibrium* is important because many cells, particularly those derived from cancers, possess ATP-dependent multidrug resistance (MDR) transporters that have evolved to expel xenobiotics (Section 8.3.4.1); indeed, rhodamine-123, which is extensively used to monitor $\Delta \psi_m$, is also used to assay MDR activity. If such an activity is present, it must first be inhibited.

The large majority of studies employ hydrophobic fluorescent cations that possess a π-orbital system to delocalise the positive charge and facilitate its permeation across the lipid bilayers (Figure 9.8). No probe is perfect, but the most reliable are tetramethylrhodamine methyl and ethyl esters (TMRM and TMRE, respectively; Figure 9.8) and

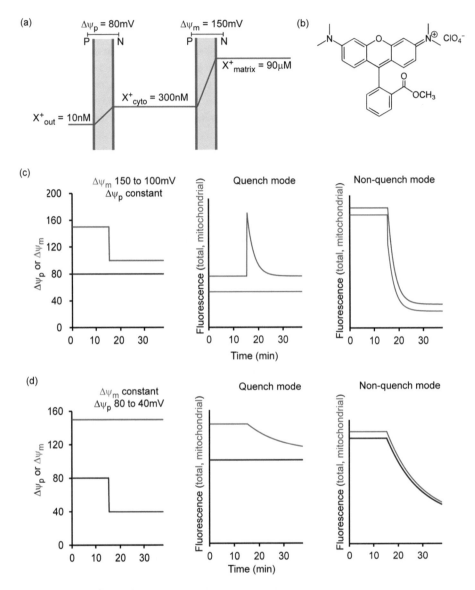

Figure 9.8   Monitoring changes in $\Delta\psi_m$ in intact cells.
(a) A membrane-permeant monovalent cation will ideally distribute to a
Nernst equilibrium across both plasma and mitochondrial membranes. (b)
Structure of TMRM. (c) Computer prediction of the change in single-cell
fluorescence with time in quench and non-quench mode when $\Delta\psi_m$ is
partially depolarised. (d) Responses to partial $\Delta\psi_p$ depolarisation.
Adapted from Ward et al. (2000).

the more hydrophilic (and thus more slowly permeant) rhodamine-123. Probes that
load unevenly and undergo variable precipitation in the mitochondria (e.g., JC-1) or
are potent inhibitors of the electron transport chain (e.g., certain cyanine dyes) must be
avoided.

Three principles govern the fluorescence response at the single-cell level:

(a) Although these are often referred to as mitochondrial membrane potential indicators, they first equilibrate across the plasma membrane, and the concentration in the matrix at equilibrium is a function of the Nernst equilibria across both membranes (Figure 9.8).

(b) Although the probes show no membrane selectivity, the vastly greater surface:volume ratio of the mitochondria compared with the cell membrane means that equilibration between cytoplasm and matrix is hundreds of times faster than that across the plasma membrane.

(c) Above a critical concentration in the matrix, the probes reversibly aggregate. The aggregates are usually nonfluorescent—that is, fluorescence quenching occurs. The design and interpretation of experiments is critically dependent on whether low (non-quenching) or higher (quench mode) probe concentrations are employed (Figure 9.8).

Practical details for these determinations have been extensively reviewed (Brand and Nicholls, 2011; Nicholls, 2012), but a few essential points are necessary in order to critically interpret the literature. Experiments may be performed at single-mitochondrial or single-cell levels of resolution. The former determines the ratio of fluorescence intensity between the mitochondria and a mitochondria-free (cytoplasmic) region of the cell. Such experiments must be performed with low (non-quenching) concentrations of probe (as, incidentally, must flow cytometry). This approach demands a high dynamic detection range because a $\Delta\psi_m$ of 150 mV implies a 300-fold gradient between matrix and cytoplasm. Mitochondria are motile within the cell, and to compensate for drift out of the focal plane, a mitochondria-targeted fluorescent protein can be used to provide a reference signal. Aspects of motility, fission and fusion, and heterogeneity can be investigated at the same time (Section 10.3).

If the overall bioenergetic status of the cell is of interest, single-cell resolution is usually sufficient. The fluorescence is a function of both $\Delta\psi_m$ and $\Delta\psi_p$ (Figure 9.8), and the equilibrium concentration of the probe X^+ in the matrix (m) relative to the external medium (e) is given by

$$[X^+]_m = [X^+]_e \times 10^{(\Delta\psi_m + \Delta\psi_p)/61}$$

[9.2]

where both potentials are given a positive sign. Note that 61 is the numerical value of 2.3RT at 37°C.

Single-cell studies can be performed in either quench mode or non-quench mode. The former is useful to detect changes in $\Delta\psi_m$ occurring during the experiment. A modest drop in $\Delta\psi_m$ results in release of some of the aggregated nonfluorescent probe from the matrix into the cytoplasm, resulting in an increase in cell fluorescence; the excess cytoplasmic probe then slowly equilibrates out of the cell to restore the initial signal (Figure 9.8c). The transient nature of the signal means that quench mode cannot be used to compare $\Delta\psi_m$ between two cell populations (e.g., in flow cytometry). Non-quench mode can in theory be used to compare two populations of cells; however, before ascribing a difference in signal to a differing $\Delta\psi_m$, it is first necessary to establish that $\Delta\psi_p$ is the same and that no change in mitochondrial amount or volume has occurred. Note that non-quench mode does not distinguish between a change in $\Delta\psi_m$ and a change in $\Delta\psi_p$ (Figure 9.8d). Inclusion of a second anionic fluorescent probe can be used to monitor $\Delta\psi_p$ and to allow the $\Delta\psi_m$ response to be distinguished (Nicholls, 2006a).

9.7.2 Mitochondrial ΔpH

Further reading: Hoek *et al.* (1980), Abad *et al.* (2004)

Until recently, there has been a tendency to ignore the ΔpH component of Δp in cellular studies (but see Hoek *et al.*, 1980). However, the availability of alkaline pH-sensitive variants of green-fluorescent protein, such as mtAlpHi (mitochondrial alkaline pH indicator), co-transfected with a relatively pH-insensitive cyan fluorescent protein as reference allows matrix pH to be determined after appropriate calibration (Figure 9.9). In order to calculate ΔpH, parallel measurements of cytoplasmic pH are required, for example, using the ratiometric probe BCECF. In the example shown in Figure 9.9, which is discussed in detail in Section 12.5, the application of a high glucose concentration to insulin-secreting cells increased ΔpH from −0.2 to −0.5 units, equivalent to an increased contribution to Δp of 18 mV. Clearly, significant errors in Δp can be introduced if this component of Δp is ignored.

Nuclear magnetic resonance (NMR) can be used to obtain a direct measurement of the pH inside and outside a cell or organelle, and it is free of some of the drawbacks inherent in the more invasive use of weak acids or bases. The basis of the technique is

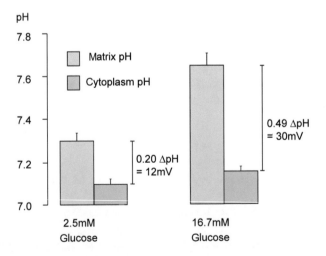

Figure 9.9   ΔpH measurements.
Matrix pH was monitored with the YFP-derived mtAlpHi targeted to the mitochondrial matrix of insulin-secreting INS-1E cells. In parallel, cytoplasmic pH was monitored with the ratiometric indicator BCECF. Increasing glucose concentrations increased matrix pH in the INS-1E cells by 0.35 units; ΔpH increased from −0.20 to −0.49 units, equivalent to an increased contribution to Δp of 18 mV.
Data from Wiederkehr et al. (2009).

that the resonance energy of the phosphorus nucleus in P_i varies according to the protonation state of the latter. Because the pK_a for $H_2PO_4^- / HPO_4^{2-}$ is 6.8, the technique can report pH values in the range 6–7.5. The NMR signal is the average for the two ionisation states because proton exchange is fast on the NMR timescale. If there is phosphate in both external and internal phases, a ΔpH can be calculated. Although the technique is used for *in vivo* monitoring (Section 9.9), the drawback for isolated cells is that NMR is a rather insensitive method and relatively thick cell suspensions are required, with attendant problems of supplying oxygen and substrates.

9.7.3 Why measure $\Delta\psi_m$ and ΔpH?

Although respiration is often the technique of choice if only one is available, measurements of potential are frequently important to remove ambiguity. One example was discussed previously. If an agent increases oligomycin-insensitive respiration, either it could be a mild uncoupler (in which case, $\Delta\psi_m$ would decrease) or it could be acting upstream to facilitate substrate delivery or electron transport chain activity (which would be associated with a mitochondrial hyperpolarisation increasing the current through the endogenous proton leak). A variant is the 'oligomycin null-point test.' Conditions can arise in dysfunctional or hypoxic cells in which the mitochondria partially depolarise, causing the ATP synthase to reverse, draining cytoplasmic, glycolytic, ATP and helping to maintain Δp by proton extrusion. Under these conditions, addition of oligomycin will result in a further depolarisation, whereas in healthy cells the inhibitor causes a hyperpolarisation. A simple experiment in quench mode can determine the direction of the oligomycin-induced $\Delta\psi_m$ change and thus establish whether the mitochondrial population in a given cell was a net producer or consumer of ATP. An example is shown in Figure 12.2.

Finally, the ability to work at single-cell or even single-mitochondrial level in combination with additional probes of cell function is central to much of the mitochondrial cell biology, which will be discussed in subsequent chapters. In this context, however, note that promising techniques are being developed to monitor respiration of single cells loaded with phosphorescent oxygen-sensitive probes. As respiration accelerates, the O_2 gradient across the cell membrane increases, and this is reflected in an increased fluorescence lifetime of the probe (Fercher *et al.*, 2010). If and when these techniques can be made to approach the precision of existing population respiration technologies, then it will become possible to truly quantify cellular bioenergetics at the single-cell level.

9.7.4 NAD(P)H and flavoprotein autofluorescence

Further reading: Duchen *et al.* (2003), Shuttleworth (2010)

NADH and NADPH are fluorescent when excited at 340–360 nm and show identical emission spectra. The $NAD^+/NADH$ pool in the cytoplasm is highly oxidised, and virtually all cellular NADH fluorescence originates from the mitochondria, particularly because its fluorescence lifetime is greatly enhanced in the matrix. NADPH fluorescence is much less enhanced, and to a first approximation, the total cellular signal can be

(a)   (b)

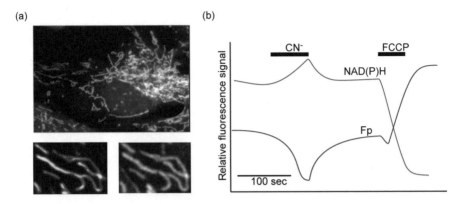

Figure 9.10   Mitochondrial autofluorescence.
(a) Endogenous NAD(P)H fluorescence from a rat cortical astrocyte excited at
351 nm showing the high mitochondrial signal. Images at the bottom show
co-localization of TMRE and autofluorescence. The faint non-mitochondrial
fluorescence may be due to cytoplasmic NADPH (Jacobson and Duchen,
2002). (b) NAD(P)H and flavoprotein autofluorescence from white adipocytes
in culture. Flavoproteins were excited at 458 nm.
Courtesy of Michael Duchen.

ascribed to matrix NADH (Figure 9.10), although the possibility of a contribution from
NADPH must be considered. The term NAD(P)H refers to this ambiguity. Flavoproteins
containing FMN or FAD cofactors fluoresce in the oxidised form when excited at 420–
480 nm. Although multiple flavoproteins contribute to the cell fluorescence, the flavins
associated with complexes I and II make a major contribution, and their reduction state
tends to mirror that of NADH (Figure 9.10). Autofluorescence can be used for *in vivo*
imaging, thus avoiding the necessity to introduce exogenous fluorescent probes.

9.7.5 ATP

The ATP content of a cell is an insensitive indicator of its bioenergetic status, other than
as a live–dead determinant. Figure 9.4 showed that the cytoplasmic ATP concentration remains virtually constant over a wide range of ΔG_p; indeed, significant changes
in ATP content generally reflect net synthesis or degradation (or cell death). ATP/ADP
ratios are better, but the complication is that the observed ratio is a mean of the cytoplasmic and matrix pools, which differ as a consequence of the electrogenicity of the
ANT (Section 9.5.1). In addition, ATP and ADP can be bound or present in nonmeta-
bolic pools (e.g., vesicles). In contrast, if a cell has creatine kinase activity, the creatine/
creatine-P ratio is an excellent way to calculate the true cytoplasmic ΔG_p (Figure 9.4).

An exception to the generalisation that cytoplasmic ATP determinations yield little
information is the pancreatic β cell, whose unique bioenergetics (Section 12.5) allow
changes in extracellular glucose to control this parameter and hence insulin secretion.
Figure 9.11 is taken from a study (Ainscow and Rutter, 2002) in which recombinant
firefly luciferase was expressed via an adenoviral vector in the cytoplasm of primary

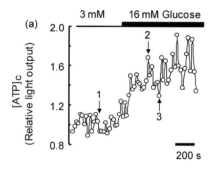
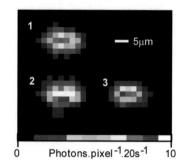

Figure 9.11   Luciferase monitoring of cytoplasmic ATP oscillations in single mouse pancreatic β-cells.
A single cell expressing cytoplasmic luciferase in the presence of luciferin was imaged as the glucose concentration was raised from 3 to 16 mM. (a) Cytoplasmic ATP levels increased and oscillations could be detected in some cells. (b) Images of the cell taken at the time points indicated in panel (a). *Data from Ainscow and Rutter (2002).*

mouse β cells. Luciferase catalyses the oxidation of its 280-Da substrate luciferin (which must be added to the incubation) in the following steps:

$$Luciferin + ATP \rightarrow luciferyl\,adenylate + pyrophosphate$$
$$Luciferyl\,adenylate + O_2 \rightarrow oxyluciferin + AMP + photons$$

The light emission is very faint (typically 3–10 photons per cell per second), and highly sensitive photon detectors are required for single-cell resolution studies.

9.8 PERMEABILISED CELLS

Further reading: Anflous *et al.* (2001), Safiulina *et al.* (2004)

The plasma membrane of cells can be permeabilised by digitonin or saponin. These detergents target the lipid rafts in the plasma membrane and disrupt the bilayer. Their utility lies in the different lipid composition of the IMM, which contains no cholesterol and so is relatively resistant. However, the OMM can be permeabilised by excess detergent and so careful titration is essential to prevent outer membrane damage and release of cyt *c*. Muscle fibres can be 'skinned' by mechanical damage to the plasma membrane, and both preparations have been employed to allow direct access to the mitochondria without having to separate the organelles. Although much of the cellular organisation remains intact, the cytoplasm is lost and replaced by the external medium. In this aspect, the preparation resembles an isolated mitochondrial incubation. In particular, the medium must be 'mitochondria-friendly,' especially in terms of a sub-micromolar Ca^{2+} concentration, at the instant of permeabilisation.

9.9 *IN VIVO* BIOENERGETICS

Further reading: Amara *et al.* (2008), Kemp and Brindle (2012)

The ultimate goal of mitochondrial physiology is to be able to monitor bioenergetic function *in vivo*. Considerable progress to this end has been made by a combination of optical and magnetic resonance spectroscopy (MRS) imaging human muscle *in vivo*. A typical protocol involves the measurement in single *in situ* skeletal muscles of basal resting values for tissue CrP (by MRS) and blood and tissue oxygenation (oxyhaemo-globin (Hb-O_2) and oxymyoglobin (Mb-O_2) respectively, by optical spectroscopy). A transient is then initiated, for example, by 15 minutes of limb ischaemia with a tourni-quet. The rates of decrease in oxygen saturation of Hb-O_2 plus Mb-O_2 are used to cal-culate the rate of mitochondria oxygen uptake, whereas the change in CrP signal allows the basal ATP turnover rate to be determined. After correction for glycolytic ATP syn-thesis, a P/O ratio (Section 4.3.7) can be calculated. The tourniquet is then removed, and the kinetics of recovery of the parameters are measured. The time constant for the recovery of CrP (remember that this can be present at more than 20 mM) is used to cal-culate the *maximal* rate of ATP synthesis in the muscle.

Measurement of P_i incorporation into ATP by ^{31}P MRS has been claimed to monitor mitochondrial ATP synthesis in muscle. However, this has been contested by the finding that a glycolytic P $\rightleftharpoons$ ATP exchange reaction occurs that is unrelated to mitochondrial ATP synthesis.

9.10 REACTIVE OXYGEN SPECIES, 'ELECTRON LEAKS'

Further reading: Murphy (2009), Handy and Loscalzo (2011)

The term *reactive oxygen species* (ROS) tends to be used indiscriminately as though all ROS molecules were equivalent. In most cells, mitochondria are responsible for the majority of the production and detoxification of ROS, although the role of cytoplasmic enzymes such as monoamine oxidase should not be forgotten. In the mitochondrial con-text, the primary ROS is the superoxide anion $O_2^{\bullet-}$, produced by the $1e^-$ reduction of O_2 in slow side reactions at multiple sites. Superoxide will spontaneously dismutate to form oxygen and hydrogen peroxide, a process that is greatly accelerated in the cell by superoxide dismutases (SODs). If nitric oxide ($^{\bullet}NO$) is present, peroxynitrite can be formed by spontaneous reaction with $O_2^{\bullet-}$, while in the presence of Fe^{2+}, H_2O_2 can pro-duce the damaging hydroxyl radical (Figure 9.12).

There are several sites in the electron transport chain where a two-electron carrier, such as ubiquinol, $FADH_2$, or $FMNH_2$, donates electrons to one-electron carriers (Fe–S centres or cytochromes). When this, or the reverse one- to two-electron transfer, occurs, the two-electron carrier must donate or accept its electrons singly, passing through an intermediate free radical form such as ubisemiquinone ($UQ^{\bullet-}$). If the thermodynamics are favourable, each of these sites is a potential source of $O_2^{\bullet-}$ in the mitochondrion by leaking a single electron to O_2. One-to-one electron transfers in the electron transfer

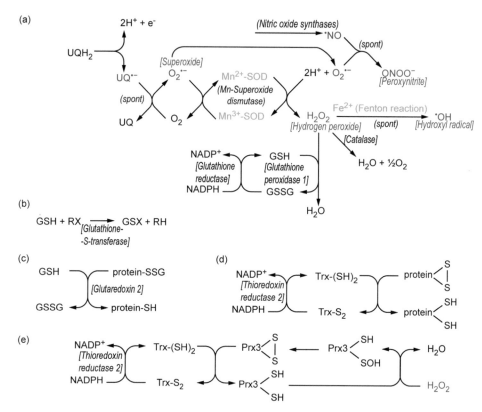

Figure 9.12 Mitochondrial ROS: generation and detoxification.
(a) Generation and metabolism of superoxide generated by 1e⁻ transfer from
the ubisemiquinone anion. Radicals are shown in red and other reactive
oxygen species in blue. spont, spontaneous process. (b–e) Thiol metabolism
in the mitochondrion. Prx3, peroxyredoxin 3; Trx, thioredoxin.

chain—for example, between Fe–S centres or cytochromes—are not sources of $O_2^{\cdot-}$,
whereas the final O_2 reduction step in complex IV (Section 5.9.2) has evolved to prevent
single-electron leakage.

The $E_{m,7}$ (Section 3.3) for the reaction $O_2 + e^- = O_2^{\cdot-}$ in the presence of $1\,M\ O_2$
is approximately $-160\,mV$. To translate this into a more relevant $E_{h,7}$ requires an estimate of the intracellular concentrations of oxygen and superoxide. Taking a value of
$25\,\mu M$ for intracellular O_2 in a tissue (equivalent to ~3% O_2) and an estimated range of
1–$250\,pM$ for $O_2^{\cdot-}$ (Murphy, 2009), this gives an estimate for $E_{h,7}$ in the range of $+150$
to $+230\,mV$. Note that these are not equilibrium concentrations because SODs continuously remove $O_2^{\cdot-}$ (Figure 9.12). Thus, sites in complexes I–III have sufficiently low
(i.e., negative) potentials to be candidate sites of $O_2^{\cdot-}$ generation. The rate of $O_2^{\cdot-}$ production at a given site is proportional to the concentration of the carrier, the fraction in
the correct redox state and the local O_2 concentration. This means that experiments performed in air, rather than under physiological intracellular O_2 tensions, may exaggerate
the role of $O_2^{\cdot-}$ generation (Figure 9.13).

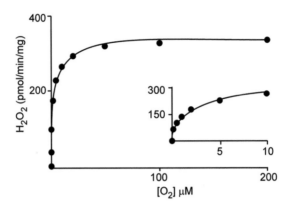

Figure 9.13   ROS generation by isolated mitochondria is a hyperbolic function of oxygen concentration.
Rat liver mitochondria respiring in the presence of succinate in the presence of varying oxygen concentrations. Note the absence of a 'hypoxic' increase in ROS. Insert: Enlarged region.
Data from Hoffman and Brookes (2009).

9.10.1 Complex I

Further reading: Pryde and Hirst (2011), Treberg *et al.* (2011)

Candidate sites in complex I for $O_2^{\bullet -}$ generation are the electron transfers from $FMNH_2$ to Fe–S centre N3 (Figure 5.10) and from centre N2 to UQ. There is debate regarding whether $O_2^{\bullet -}$ originates from the flavin site alone or whether both flavin and UQ sites contribute. The generation of $O_2^{\bullet -}$ by depolarised sub-mitochondrial particles, or by purified complex I in the absence of UQ, correlates with the redox potential of the $NADH/NAD^+$ couple (which is in equilibrium with the $FMNH_2/FMN$ couple) and has led to a proposal for a single-site (flavin) mechanism. Studies with intact mitochondria are more complicated to interpret because it is difficult to precisely control the upstream and downstream redox couples, and other $O_2^{\bullet -}$-generating sites in the mitochondrion must be controlled for. Controversy revolves around the relative super-oxide-generating abilities of normal, forward, electron transfer and so-called reversed electron transfer (RET) with succinate as substrate (Figure 4.11a). Because complex I is thermodynamically the weakest of the pumps, by which we mean that it reaches thermodynamic equilibrium with the proton electrochemical potential at a slightly lower Δp than complexes III or IV, bypassing it with succinate allows a slightly higher maximal Δp to be attained. This has the effect of re-equilibrating the redox span across complex I with Δp, resulting in an increased level of NAD^+ reduction. With isolated mitochondria, this is associated with a large increase in superoxide generation. Apparent discrepancies in the relationship between the extent of NAD(P) reduction and rate of $O_2^{\bullet -}$ production during RET, but not following rotenone addition, have led to the proposal that $O_2^{\bullet -}$ could also be produced at a rotenone-sensitive UQ site under these *in vitro* conditions of RET.

9.10.2 Complex III

Further reading: Quinlan *et al.* (2011)

The Q_p site of complex III is a potential source of $O_2^{\bullet-}$. As discussed in Chapter 5, the $E_{m,7}$ values for the two-stage oxidation of UQH_2 at the Q_p site of complex III via $UQ^{\bullet-}$ to UQ are respectively $+280$ and $-160\,\text{mV}$. This means that the $UQ^{\bullet-}/UQ$ couple is highly reducing and thermodynamically capable of donating an electron to O_2. Under normal physiological conditions, in the absence of inhibitors, the generation of $O_2^{\bullet-}$ by complex III is low compared with that of complex I, presumably because the electron from $UQ^{\bullet-}$ is rapidly transferred to cyt b_L before finding an oxygen molecule. However, the complex III inhibitor antimycin A, which binds to the inner Q_n site (Section 5.8), inhibits this pathway and the ubisemiquinone is stranded on the Q_p site, allowing it time to transfer an electron to O_2 and generate $O_2^{\bullet-}$. In contrast, myxothiazol, by preventing transfer of the first electron from UQH_2 to the Rieske protein, prevents formation of $UQ^{\bullet-}$ and inhibits complex III $O_2^{\bullet-}$ generation.

9.10.3 Other sites

Further reading: Brand (2010)

The two lipoic acid-containing oxidative decarboxylation reactions in the mitochondrion, pyruvate and 2-oxoglutarate dehydrogenases, are minor contributors to matrix $O_2^{\bullet-}$ production, as are two additional pathways that feed electrons to the UQ pool: electron-transferring flavoprotein UQ oxidoreductase, feeding electrons from β-oxidation, and glycerophosphate dehydrogenase. Complex II may also be a $O_2^{\bullet-}$ source.

9.10.4 Superoxide metabolism

Further reading: Murphy (2011)

Superoxide generated by complex I is liberated into the matrix, whereas that from complex III is believed to be released into both matrix and IMS. The efficient removal of $O_2^{\bullet-}$ is essential for survival, and estimates of the steady-state $O_2^{\bullet-}$ in cells are in the low picomolar range.

9.10.4.1 Superoxide dismutases

SODs are the first line of defence against oxidative damage and catalyse the formation of H_2O_2 and O_2 from $O_2^{\bullet-}$ (Figure 9.12). Three SODs are found in mammals. SOD1 (or CuZn-SOD) is primarily cytoplasmic, although a subfraction is present associated with the OMM, with implications for amyotrophic lateral sclerosis (Section 12.6.7). SOD2 (Mn-SOD) is in the matrix, whereas SOD3 is an extracellular CuZn isoform. Loss-of-function mutations in SOD2 are embryonic lethal.

The metal prosthetic groups undergo reversible redox changes, first accepting an electron from $O_2^{\bullet-}$, forming O_2, and then donating the electron to a second $O_2^{\bullet-}$, forming H_2O_2 (Figure 9.12):

$$O_2^{\bullet-} = O_2 + 1e^- \text{ and } O_2^{\bullet-} + 2H^+ + 1e = H_2O_2$$
$$\text{giving } 2O_2^{\bullet-} + 2H^+ = H_2O_2 + O_2$$

Hydrogen peroxide is still a ROS, and pathways have evolved for its detoxification.

9.10.4.2 Glutathione

Mitochondria maintain a glutathione pool (GSH) of 1–5 mM. GSH is synthesised in the cytoplasm and may be imported by the dicarboxylate and 2-oxoglutarate carriers. The matrix GSH pool is maintained highly reduced (E_h of approximately -280 mV; Section 3.3.5) by NADPH-linked glutathione reductase. The NADP pool (E_h of approximately -390 mV) is in turn reduced by the protonmotive force-linked transhydrogenase (Section 5.12) and the matrix NADP-linked isocitrate dehydrogenase. The large thermodynamic disequilibrium between the GSSG/GSH and $NADP^+$/NADPH couples suggests that the reductase activity may exert a major control over GSH cycling (Section 3.3.5).

Glutathione peroxidase 1 (Gpx1) detoxifies H_2O_2 or artificial pro-oxidants such as t-butylhydroperoxide (t-BuOOH) (Figure 9.12a). A second isoform (Gpx4) limits accumulation of phospholipid hydroperoxidases in the inner membrane. Glutathione-S-transferases detoxify a range of xenobiotics by forming thioethers with GSH, whereas mitochondrial glutaredoxin 2 (Grx2) catalyses the reversible formation of protein–glutathione mixed disulfides.

9.10.4.3 Thioredoxin and peroxiredoxins

Thioredoxin (Trx) is a 12-kDa protein with a redox-active dithiol at its active site, reduced by NADPH and thioredoxin reductase 2 (TR2). Thioredoxin plays a major role in maintaining protein thiol groups in the reduced state. Peroxiredoxins degrade alkyl peroxides and H_2O_2. The mitochondrial peroxiredoxin 3 (Prx3) is responsible for the breakdown of 90% of the H_2O_2 generated by the mitochondria, forming sequentially a sulfenic acid and a protein disulfide, before being reduced back to the dithiol by Trx (Figure 9.12).

9.10.5 Measurement of ROS production by mitochondria

Further reading: Dickinson *et al.* (2010), Murphy *et al.* (2011)

The literature contains estimates of the proportion of electrons leaking to form $O_2^{\bullet-}$ that vary from 0.1 to 4% for isolated mitochondria. This large discrepancy is a consequence of the range of conditions employed. The temptation is to adjust the incubation to maximise the signal or effect that is to be studied. In the case of mitochondria, this includes

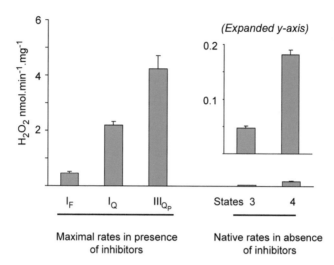

Figure 9.14   Under physiologically relevant conditions, isolated mitochondria generate $O_2^{\cdot-}$ at a small fraction of their maximal rate. Skeletal muscle mitochondria were treated with CDNB (1-chloro-2,4-dinitrobenzene) to minimise intra-mitochondrial H_2O_2 trapping. Extracellular H_2O_2 was assayed. Maximal rates for the complex I flavoprotein site (I_F) were obtained in the presence of malate and rotenone; the complex I Q-site was assayed as rotenone-sensitive H_2O_2 production in the presence of succinate and complex III Q_P site by myxothiazol-sensitive signal in the presence of succinate, malonate and antimycin A. Native rates were determined with malate plus glutamate as substrates in the absence of inhibitors and in states 3 and 4.
Data from Quinlan et al. (2012).

the use of succinate as substrate to maximise $O_2^{\cdot-}$ production from complex I and antimycin A for maximal complex III leakage. Under more physiological conditions—that is, electrons feeding into complex I in the absence of inhibitors—the production of $O_2^{\cdot-}$ is dramatically decreased (Figure 9.14).

The actual measurement of ROS is far from trivial. There are no perfect detection systems, only less bad ones. One approach is to add an $O_2^{\cdot-}$-sensitive dye such as hydroethidine or the mitochondrially targeted MitoSOX and to determine its rate of oxidation. Because of their rather inefficient competition with SODs, these dyes do not measure the rate of production of $O_2^{\cdot}$ but, rather, their reduction rate is a function of the dye concentration and the steady-state concentration of $O_2^{\cdot-}$. An increased rate of reduction of the dye therefore indicates an increased level of $O_2^{\cdot-}$. Normally, only a few percent of the $O_2^{\cdot-}$ is trapped by the dye, which has the advantage that the downstream consequences of ROS generation are not disturbed. Note that the concentration of MitoSOX in the matrix is a function of $\Delta\psi_m$ and so any plasma or mitochondrial membrane potential changes must be controlled for.

Because SODs are extremely active, the vast majority of $O_2^{\cdot-}$ is rapidly converted to H_2O_2, which freely diffuses out of the mitochondria and cell and can be trapped and detected by the oxidation of a nonfluorescent dye such as Amplex Red in combination with horseradish peroxidase. Although this can quantify the total release of H_2O_2 from the cell, this will be an underestimate of the original production because of the action of the intracellular peroxidases discussed previously. Dichlorodihydrofluorescein, frequently used to monitor H_2O_2, has many artefacts and is not recommended; however, a new generation of mitochondrially targeted boronate-based fluorescent probes for H_2O_2 show promise (Dickinson *et al.*, 2010).

9.10.6 Monitoring thiol redox potentials

Further reading: Cannon and Remington (2008), Gutscher *et al.* (2008), Morgan *et al.* (2011)

From the previous section, it is apparent that the redox poise of the glutathione couple, both in the cytoplasm and in the matrix, plays a central role in cellular metabolism. Redox-sensitive fluorescent probes have been engineered by modification of green fluorescent protein by introducing cysteines into the structure capable of reversibly forming disulfide bridges on oxidation with a concomitant change in fluorescent properties. These redox-sensitive GFPs (roGFPs) are ratiometric and equilibrate with the glutathione pool via glutaredoxin. With appropriate targeting, they can therefore be exploited to dynamically monitor the glutathione redox potentials in the cytoplasm and matrix. Recently, the approach has been extended to generation of fusion proteins between roGFP and glutaredoxin to improve the responsiveness to GSH (Figure 9.15),

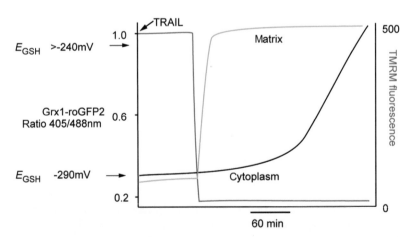

Figure 9.15   Monitoring cytoplasmic and matrix glutathione redox potentials with Grx1-roGFP constructs.
HeLa cells expressing cytoplasmic (black) or matrix targeted (orange) glutaredoxin-1 redox-sensitive GFP fusion proteins and equilibrated with TMRM (red) were exposed to TRAIL (TNF-related apoptosis-inducing ligand) to initiate the extrinsic pathway of apoptosis (Section 10.7). $\Delta\psi_m$ collapsed after approximately 1 h; the matrix glutathione pool simultaneously oxidised (increased ratio 405/488 nm). In contrast, the cytoplasmic pool slowly oxidised (from approximately −290 to −240 mV) during the next 4 h. *Adapted from Gutscher et al. (2008).*

and a fusion protein between roGFP and yeast Orp1 has been constructed to measure H_2O_2.

9.10.7 Mitochondrially targeted antioxidants

Further reading: Smith and Murphy (2011)

In addition to its role as an electron carrier in the mitochondria, ubiquinone (UQ) (and the closely related plastoquinone found in plant thylakoids; Section 6.4) can act as a 'chain breaker' antioxidant limiting free radical propagation. Attempts to administer UQ as an antioxidant supplement have had limited success because the quinone is lipophilic and will tend to dissolve in all membranes. However, if it is conjugated to a lipophilic cation, it can be targeted to the IMM and matrix in intact cells *in vivo*, and it can act as a catalytic (rather than stoichiometric) antioxidant. Mitoquinone (MitoQ) is a conjugate of UQ (lacking the isoprenoid side chain) and the triphenylphosphonium cation (Section 4.4.3) linked by a 10C hydrocarbon chain that optimally positions the quinone moiety within the hydrophobic core of the polarised IMM. Although it is unable to catalyse electron transfer in the same way as native UQ, it is an effective superoxide scavenger and extensive trials are underway to ascertain its effectiveness in a variety of clinical conditions, some of which are discussed in Chapter 12. The closely related SkQ1, a targeted plastoquinone, has been claimed to produce dramatic anti-aging effects in a variety of tissues.

9.11 REACTIVE NITROGEN SPECIES

Further reading: Radi *et al.* (2002), Erusalimsky and Moncada (2007), Brown and Borutaite (2007)

In addition to its role as a membrane-permeant second messenger, NO is a reversible inhibitor of complex IV, competing with oxygen. Its slow reduction to the non-inhibitory nitrous oxide contributes to the reversibility of the inhibition, as does the non-enzymic direct reaction of NO with O_2. It is plausible that physiological concentrations of NO decrease the apparent affinity of the complex for oxygen—and this may be of significance in the reduced oxygen environment of the intact cell—by increasing the control exerted by oxygen availability. Peroxynitrite is formed by the non-enzymatic reaction of NO with $O_2^{\cdot-}$. Although exogenously added peroxynitrite has been reported to inhibit at multiple sites in the electron transport chain, it is important to consider what physiological (or pathophysiological) concentrations would be expected *in vivo*, in view of the low-picomolar concentrations of $O_2^{\cdot-}$ believed to be present in the mitochondria. The significance of *in vitro* experiments in which peroxynitrite is added at concentrations in the range 10^{-5} to 10^{-4} M must be critically examined. Also note that the existence of a mitochondrial form of nitric oxide synthase (mtNOS) is highly contentious.

9.12 UNCOUPLING PATHWAYS, 'PROTON LEAKS'

Further reading: Nicholls (2006b), Azzu and Brand (2009)

All isolated mitochondria possess an inherent proton leak, allowing respiration to occur in the absence of ATP synthesis, whereas secondary transport processes, such as Ca^{2+} cycling, utilise the proton current and add to the overall oligomycin-insensitive current. The molecular basis of this leak has not been firmly established, although correlations suggest that MCs such as ANT may contribute. The endogenous leak is independent of that contributed by uncoupling protein(s), which is reviewed in Chapter 12 in the context of brown adipose tissue. We review these pathways here because of indications that there may be an intimate link between proton leaks (i.e., uncoupling) and electron leaks ($O_2^{\bullet-}$ generation). The concept of 'mild uncoupling'—that is, that a modest increase in proton leak can reduce $O_2^{\bullet-}$ generation by lowering $\Delta\psi_m$—is pervasive in the literature but may require more critical analysis in the light of indications that a form of neuronal preconditioning (i.e., protection; Section 12.6.2) afforded by brief exposure to low protonophore concentrations may be mediated by AMP-kinase (Section 11.3) rather than by ROS.

9.12.1 Relationships between proton leak and $O_2^{\bullet-}$

Isolated heart mitochondria oxidising succinate in state 4 maintain an elevated $\Delta\psi$ and produce relatively large amounts of $O_2^{\bullet-}$, consistent with the high reduction state of complex I by 'reversed electron transport.' Titration with protonophore produces the expected drop in $\Delta\psi$ and increase in respiration, and these are accompanied by a rapid decrease in the rate of $O_2^{\bullet-}$ generation (detected as extramitochondrial H_2O_2) (Figure 9.17a). These results raise the possibility of limiting oxidative stress by 'mild uncoupling' (i.e., a controlled increase in proton conductance), although this will inevitably be associated with a decreased maximal capacity for ATP synthesis as protons are diverted from the ATP synthase to the leak.

Intact cells cannot bypass complex I by simply utilising succinate because the substrate is generated, and fumarate removed, by NAD-linked dehydrogenases. When a similar protonophore titration experiment is performed with intact cells metabolising glucose, a rather different relationship is obtained (Figure 9.17), with a much flatter $\Delta\psi_m$ dependency for ROS. Indeed, once excess protonophore is present above that required for maximal respiration, a dramatically increased level of $O_2^{\bullet-}$ is seen, probably due to lowered NADPH reduction with consequent effect on the glutathione pool (Figure 9.12). Thus, the trade-off between lowered ROS and loss of ATP-generating capacity seems less favourable in intact cells. In fact, mitochondria have evolved a more sophisticated version of mild uncoupling through their endogenous proton leak (Section 4.5.1). The high-voltage dependency of the leak (Figure 4.12) implies that $O_2^{\bullet-}$ generation can be restricted to state 4 conditions, whereas the decrease in Δp in state 3 will result in inhibition of the leak, allowing protons to be channelled through the ATP synthases with little loss of ATP-generating capacity.

We now describe a proton leak pathway in which the loss of ATP-generating capacity (or rather the diversion of protons into the leak) is central to its mechanism, the brown fat UCP1.

9.12.2 Uncoupling protein 1

Further reading: Nicholls (2006b)

Brown adipose tissue is the seat of nonshivering thermogenesis—the ability of hibernators, cold-adapted rodents and newborn mammals in general to increase their respiration and generate heat without the necessity of shivering (Section 12.4). In extreme cases of small rodents, whole body respiration can increase up to 10-fold as a result of the enormous respiration of this tissue, which rarely accounts for more than 5% of the body weight. The tissue is innervated by noradrenergic sympathetic neurons and release of the transmitter onto an unusual class of β_3 receptors that activate adenylyl cyclase and hence hormone-sensitive lipase (Figure 9.16). This leads to the hydrolysis of the triglyceride stores, which are present in multiple small droplets, giving the cell a 'raspberry' appearance. The brown adipocytes are packed with mitochondria, whose extensive inner membranes indicate a high capacity for respiration. However, the chemiosmotic theory now poses a problem: how can the fatty acids liberated by lipolysis be oxidised by the mitochondria when control of the proton circuit is focused on the re-entry of protons into the mitochondrial matrix? The problem is compounded by the relatively low amount of ATP synthase and by the absence of any significant extramitochondrial ATP hydrolase activity.

Two solutions are possible from first principles: either the brown fat mitochondrial respiratory chain is modified so that it does not expel protons or the membrane is modified to allow re-entry of protons in the absence of ATP synthesis. The latter turns out to be the case. The mitochondrial inner membrane contains a unique 32-kDa uncoupling protein (UCP1) that binds a purine nucleotide to its cytoplasmic face and is inactive until the free fatty acid concentration in the cytoplasm starts to rise. The protein then binds a fatty acid and alters its conformation to become proton conducting. The uncoupling protein thus acts as a self-regulating endogenous uncoupling mechanism that is automatically activated in response to lipolysis, allowing uncontrolled oxidation of the fatty acids (Figure 9.16). The low conductance state is restored when lipolysis is terminated, and the mitochondria oxidise the residual fatty acids.

The physiological regulation seen in intact brown adipocytes can be mimicked with isolated mitochondria in a combined oxygen electrode/TPP$^+$ electrode chamber by the infusion of fatty acid (mimicking lipolysis) in the presence of coenzyme A, carnitine and ATP to allow the fatty acid to be activated. The increase in $C_m H^+$ correlates with the steady-state concentration of free fatty acid during the infusion and on the termination of lipolysis (Figure 9.16b).

Expression of the uncoupling protein varies in response to the adaptive status of the animal: it is present at high concentration at birth, but it is then repressed so that the mitochondria lose the protein and the capacity for nonshivering thermogenesis.

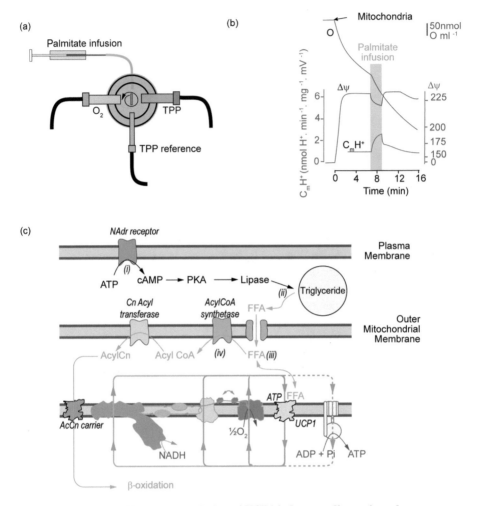

Figure 9.16   The acute regulation of UCP1 in brown adipose tissue by fatty acids.
(a) Combination TPP$^+$ and O$_2$ electrodes for the continuous monitoring of $\Delta\psi$, respiration, J$_{H^+}$ and C$_m$H$^+$ in brown adipose tissue mitochondria during infusion of the thermogenic substrate palmitate. (b) Details of an experiment in which palmitate was infused (shaded period) while respiration and $\Delta\psi_m$ were monitored. The incubation contained pyruvate and ATP, CoA and carnitine, allowing the palmitate to be activated to palmitoyl-CoA and then palmitoylcarnitine. Palmitate accumulation activates UCP1, increasing C$_m$H$^+$. At the conclusion of the infusion, the mitochondria automatically recouple as the palmitate is activated and oxidised, allowing the fatty acid to leave its binding site on UCP1. (c) Scheme applying these findings to the intact brown adipocyte. Noradrenaline binding to a β_3 receptor (i) activates the lipolytic cascade (ii), liberating free fatty acid (FFA) (iii). Although some acylcarnitine can be formed by fatty acid activation (iv), FFA accumulates as oxidation of the acylcarnitine is limited by respiratory control, until FFA binds to UCP1, activating its proton conductance. Rapid palmitoylcarnitine oxidation can now occur. On termination of lipolysis, residual fatty acids are oxidised and UCP1 reverts to its inactive state.
Data from Rial et al. (1983).

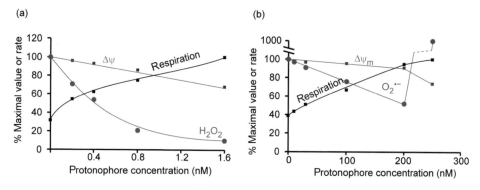

Figure 9.17   Interrelationships between proton conductance, respiration, $\Delta\psi_m$ and ROS for isolated and *in situ* mitochondria.
(a) Isolated rat heart mitochondria with succinate as substrate; titration with the protonophore SF6847 (Korshunov *et al.*, 1997). Note that this experiment was performed under conditions of high $O_2^{\cdot-}$ production (see Figure 9.14). (b) Primary rat cerebellar granule neurons in the presence of glucose; titration with FCCP (Johnson-Cadwell *et al.*, 2007).

Cold-adaptation (or, interestingly, overfeeding under certain conditions) leads to re-expression of the protein by a mechanism involving the transcriptional coactivator PGC-1α (Section 11.4). Recently, it has been recognised that brown adipose tissue is present in the adult human; we revisit the tissue in Section 12.4.

UCP1 is a member of the MC family (Section 8.2). The C-terminal domain of UCP1 shares a nine-amino acid sequence with ANT and the DNA binding domain of several transcription factors, and this has been proposed to represent the purine nucleotide binding site of the uncoupling protein (but see Section 8.2.3). UCP1 has rather low transport activity, which accounts for its presence as a major inner membrane protein in thermogenic brown adipose tissue mitochondria; however, site-directed mutagenesis at any of the three matrix loops converts the protein into a high-conductance nonselective pore, reminiscent of the permeability transition pore.

The mechanism of UCP1 has been the subject of some debate. Models must incorporate the observation that both protons (or hydroxides) and chloride are transported by the protein in a purine nucleotide-sensitive manner and that proton, but not chloride, conductance is activated by fatty acids. The simplest model is one in which the fatty acids facilitate the delivery of protons to the transport site. Interestingly, the purified, reconstituted UCP1 displays the nucleotide-sensitive but fatty acid-insensitive Cl⁻ conductance characteristic of the *in situ* protein but no significant fatty acid-dependent proton conductance. This is restored by the addition of UQ, suggesting that the coenzyme may facilitate proton delivery to the transporter.

An alternative mechanism proposes that UCP1 acts as a nonselective anion channel allowing the fatty acid anion to be transported out of the mitochondrion, completing a putative dissipative cycle including the entry of the protonated fatty acid (i.e., a classical protonophore mechanism) (Section 2.3.5). Such a mechanism can be demonstrated with a range of anion transporters in the presence of high, probably nonphysiological, levels of fatty acids. However, UCP1 is extremely sensitive to nanomolar fatty acid concentrations and also retains activity (in the absence of GDP) in the strict absence of fatty acids.

9.12.3 Novel uncoupling proteins

Further reading: Nedergaard and Cannon (2003), Nicholls (2006b), Azzu and Brand (2009), Pi *et al.* (2009)

In 1997, two genes with approximately 60% sequence identity to UCP1 were found, proposed as candidate uncoupling proteins and named UCP2 and UCP3. Later, two further genes were identified in brain, but these showed substantially less sequence similarity to UCP1. These were termed UCP4 and BMCP1 (brain mitochondrial carrier protein 1, sometimes referred to as UCP5). Of course, it is not sufficient simply to propose a function for a protein on the basis of a sequence homology, and the debate regarding their physiological function continues 15 years after their discovery. Indeed, in the 10 years since the third edition was written, UCP2 and UCP3 have been the subject of more than 1000 publications. UCP2 has a broad tissue distribution, with the highest concentration in spleen mitochondria, although even here its expression level is only 1% of that for UCP1 in brown fat. UCP3 is primarily expressed (but still at low levels) in skeletal muscle and exists in long- and short-form transcripts, with the latter lacking the sixth transmembrane domain. There is no evidence that it is thermogenic.

There is considerable uncertainty about the role of the novel uncoupling proteins (nUCPs) in the cell, or whether they are even capable of proton translocation. The starting assumption is generally that UCP2 and UCP3 share common regulatory aspects with UCP1, namely a high, nonphysiological, conductance in the absence of physiological concentrations of purine nucleoside di- or triphosphates; full inhibition in the presence of these nucleotides; and a physiologically relevant conductance induced by low concentrations of free fatty acids in the continued presence of the nucleotides. Early overexpression studies showed increased conductance but failed to demonstrate nucleotide inhibition and were ascribed to incorrect insertion or bilayer perturbation. Inhibition by GDP has therefore generally been adopted as a criterion for the presence of a functional nUCP, although it must be re-emphasised that the nucleotide-free state is nonphysiological. Thus, increases in conductance produced by agents in the absence, but not the presence, of nucleotides are of little physiological relevance. The activation of UCP2 and UCP3 proton conductance by superoxide generated exogenously by xanthine/xanthine oxidase appears to fall into this category because the $O_2^{\cdot-}$-dependent increases in conductance seen in brown fat, skeletal muscle, kidney, spleen and pancreatic β cell mitochondria are in all cases abolished by 0.5 mM GDP. Finally, putative nUCP activators should function at physiologically relevant concentrations, should make physiological sense (as in the case of fatty acids and UCP1), and a mechanism should exist to reverse any activation.

In our opinion, no nUCP study has satisfied all of these criteria. It is therefore profoundly worrying that the 'uncoupling' activity of nUCPs is sometimes taken as axiomatic. To quote from the abstract of a random paper published in November 2011, *uncoupling protein 2 (UCP2) is a mitochondrial transporter present in the inner membrane of mitochondria, and it uncouples substrate oxidation from ATP synthesis, thereby dissipating the membrane potential energy and consequently decreasing ATP production by mitochondrial respiratory chain. As a consequence of the uncoupling, UCP2 decreases the reactive oxygen species (ROS) formation by mitochondria.*

With the exception of the first sentence, all of these statements are on shaky ground. What is established in certain mouse strains is that a global deletion of UCP2 results in an oxidative shift in the redox potential of glutathione pools in blood and in tissues that would normally express the protein (Pi *et al.*, 2009). At the same time, there is a compensatory upregulation of a range of antioxidant enzymes. The effects of these changes on insulin secretion are discussed in Section 12.5.

Suggestions that UCP2 and UCP3 are modulators of the mitochondrial Ca^{2+} uniporter have not been substantiated. However, by whatever mechanism, their ability to reduce levels of ROS has been widely reported. A rather unexpected role has been proposed for UCP2, associated with its supposed protonophoric activity, to decrease insulin secretion from pancreatic β cells; however, as will be reviewed in Section 12.5, this is again contentious. One interesting feature, the significance of which is unclear, is that UCP2 and UCP3 (but not UCP1 or ANT) have a remarkably short half-life in the cell, turning over in approximately 1 h. Finally, bearing in mind that UCP1 is a member of the MC carrier family, the possibility that it acts by transporting a metabolite across the IMM should not be excluded.

9.13 THE ATP SYNTHASE INHIBITOR PROTEIN IF_1

Further reading: Campanella *et al.* (2009), Faccenda and Campanella (2012)

The ATP synthase is freely reversible, and its direction depends on the thermodynamic balance between Δp and the matrix ΔG_p. Damage to the electron transport chain, increased proton leakage, or severe hypoxia can lower Δp such that the ATP synthase reverses in the cell and starts to hydrolyse cytoplasmic ATP generated by glycolysis. Experimentally, this reversal can be detected as a decrease in $\Delta \psi_m$ upon addition of the ATP synthase inhibitor oligomycin (see Figure 12.2). Under these conditions, glycolysis is called on to service not only the entire ATP demand of the cell but also the synthase reversal. One result of this is that cells may deplete their cytoplasmic ATP to the extent that glycolysis and fatty acid oxidation, both of which require ATP, cannot proceed and the cell dies. This condition is also approached in many published experiments in which protonophores are added to cells, in which case mitochondrial ATP hydrolysis can be extremely rapid, being no longer limited by the low inner membrane proton permeability. Although the ATP depletion can be alleviated *in vitro* by the addition of oligomycin, a more subtle physiological mechanism exists in many cells, mediated by the 10-kDa inhibitor protein (IF_1; Section 7.6).

IF_1 can bind to the F_1-ATP synthase under conditions of acidic matrix pH, partially inhibiting its catalytic activity. At the molecular level, studies with *Escherichia coli* ATP synthase suggest that IF_1 acts as a 'ratchet' preventing reversal of the enzyme (Section 7.6).

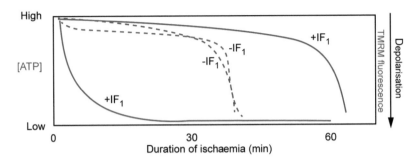

Figure 9.18   The inhibitor protein IF₁.
Schematic of the predicted time courses of ATP depletion (magenta) and $\Delta\psi_m$ (red) following ischaemia in a cell possessing maximal IF₁ activity (solid lines) and lacking IF₁ (dashed lines). The presence of the inhibitor protein prolongs the time for which the cell retains ATP, but at the cost of a rapid mitochondrial depolarisation.
Data adapted from Campanella et al. (2009).

Because an acidic matrix is normally only seen under hypoxic conditions, when the electron transport pathway is inhibited, or in the presence of a protonophore, this essentially means that IF₁ can inhibit ATP synthase reversal but is without effect on ATP synthesis (when the matrix is alkaline). The inhibition is not complete but depends on the ratio of IF₁ to the ATP synthase complex and may play an important role in limiting ATP depletion in hypoxia (Figure 9.18). Neurons generally possess higher ratios of IF₁ to F₁ than astrocytes, with the result that electron chain inhibition causes a more profound depolarisation of the former while slowing cytoplasmic ATP depletion.

10

THE CELL BIOLOGY OF THE MITOCHONDRION

10.1 INTRODUCTION

Our understanding of mitochondrial cell biology, by which we mean the dynamic aspects of the structure, life and death of the mitochondrion within the cell, has expanded enormously during the past decade. The proteins defining the structure of the mitochondrion are beginning to be defined, together with the realisation that mitochondria are intensely dynamic structures, continuously undergoing fission and fusion, interacting with the endoplasmic reticulum (ER), and being transported within the cell to sites where they are required. Finally, quality control mechanisms come into play to ensure that mitochondria with suboptimal function are eliminated or even that the entire cell is dismantled by apoptosis.

10.2 THE ARCHITECTURE OF THE MITOCHONDRION

Further reading: Frey and Mannella (2000)

The classic two-dimensional picture of the mitochondrion (see Figure 1.4) as a lozenge-shaped organelle, with relatively open connections between the cristae space and the intermembrane space between the inner and outer membranes, has been challenged as a result of high-resolution electron microscopy of serial sections and reconstruction of the three-dimensional structure of *in situ* mitochondria by computer tomography. The cristae are revealed as tortuous, frequently tubular, structures with only constricted contacts (cristae junctions with a diameter of ~30 nm) with the intermembrane space (Figure 10.1). Whereas electron microscopy of *in situ* mitochondria generally shows a so-called orthodox configuration with expanded matrix and narrow cristae, isolated mitochondria tend to display a more chaotic and variable cristae structure, although the restricted communication with the inner boundary membrane is retained. They can also show a 'condensed' conformation, where the matrix has undergone osmotic contraction

Bioenergetics. Doi: http://dx.doi.org/10.1016/B978-0-12-388425-1.00010-5

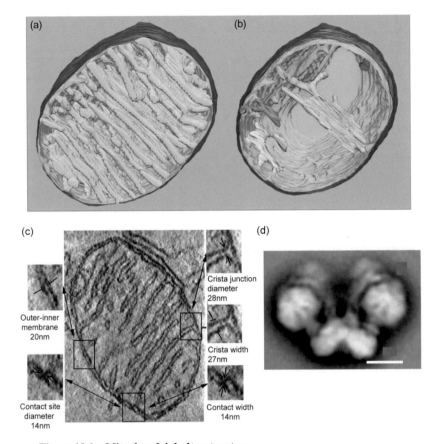

Figure 10.1   Mitochondrial ultrastructure.
(a) Three-dimensional electron tomography of a brown fat mitochondrion. Cristae are shown in yellow, the inner boundary membrane in light blue, and OMM in dark blue. (b) Four representative cristae shown in different colours. (c) Single section through the same mitochondrion showing representative dimensions of major structural features. *Reproduced with permission from Frey and Mannella (2000).* (d) Negatively stained side view of dimeric ATP synthase. *Reproduced with permission from Dudkina et al. (2010).*

due to loss of matrix ions. Such 'condensed' mitochondria may be artefacts of fixation and are probably not of physiological relevance; in particular, dehydration of the matrix (which even in the orthodox conformation may contain 500 mg of protein/ml) profoundly inhibits diffusion of substrates and coenzymes and thus respiration of NADH-linked substrates.

10.2.1 The structure of the mitochondrial inner membrane

Further reading: Pellegrini and Scorrano (2007), Rabl *et al.* (2009)

The F_oF_1 ATP synthase (Chapter 7) may play an important role in controlling the morphology of the cristae. Non-denaturing blue-native gel electrophoresis and single-particle electron microscopy of ATP synthases prepared from a variety of mitochondria indicate that they frequently occur as linked dimers in the crista membrane (CM) (Figure 10.1d). In yeast, two specific subunits, termed Su e and Su g, maintain the dimer structure. The angle between the dimers is variable from 40° to 70°, and under electron microscope long ribbons of paired F_1 heads can be seen extending for up to 80 dimers and located at the most highly curved regions of the membrane. As a result, the ribbon of dimers appears to enforce a positive curvature into the cristae membrane with a radius of approximately 17 nm. This curvature may also help in the formation of inverted submitochondrial particles by sonication (see Figure 1.4). A further protein, Fcj1, induces a negative curvature in the membrane and is concentrated at the infoldings of the inner boundary membrane (IBM) forming the narrow cristae junctions that effectively separate the CM and IBM.

Almost 95% of electron transport chain and ATP synthase complexes in beef heart mitochondria are located on the CM, and even allowing for the greater area of the CM relative to the IBM, there is still enrichment per unit area. A detailed immuno-electron microscopic analysis of the distribution of proteins between the IBM and the CM of yeast mitochondria showed that proteins involved in the insertion of nuclear and mitochondria-encoded proteins (see Figure 10.9) were concentrated in the IBM, whereas the adenine nucleotide translocator was enriched in the CM. Because adenine nucleotides continuously traffic between the mitochondria and the cytoplasm, this raises the question as to whether the crista junction creates a bottleneck for this traffic. The possible role of the creatine phosphate/creatine system to overcome this block has been discussed (Section 9.5.1.2). The distinction between the IBM and the CM is not fixed. For example, during osmotic swelling in hypotonic media, the area of the IBM increases, presumably due to recruitment from the CM, leading to rupture of the outer mitochondrial membrane (OMM).

10.2.2 The outer membrane and intermembrane space

Further reading: Shoshan-Barmatz *et al.* (2010), Herrmann and Riemer (2010), Harner *et al.* (2011)

Outer and inner membranes have distinct lipid compositions consistent with the endosymbiotic origin of mitochondria. The OMM is more closely related to ER, whereas the inner membrane contains cardiolipin but no cholesterol. The OMM must allow the passage of adenine nucleotides and metabolites from the cytoplasm without allowing leakage of intermembrane space (IMS) proteins such as cyt c. The key role is played by a 30- to 35-kDa porin, usually termed the voltage-dependent anion channel (VDAC) (Figure 10.2). The term *VDAC* originated from electrophysiological studies in which the protein was incorporated into lipid bilayers and found to change to a partially closed form on application of a field, but the terminology is confusing because there is no known mechanism to generate a physiologically relevant potential across the outer membrane. Three mammalian isoforms of VDAC have been found, and their roles have been investigated by selective knockout or RNAi approaches. A complex range of cell-specific phenotypes is found, generally associated with a reduction in the capacity of

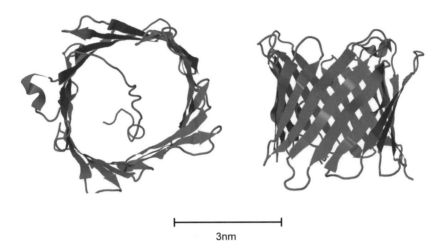

3nm

Figure 10.2   Human VDAC1.
Top and side views of human VDAC1 showing β-barrel structure. Structure:
(PDB 2K4T, Hiller *et al.*, 2008).

oxidative phosphorylation or protection against apoptosis. The best characterised iso-form is VDAC1, which forms a β-barrel structure with a central pore (diameter ~3.5 nm) and some preferential permeability to anions (Figure 10.2). There is controversy regard-ing its role in the Ca^{2+}-mediated inner membrane permeability transition pore (Section 9.4) and its ability to close under physiologically relevant conditions.

Electron microscopy reveals the presence of close contact sites, without fusion, between the IBM and the OMM (Figure 10.1). These are tight attachments that survive mitochondrial fractionation. VDAC may be concentrated at these contact sites, and evi-dence suggests that the protein assemblies that comprise the 'mitochondrial contact site complexes' are preferentially located at cristae junctions, raising the possibility that they may be capable of partially isolating the cristal space from the IMS.

10.3 MITOCHONDRIAL DYNAMICS

Further reading: Popov *et al.* (2005), Twig *et al.* (2008a)

Individual mitochondria labelled with a fluorescent cation such as TMRM (Section 9.7) or expressing a targeted fluorescent protein can easily be resolved in the light micro-scope. Mitochondria are highly dynamic structures, continuously undergoing fission (splitting into two daughter mitochondria) and fusion, and hence changing shape. The term *mitochondrion* is a combination of the Greek words for 'thread' and 'gran-ule' and was coined more than 100 years ago when the organelles were first seen under the microscope after staining tissue with Janus Green (which turns out to be a membrane-permeant cation; with hindsight, $\Delta\psi_m$ was first detected in the 19th cen-tury). Mitochondria are transported within the cell on microtubules driven by ATP-consuming motors (Section 10.4). Current research shows links between fission and

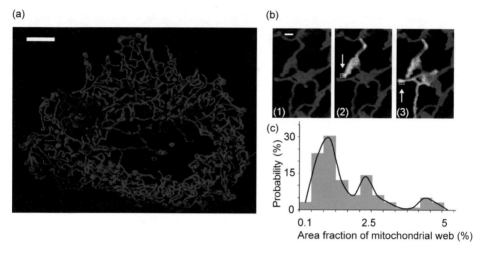

Figure 10.3 **A mitochondrial web consists of discrete abutting mitochondria.**
(a) A COS-7 cell loaded with TMRM (red) and photoactivatable mitochondria-targeted GFP (PA-GFP$_{mit}$) before photoactivation of the latter. Scale bar = 20 μm. (b) Enlarged view before and after two sequential focused photoactivations (areas indicated by arrows). Yellow, activated PA-GFP$_{mit}$. Scale bar = 2 μm. (c) Average area of a (interconnected) network as a fraction of the total cross-sectional area of the web.
Adapted from Twig et al. (2006).

fusion, mitochondrial quality control by autophagy ('mitophagy') (Section 10.6), and mitochondrial-dependent apoptosis (Section 10.7.2).

10.3.1 Discrete mitochondria versus integrated reticulum

Further reading: Twig *et al.* (2006, 2008b)

Serial confocal microscopic sections of cell bodies sometimes give the impression of a mitochondrial reticulum throughout the entire cell body (Figure 10.3). Light microscopy does not possess sufficient resolution to distinguish between an attached end-to-end 'sausage-string' network of discrete mitochondria with their individual matrices or the opposite extreme where the mitochondria have fused to produce a single giant tangled mitochondrion. Studies exploiting a photoactivatable green fluorescent protein, PA-GFP, targeted to the mitochondrial matrix have helped to resolve these alternatives with a variety of cell lines. PA-GFP is weakly fluorescent until photoisomerised by high-intensity 750-nm two-photon irradiation. A highly focused beam can excite an area of less than 0.5 μm² and can therefore photoactivate the PA-GFP within a region of single mitochondrion. The newly fluorescent GFP from this region would rapidly diffuse throughout a continuous reticulum. However, what is seen, for example, with COS-7 cells (Figure 10.3), is fluorescence that is initially limited to a small discrete volume corresponding to a single functional mitochondrial matrix. Thus, in these cells the apparent 'reticulum' is made up of a network of discrete mitochondria that at any one time each account for only approximately 1% of the complete tangle (Figure 10.3c). Similar results are obtained by irradiating a small area of a TMRE-loaded web with a sufficiently intense beam to induce phototoxicity and collapse $\Delta\psi_m$; only small regions are depolarised rather than the entire web. Indeed, a single continuum would be extremely vulnerable to global damage—a Titanic without watertight bulkheads.

In many cells, the mitochondrial morphology varies with the subcellular location. Neurons are an extreme example (Figure 10.4). Electron microscope tomography of *in situ* rat dentate gyrus neurons reveals a mitochondrial web within the cell body and extraordinarily long (up to 30 μm) filamentous branched profiles in dendrites, although it is not possible to determine whether these represent a single fused mitochondrion or a string of attached discrete mitochondria. In contrast, axonal and presynaptic mitochondria are discrete and can be as little as 200 nm in diameter (Figure 10.4). Axonal and presynaptic mitochondria have to be transported and distributed to sites where they are required along the axon. Note that the longest axons in the human body approach 1 m, or more than 20,000 times the diameter of the cell body. Axonal transport is discussed in Section 10.4.

One question that has not been fully addressed is the consequence of the existence of mitochondrial webs in many tissues for the classical isolation of mitochondria by differential centrifugation. Is there spontaneous fission to small individual mitochondria during the homogenisation and centrifugation steps? In the case of the brain, are isolated mitochondria preferentially obtained from axonal processes where they are already present as discrete entities?

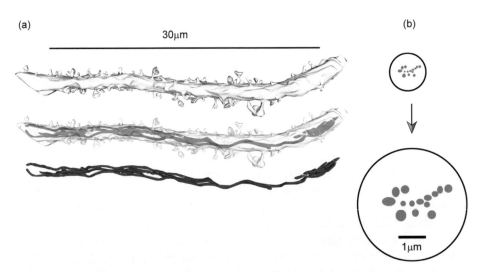

Figure 10.4   Distinct morphology of mitochondria in neuronal dendrites and presynaptic terminals.
(a) Dendritic segment from a rat dentate gyrus granule cell reconstructed by electron tomography showing branched filamentous mitochondria (blue) up to 30 μm long. Data from Popov *et al.* (2005). (b) Profiles of small discrete mitochondria in a giant presynaptic terminal (calyx of Held) shown to the same scale and enlarged 3×.
Adapted from Perkins et al. (2010).

10.3.2 Mitochondrial fission and fusion

Further reading: Twig *et al.* (2008b), Soubannier and McBride (2009)

Although the photoactivated-GFP study discussed previously shows that the mitochondrial web is composed of discrete mitochondria, the picture is highly dynamic. After photoactivation of PA-GFP in a single mitochondrion, there is a delay following which the GFP flows into an adjacent volume, consistent with a fusion event having occurred with the matrix of an adjoining mitochondrion. After a further interval, the fused mitochondrion tends to undergo fission, regenerating two discrete mitochondria (Figure 10.5). The balance between fission and fusion controls the size of individual mitochondria in a cell, allows fresh material to be incorporated into existing mitochondria, and provides a mechanism for the elimination of defective organelles.

Mitochondrial fusion involves the sequential joining of first the OMMs and then the inner mitochondrial membranes (IMMs), without permitting any leakage to the cytoplasm or between matrix and IMS (Figure 10.6). Two homologous dynamin-related GTPase proteins, mitofusins 1 and 2 (Mfn1 and Mfn2, respectively), facilitate mammalian OMM fusion. A related protein, Fzo1, acts in yeast. Mfn1 and Mfn2 are located on the OMM and have cytoplasmic N- and C-terminals that allow Mfn domains on two adjacent mitochondria to bind to each other, anchoring the two mitochondria together.

At the IBM, splice variants of the Drp (dynamin-related protein) OPA1 (optic atrophy 1) regulate inner membrane fusion in mammalian cells. Mutations in Mfn2 and OPA1 underlie the neurodegenerative disorders Charcot–Marie–Tooth neuropathy type 2a and autosomal dominant optic atrophy, respectively (Section 12.2). OPA1 exists in both long

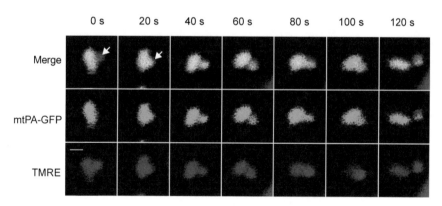

Figure 10.5   Fusion and fission between two mitochondria.
An INS1 cell expressing mitochondria-targeted PA-GFP was equilibrated with TMRE. An individual mitochondrion was photoactivated and the subsequent fusion/fission cycle with an unactivated mitochondrion (arrows) followed. Note that after fission, the two mitochondria show unequal ratios of red-to-green fluorescence, indicating that the right-hand mitochondrion is relatively hyperpolarised.
From Twig et al. (2008a).

and short isoforms, and a combination of both appears to be necessary for IMM fusion. Under stress conditions, the short isoform is favoured; this is less efficient and fusion is decreased. The mechanism by which IMM and OMM fusion is synchronised is unclear. OPA1 is also implicated in the regulation of cristae junctions.

Mitochondrial fission is controlled by the cytoplasmic dynamin-related GTPase Drp1, but it appears to be initiated at sites where ER tubules (Section 10.3.3) contact the mitochondrial outer membrane (Friedman *et al.*, 2011). Initially, the tubules appear to squeeze the mitochondrion, following which Drp1 translocates to the same location on the outer membrane. It is unclear whether this mechanism is relevant in neuronal axons. Drp1 is regulated by phosphorylation, although depending on the cell type and phosphorylation site, this can lead to either increased or decreased fission. The integral outer membrane protein Fis1 may recruit Drp1, although this is controversial, and an alternative outer membrane protein, Mff, may be more important as a receptor for Drp1. Once recruited, Drp1 molecules oligomerise to form rings around the filamentous mother mitochondria and constrict, driven by GTP hydrolysis, to give rise to two daughter mitochondria (Figure 10.6). It is unclear whether mammals possess inner membrane fission proteins.

The fission/fusion cycle plays important roles in mitochondrial quality control (Section 10.3.2). Mitochondrial fission is also important during cell division to ensure equal distribution of mitochondria between the daughter cells. Fused mitochondria may be protected against mitophagy, the selective removal of damaged mitochondria (Section 10.6). The consequences of a dysfunctional cycle are currently being investigated in the context of Parkinson's disease (Section 12.6.3).

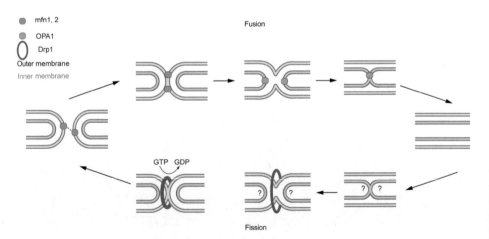

Figure 10.6   Mitochondrial fusion and fission.
Fusion and fission must occur without any leakage between compartments. Mitofusins (mfn1 and -2) on the OMM of two adjacent mitochondria bind to each other and induce OMM fusion. OPA1 on the IBM facilitates inner membrane fusion. The proteins responsible for inner membrane fission are currently unknown. OMM fission is induced by Drp1 forming a ring around the mitochondrion, which then constricts as GTP is hydrolysed.

10.3.3 Mitochondrial interactions with endoplasmic/sarcoplasmic reticulum

Further reading: Csordas *et al.* (2006), De Brito and Scorrano (2008), Giorgi *et al.* (2009)

Electron micrographs show frequent close associations between the outer membrane and the ER or SR (Figure 10.7), with a separation as low as 10–25 nm. These 'mitochondria-associated membranes' support transfer of lipids between the ER and mitochondria, and also imply that some of the mitochondrial Ca^{2+} uniporters (Section 9.4.1.1) are directly exposed to Ca^{2+} released from the through ER IP_3 or SR ryanodine receptors. As a consequence, a proportion of the released Ca^{2+} is taken up by the adjacent mitochondrion. The separation between the ER and OMM appears to be important to control this proportion and is controlled by protein tethers, disruption of which decreases the coupling between the ER and the mitochondrion (Figure 10.7). Apoptotic stimuli may narrow the gap, enhancing mitochondrial Ca^{2+} uptake and facilitating activation of the permeability transition pore. There is evidence that Mfn2, which is present on the ER membrane as well as the OMM, plays a major role in tethering the ER to the OMM. In muscle, mitochondrial uptake of a fraction of the Ca^{2+} released from the sarcoplasmic reticulum to initiate contraction has the effect of activating matrix dehydrogenases, facilitating ATP generation to power the contraction (Section 12.3.1).

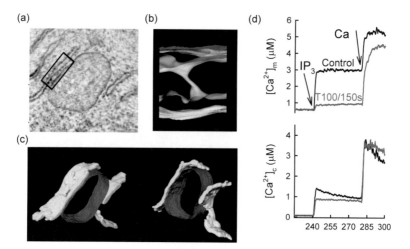

Figure 10.7 Mitochondrial ER interactions.
(a) Thin-section electron micrograph of a DT40 TKO cell showing ER adjacent to a mitochondrion. (b) Proteinaceous tether (gray) between the ER (yellow) and OMM (red). (c) Two orientations of a surface model of the same field. (d) Effect of trypsin pretreatment (100 μg/ml of trypsin for 150 s; red traces) on the IP_3-induced increase in $[Ca^{2+}]_c$ and $[Ca^{2+}]_m$ in permeabilised RBL-2H3 cells. Note the large decrease in the mitochondrial Ca^{2+} rise, ascribed to the proteolysis of the tether and consequent loss of direct Ca^{2+} transfer from ER to mitochondrion.
From Csordas et al. (2006).

10.4 TRAFFICKING OF MITOCHONDRIA

Further reading: Cai and Sheng (2009), Cai *et al.* (2011)

While the assumption that import of nuclear encoded mitochondrial proteins must occur close to the nucleus is under revision (Giuditta *et al.*, 2008), it is still essential that mitochondria are correctly distributed throughout the cell. Neurons in particular require mechanisms to transport mitochondria to synaptic sites along their axons and dendrites, where the mitochondria are required not only for ATP generation but also for cytoplasmic Ca^{2+} regulation. In the extreme case of motor neurons, these distances may be enormous: the axon of the human sciatic nerve is 1 m long—more than 20,000 times the diameter of the cell body.

Microtubule tracks provide the road along which mitochondria and other organelles are transported (Figure 10.8). Kinesin motors drive anterograde transport away from the cell body and towards the plus (+) end of the track. Mammals have variants of kinesin-1 (KIF5), with two coiled-coil heavy chains each having motor and cargo binding domains and two light chains that may contribute to cargo binding (Figure 10.8). Additional

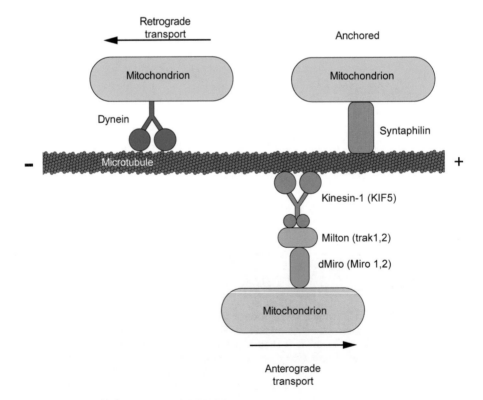

Figure 10.8 Mitochondrial transport.
Kinesin and dynein motors, respectively, drive antegrade and retrograde transport of mitochondria along microtubule tracks in neuronal axons. Syntaphilin may be involved in mitochondrial anchoring at sites of high ATP demand.

proteins required for mitochondrial recognition were first described in *Drosophila*. dMiro, an OMM protein, is a GTPase with Ca^{2+} binding domains and two mammalian orthologues, Miro1 and Miro2. Miro in turn recognises the adapter protein Milton (TRAK1 or TRAK2 in mammals), which may provide the bridge to kinesin, although this is debated.

Dynein motors drive retrograde transport (Figure 10.8). It is predicted that there is a tendency for 'new' mitochondria to be preferentially transported in an anterograde direction and 'old' or dysfunctional mitochondria to be returned to the cell body for mitophagic recycling. However, this has not been clearly demonstrated, and the factors controlling selective association of mitochondria with kinesin or dynein are still unclear.

Axonal mitochondria are small and discrete, in contrast to those in dendrites (Figure 10.4). Their transport along an axon, followed in a fluorescence microscope, is uneven, with periods of rapid motion ($1-5\,\mu m\,s^{-1}$) followed by quiescent periods or even reversal of direction (Gerencser and Nicholls, 2008), suggesting that both kinesin and dynein motors may be simultaneously attached to the mitochondria. Mitochondrial motility is inhibited in the presence of elevated Ca^{2+}, and this may play an important role in targeting mitochondria to regions of high synaptic activity and hence energy demand. Elevated Ca^{2+} mediates a change in the conformation of Miro, leading to disruption of transport. The ability of elevated Ca^{2+} to inhibit both anterograde and retrograde transport suggests that Miro also decouples dynein, although the mechanism is currently unclear. A further protein, Syntaphilin, may act as an anchor for static mitochondria (Figure 10.8).

An important interaction between mitochondrial transport and fission/fusion was revealed during investigations of the nature of the rare type 2 A Charcot–Marie–Tooth disease (CMT2A). CMT2A is caused by mutations in the mitochondrial fusion protein Mfn2. The reason why this leads to a selective degeneration of long peripheral axons was clarified when it was found that Mfn2 interacts directly with mammalian Miro and Milton proteins and hence affects mitochondrial transport.

10.5 MITOCHONDRIAL BIOGENESIS

Further reading: Diaz and Moraes (2008), Scarpulla (2011)

The quantity of mitochondria within a given cell is closely regulated. Generally, it appears that the maximal bioenergetic capacity, considering glycolysis and oxidative phosphorylation together, is sufficient to service the maximal short-term energy demand of the cell, with some safety margin. Although it might be thought that there would be benefits in having a substantial excess of mitochondria, the drawback is that the potential sources of reactive oxygen species (ROS) would increase, particularly because the mitochondria would spend much of their time close to state 4, where the high $\Delta\psi_m$ could facilitate ROS production (Section 9.10). Nevertheless, it is clear that mechanisms exist in tissues such as muscle to increase mitochondrial content in response to physical exercise (Chapter 11).

Signalling pathways exist to ensure that the cell has the appropriate number of mitochondria during cell division in response to exercise, caloric restriction, etc. The coordinate expression of mitochondrial and nuclear coded peptides requires bidirectional signalling between the mitochondrion and the nucleus. Transcription factors regulate the expression of mitochondrial genes, whereas the retrograde signalling is mediated by the cell sensing changes in bioenergetic parameters. These are reviewed in Chapter 11.

10.5.1 Protein import

Further reading: Neupert and Herrmann (2007), Endo *et al.* (2011), Dukanovic and Rapaport (2011)

Mitochondria are not created *de novo* but are generated by incorporation and assembly of proteins followed by mitochondrial fission. A total of 99% of the polypeptides required for mitochondrial function are synthesised in the cytoplasm and imported into the mitochondrion. In the case of complexes I, III and IV and the ATP synthase, nuclear and mitochondrial gene products must additionally be coordinated and assembled to produce the functional complexes. The machinery for import and sorting involves integral membrane complexes ('translocators') in the OMM and IMM, together with soluble factors in the cytoplasm, IMS and matrix (Figure 10.9). For a comprehensive description of the protein chemistry, the reader is referred to the above reviews; here, we focus on the basic principles. Much of our understanding of protein import comes from studies on *Neurospora crassa* or, in the cases of the OMM, *Saccharomyces cerevisiae*.

The TOM40 complex spanning the OMM is the major pathway for protein entry (Figure 10.9). The imported proteins are then targeted to specific complexes for delivery to one of four locations: OMM, IMS, IMM, or matrix. The mechanism by which integral proteins diffuse from the IBM to the CM is unclear, and the generic term IMM is used here. TOM40 contains at least eight polypeptides (referred to using lowercase letters as Tom70, etc.), including three receptors—Tom20, Tom22 and Tom70. The first two recognise N-terminal and internal mitochondrial targeting sequences, whereas Tom70 recognises polytopic (multiple transmembrane domain) proteins such as IMM metabolite carriers. Purified Tom40 possesses a β-barrel pore with a diameter (~20Å) sufficient to accommodate a looped polypeptide. Interestingly, the pore contains hydrophobic regions that may facilitate translocation of the unfolded proteins.

The TIM23 complex spanning the IMM transports those precursor proteins that possess an N-terminal cleavable presequence and are destined to the matrix or IMM. There may be a direct physical association between the TOM40 and TIM23 complexes to facilitate import of the hydrophobic, unfolded peptides. Unfolded polytopic proteins that will be imported via the TIM22 complex and peptides destined for the OMM (discussed later) may be protected from aggregation by a spectrum of small Tim proteins in the IMS. The TIM23 complex contains three core components and several peripheral subunits. The Tim23 peptide contains a narrow protein-conducting channel that is gated by $\Delta\psi_m$ and the presence of a transportable precursor peptide, ensuring that there is never an empty channel that could act as a proton short-circuit. Matrix proteins are threaded through Tim23 as unfolded polypeptides.

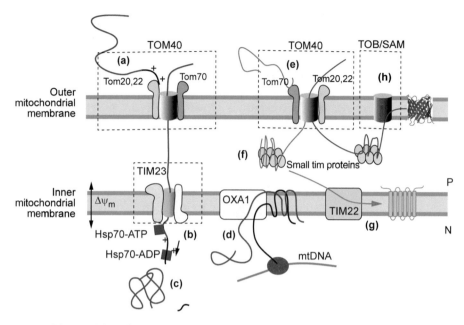

Figure 10.9   Mitochondrial protein import.
Protein precursors destined for the matrix possess positively charged leader sequences that are recognised by the Tom20 receptor of the TOM40 complex (a). (b) The peptide threads through the inner membrane TIM23 complex drawn by Δψ and pulled by a ratchet mechanism involving reversible attachment of Hsp70. (c) The imported protein folds into its functional conformation in the matrix. (d) Mitochondrial-synthesised peptides and nuclear peptides that co-assemble with them are inserted by the OXA1 complex. (e) Mitochondrial carrier proteins destined for the inner membrane lack a leader sequence and are recognised by signature sequences located at the future matrix loops. (f) Their structure is stabilised by interaction with small Tim proteins in the intermembrane space, while (g) the TIM22 complex catalyses insertion of the carrier. (h) Outer membrane β-barrel proteins such as VDAC enter via TOM40, are stabilised by small Tim proteins, and are then inserted into the OMM with the help of the TOB/SAM complex. Not shown are numerous accessory proteins and those involved in the targeting of IMS proteins.

The driving force for the translocation of the positively charged N-terminal pre-sequence across the inner membrane is $\Delta\psi_m$. However, once the presequence emerges into the matrix, an additional driving force is required. The presequence is recognised and bound by a mitochondrial heat shock protein mtHsp70, associated with the TIM23 complex. mtHsp70 is an ATPase possessing a peptide binding pocket that is open in the ATP conformation and tightly binding after the ATP is hydrolysed to ADP. Brownian motion drives the peptide into the matrix, together with a ratchet mechanism involving two molecules of mtHsp70 alternately binding to the emerging peptide and preventing the peptide from reversing. In the matrix, the N-terminal presequence is cleaved off by a matrix processing peptidase.

Integral IMM proteins, other than the metabolite carriers, which use the TIM22 complex (discussed later), are assembled into the IMM via TIM23 either by a 'stop transfer' mechanism, typically for single transmembrane domain proteins, in which direct lateral insertion occurs from the complex, or by translocation into the matrix followed by re-insertion into the IMM via an 'oxidase assembly' (OXA1) in the IMM.

The inner membrane TIM22 complex is required by many integral inner membrane proteins, including the large class of six-transmembrane metabolite carriers such as the adenine nucleotide translocator. These proteins are not synthesised with cleavable N-terminal presequences but instead have internal sequences that are recognised as targeting motifs. After recognition by the Tom70 receptor and uptake through TOM40, they are directed, with the help of small Tim proteins Tim9 and Tim10 (zinc finger proteins that recognise and bind the 'carrier signature' sequences and assist in the folding of the carrier into its three-module tertiary structure), to the TIM22 complex spanning the inner membrane. From there, they are directly inserted into the IMM by a $\Delta\psi_m$-dependent process without emerging into the matrix.

At least 40 proteins reside in the OMM, all of which are nuclear encoded. In addition to the proteins of the TOM40 complex, many of these proteins are involved in fission and fusion, apoptosis, and pore formation (e.g., VDAC, Section 10.2.2); several key metabolic pathways, such as fatty acid activation, are also located on the outer membrane. All OMM proteins have internal targeting and sorting signals. β-barrel proteins such as VDAC, Tom40 and Mdm10 (which may provide a link between the OMM and adjacent ER) are imported into the IMS by the TOM40 complex and delivered with the help of the small Tim chaperones to the OMM-located TOB/SAM complex, which is specialised for the insertion of β-barrel precursors. OMM proteins anchored to the cytoplasmic face of the IMM by short N- or C-terminal segments, including members of the Bcl-2 family (Section 10.7), are inserted into the OMM without involving import via the TOM40 complex.

A number of soluble IMS proteins contain internal disulfide bonds but are imported through TOM40 in the reduced unfolded form. The IMS has a relatively low potential redox environment, and a TIM40 'disulfide relay' system facing the IMS is required to mediate oxidative disulfide bond formation for these proteins. Key components are a disulfide carrier (Tim40/Mia40) and a sulfydryl oxidase Erv1 that is reduced by cyt c. Cytochrome c is imported into the mammalian IMS via TOM40 as the unfolded apoprotein and is then bound by HCCS (holocytochrome c synthase, formerly known as cyt c haem lyase), which recognises the N-terminus of the polypeptide, inserts the haem and releases the folded holoprotein. Unlike the unfolded apoprotein, the folded holoprotein cannot access the cytoplasm via TOM40.

10.5.2 Assembly of mitochondrial complexes

Further reading: Mimaki *et al.* (2011)

Biosynthesis of complexes I, III and IV and the ATP synthase requires the integration of the products of the nuclear and mitochondrial genomes. In mammalian mitochondria, the nuclear-encoded subunits are imported initially into the matrix via TOM23 (Figure 10.9), leader sequences are removed, and then the nuclear-encoded subunits are inserted from the matrix into the inner membrane together with the mitochondrially encoded subunits by OXA1.

Multiple assembly factors are required for each complex. Consistent with its formidable complexity, complex I requires at least 10 assembly factors, and it is expected that another 10–20 remain to be described. Some of the known assembly factors are included among the 31 'supernumerary subunits' in the mammalian complex (Section 5.6), and others are not part of the final assembled complex I but are required to stabilise assembly intermediates. Considerable progress has been made towards understanding the assembly of complex I (Figure 10.10). The basic L-shaped structure (see Figure 5.10) with 14 core subunits is conserved from bacteria to mammals. Using the human nomenclature, 7 hydrophobic peptides from mtDNA (ND1–ND6 and ND4L) form the membrane-located arm, and 7 more hydrophilic nuclear-encoded subunits (V1, V2, S1–S3, S7 and S8, each with the prefix NDUF) make up the peripheral matrix arm. The peripheral arm can be further subdivided into the N-module, responsible for binding and oxidising NADH and containing NDUFs V1, V2 and S1, and the Q-module (NDUFs S2, S3, S7 and S8) responsible for electron transfer to UQ.

Mammalian complex I assembly has been studied by a variety of approaches, including cells lacking mitochondrial DNA (ρ_0 cells; Section 12.2.2), cell lines generated with specific deletions and mutations in subunits or assembly factors, and cells derived from patients with mitochondrial diseases originating from mutations in complex I. Stalled

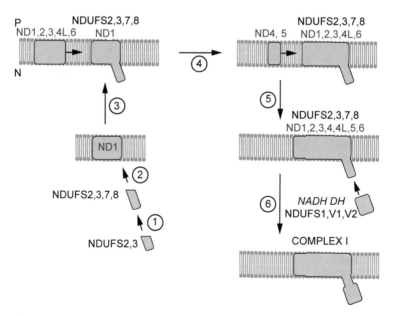

Figure 10.10 Probable assembly sequence of the core subunits of mammalian complex I.
Mitochondrially encoded subunits are shown in red; nuclear-encoded subunits are shown in black.
Adapted from Mimaki et al. (2011).

or unstable intermediate assembly states can be determined, and the incorporation of GFP tagged peptides can be monitored. Details are still being established, and there is no complete agreement, but a summary of a likely sequence of core subunit assembly is shown in Figure 10.10.

Two, 4 and 11 proteins have been identified as assembly factors for complexes II, III and IV, respectively, while the ATP synthase requires at least 3 assembly factor proteins. Mutations in assembly factors are associated with a spectrum of mitochondrial diseases (Section 12.2). Iron–sulfur clusters are present in complexes I–III and are synthesised in the matrix of mammalian mitochondria. More than 20 proteins are required for their biogenesis. Mutation in frataxin, the Fe chaperone required for Fe–S and haem synthesis, causes Friedreich ataxia (Section 12.6.5).

10.6 MITOPHAGY

Further reading: Deas *et al.* (2011), Narendra and Youle (2011), Lee *et al.* (2012)

Damaged proteins in the cell are removed either by the ubiquitin–proteasome system or by the autophagy–lysosome pathway (Figure 10.11). Note that mitochondria possess AAA protease complexes ('ATPases Associated with diverse cellular Activities') capable of degrading unfolded membrane proteins. It is likely that these function to remove excess subunits during membrane complex assembly. In these cases, the mitochondria are preserved; however, if damage is too severe, then the entire organelle can be degraded. Macroautophagy is the process in cells whereby large cytoplasmic organelles are engulfed by lysosomes and degraded. Although much research on autophagy has been performed in yeast, we restrict ourselves to mammalian systems. The pathways of general autophagy have been extensively reviewed, and we further limit discussion to mitophagy, the selective targeting of mitochondria for lysosomal degradation.

To maintain a stable and healthy mitochondrial population, the biosynthesis of fresh components and their incorporation into existing mitochondria must balance the detection, removal and destruction of damaged material. The process of mitophagy is therefore closely linked to the fission/fusion cycle because the mitochondrial network must first be disassembled, following which any defective discrete mitochondria must be identified and targeted for removal. At each cycle of fission and fusion, the mitochondrial components will be randomly sorted between the daughter organelles. If one receives a disproportionate amount of damaged components, it may display a lowered $\Delta\psi_m$ (see Figure 10.5), and observations suggest that mechanisms exist for the defective mitochondrion to exit the fission/fusion cycle and be removed.

The most convincing experiments in this context are those that exploit the natural stochastic variations in $\Delta\psi_m$ in the mitochondrial population (e.g., Twig *et al.*, 2008a). Less relevant are studies in which investigators have chronically exposed cells to high protonophore concentrations to induce global depolarisation, with the complication of cytoplasmic acidification, ATP depletion, disturbed Ca^{2+} homeostasis and the collapse of the critical pH gradient across the lysosomal membrane. Experimentally, mitophagy can also be initiated by inhibition of the TOR pathway (Section 11.6), by photo-damage,

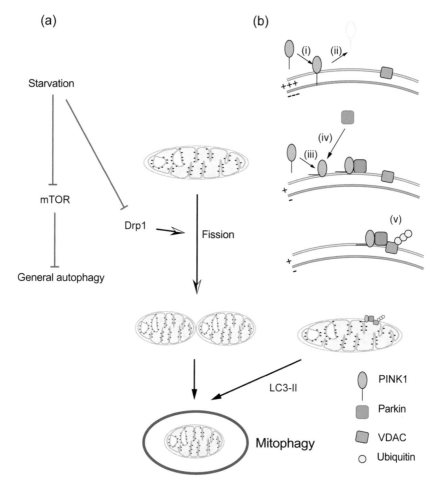

Figure 10.11   Hypotheses for mitophagy.
Prolonged mitochondrial fission can be a signal for mitophagy. During
starvation, mTOR (which normally inhibits general cellular autophagy) is
itself inhibited and thus autophagy is facilitated. However, mitochondria are
spared by the parallel inhibition of the fission protein Drp1. The kinase
PINK1 continuously translocates to the OMM (i). It appears to recognise
fully polarised mitochondria, perhaps by insertion of its N-terminal into the
TIM23 protein import pathway. Under these conditions, PINK1 is
continuously degraded by proteolysis (ii). When PINK1 associates with
partially depolarised mitochondria, it is stabilised (iii) and recruits parkin,
which proceeds to polyubiquitylate VDAC and other OMM proteins. This
provides a signal for LC3-II and related proteins to associate, targeting the
mitochondrion for mitophagy.

by excess ROS, and by enhanced mitochondrial fission. It is conceivable that small
'fissioned' mitochondria are selectively recognised by the mitophagic apparatus. Note
that mitochondria are spared from the cellular autophagy accompanying nutrient deprivation. As discussed later, under these conditions fission is inhibited, and the resulting
elongated mitochondria appear to be protected.

The phosphatidylethanolamine-conjugated protein LC3-II associates with the autophagosomal membrane and can be used in the form of a GFP construct to visualise mitophagy in the microscope by its co-localisation with a matrix-targeted fluorescent marker.

The controlled removal of defective mitochondria is of particular importance in post-mitotic cells such as neurons, and clues as to the mechanism of mitophagy have come from the study of rare forms of familial Parkinson's disease (PD) in which the process appears to be defective. PD is reviewed in Section 12.6.3; here, we discuss the normal function of the proteins revealed by this research.

PINK1 (PTEN-induced kinase 1) is a ubiquitously expressed serine/threonine kinase. Despite an N-terminal mitochondrial targeting sequence, there is a lack of consensus regarding its subcellular localisation, and various reports have localised the kinase to the cytoplasm, the IMS, the IMM and the OMM. Some of this ambiguity may be due to mislocation of an overexpressed protein. However, there is consensus that loss of PINK1 has deleterious effects on mitochondria, detected by the usual basket of parameters—decreased $\Delta\psi_m$, lowered ATP, increased ROS, decreased complex I and IV activity, increased Ca^{2+} accumulation and altered morphology.

Although there is some uncertainty regarding its physiological substrates, affinity purification shows that PINK1 can phosphorylate mitochondrial TRAP1 (heat shock protein 75) and, perhaps indirectly, the IMS protease HtrA2/Omi. The kinase interacts with Miro (Section 10.4) and may thus be implicated in mitochondrial transport. However, most research is focused on the interaction of PINK1 with a second PD-related protein, Parkin.

Parkin is a normally cytoplasmic protein with a latent E3 ubiquitin ligase activity capable of proteasome-independent ubiquitylation. Although parkin-deficient mice show a relatively mild phenotype, the protein is associated with a wide spectrum of protective pathways and effects focused on protection against mitochondrial dysfunction. There is a close interaction between PINK1 and parkin, with the latter appearing to act downstream (Figure 10.11).

We consider it regrettable that, to date, many of the high-profile studies investigating the roles of PINK1 and parkin in mitophagy have used the bioenergetically dubious approach of chronic exposure of cells to protonophore. Keeping in mind our reservations, treatment with the protonophore CCCP results in parkin recruitment to mitochondria, following which mitochondria are cleared from a subset of cells within 24 h. A more rigorous and specific means to depolarise mitochondria in cells—the combination of an electron transport inhibitor plus oligomycin (with oligomycin alone as control)—has been performed in one publication (Vives-Bauza et al., 2010). Interestingly, however, electron transport inhibition alone, which produces an approximately 20-mV mitochondrial depolarisation (the residual potential is supported by ATP synthase reversal utilising cytoplasmic ATP), did not result in significant parkin translocation, suggesting that a profound depolarisation, or one of the many side effects of the protonophore, is required.

Accepting these limitations, the present consensus is that PINK1 continually translocates to 'healthy' mitochondria but that under normal conditions is immediately degraded by the protease PARL (presenilin-associated rhomboid-like protease) or a protease sensitive to the inhibitor MG132. Protease activity is by some means dependent on a maintained $\Delta\psi_m$ by the target mitochondrion. A depolarised mitochondrion would thus accumulate PINK1 on its outer membrane. However, it is not clear how a 'dysfunctional' mitochondrion with a lowered Δp is recognised. Any mechanism has to be specific to the damaged mitochondrion, while sparing adjacent healthy ones, and so cannot respond to a released cytoplasmic signal. In addition, it must in some way be able to detect a decreased transmembrane potential across the inner membrane (or its consequences) while only being able to access the outer face of the OMM. One speculation is that in normally polarised mitochondria, the IMS N-terminal of the translocated PINK1 might engage the $\Delta\psi_m$-dependent inner membrane TIM23 import complex, holding it in a conformation accessible to the protease.

Although there is a consensus that accumulated PINK1 leads to translocation of parkin to the OMM, it is debated whether this is due to a simple binding, whether PINK1 (remember that it is a kinase) phosphorylates parkin, or whether some other proteins are involved. Once localised to the OMM, the ubiquitin–ligase activity of parkin causes the ubiquitylation of multiple outer membrane proteins, leading to their proteasomal degradation. These events precede mitophagy, and the precise trigger for the latter is currently being investigated. Because mfn1 and mfn2 are among the parkin targets, this raises the possibility that stalled, 'fissioned,' mitochondria are selective subjects for mitophagy.

In mitotic cells, starvation induces the production of autophagosomes that break down and recycle proteins and organelles to meet the cell's energy requirements. It is logical that mitochondria are protected from this process in order to retain efficient ATP generation. Starvation activates a protein kinase A-mediated phosphorylation of Drp1 (Section 10.3.2), inhibiting its translocation to the mitochondria and consequently inhibiting fission. The resulting elongated mitochondria have more cristae, show increased dimerisation of the ATP synthase (Section 10.2.1), and appear to be protected from mitophagy.

In non-mitotic neurons, dysfunction in normal mitophagy may be detrimental in two ways—either by excessive degradation of healthy mitochondria, reducing energetic capacity, or by failed mitophagy of damaged organelles, leading to a deteriorating mitochondrial quality. The evidence that this may contribute to mitochondrial deterioration in aging and neurodegenerative diseases will be developed in Chapter 12.

10.7 APOPTOSIS

Further reading: Wang and Youle (2009), Parsons and Green (2010)

Programmed cell death, or apoptosis, is the orderly shutdown of metabolism and dismantling of the entire cell with minimal leakage of cell constituents. Apoptosis occurs naturally during embryonic or immune system development. It can be induced by ligand binding to cell surface receptors (the *extrinsic pathway*), whereas a range of internal cell stresses, such as growth factor withdrawal or DNA damage, initiate the *intrinsic pathway* (Figure 10.12). Although the following section describes the main features of these apoptotic pathways, the reader is recommended to consult more specialised texts for a comprehensive description of the plethora of proteins implicated. A family of

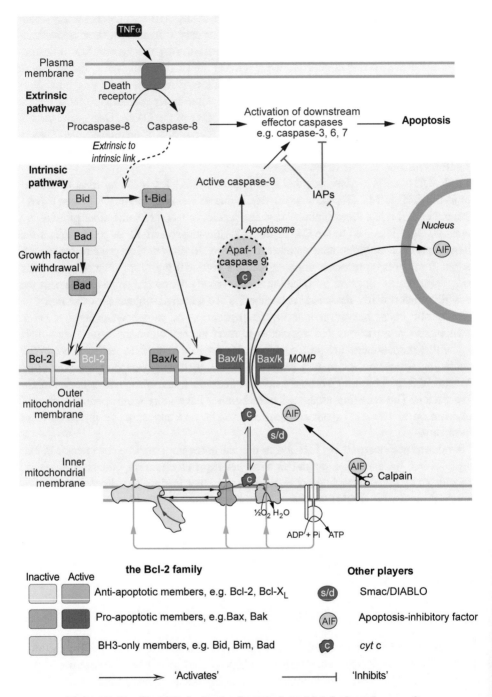

Figure 10.12   Simplified scheme of extrinsic and intrinsic pathways of apoptosis.
For details, see text. MOMP, mitochondrial outer membrane permeabilisation.

proteases, capable of selectively cleaving proteins on the C-side of Asp residues (hence 'caspases'), play a key role in the apoptotic cascade and the proteolytic digestion of the cell. Caspases exist as inactive precursor procaspases in healthy cells and are activated by carefully controlled proteolytic cleavage. Apoptosis must be contrasted to the chaotic cell death and leakage of contents that occurs in necrosis (Section 12.6.1).

10.7.1 The extrinsic pathway

The extrinsic (type 1) pathway is activated by ligand binding to 'death receptors,' including a subset of tumour necrosis factor receptors (TNFRs). The extrinsic pathway is further subdivided according to cell type: in type I cells, activation of a caspase cascade is sufficient for apoptosis, whereas in type II cells a mitochondrial amplification pathway is involved.

Extrinsic apoptosis in type I cells: TNFRs such as FAS/CD95 aggregate on the plasma membrane after ligand binding, recruiting cytoplasmic proteins and then procaspase-8. The mechanism by which the death receptors activate this 'initiator' caspase is still debated, but once activated, caspase 8 in type I cells proceeds to activate downstream 'executioner' or 'effector' caspases that are responsible for the specific cleavages characteristic of apoptosis. Caspase 3 is the dominant executioner caspase in most cells.

Extrinsic apoptosis in type II cells: Death receptor-activated caspase 8 targets and cleaves effector caspases as in type 1 cells. However, in addition, the intrinsic pathway is recruited as an amplification step.

10.7.2 The intrinsic pathway

The predominant mechanism of apoptotic cell death is via the intrinsic (type 2) pathway, which is triggered by cellular damage (e.g., growth factor withdrawal, DNA damage, heat shock, radiation, or chemotherapeutic drugs). A critical stage is the formation in the OMM of a mitochondrial outer membrane permeabilisation (MOMP) channel allowing the release from the mitochondrial IMS of pro-apoptotic proteins including cyt c, apoptosis-inducing factor (AIF), and Smac/DIABLO (Figure 10.12). Cyt c is of course the component in the electron transport chain that shuttles electrons from complex III to complex IV, so the question arises as to how the mitochondrion can continue to function after its loss from the IMS. Partial release could still retain sufficient cyt c for electron transport, and after extensive loss and inhibited electron transport, mitochondria can maintain a suboptimal $\Delta\psi_m$ by ATP synthase reversal utilising glycolytic ATP. The released cyt c forms an ATP-(or dATP-) driven 'apoptosome' together with a cytoplasmic factor 'Apaf-1' and procaspase 9. The aggregated procaspase-9 monomers proteolytically cross-activate each other, and the active caspase 9 triggers the effector caspases as in the extrinsic pathway.

Co-released Smac/DIABLO facilitates apoptosis by neutralising inhibitors of apoptosis (IAP) proteins, including the cytoplasmic X-linked inhibitor of apoptosis protein (XIAP), which would otherwise inhibit the caspase cascade. Cells with a high level of

XIAP seem to require the mitochondrial amplification pathway reviewed previously. Finally, calpains or cathepsins have access through the MOMP to cleave the membrane anchor of AIF. AIF is a flavoprotein that mediates caspase-independent apoptotic nuclear changes (chromatin condensation and large-scale DNA fragmentation) after its release from mitochondria. The normal physiological role of AIF is being studied by investigating Harlequin mutant mice, which display an 80% reduction in AIF expression and show selective neuronal degeneration and a 50% reduction in amount of complex I.

The MOMP must be carefully distinguished from the Ca^{2+} and oxidative stress-induced IMM permeability transition (mPTP) (Section 9.4), although one consequence of the latter, at least *in vitro*, is an osmotic swelling of the matrix that can lead to mechanical rupture of the OMM. The mPTP causes a catastrophic bioenergetic collapse and is primarily implicated in necrosis, although it may also occur in the final stages of apoptosis.

10.7.3 Mitochondrial outer membrane permeabilisation

Further reading: Chipuk and Green (2008), Edlich *et al.* (2011)

The composition and mechanism of formation of the MOMP channel is being intensively investigated. A simplified scheme is shown in Figure 10.12; for more details, the reader is referred to the cited reviews. The Bcl-2 family of proteins comprises more than 15 members that may be divided into three subgroups on the basis of their structure and function: anti-apoptotic, pro-apoptotic and BH_3-only members. The last subgroup contains the primary regulators of MOMP formation. In normal cells, the BH_3 proteins are inactive, either by being compartmentalised or by downregulation. Specific stresses upregulate or release distinct BH_3 proteins; for example, growth factor withdrawal allows Bad to be dephosphorylated and activated. The BH_3 protein Bid is relatively inactive until its N-terminal is cleaved, for example, by the Ca^{2+}-activated protease calpain in response to cell damage, producing a 15-kDa C-terminal fragment, truncated or t-Bid. Alternatively, t-Bid can be formed by caspase 8 cleavage of Bid, serving as the link between extrinsic and intrinsic pathways in type II cells.

At the outer membrane, there is a conflict between pro- (Bax, Bak, etc.) and anti-apoptotic (Bcl-2, Bcl-X_L, etc.) proteins. Bak and Bcl-2 are integral OMM proteins, whereas Bax is normally cytoplasmic or loosely attached to the outer face of the OMM. The pro-apoptotic proteins attempt to undergo conformational changes, promoted by t-Bid and other BH_3 proteins, to form homo-oligomers that constitute (or at least contribute to) the MOMP channel. The anti-apoptotic members inhibit this assembly but are in turn inactivated by BH_3 proteins (Bad and t-Bid in Figure 10.12). Anti-apoptotic Bcl-X_L may continuously remove Bax from the outer membrane, whereas inhibition of the mitochondrial fission protein Drp1 or overexpression of mitofusins blocks Bax insertion and oligomerisation, decreasing apoptosis.

10.7.4 Cristae remodelling and apoptosis

Further reading: Pellegrini and Scorrano (2007), Wasilewski and Scorrano (2009)

Electron tomography of mitochondria during apoptosis shows that the cristae junctions (Section 10.2) widen and individual cristae fuse, facilitating cyt c release through the MOMP. The inner membrane fusion factor OPA1 (Section 10.3.2) has a second action as an anti-apoptotic protein, once it has been released from the inner membrane following N-terminal cleavage by the intramembrane protease PARL, which will be discussed again in the context of Parkinson's disease (Section 12.6.3).

11

SIGNALLING BETWEEN THE MITOCHONDRION AND THE CELL

11.1 INTRODUCTION

Further reading: Finley and Haigis (2009)

The abundance of mitochondria varies greatly between different cell types, paralleling the requirement of the cell for oxidative phosphorylation. Many cancer-derived cell lines are sparsely populated with mitochondria and obtain much of their ATP from glycolysis (Section 12.7.1), whereas mitochondria are abundant in slow-twitch muscle fibres. Generally, cells seem to possess sufficient mitochondria to satisfy the maximum sustained ATP demand of the cell, with an additional safety margin. However, mitochondrial content can vary with exercise, diet, etc., and changes can occur with aging and aging-related diseases that adversely affect the tissue's bioenergetic capacity. Many of the signalling pathways that regulate and coordinate the expression of nuclear and mitochondrial genes have recently been elucidated and form the focus of this chapter. Figure 11.1 shows a simplified scheme of some of the major interactions between key signalling pathways. A range of retrograde signals from the mitochondria, including cytoplasmic ΔG_p, $[Ca^{2+}]_c$, hypoxia, reactive oxygen species (ROS), and $NAD^+/NADH$ redox poise, appear able to tune the expression of nuclear- and mitochondrial-encoded genes to optimise mitochondrial bioenergetic function in a changed environment. First, however, we need to review some key features of the mitochondrial genome.

11.2 THE MITOCHONDRIAL GENOME

Further reading: Falkenberg *et al.* (2007), Krishnan and Turnbull (2010); see also MITOMAP, a human mtDNA database (http://mitomap.org)

Bioenergetics. Doi: http://dx.doi.org/10.1016/B978-0-12-388425-1.00011-7

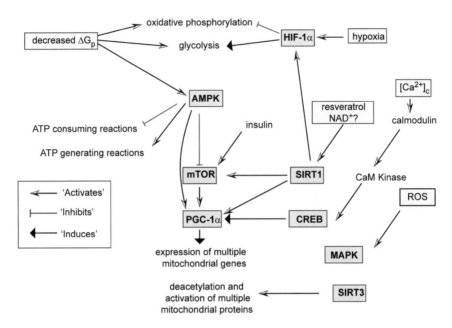

Figure 11.1   Overview of major signal transduction pathways controlling mitochondrial biogenesis and function.
AMPK, 5′-AMP-activated protein kinase; CREB, cAMP response element binding; HIF-1α, hypoxia-inducible factor 1α; MAPK, mitogen-activated protein kinase; mTOR, mammalian target of rapamycin; PGC-1α, peroxysome proliferator-activated receptor γ coactivator 1α; SIRT1 and -3, silent mating-type information regulation 2 homologues 1 and 3.

The endosymbiotic theory, whereby mitochondria are the descendants of ancient bacterial invaders into the eukaryotic cell, is generally accepted. Over the subsequent aeons, almost all of the original bacterial genes have been transferred to the nuclear genome. Mammalian mitochondria retain residual DNA encoding 37 genes, including those for 13 polypeptides—highly hydrophobic subunits of the ATP synthase and complexes I, III and IV. Although this accounts for only 1% of the total mitochondrial proteome, the coordination and integration of the protein products from the two genomes introduces major complexities.

Mammalian mitochondrial DNA (mtDNA) is a 16.568-kilobase (human) circular double-stranded structure (Figure 11.2) whose codon usage diverges slightly from that of the nuclear genome. Note that the base pair numbering for the human mtDNA runs (anticlockwise in Figure 11.2) from 001 close to the origin of the heavy chain to 16569; this is because the reference sequence (see below) was later revised when it was found that one base pair was erroneously included. To avoid renumbering all positions, a 'placeholder' was inserted at position 3107.

Twenty-two genes encode transfer RNAs (tRNAs), and another 2 encode ribosomal RNAs. These are required for translation of the mRNAs from the remaining 13 genes, which encode seven complex I polypeptides (ND1–ND6 and ND4L), cyt *b*, three

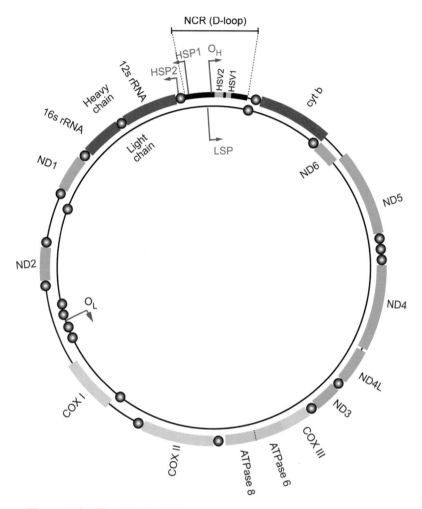

Figure 11.2   The mitochondrial genome.
The 16,568-base human mitochondrial genome encodes six complex I
peptides (ND1–ND6 and 4L), cyt *b*, three complex IV peptides (COXI–
COXIII) and two peptides of the ATP synthase F$_o$ (ATPases 6 and 8). The two
strands may be separated on a density gradient into heavy and light chains.
Both chains are coding: the light chain encodes for 8 tRNAs (balls) and the
ND6 subunit of complex I, and the heavy chain encodes the other 14 tRNAs
and remaining complex subunits. The genes for ND5 and ND6 and for
ATPase 6 and ATPase 8 overlap by a few bases.

complex IV polypeptides (COXI–III) and two polypeptides of the ATP synthase F$_o$
(ATPases 6 and 8). These proteins are very hydrophobic, and one explanation for the
retention of their genes on the mtDNA is that evolution has not chanced upon a way to
import, locate and fold them correctly. The mitochondrial genome is extraordinarily com-
pact with no introns; indeed, two pairs of genes overlap by a few bases. The complementary light chain is also transcribed, coding for one complex I subunit and several tRNAs.

mtDNA is associated with the matrix face of the inner mitochondrial membrane (IMM) and is packaged with proteins to form structures called nucleoids. The associated proteins are involved in replication and expression; they include the mitochondrial helicase (Twinkle), polymerase-γ (POLG), and transcription factor A (TFAM). The detailed mechanism of mtDNA replication is beyond the scope of this book, but we revisit the roles of POLG (Section 12.9) and TFAM (Sections 11.5 and 11.6.3).

A typical somatic cell can have 1000–10,000 copies of the mitochondrial genome. The consequence for mtDNA diseases (which are discussed in Section 12.2) is that the severity of the phenotype resulting from a deleterious mutation will depend on the extent to which the mutation is present in all mtDNA (homoplasmy) or a subpopulation (heteroplasmy). During cell division, there is no mechanism to ensure an equal partition of wild-type and mutant mtDNA between the daughter cells; as a result, the proportion of mutated mtDNA, and hence the phenotypic consequences, can differ ('segregate') widely between tissues or even individual cells. The mechanism of segregation is still not completely understood, and it differs from tissue to tissue. The way in which mutated mitochondria with diminished function evade the quality control discussed in Section 10.3.2 is clearly of central importance to the understanding of mitochondrial diseases.

11.2.1 Haplotypes

Further reading: Andrews *et al.* (1999), Soares *et al.* (2009), van Oven and Kayser (2009)

Because multiple single nucleotide polymorphisms (SNPs) are found in the human genome, it is helpful to define an arbitrary reference sequence and to express all mutations with reference to this sequence. This is known as the *Revised Cambridge Reference Sequence* (rCRS) and was obtained from an arbitrary European belonging to haplogroup H2A2A. A mitochondrial haplotype is a precise mtDNA sequence, whereas a haplogroup is a group of similar haplotypes that share a common ancestor. Because specific mutations define a haplogroup, analysis of the SNPs of an individual can assign him or her to a given haplogroup (Figure 11.3).

Single point mutations in the coding region of mtDNA that are phenotypically roughly neutral appear in the human genome at a rate of approximately 1 base per 3500 years. It is notable that this rate is at least 10 times higher than that of the nuclear genome, and this has been ascribed variously to the close proximity of the electron transport chain with its production of ROS, to the lack of protective histones, and to the more limited repair capacity. The non-coding region, sometimes called the control region (bases 16024–576), contains two sections that are highly polymorphic (i.e., associated with a high incidence of mutations). These are known as HVR1 (16024–16383) and HVR2 (57–372), and mutation analysis in these regions is commonly used for commercial genealogy investigation.

Although mutations leading to impaired function usually demonstrate heteroplasmy, haplotypes are usually assumed to be homoplasmic. There is considerable interest in investigating whether particular mitochondrial haplotypes confer any evolutionary advantage for particular populations. Generally, the phenotypic variation between haplotypes must be subtle: a spontaneous mutation leading to significant mitochondrial dysfunction would be classified as a mitochondrial disease and would, it is hoped, be

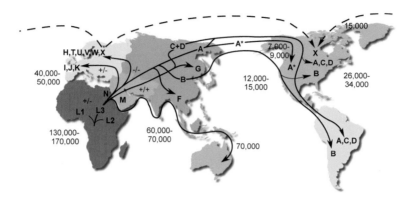

Figure 11.3   Human mtDNA migrations.
Reproduced from MITOMAP: A Human Mitochondrial Genome Database,
http://www.mitomap.org, 2013.

eliminated by quality control mechanisms. Conversely, although a haplotype may have little evolutionary advantage on its own, it may exacerbate or ameliorate simultaneous deleterious mutations (Gomez-Duran *et al.*, 2012).

11.2.2 'Mitochondrial Eve'

Further reading: Lewin (1987)

If the maternal lines of all people living today could be extrapolated backward indefinitely (mother, grandmother, great-grandmother, etc.), they would eventually converge on a single individual, the most recent (matrilineal) common ancestor (MRCA), popularly and misleadingly referred to as 'Mitochondrial Eve' and estimated to have lived in sub-Saharan Africa approximately 200,000 years ago (Figure 11.3) with a few thousand other women (whose purely matrilineal lines have subsequently died out due to their line of descent including at least one generation lacking daughters). All modern human mtDNA is descended from this individual. A clade is a group that contains an ancestor and all its descendants, so *Homo sapiens* is the mitochondrial clade defined by Mitochondrial Eve. The evolutionary tree of human mitochondrial haplogroups is exceedingly complex and can be accessed online (see http://www.phylotree.org or http://www.mitomap.org).

11.3 AMP KINASE

Further reading: Hardie (2011)

AMP-activated protein kinase (AMPK), not to be confused with 3',5'-cyclic AMP kinase, has been described as the cell's master metabolic regulator. It is important for

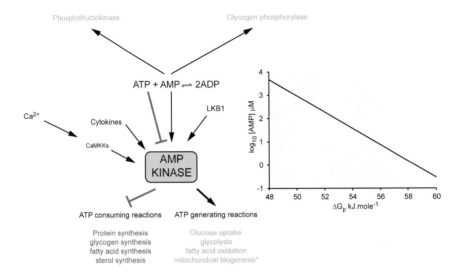

Figure 11.4   AMP kinase and AMP response to cytoplasmic phosphorylation potential (ΔGp).
(Left) Major pathways regulation by AMP and AMP kinase. Red, inhibition; green, activation; LKB 1, liver kinase B1. (Right) Calculated variation in cytoplasmic AMP concentration with ΔG_p. Note logarithmic scale.

the cell to maintain cytoplasmic ΔG_p (the free energy of the ATP = ADP + P_i reaction; Section 3.2.1) as constant as possible over a range of energy demands. Several factors contribute to this homeostasis. The basic mechanism of respiratory control (Section 4.3) means that oxidative phosphorylation increases to match demand. In addition, in many cells, Ca^{2+} activation of matrix dehydrogenases occurs under conditions of high ATP demand that are associated with elevated $[Ca^{2+}]_c$, and this has the effect of increasing the availability of NADH for the electron transport chain (Section 12.3.1). Finally, in most cells, glycolysis accelerates in an attempt to counteract any deficiencies in mitochondrial ATP production (the Pasteur effect). It would clearly be advantageous for the cell to have a sensitive means of detecting, and responding to, changes in cytoplasmic ΔG_p so that it can initiate transcriptional programmes to fine-tune its bioenergetics. We have already seen (see Figure 9.4) that the level of cytoplasmic ATP is a very insensitive indicator of subtle changes in ΔG_p, and that although creatine phosphate and creatine do respond to such changes, their high concentrations in the cell preclude their ability to interact specifically with a hypothetical regulator. Cytoplasmic ADP is better (see Figure 9.4), but by far the most sensitive indicator is AMP (Figure 11.4). Owing to the presence of adenylate kinase (or the equivalent myokinase in muscle), the close to unity equilibrium of the reaction,

$$2ADP \rightleftharpoons ATP + AMP$$

means that AMP concentrations respond very sensitively to small changes in cytoplasmic ΔG_p (Figure 11.4); note that we need a logarithmic scale to display the variation of AMP concentration over a physiologically relevant range of ΔG_p.

AMP directly activates skeletal muscle phosphorylase, enhancing glycogen break-down, and phosphofructokinase, increasing glycolysis. However, most effects of AMP are mediated via AMPK. AMPK is present in the cell in an inactive form until phos-phorylated by upstream kinases, including the tumour suppressor LKB1 (liver kinase B1) in many cells and Ca^{2+}-activated kinases in neurons and lymphocytes. Binding of AMP to the kinase inhibits its dephosphorylation and allosterically activates the enzyme. The overall result is that its activity can be enhanced 1000-fold by an increase in cyto-plasmic AMP. ATP antagonises these effects of AMP, so the kinase can act as a sensor of the AMP/ATP ratio and is thus an exquisitely sensitive cytoplasmic 'energy monitor.'

Approximately 30 downstream targets of AMPK have been identified (Figure 11.4) and can be categorised as either ATP-consuming biosynthetic reactions, which are inhibited by AMPK-dependent phosphorylation, or ATP-generating processes, which are activated. Among the latter are glucose transport and fatty acid oxidation (via inhi-bition of acetyl-CoA carboxylase, which would otherwise produce inhibitory malonyl-CoA). In addition to these acute effects, AMPK can upregulate PGC-1α (Section 11.5), either directly by phosphorylation or indirectly via SIRT1-dependent deacetylation (Section 11.8). AMPK also has an inhibitory interaction with the mTOR (mammalian target of rapamycin) pathway (Section 11.7).

'Preconditioning' is the phenomenon whereby a mild nonlethal stress can induce a hormetic (compensatory) response in the cell that induces protective mechanisms against a subsequent more severe stress (Section 12.9.3). The paradoxical protection that can be induced in neurons by low protonophore concentrations (Section 12.6.2) is in part mediated by AMPK activation.

AMPK is a pharmacological target; the nucleoside AICAR is converted in cells into an AMP analogue that activates the kinase. In addition, a range of compounds that induce a degree of mitochondrial dysfunction, and hence lead to raised AMP, can act as indirect activators. The type 2 diabetes drug metformin may function in this way to acti-vate the kinase, although this is debated (Section 12.5.2).

11.4 TRANSCRIPTION FACTORS AND TRANSCRIPTIONAL COACTIVATORS IN BIOENERGETIC CONTROL

Further reading: Baltzer *et al.* (2010), Fernandez-Marcos and Auwerx (2011), Scarpulla (2011), Blomain and McMahon (2012)

Transcription factors and nuclear receptors up- or downregulate transcription of genes by binding to adjacent enhancer or promoter regions of DNA, while fine control is exerted at a second level through the action of coactivators and corepressors that are in turn regulated by metabolic, hormonal, or environmental cues. Although a detailed description of the multiplicity of control pathways is far beyond the scope of this book (the human genome contains sequences for more than 2000 potential transcription fac-tors and 200 transcriptional coregulators), a small number of key transcription factors and coactivators play central roles in the control of mitochondrial biogenesis and func-tion and are briefly reviewed here.

Mitochondrial biogenesis requires the cell to coordinate the expression of nuclear and mitochondrial-encoded subunits. Two classes of transcription factors are therefore required, acting on the nuclear and mitochondrial genome, respectively. Binding sites for the transcription factors NRF-1 and NRF-2 ('nuclear respiratory factors') are found in the promoter regions of nuclear genes encoding many subunits of the electron transport chain, and they were originally identified as activators of cyt c and complex IV genes, respectively. Of approximately 13,000 human promoters, 690 are bound in living cells by NRF-1. Additional transcription factors controlling nuclear expression of mitochondrial proteins include ERRα (estrogen-related receptor α) and yin-yang 1, whereas peroxisome proliferator-activated receptors α and δ (PPARα and -δ) selectively control expression of fatty acid oxidation enzymes. A second group of transcription factors are targeted to the mitochondrion and control parallel expression of the 13 subunits encoded by the mitochondrial genome. These include mitochondrial transcription factors A (Tfam) and B (TFB1M and TFB2M) that act at the D-loop regulatory region of mtDNA (Section 11.2).

These transcription factors are in turn controlled by transcriptional coactivators that are themselves regulated by environmental signals. The PGC (peroxisome proliferator-activated receptor γ coactivator') family, comprising PGC-1α, PGC-1β and PRC (PGC-1α-related coactivator), between them control the expression of more than 1000 nuclear-encoded mitochondrial proteins by acting in concert with multiple nuclear transcription factors.

PGC-1α has been termed 'the master gene' in this context. Its expression is regulated by a network of upstream phosphorylation pathways responding to hormones (insulin and glucagon), cytokines, exercise and cold. Figure 11.5 shows just one of these pathways—the response to cold in muscle and brown fat. We discuss the latter in detail in Section 12.4. Once synthesised, PGC-1α is subject to a multiplicity of post-translational modifications (see the cited reviews for details). PGC-1α is phosphorylated by a range of protein kinases; that by AMPK results in its activation, implying that a low-energy state signals the need for more mitochondrial biogenesis. Overexpression of PGC-1α or PGC-1β results in mitochondrial proliferation, but surprisingly, in view of the central role played by these coactivators, knockout of either produces a viable mouse, but with decreased mitochondrial function. However, a double PGC-1α/β knockout showed profound cardiomyopathy with defective mitochondria. In fact, recent research suggests that PRC may be the member of the PGC-1 family that most powerfully controls mitochondrial gene expression. PGC-1α is acetylated and inactivated by the acetylase GCN5, and it is deacetylated (and activated) by silent information regulator 2-like 1 (SIRT1; Section 11.7). A high-energy state induces acetylation, whereas deacetylation is favoured by low-energy conditions. The regulation of SIRT1 is discussed later; however, we mention here that we have problems with the accepted view that SIRT1 activity is physiologically regulated by changes in cytoplasmic NAD^+. The distinctive role of PGC-1β is discussed in Section 12.4, in which we focus on a single tissue—brown fat.

11.5 ADAPTATIONS TO HYPOXIA

Although most experiments with isolated mitochondria or cells are performed in ambient oxygen concentrations, diffusion gradients in the lungs, bloodstream, tissue and

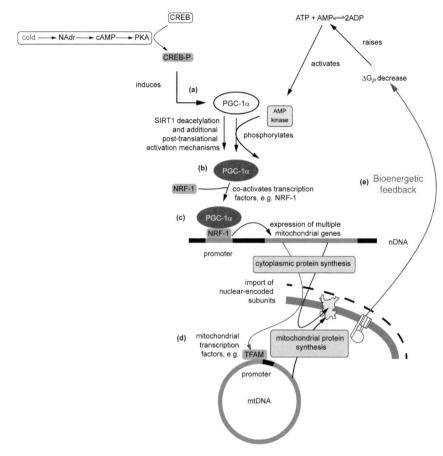

Figure 11.5   PGC-1α as a master regulator of mitochondrial biogenesis.
(a) PGC-1α can be induced by cAMP acting via CREB. In brown fat, cold can
be an initiating factor. (b) PGC-1α can be activated via a variety of post-
translational modifications, including SIRT1 deacetylation and AMP kinase-
dependent phosphorylation. (c) Active PGC-1α is a transcriptional coactivator
allowing transcription factors such as NRF-1 to promote the nuclear
expression of multiple mitochondrial genes, including (d) mitochondrial
transcription factors such as TFAM that initiate expression of the
mitochondrial genome, allowing nuclear- and mitochondrially encoded
subunits to be assembled into functional complexes. (e) An important
bioenergetic feedback mechanism under conditions of lowered ΔG_p is the
activation of AMP kinase.

finally the cell mean that the actual oxygen concentration at the *in situ* mitochondrion
can be as little as 10% of that in the atmosphere. The affinity of complex IV for O_2 is
sufficiently high for this not to impose a rate limitation on the electron transport chain,
and physiological mechanisms exist to match oxygen supply to demand, but protective
mechanisms are necessary to prevent an energetic crisis under more extreme conditions
of hypoxia, when oxygen demand may exceed supply. AMPK activation is one strategy, and another is the control exerted by hypoxia-inducible factor-1α (HIF-1α; Section

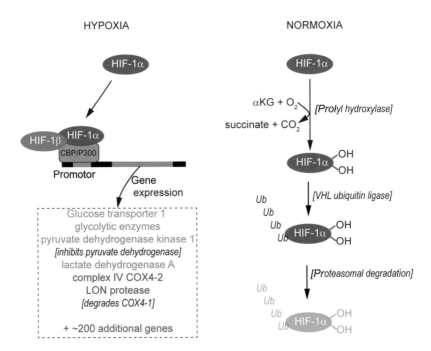

Figure 11.6   HIF-1α.
HIF-1α is continuously synthesised, but under normoxic conditions it is hydroxylated and degraded. Under hypoxic conditions, HIF-1α induces the expression of key glycolytic enzymes, inhibits pyruvate dehydrogenase by increasing PDH kinase, alters the subunit composition of complex IV and enhances the expression of approximately 200 other hypoxia-associated genes.

11.4.1). The role of the ATP synthase inhibitor protein, IF_1 (Section 7.6), in limiting ATP synthase reversal in hypoxia has already been discussed (Section 9.13).

11.5.1 Hypoxia-inducible factor

Further reading: Semenza (2007), Taylor (2008)

HIF-1 is a heterodimer composed of constitutive HIF-1α and HIF-1β subunits and acts, together with its coactivators p300 and CBP, as a transcriptional regulator of the expression of more than 200 hypoxia-dependent genes, including key bioenergetic components (Figure 11.6). The α subunit is continuously synthesised, but in the presence of oxygen, an Fe^{2+}-dependent hydroxylase (a member of a family of enzymes generally known as α-ketoglutarate-dependent oxygenases in which the splitting of α-ketoglutarate to succinate and carbon dioxide helps generate an $Fe(IV) = O$ species) inactivates HIF-1α by hydroxylation of specific proline and asparagine residues—for example,

$$Proline + \alpha\text{-ketoglutarate} + O_2 = 4\text{-hydroxyproline} + CO_2 + succinate$$

preventing it from interacting with its transcriptional coactivators and making it a target for polyubiquination by a ubiquitin ligase and subsequent proteosomal degradation.

Under conditions of limiting oxygen, the hydroxylases do not function and HIF-1α rapidly accumulates. The simplest explanation is lack of oxygen because the affinity for O_2 of complex IV (approximately $0.7\,\mu M$) is higher than that of the hydroxylases. However, more complicated hypotheses have been proposed based on an increased ROS generation under hypoxic conditions oxidising the cofactor Fe^{2+} to Fe^{3+} and thus inhibiting the enzymes. It must be emphasised that this mechanism is currently contentious; evidence for the ROS model is that myxothiazol and stigmatellin (which inhibit electron transport at complex III without generating $O_2^{\cdot-}$), but not antimycin A (which leads to high ROS generation), abolish hypoxic HIF-1α induction.

The first response to severe hypoxia is a decrease in cytoplasmic ΔG_p, with a consequent increase in AMP and activation of AMPK. The role of HIF-1α is to drive the transcription of a set of genes that facilitate the cell's more long-term adaptation to a hypoxic state. Increased transcription of the genes encoding the glycolytic pathway facilitates anaerobic glycolysis, whereas upregulation of pyruvate dehydrogenase kinase, which phosphorylates and inhibits pyruvate dehydrogenase, limits pyruvate entry into the tricarboxylic acid cycle. In the electron transport chain, HIF-1α can modify the subunit composition of complex IV, substituting the COX4-2 subunit for the more aerobic COX4-1 to optimise its ability to function under conditions of low oxygen tension.

11.6 MITOCHONDRIAL PROTEIN PHOSPHORYLATION

Further reading: Acin-Perez *et al.* (2009), Phillips *et al.* (2011), Papa *et al.* (2012), Covian and Balaban (2012)

With the exception of pyruvate dehydrogenase kinase, which phosphorylates and inhibits pyruvate dehydrogenase in response to elevated matrix ATP, NADH and acetyl-CoA, protein phosphorylation was originally thought to have little relevance for the mitochondrion. This view has changed with the development of sensitive proteomic screening techniques that have revealed an enormous number of potentially relevant mitochondrial phosphorylation sites, although the physiological significance of the large majority of these is doubtful. A recent review listed criteria that increase the likelihood that a given phosphorylation site is of functional relevance, including the following:

(a) Phosphorylations of critical conserved residues, particularly in the vicinity of active sites or redox centres, are more likely to be relevant than those at nonconserved residues in other regions of the protein.

(b) Phosphorylation of matrix-located sites should be observable in intact mitochondria without the need for disruption or the addition of exogenous kinases.

(c) Demonstration that a high mole-fraction, or occupancy, of a given residue can be phosphorylated under physiologically relevant conditions. The large majority of studies fail to quantify the extent of phosphorylation, and it is difficult to see how a low proportion of phosphorylation can affect overall function of the protein pool.

(d) Finally, demonstration that the phosphorylation state actually affects the kinetics, function, import, assembly, or degradation of the protein.

Applying these criteria places in doubt the significance of the large majority of reports on phosphorylation of inner membrane transporters—complexes I (69 reported sites), II (9 sites), III (24 sites) and the ATP synthase (67 sites). However, functional consequences of the phosphorylation of the voltage-dependent anion channel and the outer membrane TOM protein import complex have been reported, and phosphorylations of cyt c and complex IV also appear to be on firmer ground. Tyrosine phosphorylation of cyt c affects the kinetics of its interaction with complex IV, whereas isolation of mitochondria in the presence of phosphodiesterase inhibitors, tumour necrosis factor-α, or phosphatase inhibitors had an inhibitory effect on complex IV.

Although established cytoplasmic kinases and phosphatases can access proteins on the outer face of the outer mitochondrial membrane (OMM), topology demands that phosphorylation of more internal mitochondrial sites requires kinases in the same compartment—that is, the intermembrane space or matrix. That these appear to exist is shown by the ability of approximately 50 proteins to incorporate ^{32}P when intact isolated mitochondria are incubated with $\gamma^{32}P$-ATP. This raises the question as to what controls their phosphorylation (and dephosphorylation) because the vast majority of kinases lack a mitochondrial targeting sequence and so would not be predicted to be located in the matrix. An alternative approach has been to isolate mitochondrial complexes by blue native (non-denaturing) gel electrophoresis and to expose the gel to $\gamma^{32}P$-ATP. Each complex, together with the ATP synthase, was found to incorporate covalently bound ^{32}P, suggesting, controversially, that the complexes contained integral protein kinases (i.e., that they were capable of autophosphorylation).

One extensively studied system is a cAMP-dependent protein kinase A (PKA) found in mitochondrial membrane fractions, although its matrix localisation is still controversial. In this proposal, a matrix form of PKA reversibly bound to A-kinase anchor proteins associated with the IMM would be activated by cAMP derived from an HCO_3^--activated adenylyl cyclase (Figure 11.7). The implication is that changes in $O_2^{\cdot-}$ over a physiological range (10–40 mM) would signal enhanced phosphorylation of complex I.

11.7 MTOR

Further reading: Schieke *et al.* (2006), Bai *et al.* (2007), Cunningham *et al.* (2007)

mTOR (mammalian target of rapamycin) can be considered to be the mediator of 'bioenergetic plenty' for the cell and the mitochondrion. It is a ubiquitously expressed serine/threonine protein kinase in the phosphatidylinositol 3-kinase (PI3K) family that plays a central role coupling cytoplasmic nutritional and growth factor status to mitochondrial and cellular function (Figure 11.7). In particular, it is involved in the energetic balance between food intake, storage and expenditure, and it is consequently implicated in diseases such as cancer and diabetes in which this balance is disturbed (see Chapter 12). mTOR is a component in two complexes, mTORC1 (together with 'raptor') and mTORC2 (together with 'rictor'). mTOR derives its name from the ability of mTORC1 to bind, and be inhibited by, rapamycin (an immunosuppressant used in human transplantation) in complex with a protein termed FKBP12. Rapamycin is not, of

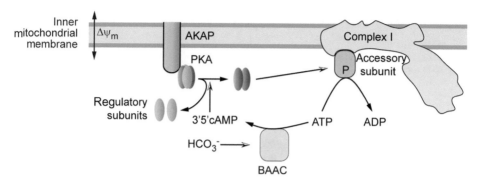

Figure 11.7 Mitochondrial protein phosphorylation.
Proposed mechanism for HCO_3^- -mediated activation of a mitochondrial form
of adenylyl cyclase (BAAC), generating matrix 3′,5′-cAMP and activating
matrix-located PKA, reversibly bound to an A-kinase anchor protein (AKAP).
Several targets have been proposed, notably accessory subunits of complex I.

course, a normal physiological component of the cell, and recent evidence suggests that
a protein on the OMM, FKB38, binds and inactivates mTORC1 until it is released and
activated by upstream regulators.

Insulin is a major upstream controller activating mTOR. A signalling cascade from
the insulin receptor—via PI3K, phosphoinositide-dependent kinase 1 (PDPK1), and
protein kinase B (AKT/PKB)—inhibits the ability of the proteins Tsc1 and Tsc2 to
activate the small GTPase Rheb, switching it to the GTP-bound state in which it activates mTORC1 (Figure 11.8). Growth factors can use additional pathways to control
mTORC1, and branched-chain amino acids (particularly leucine) activate mTORC1 and
facilitate its activation of protein synthesis.

mTOR senses the energetic status of the cytoplasmic adenine nucleotide pool via
AMPK (Section 11.3). Activation of the kinase by an increase in AMP inhibits mTOR
by a dual mechanism, both activating tsc2 and inhibiting raptor (Figure 11.8).

Downstream targets of activated mTORC1 include protein synthesis and amino acid
uptake. Protein synthesis is stimulated by blocking the translation inhibitor 4E-BP and
activating S6K1, which phosphorylates and activates the S6 ribosome. An important
negative feedback pathway mediated by mTORC1 is to enhance postprandial expression
of the appetite-suppressant hormone leptin.

The mTORC1 complex is required for the coactivator PGC-1α to bind to the transcription factor ying-yang 1 and initiate nuclear expression of multiple mitochondrial
genes. Consistent with this, blocking mTOR function with rapamycin results in mitochondria with decreased bioenergetic function, although there is controversy regarding whether this is due to decreased transcription or to a direct effect. mTOR signalling
plays a central role in current theories of aging (Section 12.9.4), and rapamycin has
been extensively investigated as a caloric-restriction mimetic, although this is complicated by the finding that rapamycin will also inhibit mTORC2 and that this is associated
with impaired glucose homeostasis and insulin resistance.

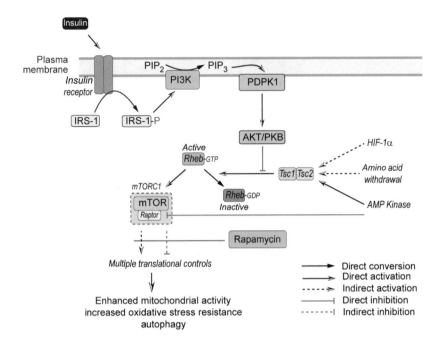

Figure 11.8   Simplified scheme of major signalling pathways acting via mTOR.
Note that insulin protects Rheb from inactivation and thus enhances mTORC1 activity, whereas a range of metabolic restrictions (hypoxia, amino acid withdrawal and increased AMP) have an opposing effect. Tsc is the tuberous sclerosis protein. Additional components and interactions exist (Evans *et al.*, 2011).

11.8 SIRTUINS AND MITOCHONDRIAL FUNCTION

Further reading: Verdin *et al.* (2010), Giralt and Villarroya (2012), Nogueiras *et al.* (2012), He *et al.* (2012), Anderson and Hirschey (2012)

Sirtuins are proteins originally associated with an increased life span in yeast, flies and worms. There are seven mammalian orthologues (SIRT1–SIRT7). With the possible exception of SIRT4, each catalyses the deacetylation of acetylated protein lysine residues in an NAD^+-dependent reaction that generates nicotinamide, 2'-*O*-acetyl-ADP-ribose, and the deacetylated substrate (Figure 11.9). Note that cells also contain acetyl-lysine deacetylases (histone deacetylases or HDACs) that catalyse direct hydrolysis. SIRT4 is an NAD^+ ADP-ribosyltransferase (transferring an ADP-ribose onto a protein amino acid residue with the release of nicotinamide). SIRT1, -6 and -7 are nuclear; SIRT2 is cytoplasmic; and SIRT3, -4 and -5 are mitochondrial.

SIRT1 acts as a nuclear HDAC, but the other orthologues have wide substrate specificities. For a protein to be a substrate for deacetylation, it must first be acetylated; however, little is known about the mechanism of acetylation in the mitochondrion, although

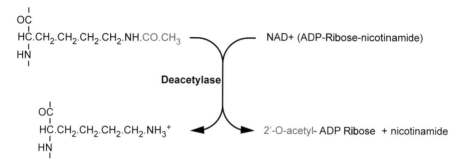

Figure 11.9   Protein deacetylation by sirtuins.
Note that SIRT4 ADP ribosylates, rather than deacetylates, proteins.

a candidate mitochondrial acetylase, GCN5, utilising acetyl-CoA, has recently been described (Scott *et al.*, 2012).

Current bioenergetic research has several goals:

(a) To investigate how nuclear-located SIRTs are controlled and affect the transcription of mitochondrial and metabolic genes: SIRT1 deacetylates a range of transcription factors and coactivators that are central to bioenergetics. The consensus is that SIRT1 acts as a sensor of nutrient availability. In this context, it is generally stated that SIRT1 senses NAD^+ levels and thus responds to conditions of substrate deprivation. However, this is problematic to metabolism experts. Typical estimates for the free NAD^+/NADH ratio in the cytoplasm, calculated from the pyruvate/lactate ratio, are approximately 500:1. Note that whole tissue estimates (e.g., Canto *et al.*, 2009) include the large, and much more reduced, mitochondrial nucleotide pool plus all the bound molecules of NAD^+ and NADH. Thus, imperceptible changes in the concentration of NAD^+ would be expected in all but the most catastrophic conditions, such as potentially pathological conditions in which the entire NAD pool is depleted, for example, by PARP (poly(ADP-ribose) polymerase 1) activation (Section 12.6.2.1). Contrast this with AMPK, the concentration of whose activator (AMP) is exquisitely sensitive to energy status (Figure 11.4). There is clearly a problem here that needs to be resolved. However, by whatever mechanism, SIRT1 responds to caloric restriction or fasting by inhibiting TORC2 and thus decreasing gluconeogenesis, activating PGC-1α and increasing fatty acid oxidation, facilitating insulin secretion and glucose homeostasis. A polyphenol component of red wine, resveratrol, activates SIRT1, and we revisit this compound in the context of dietary restriction and aging (Section 12.9). The full complexity of SIRT1 regulation is beyond the scope of this book, and the reader is referred to the previously cited reviews.

(b) To identify acetylated mitochondrial proteins: at least 130 mitochondrial proteins, including 70 related to metabolism, have been found by mass spectrometry to have acetylated lysine residues. These include subunits of complexes I (at no less than 17 sites), II and III, ATP synthase and several metabolic enzymes. However, the caveats discussed in the context of mitochondrial protein phosphorylation (Section 11.5) are equally applicable here.

(c) To determine which proteins are substrates for specific mitochondrial SIRTs and to establish the bioenergetic and physiological consequences (if any) of their acetylation status: the metabolic consequences of mitochondrial protein acetylation status are currently being investigated. Knockout mice lacking SIRT4 and SIRT5 show no detectable change in mitochondrial acetylation, suggesting a limited range of targets. The only known substrates of SIRT4 are ANT2 and glutamate dehydrogenase, with ADP-ribosylation of the latter being inhibitory. SIRT5 deacetylates and activates carbamoyl phosphate synthetase 1, the rate-controlling enzyme for urea synthesis.

In contrast to the other isoforms, SIRT3 knockout produces a widespread mitochondrial hyperacetylation. Although the knockout mice show no dramatic phenotype, a number of metabolic alterations have been reported to be associated with the resulting hyperacetylation that include electron transport chain complexes, several enzymes involved in fatty acid synthesis and catabolism, isocitrate dehydrogenase 2 and SOD2, while the mitochondria show reduced ATP synthesis and fatty acid oxidation. Of more physiological relevance, the overall mitochondrial acetylation state is responsive to caloric restriction (decreased acetylation), high-fat diet and ethanol metabolism (increased acetylation in liver). Expression of SIRT3 is controlled by PGC-1α, acting via the ERRα nuclear receptor (Section 11.4).

SIRT3 has also been proposed to be an upstream regulator of AMPK by deacetylating, and hence activating, LKB1, although this has to be reconciled with an apparent different compartmentation of the two kinases. SIRT3$^{-/-}$ mice show a 50% reduction in tissue ATP levels relative to wild-type (Ahn *et al.*, 2008), but as emphasised in Section 9.7.5, this does not necessarily indicate any bioenergetic dysfunction but could be due simply to an altered adenine nucleotide pool size. However, a significant reduction in state 3_{ADP} respiration by isolated liver mitochondria was consistent with a decreased complex I activity, while the activities of complexes III and IV and the ATP synthase have each been reported to be altered by SIRT3 deacetylation.

11.9 REDOX SIGNALLING AND OXIDATIVE STRESS

Further reading: Droge (2002), Handy and Loscalzo (2011), Ray *et al.* (2012)

The terms *ROS* (reactive oxygen species) and *oxidative stress* are cited with increasing frequency in studies of mitochondrial and cellular dysfunction associated with aging and pathological states, as well as a range of signalling processes. However, unless the precise ROS is identified, it is difficult to evaluate the study and test the proposed hypothesis. Common ROS include $O_2^{\cdot-}$, H_2O_2, $ONOO^-$ (peroxynitrite), $NO^{\bullet}$ (nitric oxide), singlet oxygen and $OH^{\bullet}$ (hydroxyl radical), each with distinctive pathways of formation, reactivity and degradation. Oxidative stress also tends to be loosely defined as a condition under which there is a potentially damaging increase in the steady-state concentration of one or more of the previous ROS, or sometimes as a condition under which the cytoplasmic redox potential of a couple such as $NADP^+/NADPH$ or disulfide/dithiol undergoes a shift to a more oxidised potential. Oxidative damage in turn can be assessed in terms of modifications to proteins (e.g., formation of protein carbonyl, 3-nitrotyrosine, or 4-hydroxynonenal adducts), lipids (lipoperoxidation), or DNA (e.g., formation of 8-hydroxydeoxyguanosine bases in mtDNA). Finally, there is no single mode of action of antioxidants: they can be enzymatic (e.g., catalase), catalytic (e.g., MitoQ), stoichiometric, hydrophilic, or lipophilic and are generally specific for a given ROS.

Given this bewildering complexity, a checklist has been proposed to help assess a possible role of ROS in a biological process (Murphy *et al.*, 2011). Which ROS is responsible? Does the mechanism make chemical sense? Is the ROS in the right place and at a sufficient concentration? Can you modulate the response by altering the level of the particular ROS? To which we can add, 'Can pathways be modified by the addition of exogenous ROS at physiologically relevant concentrations?' In evaluating the literature, it is important to bear in mind that the maximal intracellular concentration of H_2O_2 probably does not exceed 10–50 μM and that phenomena requiring the addition of millimolar concentrations of the ROS may be of limited physiological relevance. Similar caution must be adopted in interpreting studies in which exogenous generators of $O_2^{\cdot-}$ are employed because these can generate levels of the ROS far higher than those attained *in vivo*. Answers to these questions are important in order to decide where the boundary lies between oxidative damage and the potentially beneficial role of specific ROS in signalling between the mitochondrion and the cell.

The effect of ROS trapping agents can help to establish the cause–effect relationship between ROS and cell damage. Thus, the large increase in the level of $O_2^{\cdot-}$ that is seen in neurons undergoing delayed Ca^{2+} deregulation (see Figure 12.6) as a consequence of pathological NMDA receptor activity is ascribed to a downstream consequence of cell damage because trapping agents do not protect the cell.

ROS are implicated in a wide range of signalling cascades. In addition to the mitochondrion as a source (and sink) for ROS, activated macrophages and neutrophils employ NADPH oxidase to produce superoxide, generating H_2O_2 concentrations in their vicinity that can reach 100 μM. In nonphagocytic cells, NAD(P)H oxidase isoforms generate $O_2^{\cdot-}$ (and thus H_2O_2) levels that are implicated in a wide range of signal transduction pathways, including mitogen-activated protein kinase (MAPK) cascades (Figure 11.1), activation of antioxidant defences, stress-induced increases in pro-apoptotic proteins such as p66Shc, and PI3K signalling. In this last example, ROS facilitate growth factor and insulin-mediated signalling, both by facilitating receptor autophosphorylation (via activation of NADPH oxidase) and by facilitating thiol group oxidation, and consequent inactivation, of PTEN (phosphatase and tensin homologue), which would otherwise function to inhibit growth factor signalling via PI3K.

The dominant effect of added H_2O_2 is to act as substrate for glutathione peroxidase and peroxiredoxin, leading to an oxidative shift in the thiol redox potentials for the glutathione and thioredoxin pools. This in turn facilitates the formation of protein disulfides and may additionally lead to oxidation of the cytoplasmic $NADP^+/NADPH$ pool. Pro-oxidants such as *t*-butylhydroquinone have similar actions. The adenine nucleotide translocator contains redox-active thiol groups, and their oxidation with the formation of intramolecular disulfide bonds results in inhibited transport activity. Another mitochondrial protein that is very sensitive to oxidative stress ($O_2^{\cdot-}$) is aconitase, which undergoes loss of its iron and inactivation following oxidation of its Fe–S centre. The process is reversible following Fe–S centre reduction and reincorporation of Fe^{2+}.

Although mitochondria are generally considered the main source of ROS, efficient trapping systems (SOD, glutathione peroxidase, etc.) mean that their steady-state concentrations (the key parameter for signal transduction) are low under physiologically relevant conditions (see Figure 9.14). Experiments in which these levels are enhanced by the addition to cells of electron transport chain inhibitors such as rotenone or antimycin A should be interpreted with caution, particularly if the dramatic effects on cellular bioenergetics are not controlled for. Endogenous H_2O_2 levels can be increased in appropriate cells by monoamine oxidase substrates and lowered by ectopic expression of catalase. The ability of 'mild uncoupling' to decrease steady-state levels of HCO_3^- and H_2O_2 in cells is, as we have previously discussed (Section 9.12), contentious.

12 MITOCHONDRIA IN PHYSIOLOGY AND PATHOLOGY

12.1 INTRODUCTION

'Mitochondrial' hypotheses are currently being investigated in a wide range of disorders. The field tends to advance by the discovery of a new pathway or component, followed by a broad testing of hypotheses that a defect in the pathway may have implications for a particular disease state. How many of these hypotheses will stand the test of time before the next edition of *Bioenergetics* cannot be predicted; however, based on the experience of the previous decade, it is likely that 'unknown unknowns' (i.e., defects in pathways that we do not currently know exist) may hold the key to many of the pathologies. For example, one of the least understood aspects of mitochondrial diseases is their tissue specificity: why do certain mutations in complex I in LHON (Leber's hereditary optic neuropathy) and mutations in genes for the mitochondrial fusion protein OPA1, underlying dominant optic atrophy, both target retinal ganglion cells? Finally, we currently have little insight into one of the most important questions: why is age the risk factor in cancers, stroke, heart attack and neurodegenerative diseases?

The complexity of the problems and the importance to society of the conclusions from the research in this area demand that the highest standards of bioenergetic analysis, which we have tried to review in previous chapters, are applied. Unfortunately, this is not always the case. In studying the primary literature, the reader is encouraged to apply the criteria we have outlined, for example, in Chapter 9, to assess the weight that should be applied to conclusions the authors draw. We have included a limited selection of studies from the primary literature as illustrations of the range of bioenergetic techniques that can be applied.

12.2 MITOCHONDRIAL DISEASES

Further reading: Duchen and Szabadkai (2010), Schapira (2012)

A mitochondrial disease can be defined as any disorder in which the primary lesion is a dysfunction or abnormal behaviour of the organelle. Many mitochondrial diseases are

Bioenergetics. Doi: http://dx.doi.org/10.1016/B978-0-12-388425-1.00012-9

genetic, including inherited deleterious mutations in mitochondrial DNA (mtDNA), which can be caused by deletions, inversions, duplications, point mutations, or rearrangements, all of which may affect genes encoding complex subunits or transfer RNAs (tRNAs). In addition, mutations occur in nuclear-encoded mitochondrial genes that may affect structural genes, mtDNA replication, or repair. Finally, many aging-related diseases, such as type 2 diabetes and Parkinson's and Alzheimer's diseases, have aspects associated with altered mitochondrial function.

12.2.1 mtDNA mutations

Further reading: Shoubridge and Wai (2008), Krishnan and Turnbull (2010), Lax *et al.* (2011), Craigen (2012), Ghezzi and Zeviani (2012)

Phenotypically neutral or mildly advantageous mtDNA polymorphisms (haplotypes) were reviewed in Section 10.5.1. mtDNA has a mutation rate 10 times higher than that of the nuclear genome, and deleterious somatic mutations (point mutations or large-scale deletions of up to 8 kb) can arise during life, exacerbated by the exposure of mtDNA to reactive oxygen species (ROS) from the adjacent electron transport chain, limited mtDNA repair mechanisms, and lack of protective histones. Mitochondrial theories of aging are discussed in Section 12.9.

It is not clear how deleterious mtDNA mutations escape mitochondrial quality control (Section 10.3.2) and the oocyte bottleneck (Section 12.2.2), but more than 200 pathogenic mutations have been described, with a total prevalence of approximately 1 in 10,000 of the adult population. Almost all mutations are functionally recessive, with their phenotype depending on the extent to which their degree of heteroplasmy (the fraction of mutant rather than wild-type mtDNA in a cell) exceeds a critical threshold.

It is notable that many mtDNA mutations are referred to as 'encephalomyopathies' because brain and muscle are the tissues whose mitochondria can be called upon to produce bursts of high activity and could therefore be preferentially compromised by a restriction in mitochondrial function. Mutations can occur in protein coding genes, but a disproportionate number are found in tRNA genes and thus uniformly affect synthesis of all 13 mt-DNA-encoded proteins.

The most common pathogenic mtDNA point mutation is A3243G in the leucine tRNA gene. It manifests as a mitochondrial myopathy and encephalopathy with lactic acidosis and stroke-like episodes (MELAS). A point mutation 8344 A>G in the lysine tRNA gene leads to MERRF (myoclonic epilepsy ragged red fibres, which are muscle fibres with proliferating dysfunctional mitochondria). LHON is caused by point mutations in one of three loci all involving complex I subunits. It is unclear why the pathology is almost entirely restricted to the retinal ganglion cells in the optic nerve. The so-called common deletion in mtDNA covers a 4977-bp region (8482–13459), affecting the expression of several tRNAs and polypeptides. As with other mutations, the phenotype depends on the degree of heteroplasmy, and above the threshold leads to Kearns–Sayre syndrome.

12.2.2 Oocytes and generational quality control

Further reading: Shoubridge and Wai (2007, 2008), Stewart *et al.* (2008), Jokinen and Battersby (2012)

Mammalian mtDNA is inherited maternally because any sperm mitochondria fail to survive in the ovum after fertilisation. Mature oocytes contain 100,000 copies of mtDNA and almost as many mitochondria. However, primordial germ cells (PGCs) in the embryo initially contain as few as 10–100 mitochondria, so only 0.01% of the mtDNA in the parental oocyte will be passed to the second generation. This does not necessarily imply the destruction of 99.99% of the oocyte mitochondria but, rather, suggests that mitochondrial replication is suppressed at this stage so that the 100,000 mitochondria from the oocyte are apportioned to approximately 10,000 cells (including PGCs) in the early embryo before replication resumes. This bottleneck facilitates the segregation of mtDNA in subsequent generations by increasing the chances that mature oocytes will be homoplasmic. Thus, most individuals have homoplasmic mtDNA, although the variation between individuals is typically up to 50 nucleotides.

In order to investigate whether mechanisms exist to prevent a progressive build-up of deleterious mitochondrial mutations in succeeding generations, 'mutational meltdown' experiments have been performed in which the fate of mtDNA mutations was followed over subsequent generations. The mtDNA mutator mouse expresses a mutation in the *PolgA* catalytic subunit of mitochondrial DNA polymerase affecting its proofreading ability and therefore has a high level of mtDNA mutations. Homozygous mutated females were bred with wild-type males, and heterozygous female progeny were used to establish wild-type lines to follow the fate of the original mutations over generations. A strong 'purifying' selection was found against deleterious mutations, as opposed to synonymous changes (which did not affect the amino acid sequence), suggesting some mechanism for functional testing of the mitochondria in the maternal germline. How so many disease-causing mtDNA mutations escape this quality control is not clear.

12.2.3 Cybrids

Further reading: Swerdlow (2011), Vithayathil *et al.* (2012)

A mitochondrial dysfunction can result from a combination of environmental (cytoplasmic) factors, mutations in nuclear genes, and mutations in mtDNA. Cybrids provide a means to selectively investigate the consequences of mtDNA mutations in a wild-type cellular environment. The first stage in cybrid creation is the generation of $\rho 0$ (rho-zero) cells depleted of their mtDNA by prolonged exposure to cationic DNA mutagens such as ethidium, which are concentrated in the mitochondria by $\Delta\psi_m$ (Figure 12.1). The $\rho 0$ cells retain mitochondrial 'ghosts' that lack both electron transport chain activity and a functional ATP synthase (due to the lack of two F_o subunits). However, the F_1 subunits remain able to hydrolyse glycolytic ATP; this involves the adenine nucleotide translocator (ANT) taking in ATP^{4-} and releasing ADP^{3-} (see Figure 9.4). This electrogenic

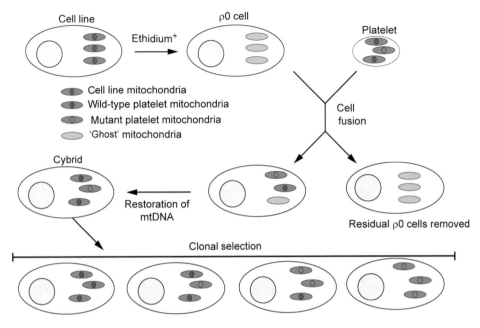

Figure 12.1   ρ0 cells and cybrids.
A cybrid is a cell in which mtDNA is selectively depleted, leading to a ρ0 (rho-zero) cell with 'ghost' mitochondria lacking electron transport and a functional ATP synthase (but retaining ATPase activity). The cell is then repopulated with mitochondrial from an anucleate cell (e.g., a platelet). If the platelet's mitochondria are heteroplasmic, this allows clones to be produced with differing degrees of heteroplasmy.

exchange can maintain a suboptimal $\Delta\psi_m$, which seems to be necessary for ρ0 cell growth.

Fusion of ρ0 cells with cells containing mitochondria but lacking a nucleus (e.g., fibroblast cytoplasts or blood platelets) creates a cybrid cell line (Figure 12.1). If the platelets' mtDNA is heteroplasmic, then clones can be isolated with differing heteroplasmic ratios, and these can be exploited to investigate thresholds at which a bioenergetic phenotype becomes apparent. If a cybrid retains aspects of the bioenergetic phenotype of the disease, then the indication is that a mtDNA mutation plays some role in the disease.

Figure 12.2 shows part of a detailed bioenergetic analysis of a related MELAS-type mutation, A13528G, affecting the ND5 gene and is shown as an example of how we believe such studies should be conducted. Patient fibroblasts in a high KCl medium (to minimise any contribution from $\Delta\psi_p$) showed decreased TMRM fluorescence (non-quench mode; Section 9.7.1) indicative of partial mitochondrial depolarisation (or, it should be noted, decreased mitochondrial content), together with increased lactate production due to compensatory glycolysis. These bioenergetic defects were retained when cybrids were created with the patient's mitochondria, confirming the mitochondrial origin of the bioenergetic defect. A cell respiratory control (CRC; Section 9.6) analysis of

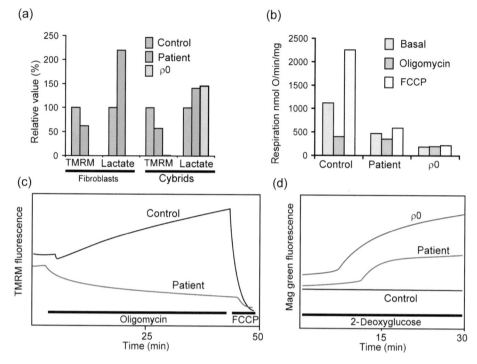

Figure 12.2   Bioenergetic analysis of fibroblasts and cybrids from a patient with a mutation in the complex I ND5 gene.
(a) TMRM fluorescence (under non-quench conditions) and lactate accumulated during 48 h were determined for fibroblasts from the patient and a control subject and for cybrids containing mitochondria from the patient, the control subject, and the empty $\rho 0$ cells. Note decreased $\Delta\psi_m$ and increased reliance on glycolysis. (b) Cybrid cell respiratory control experiment. Note decreased spare respiratory capacity. (c) Oligomycin null-point assay. Control cybrids hyperpolarise after oligomycin, whereas the patient cybrids depolarise, indicating that their ATP synthase was running in reverse. (d) Mag Green detects the increase in free cytoplasmic Mg^{2+} when cytoplasmic ATP is hydrolysed to ADP. 2-Deoxyglucose restricts glycolysis. The greater reliance of patient cybrids (and particularly $\rho 0$ cells) on glycolysis means that cytoplasmic ATP is depleted sooner.
Adapted from McKenzie et al. (2007).

the intact cybrids showed a dramatic decrease in basal and FCCP-stimulated respiration with little change in proton leak. An 'oligomycin null point' analysis (Section 9.7.3) showed that the ATP synthase in the patient's cybrids was running in reverse, hydrolysing cytoplasmic ATP, because addition of oligomycin depolarised $\Delta\psi_m$, contrasting with a hyperpolarisation in the control. Inhibition of glycolysis by addition of 2-deoxyglucose produced a more rapid depletion of residual ATP in the patient's cybrids, consistent with their greater reliance on glycolysis. This assay utilised an Mg^{2+}-selective indicator (Mag Green) and is based on the lower ability of ADP to chelate Mg^{2+} compared to ATP. Finally, with digitonin-permeabilised cells (Section 9.8), state 3 respiration was much more depressed with complex I substrates than with succinate, indicative of a complex I locus (not shown).

12.2.4 Nuclear mutations

Further reading: Trifunovic *et al.* (2004), Lenaers *et al.* (2009), Copeland (2012)

Because 99% of mitochondrial genes are nuclear encoded, it is not surprising that many genetic mitochondrial diseases have a nuclear, rather than mitochondrial, origin. mtDNA is replicated and repaired by DNA polymerase γ encoded by *POLG* and *POLG2* genes. Approximately 200 *POLG* mutations associated with mitochondrial disease (mostly forms of progressive external ophthalmoplegia, an eye movement disorder) have been reported mapped to its polymerase domain, resulting in decreased catalytic activity or defective proofreading. An mtDNA 'mutator' mouse has been constructed that is homozygous for the knockin of an error-prone mtDNA polymerase with defective proofreading ability. The mice have a three- to fivefold increase in mtDNA point mutations and show bioenergetic deficiencies and a spectrum of premature aging phenotypes. The helicase that separates the mtDNA strands for replication is encoded by the *TWINKLE* gene, and more than 30 pathogenic mutations in this protein have been reported. Further pathological mutations are found in the transport and metabolism of precursors for the four deoxynucleotide triphosphates required for mtDNA synthesis.

Charcot–Marie–Tooth disease type 2A (CMT2A) is one of a class of neuropathies characterised by a loss of function and degeneration of long peripheral axons. CMT2A is caused by mutations in the outer mitochondrial membrane (OMM) fusion protein mitofusin2 (Mfn2). Neurons cultured from Mfn2 knockout mice show defects in mitochondrial trafficking, and it is a defective interaction with Miro and Milton (Section 10.4), rather than an alteration in the fission/fusion cycle, that may inhibit the trafficking of mitochondria to the extremities of these enormously long axons. Mutations in Opa1 (optic atrophy 1), which regulates inner mitochondrial membrane (IMM) fusion (Section 10.3.2), lead to a dominant inherited optic neuropathy that selectively affects the retinal ganglion cells that form the optic nerve and can lead to blindness. The most pronounced bioenergetic effect of this mutation is an increase in proton leak (Loiseau *et al.*, 2007).

12.3 THE HEART

With the exception of the brain, no organ is more critically dependent on continuous mitochondrial function in response to variable energy demand than the heart. Cardiac output can increase from 5 to 30 l of blood min^{-1} during severe exercise. Ca^{2+}-mediated signalling to the mitochondria allows this to occur with little or no drop in cytoplasmic ΔG_p as long as coronary blood supply is unimpeded. However, a transient blockage (heart attack) initiates a sequence of events in which mitochondrial dysfunction plays a central role.

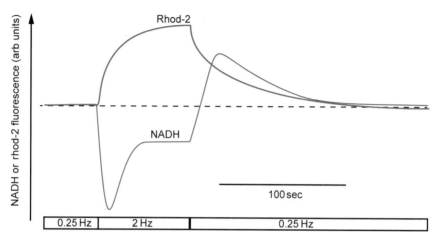

Figure 12.3   NADH redox state in the beating heart.
Rat heart tissue was loaded with the Ca^{2+} indicator rhod-2-AM, which
primarily locates to the mitochondria by virtue of its positive charge (see
Figure 9.2). Rhod-2 and endogenous NADH fluorescence were monitored in
parallel as the stimulation frequency was changed from 0.25 to 2 Hz. Note the
progressive rise in the presumed mitochondrial component of the rhod-2
signal, consistent with an increase in $[Ca^{2+}]_m$, correlating with the partial
recovery of the NADH fluorescence, indicating Ca^{2+}-dependent activation of
mitochondrial metabolism. Note also the overshoot of NADH reduction as the
stimulation frequency is restored to 0.25 Hz.
After Brandes and Bers (2002).

12.3.1 Bioenergetic tuning to altered workload

Further reading: Griffiths (2009), Liu and O'Rourke (2009), Balaban (2009b)

In Chapter 3, we introduced the idea that bioenergetic flux is a function of the thermodynamic disequilibrium between electron transport and Δp and between Δp and ΔG_p.
Thus, increasing ATP turnover in isolated mitochondria—for example, with glucose plus
hexokinase—is accompanied by a decrease in the value of both parameters. In isolated
models, this drop is not problematic; however, in the working heart, in which cardiac output can increase sixfold between rest and vigorous exercise, such a drop in ΔG_p could
seriously compromise heart function, in terms of both contraction and Ca^{2+} regulation. In
practice, ^{31}P nuclear magnetic resonance spectroscopy of the working heart shows little
or no drop in ΔG_p with work output. Similarly, studies measuring NAD(P)H fluorescence
(which responds to changes in Δp; see Figure 9.10) in the isolated perfused heart also
show that compensatory mechanisms exist to minimise changes in the face of increased
workload (Figure 12.3). Although the creatine kinase system can buffer ΔG_p during
short-term burst activity, the mechanism for this energetic homeostasis must lie upstream.
    Increased cytoplasmic free Ca^{2+} ($[Ca^{2+}]_c$) is the main signal tuning the cardiac
mitochondria to an increased workload. During excitation–contraction coupling in the

heart, a small Ca^{2+} entry via plasma membrane L-type voltage-activated Ca^{2+} channels (VACCs) triggers a much larger Ca^{2+} release from the sarcoplasmic reticulum T-tubules to trigger contraction. In heart cytoplasm, $[Ca^{2+}]_c$ varies from approximately 150 nM in diastole (dilation) to approximately 1 μM in systole (contraction). Because the T-tubules are in close apposition to the mitochondria, the mitochondrial Ca^{2+} uniporter will see a high localised Ca^{2+} concentration, allowing a proportion of the released Ca^{2+} to be accumulated by the mitochondria (Figure 10.7), where it activates key metabolic enzymes: NAD-dependent isocitrate dehydrogenase, α-ketoglutarate dehydrogenase, and pyruvate dehydrogenase (the latter is activated indirectly as a result of Ca^{2+}-dependent activation of pyruvate dehydrogenase phosphatase). This $[Ca^{2+}]_m$-mediated activation of the dehydrogenases, together with activation of the glutamate/aspartate transporters aralar/citrin (Section 9.5.3), helps to maintain the matrix NADH/NAD^+ ratio. An additional less characterised locus of Ca^{2+} activation appears to be the ATP synthase. Although direct addition of Ca^{2+} to isolated mitochondria does not affect the kinetics of the synthase, in the intact heart Ca^{2+}-triggered post-translational modification has been suggested.

The question as to whether mitochondrial Ca^{2+} transport is sufficiently rapid to respond to these oscillations on a beat-to-beat basis (the mouse heart rate is ~500 beats/min) or whether the matrix sees merely a time-averaged increase in $[Ca^{2+}]_m$ during increased activity is controversial (O'Rourke and Blatter, 2009). Although several matrix-targeted Ca^{2+} indicators exist (see Figure 9.2), the possibility that oscillations originate from a fraction miss-targeted to the cytoplasm has to be rigorously excluded. Note that isolated cardiac mitochondria have rather low mitochondrial Ca^{2+} uniporter (MCU; Section 9.4.1.1) activity relative to other mitochondria.

12.3.2 Mitochondria and cardiac ischaemia/reperfusion injury

Further reading: Solaini and Harris (2005), Halestrap and Pasdois (2009), Griffiths (2012)

A myocardial infarction (heart attack) is caused by a blood clot in a coronary artery supplying the left ventricle with fuel and blood, and it results in an area of ischaemia. Glycolysis cannot adequately compensate for the failure of oxidative phosphorylation, in part because the intracellular acidification resulting from lactate production inhibits phosphofructokinase. The decrease in ATP inhibits the Na^+/K^+-ATPase, allowing $[Ca^{2+}]_c$ to rise, while some of the concomitant AMP may be degraded, depleting the intracellular adenine nucleotide pool and generating xanthine, which has been proposed to be a source of superoxide via xanthine oxidase even under these very low oxygen conditions. Despite the high $[Ca^{2+}]_c$, low ATP, and oxidative stress, the permeability transition does not appear to be activated during the actual ischaemic period, probably as a result of the low cytoplasmic pH.

To limit cell necrosis, it is essential to restore blood flow as soon as possible. However, reperfusion can increase the damage ('reperfusion injury'), and there is convincing evidence for the involvement of the mitochondrial permeability transition pore (mPTP) in this process (Figure 12.4). Paradoxically, the return of oxygen is the time of

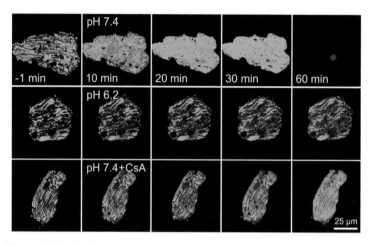

Figure 12.4   Activation of the permeability transition in a model of cardiac reperfusion injury.
Rat cardiac myocytes were loaded with fluorescent calcein, under conditions in which it predominantly locates to the mitochondria, and subjected to 3 h ischaemia at pH 6.2, followed by reperfusion at pH 7.4, 6.2, or 7.4 in the presence of cyclosporine A (CsA). Red fluorescing propidium iodide detects loss of cell viability. Note that reperfusion at pH 7.4 is accompanied by release of calcein into the cytoplasm, indicating activation of the permeability transition, but this is delayed or prevented by acidic reperfusion or the presence of CsA.
From Kim et al. (2006), courtesy of John Lemasters.

greatest damage to the tissue; return to a neutral pH, further ROS production (due to the combination of high O_2 and a highly reduced electron transport chain), and uptake into the mitochondria of Ca^{2+} accumulated in the cytoplasm all contribute toward mPTP activation. Evidence that the mPTP is activated during reperfusion has been obtained with perfused hearts using [^{3}H]-deoxyglucose, which is phosphorylated in the cytoplasm to deoxyglucose-6-phosphate. This in turn is not permeable across the IMM unless a permeability transition pore is formed. Mitochondria are isolated in the presence of EGTA to reseal the pore contained the label, indicating that the permeability transition had been activated. Alternatively, mitochondria within cardiac myocytes can be loaded with the fluorescent probe calcein. This is released into the cytoplasm if the permeability pore is activated (Figure 12.4). Consistent with these findings, cyclosporin A (CsA), an inhibitor of the mPTP, or genetic ablation of its target, cyclophilin D, in animal models provides some protection against reperfusion injury. However, the usefulness of CsA is limited by its ability to inhibit calcineurin in the cytoplasm and generate a generalised hyperphosphorylated state. To counter this, more selective analogues have been developed.

A word is apposite about supposed MCU inhibitors in these studies. The hexavalent cation, ruthenium red (RuRed), was originally produced as a sialic acid stain for the plasma membrane. Polycations (RuRed, Mg^{2+}, and La^{3+}) restrict mitochondrial Ca^{2+} transport by altering the surface charge in the vicinity of the transporter; thus, RuRed is highly potent with isolated mitochondria. The assumption is frequently made that RuRed enters intact cells to inhibit the MCU, and it has been reported to protect myocytes and hearts against reperfusion damage at low micromolar concentrations; however, it is more likely that it acts at the plasma membrane to limit Ca^{2+} entry into the cell. The more potent derivative of RuRed, Ru360, is often used uncritically as a cell-permeant MCU inhibitor, but careful parallel analysis of $[Ca^{2+}]_c$ and $[Ca^{2+}]_m$ in a variety of cells, including rat cardiac myoblast cells, has shown that a massive exposure to Ru360 ($10\,\mu M$ for $12\,h$) also failed to inhibit mitochondrial Ca^{2+} uptake in intact cells (Hajnoczky et al., 2006). Studies of reperfusion injury interpreting these inhibitors as acting at the MCU should be re-evaluated.

Short periods of ischaemia followed by reperfusion can protect the heart (and brain, Section 12.6.2) against a subsequent more prolonged ischaemic attack. The mechanism of this 'ischaemic preconditioning' is debated but may involve increased defences against oxidative stress, AMP kinase (AMPK) activation, or the protective effect of HIF-1α (Section 11.5.1). Despite the controversy surrounding the very existence of a mitochondrial ATP-inhibitable K^+ channel (mK_{ATP}; Section 9.3), the role of putative channel activators as cardioprotective agents has been extensively investigated, with proposals that preconditioning may involve opening of the mK_{ATP} in response to the decrease in cytoplasmic ATP, resulting in swelling of the mitochondrial matrix. High levels of ROS have also been implicated in channel opening *in vitro* but may not be physiologically relevant. It must be emphasised that the way in which an increase in matrix volume is cardioprotective is far from clear.

12.4 BROWN ADIPOSE TISSUE AND TRANSCRIPTIONAL CONTROL

Further reading: Sears et al. (1996), Scarpulla (2008, 2011), Ravussin and Kozak (2009), Kajimura et al. (2010), Bartelt and Heeren (2012)

The bioenergetics of brown adipose tissue (BAT) and the role of the uncoupling protein UCP1 were reviewed in Section 9.12.2. BAT has become the tissue of choice in which to test and develop theories on the transcriptional control of cellular differentiation and bioenergetic function. There are several reasons for this. First, UCP1 confers unique bioenergetic properties on the mitochondria and the tissue. Second, cold adaptation or chronic noradrenergic stimulation of rodents produces a remarkable proliferation of mitochondria in the tissue and an even more dramatic increase in UCP1, sufficient for the content of UCP1 per mitochondrion to increase almost 10-fold over the level in the thermoneutral animal. Third, BAT has proven to be an invaluable model in which to investigate cellular differentiation. Finally, there is great excitement regarding the possible role of BAT deposits in the adult human in relation to obesity. Each of the transcriptional controls reviewed in Chapter 11 has been investigated (and, in some cases, originated) in BAT, which therefore serves as an interesting case study.

Classic brown adipocytes in adult rodents originate from foetal progenitor cells able to differentiate into either muscle cells or interscapular BAT (iBAT). In addition, currently undefined pathways allow 'brown-like' (beige) adipocytes ('wBAT') containing UCP1 to be formed independently within white fat depots in response to β stimulation.

The brown preadipocyte cell line, H1B-1B (Sears *et al.*, 1996), can be induced to differentiate into UCP1-containing brown adipocytes in the presence of insulin and triiodothyronine. The nuclear transcription factor PPARγ was found to be necessary but not sufficient for this differentiation, and analysis of a brown fat cDNA library later identified the additional factor as PGC-1α (Section 11.4). Although PGC-1α is induced by cold adaptation and plays a central role in cold-adapted mitochondrial proliferation and UCP1 expression in BAT, depletion of the coregulator still allows brown adipocytes to be formed, indicating that other factors control cell differentiation. The transcriptional regulator PRDM16 has been shown to play this key role in the generation of brown adipocytes from precursor cells.

Adaptive thermogenesis, the increase in thermogenic capacity of the tissue in response to cold, is mediated predominantly by $β_3$-adrenergic signalling via the CREB pathway, leading to activation of PGC-1α (Figure 12.5). However, although this is necessary, it appears not to be sufficient for maximal induction of thermogenic capacity because the response is blunted in SIRT3$^{-/-}$ mice. SIRT3 is highly expressed in BAT and is further upregulated by cold exposure by a PGC-1α-dependent mechanism. SIRT3 may in turn act in a positive feedback increasing PGC-1α expression. Deletion of the gene for poly(ADP ribose) polymerase (PARP), which consumes NAD^+ in response to DNA damage (Section 12.6.2.1), resulted in raised NAD^+ levels, SIRT1 activation and increased mitochondrial biogenesis in BAT.

Recently, it has been recognised that BAT is present in the adult human. Positron emission tomography (PET) scanning in the presence of 2-^{18}F-fluorodeoxy-D-glucose has been used by oncologists to identify tumours with high glycolytic rates. Puzzling noncancerous areas in some patients have been identified as BAT, previously thought to be restricted to neonates (Cypess *et al.*, 2009). This has revived interest in the possibility that the high oxidative capacity of the tissue may in some cases contribute to weight homeostasis in the adult.

12.5 MITOCHONDRIA, THE PANCREATIC β CELL AND DIABETES

More than 8% of the U.S. population has some form of diabetes. Type 1 diabetes (T1D) is usually a consequence of the autoimmune destruction of β cells, although what causes this is unclear. T1D accounts for 5–10% of all cases and is treated with insulin therapy. Type 2 diabetes (T2D) is the most rapidly advancing disease in the developed world. It is characterised by a high blood glucose level associated with a combination of insulin resistance in target tissues (particularly muscle and adipose tissue), increased glucose output from the liver, and decreased insulin secretion by the β cells. The majority of T2D cases were previously considered to be sporadic, with obesity as a primary cause; however, it is now recognised that genetic variants play a major role in defining an individual's risk factor in combination with lifestyle. Exercise and dietary control are the initial forms of treatment. Failure to control T2D results in a greatly increased risk of cardiovascular disease, stroke, blindness, kidney failure and the need for lower limb amputation. We first review the central role played by mitochondria in the normal physiological mechanism of insulin secretion.

12.5.1 Glucose-stimulated insulin secretion

Further reading: Rutter (2004), Jitrapakdee *et al.* (2010), Maechler *et al.* (2010)

Although the outline pathway for glucose-stimulated insulin secretion (GSIS) has long been established (Figure 12.5a), there is continuing controversy concerning the details. Pancreatic β cells possess a low-affinity glucose transporter, (GLUT1 in the human and GLUT2 in the mouse), allowing glucose transport into the cell to respond to physiological fluctuations in blood glucose. Once in the cytoplasm, glucose is phosphorylated by the low-affinity type 4 hexokinase (glucokinase), rather than the higher affinity hexokinase, allowing glycolytic activity in turn to vary with external glucose concentration (Figure 12.5a).

Unlike most other cells, β cells possess low levels of both lactate dehydrogenase and the plasma membrane monocarboxylate transporter, with the result that pyruvate generated by glycolysis must be oxidised by the mitochondrion, without the option, under conditions of deficient mitochondrial ATP synthesis, of being reduced and exported as lactate. The absence of these so-called 'disallowed' proteins is of critical importance because it disables the Pasteur effect (Section 12.7.1) whereby fluctuations in oxidative phosphorylation can be compensated for by increased glycolysis. It also means that the cells are insensitive to circulating levels of pyruvate and lactate. As in other cells, the α-glycerophosphate and malate/aspartate shuttles (Section 9.5.2) allow the cytoplasmic NADH generated by glycolysis to be oxidised by the mitochondria.

β cell mitochondria are distinctive in possessing a high endogenous proton conductance (Affourtit and Brand, 2005), one result of which is that substrate availability exerts a high control over Δp (Section 4.8). Interestingly, both the $\Delta\psi_m$ and ΔpH components of Δp are enhanced by elevated glycolysis (Figures 12.5b and 12.5c). The effect is to raise the cytoplasmic ATP/ADP ratio. The cells have a high ATP turnover, although the ATP-utilising pathways are ill-defined. The overall result of this distinctive bioenergetic behaviour is that cytoplasmic ATP/ADP ratios are uniquely responsive to changes in glucose availability. According to the canonical pathway, the K_{ATP} channel complex at the plasma membrane is inhibited by ATP (in the Mg-free ATP^{4-} form) and activated by ADP (in the Mg form). As a result, the high cytoplasmic ATP/ADP ratio will depolarise $\Delta\psi_p$, allowing VACCs to open, with the resulting Ca^{2+} entry triggering insulin secretion.

This is only part of the story, however, because insulin secretion is not continuous but, rather, oscillates with a periodicity of a few minutes (Figure 12.5e). Complex upstream oscillations in $\Delta\psi_p$ and $[Ca^{2+}]_c$ (Figure 12.5d), as well as in respiration and glycolysis, have been described. $[Ca^{2+}]_c$ measurements show clusters of fast oscillations with a period of a few seconds, superimposed on slow oscillations with a period of 2–4 min, the latter corresponding to the insulin secretion pattern. The mechanisms underlying these oscillations are still being resolved, with the slow component being ascribed to metabolic (glycolytic) factors and the fast components to ion channel and Ca^{2+} effects. Insulin secretion in response to a continuous elevation of glucose is biphasic; although the underlying mechanism is not entirely clear, one possibility is that the first phase is due to the release of insulin vesicles closely apposed to the plasma membrane, while the vesicles responsible for the second phase may first require translocation to the release site.

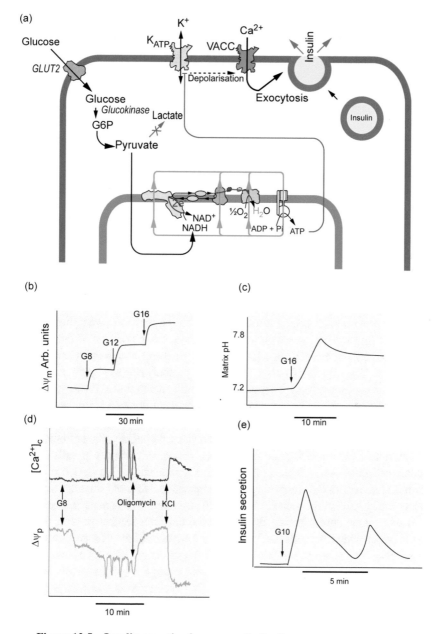

Figure 12.5   Insulin secretion by pancreatic β cells.
(a) Schematic of the basic mechanism for glucose-stimulated insulin
secretion. GLUT2, glucose transporter 2; K_{ATP}, ATP-sensitive K^+ channel;
VACC, voltage-activated Ca^{2+} channel. (b) Glucose-induced mitochondrial
hyperpolarisation in single β cells (from Ainscow and Rutter, 2002). (c)
Glucose-induced matrix alkalinisation in INS-1E cells (from Wiederkehr
et al., 2009). (d) Glucose-induced $[Ca^{2+}]_c$ and $\Delta\psi_p$ oscillations in INS-1
832/13 cells (from Goehring *et al.*, 2012). (e) Glucose-stimulated insulin
secretion from MIN-6 cells (from Shigeto *et al.*, 2006). G8, etc.: glucose
concentration (mM).

Apparent anomalies between $[Ca^{2+}]_c$ signals and insulin secretion have led to the concept of mitochondria-derived metabolic 'coupling factors' that modulate insulin secretion in parallel with the previously mentioned pathway. Pyruvate entering the mitochondrion is metabolised by both pyruvate dehydrogenase and pyruvate carboxylase. The latter pathway is anaplerotic; that is, it produces a net increase in citric acid cycle intermediates, thus providing metabolites for export to the cytoplasm, where they can be used for the net formation of the amino acids required for the synthesis of insulin and other proteins. Numerous investigations with primary β cells or insulin-secreting cell lines have identified metabolic factors that influence insulin secretion. The generation of cytoplasmic NADPH is a common feature of a number of metabolic cycles connecting the matrix and cytoplasmic pools of pyruvate, malate, citrate, or isocitrate that have been reported to enhance GSIS. Glutamate has also been proposed as a coupling factor. However, to date, no clear mechanisms have been advanced to explain these findings. Resolution will probably require primary bioenergetic effects to be eliminated, followed by an investigation of plasma membrane ion fluxes and potential changes, before individual coupling factors can be explained by bioenergetics, ion fluxes, or direct effect on exocytosis per se.

Fatty acids are major energy sources for β cells under basal glucose conditions, but they do not stimulate insulin secretion until glucose is elevated. The mechanism that must exist to prevent ATP generation linked to β-oxidation from swamping the glucose control of insulin secretion is not entirely clear. Thus, acute addition of fatty acids does not induce insulin secretion from islets in 3 mM glucose but enhances that evoked by 15 mM glucose. The effects of fatty acids may be partially due to signalling, in addition to bioenergetics, because they interact with plasma membrane GPR40 receptors that facilitate insulin secretion. Chronic exposure to elevated fatty acid is implicated in β cell failure, blunting the cells' response to glucose.

GSIS can be investigated at a variety of levels from the whole animal down. Those of most bioenergetic relevance include intact islets, dissociated primary β cells, and clonal insulinoma cell lines. As with most physiological investigations, the degree of molecular precision decreases with the complexity of the preparation. Islets of Langerhans are clusters of cells, typically ~500 in the mouse, where insulin-secreting β cells make up the core of the islet and account for approximately 60–80% of the total cell number. α cells, δ cells, and pancreatic polypeptide (PP) cells are found on the periphery of the islet and are responsible for the secretion of glucagon, somatostatin and PP respectively. Islet β cells are coupled to each other via connexin gap junctions, allowing them to respond synchronously to glucose.

Single dissociated β cells allow more detailed bioenergetic analysis, while clonal cell lines, apart from their ease of preparation, possess the plasma membrane monocarboxylate transporter and thus allow glycolysis to be bypassed by exogenous pyruvate. Interestingly, clonal cells metabolising exogenous pyruvate still show plasma membrane potential oscillations (Goehring et al., 2012), showing that these can be dissociated from glycolysis (Figure 12.5d).

12.5.2 Type 2 diabetes

Further reading: Morino *et al.* (2006), Szendroedi *et al.* (2012), Newsholme *et al.* (2012), Supale *et al.* (2012)

In the healthy individual, insulin secreted in response to a glucose load works to restore basal blood levels of the sugar by enhancing uptake, primarily into skeletal muscle, decreasing hepatic gluconeogenesis, inhibiting adipose tissue lipolysis, and facilitating a switch from fatty acid oxidation to glycolysis as a primary fuel by inhibiting glucagon secretion by the α cells of the pancreas. The coordinated actions of insulin and glucagon help to maintain plasma glucose concentrations in healthy individuals in a range from 3.6 to 5.8 mM. 'Metabolic syndrome' is defined as a combination of three or more of the following factors: abdominal obesity (visceral fat), elevated blood pressure, elevated triglycerides, low high-density lipoprotein cholesterol and hyperglycaemia. It currently affects approximately 30% of the U.S. population and is associated with insulin resistance, the failure of target tissues to respond to the hormone. Insulin resistance often progresses to full T2D with deficient glucose control of insulin secretion.

The molecular mechanisms underlying insulin resistance are being intensively investigated. In the liver, insulin resistance correlates with intracellular lipid accumulation and means that insulin fails to stimulate glycogen synthesis or to inhibit gluconeogenesis. There is strong evidence for a defect in insulin-stimulated glucose transport in skeletal muscle caused by fatty acid-induced failure of insulin receptor phosphorylation of its immediate target IRS-1 (see Figure 11.7) correlating with intramuscular triglyceride content and circulating plasma free fatty acids. A strong correlation is also found between insulin resistance and the amount of visceral fat. These deep fat depots not only store and release fatty acids but also are the source of hormones such as leptin, resistin and adiponectin that affect energy metabolism. Counterintuitively, plasma adiponectin levels decrease in obesity, and administration of the hormone decreases insulin resistance. A role of the hormone in the control of skeletal muscle mitochondrial biogenesis has recently been proposed (Iwabu et al., 2010), mediated by $[Ca^{2+}]_c$ elevation, AMPK activation and consequent phosphorylation and activation of PGC-1α. Activated AMPK can also increase carnitine palmitoyltransferase 1 (which may be limiting fatty acid entry into mitochondria) and thus facilitate fatty acid oxidation (McCarty, 2005).

Defective mitochondrial biogenesis in T2D is consistent with magnetic resonance scans (MRS) of skeletal muscle of lean insulin-resistant subjects. These show a decrease in mitochondrial ATP synthesis capacity together with a large increase in triglyceride content. A DNA microarray expression profile of 22,000 genes (Mootha et al., 2003), grouped into metabolically related sets, was used to compare expression levels in skeletal muscle biopsy samples from subjects with normal glucose tolerance and with T2D. The most consistent change was an average 20% decrease in the expression of a set of 106 'oxidative phosphorylation' genes. Similar results were seen with samples from prediabetic impaired-glucose tolerance patients with normal fasting glucose, showing that the change preceded hyperglycaemia. A subset of these genes appeared to be coregulated in three mouse tissues, skeletal muscle, heart and brown adipose tissue—the principal sites of insulin-controlled glucose uptake. PGC-1α was decreased by 20% in the diabetic muscle, and overexpression of the coactivator in mice strongly upregulated the same subset. Two genes that have been implicated are both transcription partners of PGC-1α.

Although changes in mitochondrial biogenesis in skeletal muscle and other tissues may correlate with insulin sensitivity, such as in aging, a causal linkage is far from evident. Acute elevations in circulating fatty acids at fasting insulin levels induce insulin resistance with no apparent effect on mitochondrial function. However, a combination of decreased oxidative capacity and increased circulating fatty acids may account for the intramuscular accumulation of triglycerides and lipid intermediates associated with insulin resistance and T2D.

Thiazolidinediones such as pioglitazone are activators of PPARγ, increase expression of PGC-1α and mitochondrial biogenesis in skeletal muscle, and were formerly used in the treatment of T2D, but have been withdrawn for safety reasons. Metformin, which is the most widely prescribed drug for T2D, inhibits hepatic gluconeogenesis and acts synergistically with insulin in restoring basal plasma glucose levels (Viollet *et al.*, 2012). Despite intensive research, its precise mechanism of action is still controversial. High concentrations of metformin *in vitro* weakly inhibit complex I and raise hepatocyte cytoplasmic AMP levels sufficiently to activate AMP kinase and inhibit gluconeogenesis. At first sight, it is not clear how this could occur *in vivo* without a potentially damaging systemic restriction of ATP-generating capacity and potential long-term toxicity. We shall later review the evidence that spare respiratory capacity is an important parameter for neuronal survival (Section 12.6.2) and that complex I inhibitors in the central nervous system can cause dopaminergic neuronal loss and hence the symptoms of Parkinson's disease (Section 12.6.3). Possible resolutions to this apparent paradox are selective uptake of metformin into the liver and the principle of mitohormesis (Section 12.9.3), whereby mild bioenergetic stresses act via AMPK and PGC-1α to increase mitochondrial biogenesis. However, whether AMPK is even the relevant target, direct or indirect, for metformin is also debated because the drug can inhibit gluconeogenesis in mice lacking liver AMPK (Foretz *et al.*, 2010).

12.5.2.1 β cell failure and T2D

Further reading: Prentki and Nolan (2006), Mulder and Ling (2009)

Insulin resistance in target tissues, leading to chronically elevated plasma glucose, means that the β cells will continuously release insulin, leading to sustained hyperinsulinaemia. Full T2D, however, only develops with the onset of β cell dysfunction, when the cells are no longer able to sustain the hypersecretion of insulin, despite an initial increase in β cell mass and upregulation of insulin gene expression. Chronically elevated circulating glucose and fatty acid levels both contribute to β cell dysfunction, hence the term glucolipotoxicity. Insulin secretion is also blunted by prolonged exposure of β cells to elevated fatty acid levels, although suggestions of fatty acid-induced mitochondrial uncoupling are not convincing (Ravnskjaer *et al.*, 2005).

Although some apoptosis is seen in T2D, a functional deficit is also involved. The C57BL/6J mouse strain becomes severely insulin resistant when fed a high-fat diet, and this allows the reaction of the β cells to this hyperglycaemic hyperinsulinaemic condition to be studied. The most dramatic change in the isolated islets is a switch from glucose to alternative mitochondrial substrates including palmitate and glutamine, with a consequent deregulation of GSIS. Interestingly, a substrain of these mice that show defective glucose regulation on a normal diet carry a mutation that inactivates the nicotinamide nucleotide transhydrogenase (Section 5.11). The resultant oxidative shift in the $NADP^+/NADPH$ redox potential affects insulin secretion, either by the redox consequences of impaired glutathione reduction or as a 'coupling factor.' It is clearly important to establish the precise strain used in any publication using C57BL/6J mice.

12.5.2.2 A role for UCP2?

Further reading: Zhang *et al.* (2001), Pi *et al.* (2009)

The functions of the 'novel uncoupling proteins' UCP2 and UCP3 remain contentious 15 years after their discovery. We have reviewed (Section 9.12.3) the evidence that these mitochondrial carriers may function by some mechanism other than by increasing membrane proton conductance. UCP2 knockout mice have been reported to be more efficient in secreting insulin and maintained lower blood glucose levels. The interpretation was predicated on the supposed protonophoric activity of UCP2, such that in wild-type mice the uncoupling pathway would lower the ATP/ADP ratio at a given glucose concentration and thus restrict GSIS. Inhibition or ablation of β cell UCP2 would thus improve insulin secretion as a means of treating T2D. In contrast, a subsequent study in which the mice were more exhaustively back-crossed into three strains showed an opposite effect of the knockout—decreased GSIS in the knockouts accompanied by an oxidative shift in glutathione redox potential and an upregulation of antioxidant pathways. However the controversy is finally resolved, it is important not to take as axiomatic that UCP2 acts as a protonophore in this and other systems.

12.6 MITOCHONDRIA AND THE BRAIN

The human brain accounts for only 2% of body weight but consumes almost 20% of total oxygen, largely to maintain the basal and stimulated ion gradients across the plasma membranes of the neurons and glia. Together with the heart, the brain has the most stringent requirement for unimpaired oxidative phosphorylation, and possible mitochondrial bioenergetic deficiencies are being investigated in each of the major neurodegenerative diseases. The technical problems are considerable: the 'gold standard' is the noninvasive investigation of altered metabolism in human subjects by functional magnetic resonance imaging (fMRI) to monitor blood flow or by PET to measure glucose metabolism using ^{18}F-deoxyglucose. Animal models, of course, allow much more intervention; however, with the exception of stroke and traumatic brain injury, it is difficult to generate animal models that accurately reproduce all aspects of the human disease. Transgenic mice rely on the incorporation of mutations associated with the relatively rare examples of familial Alzheimer's and Parkinson's diseases, and even where the underlying mutation is firmly established, as in the case of polyQ diseases such as Huntington's disease, there is a variety of genetic models with differing degrees of expression of the human disease phenotype. One problem is that the human diseases are age-related and develop over several years, whereas the life span of a mouse requires that this time course is compressed into a matter of months.

Primary neural cells can provide valuable information, with the limitation that the former must be prepared from neonatal animals, whereas the human diseases are associated with aging. Cell lines are extensively employed, but it must be borne in mind that these are transformed cells with altered dependencies on oxidative phosphorylation and glycolysis. Isolated brain mitochondrial preparations suffer from extreme heterogeneity even if prepared from specific brain regions; in addition to the multiplicity of neuronal subtypes, the location within the neuron (soma, terminals, and dendrites) may affect function (see Figure 10.4), while mitochondria from neurons and glia are known to have different properties.

12.6.1 Neurodegeneration

Further reading: Karbowski and Neutzner (2012), Correia *et al.* (2012), Martin (2012)

In the following sections, we review some of the current evidence linking mitochondrial dysfunction to the major neurodegenerative disorders. As the range of factors known to control normal mitochondrial function increases, hypotheses as to possible malfunctions expand in parallel. Thus, in addition to 'classic' targets such as mutated mtDNA, insufficient ATP generation, excess Ca^{2+} accumulation, or damaging ROS generation, novel targets such as fission and fusion, axonal trafficking, biogenesis, protein import, and assembly are all being tested as possible neurotoxic loci. Much remains to be elucidated; in particular, the factors that underlie the tissue specificities of individual neurodegenerative diseases remain largely obscure.

12.6.2 Mitochondria, stroke and glutamate excitotoxicity

Further reading: Stavrovskaya and Kristal (2005), Nicholls (2008a), Perez-Pinzon *et al.* (2012)

Approximately 80% of the nerve terminals in the mammalian brain utilise glutamate as neurotransmitter. Before this discovery, neurotransmitters appeared to be distinct molecules (acetylcholine, dopamine, GABA, etc.) that had evolved to perform just this role. That such a ubiquitous amino acid can perform such a specialised role is dependent on its precisely controlled compartmentation, and this in turn is critically dependent on mitochondrial function. Glutamate in the neuronal cytoplasm is in the range 1–10 mM; this contrasts with approximately 1 μM in the extracellular medium. This large concentration gradient is maintained by plasma membrane excitatory amino acid transporters (EAATs) on both neurons and surrounding glia. EAATs are driven by the Na^+-electrochemical gradient ($\Delta\tilde{\mu}_{Na^+}$) and are therefore dependent on cytoplasmic ATP supply to the Na^+/K^+-ATPase. ATP is also required for the synaptic vesicle vesicular (V) ATPase to further accumulate glutamate into the lumen of the vesicles.

A stroke (focal ischaemia) deprives a brain region of oxygen, immediately inhibiting electron transport in neurons and glia. The capacity of glycolysis to compensate for the loss of oxidative phosphorylation is severely limited and cytoplasmic ATP levels rapidly collapse. The failing Na^+/K^+-ATPase cannot maintain $\Delta\tilde{\mu}_{Na^+}$, and there is a catastrophic release of glutamate that diffuses out from the ischaemic core. Glutamate is the agonist for a variety of ion channel and G protein-coupled receptors that have evolved to respond to brief (millisecond) exposures to the neurotransmitter. One group of these, N-methyl-D-aspartate (NMDA) receptors, play a central role in learning and memory, but their properties, allowing both Na^+ and Ca^{2+} into cells and remaining active for as long as they are exposed to glutamate, pose a threat to the survival of neurons in the partial oxygenated 'penumbra' surrounding the ischaemic core.

The bioenergetic consequences of chronic NMDA receptor activation are summarised in Figure 12.6. Although NMDA receptors typically have a 10:1 selectivity for Ca^{2+} over Na^+, the concentration of the latter ion in the medium is 100 times greater, with the result that Na^+ entry into the cytoplasm exceeds that of Ca^{2+}. The deleterious effects of the Na^+ entry are twofold. First, it imposes an enormous energy demand on the mitochondria, which must supply ATP to drive the fully activated Na^+/K^+-ATPase as it attempts to extrude the Na^+ from the cytoplasm. Second, the net uptake of Na^+ is balanced by a net loss of K^+, which can lead to spreading depression (discussed later). At the same time, Ca^{2+} entry through the NMDA receptor rapidly raises $[Ca^{2+}]_c$ above the set point of approximately $0.5\,\mu M$, at which mitochondria become net accumulators of the cation (Section 9.4.1). Ca^{2+} uptake into the matrix imposes an additional demand on the proton circuit. Finally, if the Ca^{2+} loading of the matrix exceeds the capacity of the mitochondrion, the permeability transition pore mPTP is activated, $\Delta\psi_m$ collapses, and the neuron initiates a necrotic death.

Oxidative stress was previously thought to be an early event in glutamate excitotoxic neuronal cell death. However, the concentration of $O_2^{\cdot-}$ only rises after $\Delta\psi_m$ collapses as a consequence of Ca^{2+} overload and the initiation of necrotic cell death. It is apparent that a critical factor in this life-or-death struggle of the neuron is its 'spare respiratory capacity' to supply the proton current required by these processes. In acute cell culture experiments, any factor that decreases the maximal capacity of the mitochondria to generate ATP (partial electron transport inhibition, oxidative damage to the ANT, or mild uncoupling) potentiates the lethality of parallel pathological glutamate exposure. In contrast, a subtle uncoupler-induced increase in C_mH^+ in the absence of glutamate can initiate a delayed protective preconditioning by activating AMPK (Weisova et al., 2012). We previously discussed this concept of hormesis, whereby an intermittent nonlethal stress can induce pathways that protect the cell against a subsequent more extreme stress, in the context of cardiac ischaemia/reperfusion (Section 12.3.2). In these neuronal studies, AMPK activation was accompanied by an increase in GLUT3 activity and an increased $\Delta\psi_m$ and cellular ATP following washout of the protonophore (Figure 12.7).

12.6.2.1 PARP and NAD⁺ depletion

Further reading: Moroni (2008), Alano et al. (2010), Bai and Canto (2012)

PARPs (poly(ADP-ribose) polymerases) catalyse the transfer of ADP ribose units from NAD^+ to protein amino acid side chains, generating poly(ADP ribose) polymers. In the nucleus, they play a role in gene expression, cell cycle control and genomic stability. Seventeen human PARPs are currently known, with the ubiquitously expressed PARP-1 being the most abundant. The predominantly nuclear PARP-1 possesses a DNA binding domain that can recognise DNA strand breaks, and interaction with these can increase

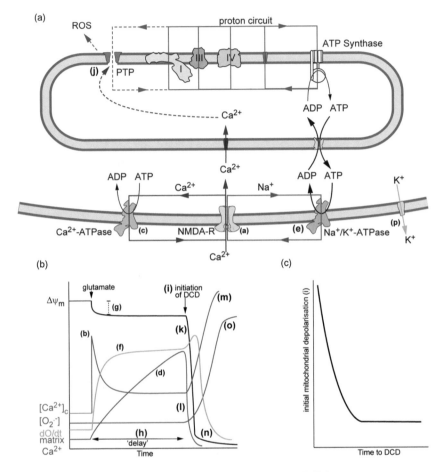

Figure 12.6   Bioenergetic consequences of chronic NMDA receptor activation: excitotoxicity.

(A and B) NMDA receptor activation (a) results in the entry of Na^+ and Ca^{2+} into the cell. $[Ca^{2+}]_c$ spikes (b) and then partially recovers as the receptor partially desensitises and the Ca^{2+}-ATPase (c) is activated to expel Ca^{2+}. $[Ca^{2+}]_c$ plateaus above the mitochondrial set point (Section 9.4.1) and so there is a net accumulation of Ca^{2+} into the matrix (d). The Na^+/K^+-ATPase (e) is activated to expel the Na^+ entering the cytoplasm, and the energy demand of the two ATPases and the matrix Ca^{2+} uptake results in a large increase in respiration (f) and a slight mitochondrial depolarisation (g). After a stochastic delay (h), the combination of matrix Ca^{2+} accumulation and ATP insufficiency can lead to delayed Ca^{2+} deregulation (DCD) (i) and activation of the PTP (j): $\Delta\psi_m$ collapses (k); Ca^{2+} is released back into the cytoplasm (l) and $[Ca^{2+}]_c$ increases (m); respiration fails due to the release of cytochrome c (n) and a large increase in the level of reactive oxygen species (o) occurs as a downstream consequence of PTP opening and DCD. Accumulation of Na^+ in the cytoplasm is balanced by a loss of K^+ (p), which can raise the external K^+ concentration and contribute to spreading depression. (C) There is an inverse relationship between the survival time of neurons prior to DCD and the extent of the initial $\Delta\psi_m$ depolarisation. *Data from Ward et al. (2000).*

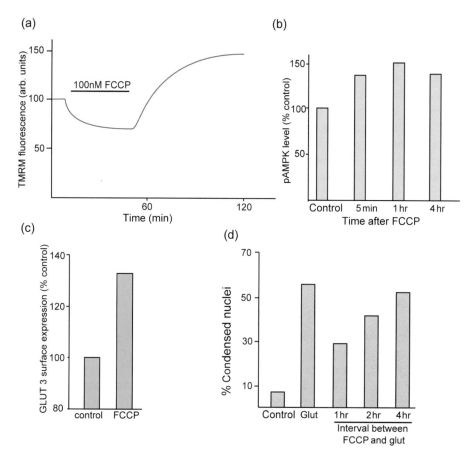

Figure 12.7   AMPK mediates neuronal preconditioning induced by mild uncoupling.
(a) Cerebellar granule neurons were briefly exposed to low (100 nM) FCCP. TMRM (non-quench mode) monitored $\Delta\psi_m$. Note that the *in situ* mitochondria hyperpolarise after removal of the protonophore. (b) 100 nM FCCP for 30 min increases levels of phosphorylated (active) AMPK; the effect persists for at least 4 h. (c) Increased surface expression of the glucose transporter GLUT3 1 h after the same FCCP pretreatment. (d) Cell death 24 h after 10-min exposure to glutamate/glycine to induce excitotoxicity; prior mild uncoupling is protective.
Data from Weisova et al. (2012).

basal PARP-1 activity by 500-fold. PARP-1 is pathologically overactivated in multiple brain cell types following ischaemia, and this seems to contribute to cell death in the ischaemic penumbra because knockout of the protein or addition of PARP inhibitors reduces the infarct volume in stroke models.

Although there are multiple hypotheses concerning the toxic effects of PARP-1 activation in stroke, one of the most convincing is that the massive overactivation of this enzyme depletes cytoplasmic NAD^+ to a level where bioenergetic function is impaired. Within minutes of DNA damage, the cytoplasmic $NAD^+ + NADH$ pool can be decreased to 10–20% of normal, with severe consequences for glycolysis. There is controversy regarding whether the bioenergetic failure accompanying PARP-1 activation can be purely ascribed to NAD^+ depletion or whether PARP-1 has direct effects on mitochondria, perhaps as a consequence of an effect on some matrix-located enzyme. Depletion of NAD^+ in neurons transfected with another NAD^+-consuming enzyme, NAD^+ glycohydrolase, reproduces the phenotype. Exogenous pyruvate can still be utilised by the mitochondria in these cells and prevents the bioenergetic collapse, showing that matrix $NAD^+ + NADH$ content is retained. Surprisingly, the addition of high NAD^+ concentrations to the medium results in uptake of the nucleotide into the cytoplasm of cultured neurons, preventing or reversing the bioenergetic deficits.

Although we have reservations about normal physiological fluctuations in cytoplasmic NAD^+ to regulate SIRT1 (Section 11.8), the profound depletion induced by PARP-1 activation does inhibit the deacetylase. Notably, no effect is seen on the mitochondrial located SIRT3.

12.6.2.2 Spreading depression

Further reading: Zhou *et al.* (2010), Carlson *et al.* (2012)

The 'penumbra' surrounding a focal infarct into which excitotoxic glutamate diffuses is a few millimetres wide regardless of brain size; thus, while protecting this zone produces an improved outcome in rodent models, only a limited effect would be predicted in the human brain. In contrast, 'spreading depression' (SD), characterised by neuronal swelling, collapse of $\Delta\psi_p$ and silencing of brain activity, can spread in the human brain at several millimetres per minute following brain haemorrhage, traumatic brain injury, stroke, or transient global ischaemia (heart attack). Because of practical difficulties in detecting SD in human patients using electroencephalograms, it has only recently been recognised that it is common following brain damage. The ionic and bioenergetic changes associated with SD have many of the characteristics of glutamate excitotoxicity, including massive glutamate release, $[Ca^{2+}]_c$ elevation and a huge energy demand that can lead to bioenergetic insufficiency, mitochondrial depolarisation and activation of the permeability transition. An unusual feature of NMDA receptors is that extracellular Mg^{2+} blocks channel activity, even in the presence of glutamate, unless $\Delta\psi_p$ is depolarised. In SD, K^+ ions, released in proportion to the Na^+ ions accumulated into the cytoplasm, may diffuse to adjacent neurons, depolarising them and, in combination with glutamate, initiating a chain reaction of spreading depression. Indeed, SD can be initiated *in vivo* and in slice models by the focal addition of KCl. Thus, the lethal combination of K^+ and glutamate release may initiate a self-propagating wave of neuronal failure far beyond the original infarct or cell damage.

12.6.3 Mitochondria and Parkinson's disease

Further reading: Winklhofer and Haass (2010), Pilsl and Winklhofer (2011), Swerdlow (2011), Rochet *et al.* (2012), Exner *et al.* (2012), Hauser and Hastings (2012)

With a lifetime risk of developing the disease of approximately 1 in 40, Parkinson's disease (PD) is second only to Alzheimer's disease among age-related neurodegenerative

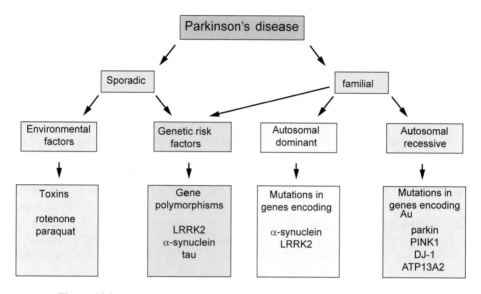

Figure 12.8
Multiple factors affect susceptibility to Parkinson's disease.
After Pilsl and Winklhofer (2011).

disorders. The main motor disorders, slowing of movement (bradykinesia) rigidity and tremor, can be ascribed to the loss of dopaminergic neurons with their cell bodies in the substantia nigra pars compacta and projecting into the striatum. However, other pathways can also be affected, leading to cognitive impairment. Pathology reveals protein aggregates (Lewy bodies) in surviving cell bodies and processes, enriched with aggregated forms of the presynaptic protein α-synuclein. The majority of PD cases are sporadic, although the contribution of genetic and environmental factors is still unclear (Figure 12.8). Approximately 10% of PD is familial and currently associated with mutations in five or six *PARK* genes, coding for proteins (α-synuclein, parkin, PINK1, DJ-1, LRRK2, and perhaps ATP13A2) that are either mitochondria-located or capable of association with mitochondria.

Mitochondrial dysfunction has played a central role in PD research since the notorious cases in the 1970s and 1980s when an illegally produced meperidine analogue contaminated with 1-methyl-4-phenyl-1,2,3,6-tetrahydropyridine (MPTP) produced symptoms closely analogous to PD. MPTP crosses the blood–brain barrier and is oxidised to MPP$^+$ (1-methyl-4-phenyl-pyridinium) by glial monoamine oxidase B (MAO-B). MPP$^+$ is a substrate for the dopamine (DA) transporter on the plasma membrane of DA neurons. As a hydrophobic cation, it is further accumulated into the mitochondrial matrix, in much the same way as $\Delta\psi_m$ indicators such as TMRM, where it inhibits complex I sufficiently to restrict respiratory capacity and enhance ROS generation, although the relative importance of these two parameters in subsequent cell death is debated. Note that in inverted submitochondrial particles, MPP$^+$ is a weak inhibitor of complex I, with a K_i of approximately 10^{-4} M. However, the combined concentrative capacity of the plasma membrane transporter and the mitochondrial membrane potential in intact cells implies that nanomolar concentrations of extracellular MPP$^+$ are toxic *in vivo*.

Because animals do not get PD, experimental models must be constructed that attempt to reproduce relevant aspects of the human disease (Terzioglu and Galter, 2008). In addition to MPP$^+$, 6-OH-dopamine is specifically accumulated into, and destroys, DA neurons. Systemic nonselective toxins (e.g., rotenone and paraquat) attempt to exploit a differential susceptibility of DA neurons. Genetic modifications in *PARK* genes can include the generation of mice with null or point mutations to model a loss of function in PINK1, parkin, or DJ-1. More selective genetic manipulation can be performed with site-specific recombinase technology. Cre recombinase expression directed by the DA transporter promoter can be used to introduce mutations specifically into DA neurons.

Clearly, a model in which DA neurons have already been destroyed is of limited interest for a bioenergetic study. The more subtle modifications each test a hypothesis—for example, that PINK1 impacts on mitophagy (Section 10.6). If it is accepted that DA mitochondrial bioenergetic impairment (by whatever mechanism) is causal in the pathology, then subtle DA neuron-selective restriction of mitochondrial function can be informative. The MitoPark mouse (Galter *et al.*, 2009) has been engineered for the specific deletion of the mitochondrial transcription factor Tfam in midbrain DA neurons. The mice are normal until adolescence, but then they develop progressive motor disabilities resembling human PD and responsive to L-DOPA. The slow progression of symptoms allows the model to be used to test potential therapies directed at slowing the progression of the human disease. The mouse adds to the body of evidence that a decline in mitochondrial function in the DA neurons is strongly implicated in human PD, but it does not answer the central question regarding the nature of the factors that induce the decline in function in the first place.

12.6.3.1 Mitochondrial dysfunction and sporadic PD

Further reading: Martinez and Greenamyre (2011, 2012)

In the MPP$^+$ model, DA neurons are selectively vulnerable by virtue of their plasma membrane DA transporter. However, careful systemic titration of rats with the complex I inhibitor rotenone can reproduce some of the pathology of PD. This suggests that DA neurons may be inherently more susceptible to bioenergetic compromise, although a cell respiratory control study of mouse dopaminergic and nondopaminergic synaptosomes revealed no significant difference in basal, oligomycin-insensitive, or maximal respiratory rates or sensitivity to stress (Choi *et al.*, 2011).

Postmortem investigations of human PD brains revealed an approximately 30% deficiency in complex I activity in the substantia nigra. No other brain region was affected, and the result could not be ascribed to the patients' long-term treatment with L-DOPA because treated non-parkinsonian controls showed no deficit. Some complex I deficit can be detected in PD patients' platelets, but results are too variable to allow this to be used as a biomarker. Cybrid technology (Section 12.2.2) has been used to remove the background variability and to establish whether mtDNA variation might play a role in PD. However, although there have been several reports of complex I deficiency being retained in cybrids from PD platelets (see Swerdlow, 2011), no consistent mutation has been identified and this remains a controversial field.

12.6.3.2 Familial PD: parkin and PINK1

Further reading: Chu (2010), Deas *et al.* (2011)

The normal roles of two proteins, the E3 ubiquitin ligase parkin and the ubiquitously expressed protein kinase PINK1, encoded by *PARK* genes, have already been discussed in the context of mitochondrial fission and mitophagy (Section 10.3.2). Approximately 30 pathogenic mutations in PINK1 have been described that impair its kinase activity and abrogate its neuroprotective effect. A spectrum of mitochondrial dysfunction has been observed in isolated mitochondria and cell culture (decreased $\Delta\psi_m$, complex I and IV activities, ATP production, increased ROS, etc.) as well as impaired axonal transport. PINK1 knockout mice show a 10–15% decrease in state 3 respiratory capacity of their isolated striatal mitochondria (Gautier *et al.*, 2008), while PINK1-deficient mouse embryonic fibroblasts have slight electron transport chain defects (Amo *et al.*, 2011). Despite this, the mice have a mild phenotype with no gross morphological changes. This is in contrast to severe defects found in corresponding *Drosophila* models.

More than 100 different mutations have been described in parkin associated with juvenile-onset autosomal recessive PD. Wild-type parkin has a neuroprotective capacity that can be enhanced by overexpression and is associated with proteosome-independent ubiquitination. However, parkin knockout mice show few signs of dopaminergic degeneration. Interestingly, it has been suggested that parkin misfolding may be associated with some apparently sporadic forms of PD. The involvement of wild-type PINK1 and parkin in mitophagy (Section 10.6) has led to the suggestion that defective removal of damaged mitochondria in the mutants may underlie the bioenergetic dysfunction. However, it is possible that parkin plays separate roles in mitophagy and neuroprotection. Thus, a new target for parkin-mediated ubiquitination has been recently proposed—PARIS, a transcriptional repressor of PGC-1α expression (Shin *et al.*, 2011). Parkin knockout leads to overexpression of PARIS associated with repression of PGC-1α and loss of DA neurons.

12.6.3.3 α-Synuclein, DJ-1 and LRRK2

Further reading: Rochet *et al.* (2012), Wang *et al.* (2012), Tsika and Moore (2012), Exner *et al.* (2012)

Mutations in the genes for LRRK2 (leucine-rich repeat kinase 2) and α-synuclein are linked to autosomal dominant forms of PD. For these gain-of-function mutations, transgenic models can be constructed by insertion of extra copies into the genome or by viral gene delivery. The physiological function of the presynaptic α-synuclein, which exists in multiple isoforms with 98–140 residues, is imperfectly understood, but it may play a role in the control of synaptic vesicle exocytosis or recycling. Point mutations and mutations that increase the level of the wild-type protein are associated with early onset familial PD. Because Lewy bodies contain fibrillar α-synuclein, one hypothesis is that the mutations enhance formation of these neurotoxic aggregates. Several studies report mitochondrial dysfunction associated with α-synuclein overexpression. There is also evidence that α-synuclein can be transmitted from cell to cell, contributing to the spread of the pathology. Molecular chaperones such as Hsp70 and DJ-1 may limit α-synuclein aggregation or the toxicity of the aggregates.

DJ-1 is a small, ubiquitously expressed, 20-kDa protein that localises to the mito-chondrion under conditions of oxidative stress. Mutations in DJ-1 cause a rare early onset autosomal recessive form of PD. Wild-type DJ-1 protects cells from oxidative stress and favours mitochondrial elongation. In contrast, mutated DJ-1 or wild-type DJ-1 deficiency increases cell vulnerability to oxidative stress and facilitates mitochon-drial fragmentation. This suggests that wild-type and mutant DJ-1 may interact in dif-ferent ways with the fission/fusion machinery. Expression of the fission protein Drp-1 (Section 10.3.2) is decreased in cells overexpressing wild-type, but not mutant, DJ-1. The protein possesses a redox-sensitive Cys residue that can be oxidised to a sulfinic acid under conditions of oxidative stress, and it has been proposed that wild-type DJ-1 functions as a redox sensor and may be responsible for the increased mitochondrial fragmentation that occurs under conditions of oxidative stress.

LRRK2 contains both protein kinase and GTPase domains, but apart from auto-phosphorylation, the physiological targets are still not established. Overexpression of mutated LRRK2 decreases neurite length and branching in primary neuronal cultures, and it has been proposed that the kinase activity of LRRK2 regulates aspects of microtu-bule assembly and cytoskeletal organisation. LRRK2 can also associate with the OMM, and mice bearing the mutated kinase show an altered mitochondrial morphology in DA neurons, suggesting a defect in mitophagy. LRRK2 is ubiquitously expressed, and fibro-blasts from patients with mutated LRRK2 have been analysed.

12.6.4 Mitochondria and Huntington's disease

Further reading: Fan and Raymond (2007), Gil and Rego (2008), Damiano *et al.* (2010)

Huntington's disease (HD) is an incurable autosomal dominant neurodegenerative disor-der caused by expansion of a CAG repeat in the ubiquitously expressed huntingtin (htt) gene (*Hdh*), leading to an expanded polyglutamine sequence in the protein. The normal function of htt is not known, although homozygous knockout of the gene is embryonic lethal. CAG repeats beyond 36 are associated with the human disease, whose symptoms include uncontrolled gait, psychiatric disturbances and dementia. HD is characterised by a massive loss of primarily GABAergic neurons in the striatum.

In vivo PET and MRS imaging of human patients are consistent with impaired brain metabolism (Saft *et al.*, 2005). *In vivo* ^{31}P-MRS of skeletal muscle of HD patients showed slowed recovery of phosphocreatine after aerobic exercise (Section 9.9). Deficits can be seen in subjects with asymptomatic mutations, suggesting that an energy deficit could be an early causative event in the disease. A possible upstream mitochon-drial involvement has been suggested by the finding that basal ganglia (which include the striatum) are particularly susceptible to partial complex II inhibition by the irrevers-ible inhibitor 3-nitropropionic acid (3-NPA) or the competitive inhibitor malonate. The resulting chronic energetic limitation and pathology reproduces some of the symptoms characteristic of HD in humans, primates and rodents.

A variety of animal and cell culture models have been engineered to express mutant htt (Menalled and Chesselet, 2002). Transgenic mouse models, in which all or part of the mutant gene is randomly inserted into the genome, express mutant htt in addition to the wild-type protein. R6/2 mice express exon 1 of human htt containing long, approximately 150 CAG repeats; this produces a severe phenotype. Transgenic models expressing the entire gene, located on either a bacterial (BACHD) or yeast (YAC128) artificial chromosome, have a milder phenotype. Knockin mice, in which the mutation is introduced into the mouse native *Hdh* gene, express htt under the control of its natural promoter and can be either homozygous or heterozygous for the mutation. 150 CAG repeat knockin mice show mild, late-onset phenotypes and limited neuronal death.

The reader of the primary literature will be excused for finding the literature on possible bioenergetic changes in mitochondria from these diverse models confusing and frequently contradictory. The inherent heterogeneity of isolated brain mitochondrial preparations, even when prepared from defined brain regions of mutant mice, limits their usefulness. Problems are compounded by the technical difficulties in preparing minute amounts of undamaged mitochondria from mouse striata. Indeed, one rarely sees data on classic respiratory control ratios with such preparations.

Although 3-NPA can model the downstream consequences of restricted complex II activity, the question regarding whether HD is associated with a selective decrease in activity of the complex remains controversial. There is controversial literature reporting that mutant htt (full length or exon 1) reduces the capacity of isolated mitochondria to accumulate Ca^{2+} before the mPTP is activated. Experiments consist of the preparation of mitochondria from the mutant mice, the direct addition of recombinant htt to normal mitochondria, or, in one case, the preparation of mitochondria from lymphoblast cell lines that retain patient mutant htt. The quality of the data varies, and a critical analysis must ask how closely these experiments might reproduce the conditions pertaining *in vivo* within striatal neurons.

A powerful test of the hypothesis that defective mitochondrial Ca^{2+} handling contributes to the pathology of HD was performed with mice crossed between R6/2 and cyclophilin D (CypD) knockout (Perry *et al.*, 2010). CypD is a component of the mitochondrial permeability transition pore, and so its deletion greatly enhances the capacity of mitochondria to accumulate Ca^{2+} prior to mPTP activation (Figure 12.9c). The prediction would be that R6/2:CypD$^{-/-}$ mice would show resistance to the pathology generated by the presence of the 2765Q exon 1 of htt. However, no behavioural or pathological differences were detected, leading to the conclusion that enhancing mitochondrial Ca^{2+} buffering capacity is not beneficial in this model.

The slow stochastic loss of striatal neurons in HD raises the possibility that glutamate excitotoxicity (Section 12.6.2) may be implicated. There is compelling evidence for this: NMDA receptor agonists injected into the striatum reproduce the pattern and selectivity of neurodegeneration seen in HD, and 3-NPA increases toxicity, consistent with the spare respiratory capacity concept, which implies that any factor that decreases neuronal ATP-generating capacity will increase the susceptibility to excitotoxic death. The suggestive evidence from 3-NPA and malonate models that HD neurons might display defective bioenergetics was tested in a cell respiratory control study (Oliveira *et al.*, 2007) comparing primary striatal neurons from wild-type and Hd$^{150/+}$ heterozygous knockin mice (Figure 12.10a). Both preparations showed a remarkably high spare

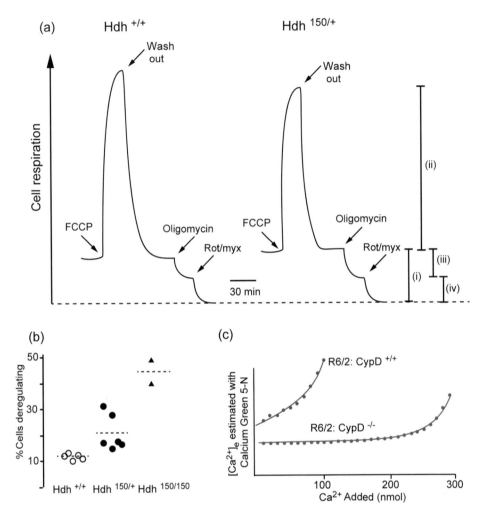

Figure 12.9   Cell respiratory control and mitochondrial Ca^{2+} handling in genetic models of Huntington's disease.

(a) Neurons cultured from mice heterozygous for 150 repeat polyQ huntingtin (Hdh$^{150/+}$) and wild-type littermates (Hdh$^{+/+}$) were incubated in the cell respirometer (Figure 9.6). Basal respiration (i), spare respiratory capacity (ii), ATP turnover (iii), and proton leak (iv) were estimated. The heterozygotes retain robust mitochondrial bioenergetic function. (b) Despite this, neurons from Hdh$^{150/+}$ and homozygous Hdh$^{150/150}$ mice show enhanced delayed Ca^{2+} deregulation (the failure to restore basal [Ca^{2+}]$_c$ after brief exposure to glutamate). Data from Oliveira *et al.* (2007). (c) Isolated cortical mitochondria from R6/s HD model mice with cyclophilin D (CypD) knockout show greatly enhanced Ca^{2+} accumulation capacity, judged by their ability to restore a low [Ca^{2+}]$_e$ (monitored by the low-affinity extramitochondrial indicator Calcium Green-5N). Despite this, they show no behavioural or pathological rescue (see text). *Adapted from Perry et al. (2010).*

respiratory capacity during brief exposure to FCCP; thus, at least under these relatively mild conditions (neonatal, heterozygous 150Q knockin mice), no dramatic bioenergetic phenotype is seen. However, electrophysiological alterations in NMDA receptor activity have been observed in HD models, and the recovery of $[Ca^{2+}]_c$ to basal levels after glutamate exposure is slowed. Delayed Ca^{2+} deregulation, which initiates irreversible excitotoxicity (Section 12.6.2), is caused by a combination of mitochondrial Ca^{2+} overload and energetic insufficiency. Apparent contradictions between mitochondrial Ca^{2+} overload in excitotoxicity and the finding that CypD knockout does not protect R6/2 mice need to be resolved.

In parallel with advances in the understanding of factors controlling mitochondrial cell biology, investigation into mitochondrial aspects of HD has expanded to test these novel loci. The fusion/fission cycle, mitochondrial axonal trafficking and PGC-1α control of biogenesis are all being actively investigated (Chaturvedi *et al.*, 2009).

12.6.5 Friedreich's ataxia

Further reading: Santos *et al.* (2010), Vaubel and Isaya (2012)

Friedreich's ataxia is an autosomal recessive disease affecting the central and peripheral nervous system and the heart. It is caused by a mutation in the gene for frataxin, a widely expressed mitochondrial matrix protein that is required for Fe–S centre synthesis, although its precise role in Fe transport, storage, or Fe–S assembly is not fully understood. The mutation most commonly found is the presence of GAA repeat expansions in an intron, resulting in decreased expression of the protein. The consequence of this is multiple deficiencies in Fe–S centres in mitochondria and throughout the cell. Dysfunctional frataxin results in excessive accumulation of iron and its carrier ferritin in the mitochondrial matrix and cytoplasmic iron depletion. The former may expose the mitochondrion to dangerous Fenton chemistry that can lead to hydroxyl radical formation (see Figure 9.12), although the contribution of oxidative stress to the pathology of Friedreich's ataxia is currently debated. There is currently no effective therapy, although the free radical scavenger idebenone is beneficial.

12.6.6 Mitochondria and Alzheimer's disease

Further reading: Rapoport (2003), Diana *et al.* (2011), Eckert *et al.* (2011)

Alzheimer's disease (AD) is the most common neurodegenerative disorder. It is characterised by memory loss, dementia and a pathology consisting of extracellular plaques containing β-amyloid peptide and intracellular neurofibrillary tangles containing phosphorylated tau protein. As with other neurodegenerative diseases, the 'gold standard' for investigation is the functional imaging of human patients. Early bioenergetic effects, preceding psychological or anatomical changes, have been monitored *in vivo* using PET and fMRI, and these reveal a decrease in cerebral blood flow and glucose uptake consistent with a decreased metabolic demand, although in the early stages brain metabolism can be activated almost normally by a recognition task. The decrease in glucose utilisation in AD patients is to some extent compensated by an increase in the utilisation of alternative fuels, notably ketone bodies. Thus, a ratio of glucose to alternative fuel that is 29:1 in age controls can be reduced to 2:1 in patients with incipient AD.

Note that a mitochondrial bioenergetic defect at this stage with retained energy demands might be expected to result in an increase in compensatory glycolysis and hence raised glucose utilisation. AD is accompanied by a progressive loss of synaptic contacts, detected by the loss of synaptophysin, and it is clearly important to establish whether this is a cause or an effect of the decreased energy demand. The former is suggested by a comparable reduction in expression of bioenergetic markers in primate models of sensory deprivation. Note that a decrease in bioenergetic capacity following synapse loss could result in insufficient capacity to deal with the energetic demands of subsequent glutamate excitotoxicity.

Postmortem studies from the mid temporal cortex of AD brain show a 50% reduction in the levels of mtRNA for complex I and IV subunits, with no change in that for β-actin. The motor cortex, which is relatively spared in AD, was not affected. At the protein level, several independent studies have reported modest reductions in cytochrome oxidase activity. Of course, apart from the debris (plaques and tangles), postmortem studies only allow analysis of the neurons that survived until death, rather than those that died during the progression of the disease.

Although there is an association between mitochondrial dysfunction and plaque and tangle formation, it is still not clear which is the upstream cause; that is, does mitochondrial dysfunction cause the pathology, or does the pathology result in mitochondrial dysfunction? Both schools of thought are currently represented in the literature and are briefly reviewed here. However, the two are not mutually exclusive because positive feedback between bioenergetics and pathology is likely. Finally, any hypothesis must ultimately explain not only why AD, in common with most neurodegenerative disorders, is age related but also why it takes years to progress and why the majority in old age still manage to escape the disease.

12.6.6.1 β-Amyloid effects on mitochondria

Further reading: Du *et al.* (2010), Eckert *et al.* (2011), Pagani and Eckert (2011)

Amyloid precursor protein (APP) is an integral glycoprotein ubiquitously expressed in the plasma membrane; its normal function is unclear. As with other integral plasma membrane proteins, APP is synthesised in the endoplasmic reticulum, trafficked through the Golgi to the cell surface, and undergoes endosomal retrieval. APP can be cleaved by three secretases—α, β and γ. Sequential cleavage by β- and γ-secretases liberates β-amyloid peptides of varying length, particularly $A\beta_{40}$ and $A\beta_{42}$. The catalytic subunit of γ-secretase is the aspartyl protease presenilin, mutations in which are associated with one form of familial AD. Depending on the location of processing, Aβ can be liberated either inside or outside the cell. Although the role of the extracellular insoluble plaques has been most intensively investigated, there are indications that soluble cytoplasmic Aβ (particularly $A\beta_{42}$) may interact directly with mitochondria, perhaps being imported via the TOM and TIM complexes. Although we do not expect that all the following targets will prove to be relevant to the human disease, Aβ *in vitro* has been shown to impact on mitochondrial function by inhibiting enzymes of the tricarboxylic acid (TCA) cycle, complex IV, the outer membrane fission protein Fis1 (enhancing mitochondrial fission), cyclophilin D (facilitating the permeability transition), and ABAD (amyloid-binding alcohol dehydrogenase). Increased mitochondrial ROS has also been proposed. Total and mitochondria-associated $A\beta_{40}$ and $A\beta_{42}$ levels increase with age and are further enhanced in transgenic mice expressing the APP Swedish mutation (K595N/M596L).

As with other neurodegenerative disorders, a major challenge is to create an animal model that reproduces the pathology and development of the disease, ideally by targeting the critical upstream initiating sites. Current models have been developed by reproducing mutations responsible for rare inherited forms of the human disease. Although detailed description of these constructs is beyond the scope of this book, mutations in amyloid precursor (APP), presenilin and tau genes have been explored separately and in combination.

One way to disentangle cause and effect is to study the temporal relationships between changes in mitochondrial function, intraneuronal and extracellular accumulation of β-amyloid, neurofibrillary tangles, cell death and cognitive changes. Proteomic studies based on the level of expression of mitochondrial proteins are complicated because although a decrease may indicate a restricted bioenergetic capacity, an increased expression can be interpreted in a similar manner as a compensatory 'hormetic' response to a bioenergetic defect, as occurs in MERFF (Section 12.2.1). Also, values related to mitochondrial protein will obscure changes in total mitochondrial content.

Triple transgenic mice (3xTg-AD) incorporating a knockin mutated presenilin 1 gene (M146V) and transgene constructs for the human APP Swedish mutation and human tau (P301L) have become popular models, showing progressive cognitive and pathological changes broadly consistent with the human disease. Although many other mouse models exist, we focus on these to review some of the bioenergetic approaches that are being taken; however, it must be appreciated that the chances of a human patient presenting with all three mutations is infinitesimal.

It is important to establish the sequence of events during the development of pathology in this and other mouse models. Proteomic analysis is ambiguous; at 6 months, blue native gels reveal increased expression of ATP synthase subunits and decreased complex I and IV subunits, while analysis of denatured proteins shows a general increase in a range of mitochondrial proteins. Mitochondrial 'dysfunction' and oxidative stress are apparent by 3 months, after which intraneuronal β-amyloid and memory deficits (at 4 months), extracellular β-amyloid plaques and synaptic dysfunction (at 6 months) and finally neurofibrillary tangles (at 12 months) appear sequentially. However, a limitation of many studies is the way in which 'dysfunction' is defined and quantified; thus, in one study, dysfunction was defined as a statistically significant decrease in pyruvate dehydrogenase E1α subunits normalised to β-actin and increased oxidative stress by lipid peroxide content of isolated mitochondria. Cultured embryonic hippocampal neurons also displayed decreased spare respiratory capacity. However, this and related studies do not prove that the observed changes in mitochondrial function are causative of the pathology.

Mitochondria isolated from specific brain regions have a highly heterogenous origin—from different types of neurons, from glia, and from cell bodies, dendrites, or axons. As a result, interpretation is difficult, while real changes in specific locations can be swamped by the background. Synaptosomes (Section 9.6), while still retaining transmitter heterogeneity, can be obtained from animals of any age and allow presynaptic function to be assessed. However, there is no consensus regarding the existence of a significant bioenergetic dysfunction in synaptosomes prepared from AD-model transgenic mice. Although a slight decline in respiratory control ratio was detected in mitochondria further isolated from cortical synaptosomes of 'Tg mAPP' mice over-expressing a mutant form of human APP (Du *et al.*, 2010), in a separate study (Choi *et al.*, 2012) no difference was detected in the bioenergetic parameters of three other AD-model transgenics after exhaustive control for synaptosomal purity.

12.6.6.2 *Mitochondria as upstream initiators in transgenic models*

Further reading: Yao *et al.* (2009), McManus *et al.* (2011), Diana *et al.* (2011)

A powerful approach to establish cause and effect is to determine whether an interven-tion to prevent or delay mitochondrial-related changes in transgenic AD models blocks the development of AD-related pathology. Mitochondria-targeted antioxidants such as MitoQ and SkQ1 (Section 9.10.5) have been reported to have dramatic effects on a range of age-related pathologies. At 6 months, 3xTg-AD mice display substantial cognitive deficits, measured by an increased time to locate a hidden platform in the Morris water maze. A parallel group that had been treated with MitoQ for 4 months showed no deficit. An increase in a variety of oxidative stress and apoptotic markers was also prevented by MitoQ treatment (Figure 12.10), as was a loss of synapses (measured as a decrease in the presynaptic synaptophysin). Importantly, the increase in β-amyloid (1–42) in the transgenic was not seen in the treated mice. Taken together, these and related results sup-port a model in which mitochondria play a role in mediating the translation of upstream events initiated by this combination of mutations into cognitive deficits and pathology.

12.6.7 **Amyotrophic lateral sclerosis**

Further reading: Pedrini *et al.* (2010), Duffy *et al.* (2011), Sack (2011)

Amyotrophic lateral sclerosis (ALS) is a fatal disease characterised by a progressive loss of motor neurons. Although most cases are sporadic, approximately 20% are the result of an autosomal dominant mutation in the gene for superoxide dismutase 1 (SOD1) associ-ated with a toxic gain of function, the precise nature of which is still debated. The pres-ence of swollen vacuolated mitochondria suggests a mitochondrial component to the disease, and this is strengthened by the association of a subfraction of the mutant SOD1 (mSOD1) with the OMM or the intermembrane space. Studies with animal models have resulted in various hypotheses to relate this localisation to the pathology, including decreased electron chain capacity, damage to the protein import machinery, association of mSOD1 with Bcl-2 leading to enhanced apoptosis, Ca^{2+} overload as a consequence of altered plasma membrane channel activity, PTP activation and increased ROS.

The changes in mitochondrial morphology occur early in the progression of pathol-ogy in the animal model, before symptoms are apparent. Increased mitochondrial fission and defective axonal mitochondrial transport (specifically anterograde) have also been reported. The reader will appreciate from this plethora of potential mechanisms that the primary cause of the disease may still be elusive.

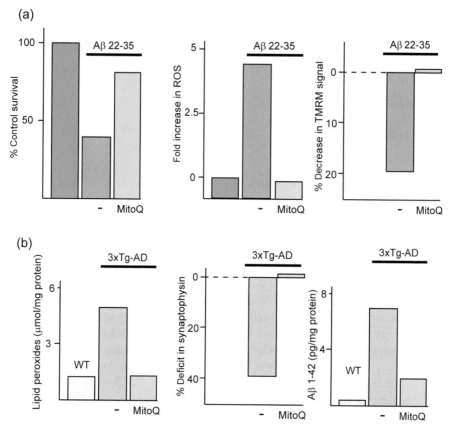

Figure 12.10   The mitochondria-targeted antioxidant MitoQ protects cultured neurons *in vitro* against β-amyloid and prevents early *in vivo* pathology in a transgenic mouse model of AD.
(a) 1 nM MitoQ protects cultured mouse cortical neurons exposed to amyloid-β 22–35 against cell death, increased ROS and decreased $\Delta\psi_m$. (b) Chronic treatment of young female 3xTg mice with MitoQ prevents an increase in lipid peroxidation, loss of synapses (synaptophysin) and accumulation of Aβ 1–42.
Data from McManus et al., 2011.

12.7 MITOCHONDRIA AND CANCER

Further reading: Ralph *et al.* (2010), Cairns *et al.* (2011)

Cancer cells are characterised by uncontrolled cell growth, restricted apoptosis and a potent anabolic metabolism. Transformation to a malignant cell involves activation of oncogenes and/or repression of tumour-suppression genes, accompanied by a reprogramming of metabolism and creation of a new microenvironment. Areas where mitochondria impinge on oncology are the aberrant partition of energy metabolism between glycolysis and oxidative phosphorylation (the Warburg and Crabtree effects), the interactions between tumour suppressors such as p53 and mitochondrial-dependent apoptosis and metabolic reprogramming. One consequence of the realisation that functional mitochondria are essential for tumour progression has been the development of 'mitocans,' drugs for cancer therapy targeted against mitochondria.

12.7.1   The Warburg and Crabtree effects

Further reading: Vander Heiden *et al.* (2009), Jose *et al.* (2010), Diaz-Ruiz *et al.* (2011)

Cancer cells characteristically upregulate glycolysis and downregulate oxidative phosphorylation so that ATP synthesis is split roughly equally between the two pathways, rather than the greater than 90% attributable to the latter in most cells. The low ATP yield from aerobic glycolysis (2 ATP/glucose) contrasting with oxidative phosphorylation (>30 ATP/glucose) means that cancer cells have a high rate of glucose utilisation, and this is exploited in PET scanning with ^{18}F-deoxyglucose to localise tumours.

When this phenomenon was first described (Warburg *et al.*, 1927), the only analogy was in yeast, in which oxygen represses glycolysis and enhances mitochondrial biogenesis and function, and so the cancer cell's repression of oxidative phosphorylation in the presence of oxygen was termed 'aerobic glycolysis' and became known as the Warburg effect. The advantages conferred on the cancer cell are still unclear, but they may relate to the ability of a tumour to proliferate in the absence of oxygen, to a decreased susceptibility to mitochondria-dependent apoptosis, or to the need of the cancer cell to accumulate nutrients for rapid proliferation.

Because of the heterogeneity of cancer cells, there is no single mechanism for the Warburg effect, and multiple control points for glycolysis have been implicated, including overexpression of high-affinity glucose transporters and high-activity (or unregulated) isoforms of hexokinase (particularly hexokinase II, which is associated with the outer membrane of tumour mitochondria and may preferentially use mitochondrial-generated ATP), phosphofructokinase, pyruvate kinase and lactate dehydrogenase. Conversely, possible mechanisms for the limitation of mitochondrial metabolism include inhibition of pyruvate dehydrogenase by overexpression of pyruvate dehydrogenase kinase mediated by HIF-1α, direct effects on electron transport activity, and an inhibition of VDAC restricting transport between matrix and cytoplasm. All of these proposals remain speculative, however, and mitochondria isolated from cancer cells appear normal.

It must be emphasised that many cancer cells do not show a Warburg effect under all conditions. Indeed, in some cases, transformation can involve a switch from glycolytic to oxidative metabolism (Figure 12.11). Slowly proliferating tumour cells rely more on oxidative phosphorylation than rapidly growing cells, and many tumour cells can grow in the presence of 'nonglycolytic' substrates such as galactose plus glutamine or in the presence of limiting glucose, relying on oxidative phosphorylation but rapidly switching to a glycolytic mode when glucose is increased. This short-term reversible switch is known as the Crabtree effect. Note that as little as 5 mM glucose can activate the Crabtree effect and that this may contribute to an artefactual glycolytic mode of cancer

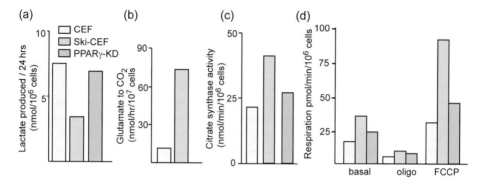

Figure 12.11   PPAR-γ mediates an oncogene-induced oxidative shift in fibroblasts.
Chicken embryonic fibroblasts (CEF) were transformed with the *ski* oncogene. In contrast to most other oncogene transformed cells, ski-transformed CEFs (ski-CEF) do not display a Warburg effect but, rather, become more oxidative. This bioenergetic reprogramming involves PPARγ and is blocked in PPARγ knocked-down ski-CEF cells. (a) Lactate production is decreased by transformation in a PPARγ-dependent manner, (b) glutamate oxidation is enhanced by transformation, (c) citrate synthase (a 'housekeeping' mitochondrial protein) is increased, and (d) CRC experiments show increased spare respiratory capacity after transformation. *Adapted from Ye et al. (2011).*

cells cultured in the presence of excessive glucose. The mechanism of the Crabtree effect is unclear.

12.7.2 Transcription factors and metabolic reprogramming

Further reading: Vaseva and Moll (2009), Jose *et al.* (2010), Cairns *et al.* (2011)

p53 has been described as the guardian of the genome. Together with related transcription factors, it primarily regulates the expression of multiple genes regulating the cell cycle. In normal cells, short-term glucose deprivation is a signal to inhibit cell division, and this is mediated via activation of AMPK, which phosphorylates and activates p53. p53 additionally antagonises the Warburg effect by inducing 'TIGAR' (TP52-induced glycolysis and apoptosis regulator), which leads to partial inhibition of phosphofructo-kinase (and hence glycolysis) and consequent activation of the pentose phosphate pathway (increasing cytoplasmic $NADP^+$ reduction and resistance to oxidative stress). At the same time, p53 promotes expression of the complex IV assembly factor 'synthesis of cytochrome c oxidase 2' (SCO2). In cancer cells with disrupted p53, the glycolytic limitation discussed previously is removed, and complex IV expression is reduced.

p53 has the ability independent of transcription or translation to induce apoptosis by translocating to the mitochondria in response to stresses such as nuclear DNA damage, oncogene deregulation and general oxidative damage. Once associated with the OMM, it will interact with pro- and anti-apoptotic members of the Bcl-2 family to induce outer membrane permeabilisation and release of cyt *c*. Specifically, wild-type p53 neutralises the anti-apoptotic Bcl-xL and Bcl-2, releasing tBid and Bax from inhibition (Section 10.7) and the pro-apoptotic Bak from its inhibitory interaction with Mcl-1. In cancer cells with defective p53, these pro-apoptotic protective mechanisms would be absent.

The PI3K–AKT1–mTOR pathway (Figure 11.7) is altered in many cancers. AKT1 activates glycolytic capacity by multiple mechanisms and also indirectly activates mTOR, which in turn stimulates lipid and protein biosynthesis for cell growth, as well as facilitating the stabilisation of the transcription factor HIF-1α. HIF-1α has been discussed (Section 11.6.1) in the context of a hypoxic adaptation favouring glycolysis at the expense of oxidative phosphorylation. Its continual degradation under normoxic conditions by the sequential actions of prolyl hydroxylase and the ubiquitin ligase VHL can be disrupted in tumours by mutations in VHL and also by TCA cycle mutations in succinate dehydrogenase and fumarate hydratase, leading to accumulation of succinate, which in turn might cause product inhibition of the α-ketoglutatate-dependent prolyl hydroxylase (Chapter 11). Together, these stabilise HIF-1α, leading to a 'hypoxic' response in the tumour, consistent with the upregulation of glycolysis and inhibition of mitochondrial activity.

One action of HIF-1α is to facilitate the activation of pyruvate dehydrogenase kinase (PDK) and thus the inhibition of pyruvate dehydrogenase. The PDK inhibitor dichloracetate is being investigated as a means of reversing this inhibition. The transcription factor c-Myc, encoded by the MYC oncogene, cooperates with HIF-1α to increase the expression of glycolytic enzymes. It also enhances mitochondrial glutamine metabolism, which may be important for an anapleurotic role of the TCA cycle in supplying amino acid intermediates for biosynthesis in the tumour. A surprising finding is that some tumours express a mutated form of mitochondrial or cytoplasmic NADP-specific isocitrate dehydrogenase (NADP-ICDH) that catalyses a novel reaction, reducing α-ketoglutarate to 2-hydroxyglutarate, the accumulation of which appears to facilitate tumour growth by a currently unknown mechanism.

12.7.3   The contribution of mtDNA mutations

Further reading: Ericson *et al.* (2012)

Genetic analyses of the mtDNA in human cancers frequently reveal the presence of clonal mutations arising from a random point mutation that confers a proliferative advantage, leading to its homoplasmic presence in the tumour. mtDNA mutates at a rate 10–20 times faster than nDNA, and the implicit assumption has been that cancer cells display increased random mtDNA mutagenesis paralleling the large increase in the frequency of such mutations seen in their nDNA, where such genomic instability facilitates the development of a neoplastic phenotype. However, this has been thrown in to doubt by the finding that the frequency of random nonclonal mtDNA point mutations is actually lower in a range of human colorectal cancers than in adjacent healthy tissue. This implies that increased mtDNA random mutation not only might fail to facilitate cancer progression but also may even provide a brake.

12.7.4 Targeting mitochondria and glycolysis in cancer therapy

Further reading: D'Souza *et al.* (2011), Cardaci *et al.* (2012)

As with all cancer therapy, a central goal is to discover and exploit xenobiotics that selectively kill cancer cells without causing systemic damage, either by targeting a process that is unusually critical for cancer cell survival or by the selective accumulation within the cancer cell (and ideally its mitochondria) of a compound that displays general toxicity. Several currently approved and experimental anticancer drugs target the cell's apoptotic machinery by activating the mitochondria-dependent intrinsic pathway, either by activating the permeability transition pore or by interacting with anti-apoptotic proteins such as Bcl-2 (Section 10.7). Unfortunately, systemic effects limit the effectiveness of most, if not all, of the current drugs, and much effort is being devoted to selective targeting of drugs to cancer cells and further to their mitochondria.

Those familiar with imaging $\Delta\psi_m$ in cells equilibrated with probes such as rhodamine-123 (Section 9.7.1) know that it is essential to reduce the exciting light intensity as much as possible to avoid phototoxicity due to the production of singlet oxygen. Photodynamic therapy exploits this sensitivity by photo-irradiation of superficial tumours loaded with appropriate dyes. Conjugation with a triphenylphosphonium group targets the dye to the mitochondria, and TPP-porphyrins are being investigated for mitochondria-targeted photodynamic therapy. A new class of mitochondria-targeted mitochondrial drugs is the gamitrinibs, synthesised by combinatorial chemistry to combine a mitochondrial targeting group (TPP or related cationic group) via a linker to geldanamycin, an inhibitor of Hsp90, which is an abundant chaperone ubiquitously expressed in normal cells and overexpressed in many cancer cells. One pool of Hsp90 is found in the mitochondrial matrix in cancer cells, and together with its homologue TRAP1, it assists in folding and assembly of matrix proteins. Accordingly, their inhibition induces mitochondrial dysfunction culminating in the permeability transition and apoptotic cell death. Nontargeted Hsp90 inhibitors have been only modestly effective in clinical trials, but the targeted gamitrinibs have shown much enhanced potency in *in vitro* experiments. In addition to lipophilic cations, mitochondrial targeting sequences have been engineered onto pro-apoptotic peptides and liposomes, and solid nanoparticles are being investigated as delivery and targeting vehicles.

The high dependency of cancer cells on aerobic glycolysis raises the possibility that drugs targeting components of the glycolytic pathway may show selective toxicity. A promising candidate is the alkylating agent 3-bromopyruvate, which has a number of targets, including hexokinase II and glyceraldehyde–phosphate dehydrogenase (GAPDH), and has effectiveness in animal models of fast-growing cancers.

12.8 STEM CELLS

Further reading: Okita and Yamanaka (2011), Folmes *et al.* (2012)

Stem cells share some of the bioenergetic and metabolic demands of cancer cells in that they must be equipped with the anaplerotic pathways required for the biosynthesis

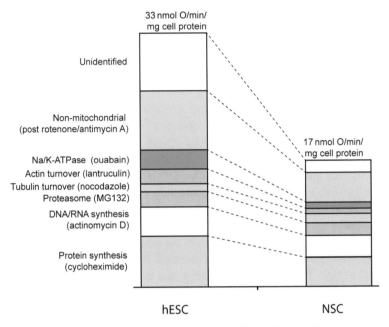

Figure 12.12   Bioenergetic changes during differentiation of human embryonic stem cells into neural stem cells.
Respiration rates (normalised to cell protein) for human embryonic stem cells (hESC) and neural stem cells (NSC). The contribution of each ATP-requiring process in the cell is determined from the decrease in respiration on adding a specific inhibitor of that process.
Adapted from Birket et al. (2011).

of cell components and rapid cell growth. 'Stemness' is therefore associated with a Warburg-like reliance on glycolysis and pentose–phosphate pathway activity, with mitochondria playing an anaplerotic role. However, reported changes in mitochondrial bioenergetic function during stem cell differentiation are not uniform: undifferentiated human mesenchymal stem cells were reported to be highly glycolytic, with an increased mitochondrial respiration during osteogenic differentiation (Chen *et al.*, 2009), whereas I6 hESCs were found to have a high initial dependency on oxidative phosphorylation and decreased mitochondrial capacity as they differentiated into neural stem cells (Birket *et al.*, 2011) (Figure 12.12). Clearly, it is not possible to generalise at this time.

   The relative ATP demand for key cellular processes can be estimated from the acute drop in respiration after adding specific inhibitors of individual processes (Figure 12.12). It is notable that protein and nucleic acid synthesis are major energy utilisers, both for hESCs derived from the blastocyst and for neural stem cells obtained from the adult animal. The introduction of inducible pluripotent stem cells from adult fibroblasts raises the interesting question as to whether they retain the age-related accumulation of mtDNA mutations or whether some mechanism similar to that in the fertilised ovum (Section 12.9.2) allows for the rejuvenation of the mitochondrial population.

12.9 **MITOCHONDRIAL THEORIES OF AGING**

Further reading: Sanz *et al.* (2010), Vendelbo and Nair (2011), Lee and Wei (2012)

We must start with some definitions. Aging must not be confused with longevity. *Aging* is a term for the complex basket of physiological changes that occur with chronological age, altering appearance, increasing the risk of neurodegenerative diseases, cancer, etc. Longevity (or life span) can be defined as the median life expectancy at birth. Although the latter has doubled in the past 150 years (from 38 to 75 years for white males in the United States) largely due to the conquering of infectious diseases, a healthy 70-year-old today is not physiologically or cognitively greatly different from a healthy 70-year-old in the 19th century. A third parameter is 'health span,' the proportion of one's life that can be enjoyed without the encroachment of aging-related diseases such as cancer or neurodegenerative disorders.

Aging research is being carried out from yeast, via the soil nematode *Caenorhabditis elegans*, the fruit fly *Drosophila melanogaster*, and the mouse and related rodents, to human. Although some interventions are sufficiently universal to be applicable at all levels, others, such as the ability of *C. elegans* to thrive following complete electron transport inhibition, are clearly less relevant to higher organisms.

12.9.1 **The mitochondrial free radical theory of aging**

Further reading: Jang and Remmen (2009), Ristow and Zarse (2010), Hekimi *et al.* (2011)

In its simplest form, the mitochondrial free radical theory of aging (MFRTA) proposed more than 50 years ago by Harman (Harman 1956) states that ROS generated by the normal functioning of mitochondria leads to the accumulation of cellular oxidative damage, disturbing cellular function, which in turn is responsible for the aging process. A direct linkage between mitochondrial ROS (mtROS) and aging implied by the theory generates a number of testable predictions: that longer lived individuals should produce less mtROS (or more precisely should display lower steady-state concentrations of mtROS); that decreasing mtROS production (or increasing the efficiency of mtROS trapping before damage can occur) should slow aging; and, conversely, that a chronic increase in mtROS levels should accelerate aging.

Despite a vast amount of research, evidence for a correlation between lowered ROS and increased life span is at best fragmentary and frequently contradictory; in particular, antioxidants have been singularly ineffective in increasing life span in these models. One high-profile exception has been the transgenic 'mitochondrial catalase mouse,' which shows a 5-month increase in median life span. Catalase is normally peroxisomal, and its ectopic expression in the matrix will have the effect of complementing glutathione peroxidase (Figure 9.12) to reduce levels of matrix H_2O_2.

An early error was the linking of MFRTA to the 'rate of living hypothesis,' attempting to rationalise the fact that larger animals tend to live longer and have lower specific metabolic rates by assuming that ROS generation was linked to the rate of metabolism—the more O_2 used, the more O_2^- is formed. It should be apparent from previous chapters that the inverse occurs: when electrons pass rapidly through potential ROS-generating sites in complex I or III, there is a decreased probability that they will have sufficient time to leak to oxygen. Somatic mtDNA mutations accumulate with age and at the same time there is a decline in respiratory chain function. However, these cannot be causally related because the highest levels of mutations or deletions (0.5–2%) are below the threshold that would produce a bioenergetic deficiency. It remains possible that deleterious mutations can accumulate clonally in individual cells and result in respiratory deficiency in those cells.

12.9.2 Mitohormesis

Further reading: Caldeira da Silva *et al.* (2008), Blagosklonny (2011)

While the MFRTA proposes that increased ROS drives the functional deterioration that is aging, an opposing view is gaining credence; namely that moderate levels of ROS are ultimately protective by inducing antioxidant defences that outweigh any initial damage. This is an example of the concept of 'mitohormesis,' in which a mild survivable restriction of mitochondrial bioenergetic capacity activates compensatory pathways, for example, via PGC1α and/or AMPK to increase mitochondrial biogenesis. Mitohormesis may underlie apparently paradoxical findings that metformin, which is proposed to act as a weak inhibitor of complex I, is an effective drug in T2D to increase hepatic mitochondrial bioenergetic capacity and inhibit gluconeogenesis (Section 12.5.2.1) or that 2,4-dinitrophenol, the classic protonophore explored so catastrophically as a slimming agent in the 1930s, may under carefully controlled conditions reduce oxidative damage and slightly extend life span in mice. The concepts of hormesis and MFRTA are to some extent mutually contradictory: if damage causes aging, then additional imposed stresses should accelerate the aging process.

12.9.3 Dietary restriction and the TOR pathway

Further reading: Finley and Haigis (2009), Kapahi *et al.* (2010), Baltzer *et al.* (2010), Evans *et al.* (2011)

An extensive list of genes have been shown to affect life span in model organisms, many of which impact on the TOR pathway. The insulin, growth factor, and nutrient-activated TOR pathway (see Figure 11.7) drive protein synthesis and cellular hypertrophy, and it has been proposed (e.g., Blagosklonny, 2011) that aging and age-related disease may be associated with a supply/demand imbalance in the pathway that can be alleviated by restricting its activity by a variety of interventions, including caloric restriction (CR), physical activity, or agents such as metformin, resveratrol, or rapamycin. Because inhibition of the TOR pathway increases life span, this suggests that the TOR pathway has a role in limiting life span, at least in yeast, worms and flies.

   Dietary restriction (DR), without a deficiency in essential nutrients, is the most effective means to increase median and maximal life span from worms to mammals. The literature may be difficult to interpret because of the loose definitions of DR (restriction of a class of nutrient) and CR (general caloric restriction) and variability in its extent, duration, the composition of the residual diet, the diurnal feeding pattern, etc. (Baltzer et al., 2010). In model organisms, the effects of DR can primarily be ascribed to inhibition of the nutrient sensor TOR by multiple pathways, including (1) a decrease in cytoplasmic ΔG_p and consequent increase in AMP activating AMPK (see Figure 11.7) and (2) decreased availability of amino acids, particularly glutamine, which acts synergistically with essential amino acids to maintain TOR activity. In rodents, restricting the intake of single essential amino acids such as methionine or tryptophan can result in an extension of life span comparable to global CR. TOR activity may also be restricted by inhibiting the insulin/insulin-like growth factor-1/AKT pathway (see Figure 11.7). There is controversy regarding whether CR increases mitochondrial biogenesis (Hancock et al., 2011).

Under conditions of DR, there is a metabolic shift from growth and reproduction to somatic maintenance (essential survival functions). There is convincing evidence in yeast, C. elegans and Drosophila that decreased TOR activity slows aging and extends life span. However, in mammals, the evidence is more equivocal. While TORC1 deletion is embryonic lethal by disrupting development, rapamycin can extend the life span of old mice by 10–15%. However, whereas DR improves insulin sensitivity and glucose tolerance, rapamycin can have the opposite effect, increasing hepatic gluconeogenesis. This has been ascribed to the ability of rapamycin also to inhibit mTORC2, which is required for insulin-mediated inhibition of hepatic gluconeogenesis.

Is there a mitochondrial component to the life span extension ascribed to TORC1 inhibition? Because mTOR facilitates PGC-1α activity and hence mitochondrial biogenesis, disruption of mTOR should result in decreased mitochondrial capacity. Indeed, skeletal muscle-specific deletion of the raptor component of mTORC1 (see Figure 11.7) results in downregulated expression of PGC-1α, reduced oxidative capacity, and premature death (Bentzinger et al., 2008). Part of the problem in extrapolating from model organisms to mammals is that the former are less dependent on mitochondrial function. Indeed, the life span of the facultative anaerobe C. elegans is extended by the clk-1 mutation, which blocks UQ biosynthesis, and by approximately 40 'Mit' mutations affecting electron transport or ATP synthase (Butler et al., 2010).

Telomeres are the TTAGG repeats that cap the ends of chromosomes to prevent these from being recognised as DNA damage. Telomeres shorten with each replication, and this can lead ultimately to cell senescence. Teleomere research has tended to be parallel but independent from bioenergetics in the aging context, but it has been suggested that residual telomeres may activate the DNA damage response, inducing p53, which binds to the promoters of PGC-1α and β, thus repressing their expression (Sahin and DePinho, 2012). This could in turn result in impaired mitochondrial biogenesis, increasing levels of ROS that would act as a positive feedforward to accelerate DNA damage, overriding the normal hormetic response to activate protective antioxidant pathways. A corollary of this hypothesis is that factors that inhibit this vicious cycle, such as enhanced telomerase expression, attenuation of p53 activity, or enhanced PGC-1α activity, should delay the onset of aging-related bioenergetic decline. Sirtuins, which can deacetylate both p53 (decreasing its activity) and PGC-1α (increasing its activity), are being intensely investigated in this context, as is the polyphenol resveratrol as a SIRT1 activator. However, the mechanism of action of resveratrol is currently contentious, and the reader is advised to evaluate the literature critically.

12.10 CONCLUSIONS

The ability of mitochondrial research to make a positive impact on medical research into disease and aging is highly dependent on the quality of the underlying science. In this book, we have attempted to review the current status of bioenergetic research, from the crystal structure of individual complexes to the *in vivo* condition. We retain our concern that the quality of much applied bioenergetic research is deficient, and a primary motivation in writing this edition is to provide a one-stop source for researchers who share our concerns.

REFERENCES

Abad, M.F.C., Di Benedetto, G., Magalhaes, P.J., Filippin, L., Pozzan, T., 2004. Mitochondrial pH monitored by a new engineered green fluorescent protein mutant. J. Biol. Chem. 279, 11521–11529.

Abrahams, J.P., Leslie, A.G., Lutter, R., Walker, J.E., 1994. Structure at 2.8Å resolution of F1-ATPase from bovine heart mitochondria. Nature 370, 621–628.

Abramson, J., Svensson, E., Byrne, B., Iwata, S., 2001. Structure of cytochrome c oxidase: a comparison of the bacterial and mitochondrial enzymes. Biochim. Biophys. Acta 1544, 1–9.

Abramson, J., Smirnova, I., Kasho, V., Verner, G., Kaback, H.R., Iwata, S., 2003. Structure and mechanism of the lactose permease of *Escherichia coli*. Science 301, 610–615.

Acin-Perez, R., Salazar, E., Kamenetsky, M., Buck, J., Levin, L.R., Manfredi, G., 2009. Cyclic AMP produced inside mitochondria regulates oxidative phosphorylation. Cell Metab. 9, 265–276.

Adachi, K., Nishizaka, T., Kinosita, K., 2012. Rotational catalysis by the ATPase. Comp. Biophys. 8, 266–288.

Affourtit, C., Brand, M.D., 2005. Stronger control of ATP/ADP by proton leak in pancreatic beta cell than skeletal muscle mitochondria. Biochem. J. 393, 151–159.

Ahn, B.H., Kim, H.S., Song, S., Lee, I.H., Liu, J., Vassilopoulos, A., et al., 2008. A role for the mitochondrial deacetylase Sirt3 in regulating energy homeostasis. Proc. Natl. Acad. Sci. USA 105, 14447–14452.

Ainscow, E.K., Rutter, G.A., 2002. Glucose-stimulated oscillations in free cytosolic ATP concentration imaged in single islet beta-cells: evidence for a Ca^{2+}-dependent mechanism. Diabetes 51 (**Suppl 1**), S162–S170.

Alano, C.C., Garnier, P., Ying, W., Higashi, Y., Kauppinen, T.M., Swanson, R.A., 2010. NAD^+ depletion is necessary and sufficient for poly(ADP-ribose) polymerase-1-mediated neuronal death. J. Neurosci. 30, 2967–2978.

Albury, M.S., Elliott, C., Moore, A.L., 2009. Towards a structural elucidation of the alternative oxidase in plants. Physiol. Plant. 137, 316–327.

Amara, C.E., Marcinek, D.J., Shankland, E.G., Schenkman, K.A., Arakaki, L.S., Conley, K.E., 2008. Mitochondrial function *in vivo*: spectroscopy provides window on cellular energetics. Methods 46, 312–318.

Amo, T., Sato, S., Saiki, S., Wolf, A.M., Toyomizu, M., Gautier, C.A., et al., 2011. Mitochondrial membrane potential decrease caused by loss of PINK1 is not due to proton leak, but to respiratory chain defects. Neurobiol. Dis. 41, 111–118.

Amunts, A., Drory, O., Nelson, N., 2007. The structure of a plant photosystem I supercomplex at 3.4Å resolution. Nature 447, 58–63.

Amunts, A., Toporik, H., Borovikova, A., Nelson, N., 2010. Structure determination and improved model of plant photosystem I. J Biol. Chem. 285, 3478–3486.

Anderson, K.A., Hirschey, M.D., 2012. Mitochondrial protein acetylation regulates metabolism. Essays Biochem. 52, 23–35.

Andrews, R.M., Kubacka, I., Chinnery, P.F., Lightowlers, R.N., Turnbull, D.M., Howell, N., 1999. Reanalysis and revision of the Cambridge reference sequence for human mitochondrial DNA. Nat. Genet. 23, 147.

Anflous, K., Armstrong, D.D., Craigen, W.J., 2001. Altered mitochondrial sensitivity for ADP and maintenance of creatine-stimulated respiration in oxidative striated muscles from VDAC1-deficient mice. J. Biol. Chem. 276, 1954–1960.

Arco, A.D., Satrustegui, J., 2005. New mitochondrial carriers: an overview. Cell Mol. Life Sci. 62, 2204–2227.

Arnou, B., Nissen, P., 2012. Structure–function relationships in P-type ATPases. Compr. Biophys. 8, 10–34.

Azzu, V., Brand, M.D., 2009. The on–off switches of the mitochondrial uncoupling proteins. Trends Biochem. Sci. 35, 298–307.

Bai, P., Canto, C., 2012. The role of PARP-1 and PARP-2 enzymes in metabolic regulation and disease. Cell Metab. 16, 290–295.

Bai, X., Ma, D., Liu, A., Shen, X., Wang, Q.J., Liu, Y., et al., 2007. Rheb activates mTOR by antagonizing its endogenous inhibitor, FKBP38. Science 318, 977–980.

Baker, L.A., Watt, I.N., Runswick, M.J., Walker, J.E., Rubinstein, J.L., 2012. Arrangement of subunits in intact mammalian mitochondrial ATP synthase determined by cryo-EM. Proc. Natl. Acad. Sci. USA 109, 11675–11680.

Balaban, R.S., 2009a. Domestication of the cardiac mitochondrion for energy conversion. J Mol. Cell. Cardiol. 46, 832–841.

Balaban, R.S., 2009b. The role of Ca^{2+} signaling in the coordination of mitochondrial ATP production with cardiac work. Biochim. Biophys. Acta Bioenerg. 1787, 1334–1341.

Baltzer, C., Tiefenbock, S.K., Frei, C., 2010. Mitochondria in response to nutrients and nutrient-sensitive pathways. Mitochondrion 10, 589–597.

Baniulis, D., Yamashita, E., Zhang, H., Hasan, S.S., Cramer, W.A., 2008. Structure–function of the cytochrome b_6f complex. Photochem. Photobiol. 84, 1349–1358.

Baradaran, R., Berrisford, J.M., Minhas, G.S., Sazanov, L.A., 2013. Crystal structure of the entire respiratory complex I. Nature 494, 443–448.

Bartelt, A., Heeren, J., 2012. The holy grail of metabolic disease: brown adipose tissue. Curr. Opin. Lipidol. 23, 190–195.

Bason, J.V., Runswick, M.J., Fearnley, I.M., Walker, J.E., 2011. Binding of the inhibitor protein IF(1) to bovine F(1)-ATPase. J. Mol. Biol. 406, 443–453.

Baughman, J.M., Perocchi, F., Girgis, H.S., Plovanich, M., Belcher-Timme, C.A., Sancak, Y., et al., 2011. Integrative genomics identifies MCU as an essential component of the mitochondrial calcium uniporter. Nature 476, 341–345.

Baumgartner, H.K., Gerasimenko, J.V., Thorne, C., Ferdek, P., Pozzan, T., Tepikin, A.V., et al., 2009. Calcium elevation in mitochondria is the main Ca^{2+} requirement for mitochondrial permeability transition pore (mPTP) opening. J. Biol. Chem. 284, 20796–20803.

Ben-Shem, A., Frolow, F., Nelson, N., 2003. Crystal structure of plant photosystem I. Nature 426, 630–635.

Bentzinger, C.F., Romanino, K., Cloetta, D., Lin, S., Mascarenhas, J.B., Oliveri, F., et al., 2008. Skeletal muscle-specific ablation of raptor, but not of rictor, causes metabolic changes and results in muscle dystrophy. Cell Metab. 8, 411–424.

Berrisford, J.M., Sazanov, L.A., 2009. Structural basis for the mechanism of respiratory complex I. J. Biol. Chem. 284, 29773–29783.

Berry, R.M., Sowa, Y., 2012. The rotary bacterial flagellar motor. Compr. Biophys. 8, 50–71.

Biegel, E., Schmidt, S., Gonzalez, J.M., Muller, V., 2011. Biochemistry, evolution and physiological function of the Rnf complex, a novel ion-motive electron transport complex in prokaryotes. Cell Mol. Life Sci. 68, 613–634.

Birket, M.J., Orr, A.L., Gerencser, A.A., Madden, D.T., Vitelli, C., Swistowski, A., et al., 2011. A reduction in ATP demand and mitochondrial activity with neural differentiation of human embryonic stem cells. J. Cell. Sci. 124, 348–358.

Blagosklonny, M.V., 2011. Hormesis does not make sense except in the light of TOR-driven aging. Aging 3, 1051–1062.

Blomain, E.S., McMahon, S.B., 2012. Dynamic regulation of mitochondrial transcription as a mechanism of cellular adaptation. Biochim. Biophys. Acta 1819, 1075–1079.

Bowler, M.W., Montgomery, M.G., Leslie, A.G., Walker, J.E., 2007. Ground state structure of F_1-ATPase from bovine heart mitochondria at 1.9Å resolution. J. Biol. Chem. 282, 14238–14242.

Brand, M.D., 1997. Regulation analysis of energy metabolism. J. Exp. Biol. 200 (**Pt 2**), 193–202.

Brand, M.D., 2010. The sites and topology of mitochondrial superoxide production. Exp. Gerontol. 46, 466–472.

Brand, M.D., Nicholls, D.G., 2011. Assessing mitochondrial dysfunction in cells. Biochem. J. 435, 297–312.

Brand, M.D., Chien, L.F., Ainscow, E.K., Rolfe, D.F., Porter, R.K., 1994. The causes and functions of mitochondrial proton leak. Biochim. Biophys. Acta 1187, 132–139.

Brandes, R., Bers, D.M., 2002. Simultaneous measurements of mitochondrial NADH and Ca^{2+} during increased work in intact rat heart trabeculae. Biophys. J. 83, 587–604.

Bricker, D.K., Taylor, E.B., Schell, J.C., Orsak, T., Boutron, A., Chen, Y.C., et al., 2012. A mitochondrial pyruvate carrier required for pyruvate uptake in yeast, *Drosophila*, and humans. Science 337, 96–100.

Brierley, G.P., Baysal, K., Jung, D.W., 1994. Cation transport systems in mitochondria: Na^+ and K^+ uniports and exchangers. J. Bioenerg. Biomembr. 26, 519–526.

Brini, M., Carafoli, E., 2009. Calcium pumps in health and disease. Physiol. Rev. 89, 1341–1378.

Brown, G.C., Borutaite, V., 2007. Nitric oxide and mitochondrial respiration in the heart. Cardiovasc. Res. 75, 283–290.

Buckel, W., Thauer, R.K., 2013. Energy conservation via electron bifurcating ferredoxin reduction and proton/Na(+) translocating ferredoxin oxidation. Biochim. Biophys. Acta 1827, 94–113.

Busch, A., Hippler, M., 2011. The structure and function of eukaryotic photosystem I. Biochim. Biophys. Acta 1807, 864–877.

Butler, J.A., Ventura, N., Johnson, T.E., Rea, S.L., 2010. Long-lived mitochondrial (Mit) mutants of *Caenorhabditis elegans* utilize a novel metabolism. FASEB J. 24, 4977–4988.

Cai, Q., Sheng, Z.H., 2009. Mitochondrial transport and docking in axons. Exp. Neurol. 218, 257–267.

Cai, Q., Davis, M.L., Sheng, Z.H., 2011. Regulation of axonal mitochondrial transport and its impact on synaptic transmission. Neurosci. Res. 70, 9–15.

Cairns, R.A., Harris, I.S., Mak, T.W., 2011. Regulation of cancer cell metabolism. Nat. Rev. Cancer 11, 85–95.

Caldeira da Silva, C.C., Cerqueira, F.M., Barbosa, L.F., Medeiros, M.H., Kowaltowski, A.J., 2008. Mild mitochondrial uncoupling in mice affects energy metabolism, redox balance and longevity. Aging Cell 7, 552–560.

Callaghan, R., Geroge, A.M., Kerr, I.D., 2012. Molecular aspects of the translocation process by ABC Proteins. Compr. Biophys. 8, 146–172.

Campanella, M., Parker, N., Tan, C.H., Hall, A.M., Duchen, M.R., 2009. IF(1): setting the pace of the F(1)F(o)-ATP synthase. Trends. Biochem. Sci. 34, 343–350.

Cannon, M.B., Remington, S.J., 2008. Redox-sensitive green fluorescent protein: probes for dynamic intracellular redox responses. A review. Methods Mol. Biol. 476, 51–65.

Canto, C., Gerhart-Hines, Z., Feige, J.N., Lagouge, M., Noriega, L., Milne, J.C., et al., 2009. AMPK regulates energy expenditure by modulating NAD$^+$ metabolism and SIRT1 activity. Nature 458, 1056–1060.

Cardaci, S., Desideri, E., Ciriolo, M.R., 2012. Targeting aerobic glycolysis: 3-Bromopyruvate as a promising anticancer drug. J. Bioenerg. Biomembr. 44, 17–29.

Carlson, A.P., Carter, R.E., Shuttleworth, C.W., 2012. Vascular, electrophysiological, and metabolic consequences of cortical spreading depression in a mouse model of simulated neurosurgical conditions. Neurol. Res. 34, 223–231.

Chalmers, S., Nicholls, D.G., 2003. The relationship between free and total calcium concentrations in the matrix of liver and brain mitochondria. J. Biol. Chem. 279, 19062–19070.

Chance, B., Williams, G.R., 1955. Respiratory enzymes in oxidative phosphorylation: III. The steady state. J. Biol. Chem. 217, 409–427.

Chaturvedi, R.K., Adhihetty, P., Shukla, S., Hennessy, T., Calingasan, N., Yang, L., et al., 2009. Impaired PGC-1alpha function in muscle in Huntington's disease. Hum. Mol. Genet. 18, 3048–3065.

Chen, C.T., Hsu, S.H., Wei, Y.H., 2009. Upregulation of mitochondrial function and antioxidant defense in the differentiation of stem cells. Biochim. Biophys. Acta 1800, 257–263.

Cherepanov, D.A., Mulkidjanian, A.Y., Junge, W., 1999. Transient accumulation of elastic energy in proton translocating ATP synthase. FEBS. Lett. 449, 1–6.

Chipuk, J.E., Green, D.R., 2008. How do BCL-2 proteins induce mitochondrial outer membrane permeabilization? Trends Cell Biol. 18, 157–164.

Choi, S., Gerencser, A.A., Nicholls, D.G., 2009. Bioenergetic analysis of isolated cerebrocortical nerve terminals on a microgram scale: spare respiratory capacity and stochastic mitochondrial failure. J. Neurochem. 109, 1179–1191.

Choi, S.W., Gerencser, A.A., Lee, D., Rajagopalan, S., Nicholls, D.G., Andersen, J.K., et al., 2011. Intrinsic bioenergetic properties and stress-sensitivity of dopaminergic synaptosomes. J. Neurosci. 31, 4524–4534.

Choi, S.W., Gerencser, A.A., Ng, R., Flynn, J.M., Melov, S., Danielson, S.R., et al., 2012. No consistent mitochondrial bioenergetic defects in presynaptic nerve terminals isolated from mouse models of Alzheimer's disease. J. Neurosci. 32, 16775–16784.

Chu, C.T., 2010. Tickled PINK1: mitochondrial homeostasis and autophagy in recessive parkinsonism. Biochim. Biophys. Acta 1802, 20–29.

Cogdell, R.J., Gall, A., Kohler, J., 2006. The architecture and function of the light-harvesting apparatus of purple bacteria: from single molecules to *in vivo* membranes. Q. Rev. Biophys. 39, 227–324.

Copeland, W.C., 2012. Defects in mitochondrial DNA replication and human disease. Crit. Rev. Biochem. Mol. Biol. 47, 64–74.

Correia, S.C., Santos, R.X., Perry, G., Zhu, X., Moreira, P.I., Smith, M.A., 2012. Mitochondrial importance in Alzheimer's, Huntington's and Parkinson's diseases. Adv. Exp. Med. Biol. 724, 205–221.

Covian, R., Balaban, R.S., 2012. Cardiac mitochondrial matrix and respiratory complex protein phosphorylation. Am. J. Physiol. Heart Circ. Physiol. 303, H940–H966.

Craigen, W.J., 2012. Mitochondrial DNA mutations: an overview of clinical and molecular aspects. Methods Mol. Biol. 837, 3–15.

Cramer, W.A., Hasan, S.S., Yamashita, E., 2011. The Q cycle of cytochrome *bc* complexes: a structure perspective. Biochim. Biophys. Acta 1807, 788–802.

Crofts, A.R., 1993. Peter Mitchell [Obituary]. Photosynth. Res. 35, 1–4.

Crofts, A.R., 2004. The cytochrome bc_1 complex: function in the context of structure. Annu. Rev. Physiol. 66, 689–733.

Crofts, A.R., Hong, S., Zhang, Z., Berry, E.A., 1999. Physicochemical aspects of the movement of the Rieske iron sulfur protein during quinol oxidation by the bc_1 complex from mitochondria and photosynthetic bacteria. Biochemistry 38, 15827–15839.

Crofts, A.R., Holland, J.T., Victoria, D., Kolling, D.R., Dikanov, S.A., Gilbreth, R., et al., 2008. The Q-cycle reviewed: how well does a monomeric mechanism of the $bc(1)$ complex account for the function of a dimeric complex? Biochim. Biophys. Acta 1777, 1001–1019.

Crompton, M., Heid, I., 1978. The cycling of calcium, sodium, and protons across the inner membrane of cardiac mitochondria. Eur. J. Biochem. 91, 599–608.

Cross, R.L., 1981. The mechanism and regulation of ATP synthesis by F1-ATPases. Annu. Rev. Biochem. 50, 681–714.

Cruz-Gallardo, I., Diaz-Moreno, I., Diaz-Quintana, A., De la Rosa, M.A., 2012. The cytochrome f-plastocyanin complex as a model to study transient interactions between redox proteins. FEBS Lett. 586, 646–652.

Csordas, G., Renken, C., Varnai, P., Walter, L., Weaver, D., Buttle, K.F., et al., 2006. Structural and functional features and significance of the physical linkage between ER and mitochondria. J. Cell Biol. 174, 915–921.

Cunningham, J.T., Rodgers, J.T., Arlow, D.H., Vazquez, F., Mootha, V.K., Puigserver, P., 2007. mTOR controls mitochondrial oxidative function through a YY1-PGC-1alpha transcriptional complex. Nature 450, 736–740.

Cypess, A.M., Lehman, S., Williams, G., Tal, I., Rodman, D., Goldfine, A.B., et al., 2009. Identification and importance of brown adipose tissue in adult humans. N. Engl. J. Med. 360, 1509–1517.

Damiano, M., Galvan, L., Deglon, N., Brouillet, E., 2010. Mitochondria in Huntington's disease. Biochim. Biophys. Acta 1802, 52–61.

Dang, S., Sun, L., Huang, Y., Lu, F., Liu, Y., Gong, H., et al., 2010. Structure of a fucose transporter in an outward-open conformation. Nature 467, 734–738.

Darrouzet, E., Cooley, J.W., Daldal, F., 2004. The cytochrome $bc(1)$ complex and its homologue the $b(6)f$ complex: similarities and differences. Photosynth. Res. 79, 25–44.

Daum, B., Nicastro, D., Austin, J., McIntosh, J.R., Kuhlbrandt, W., 2010. Arrangement of photosystem II and ATP synthase in chloroplast membranes of spinach and pea. Plant. Cell 22, 1299–1312.

Davey, G.P., Peuchen, S., Clark, J.B., 1998. Energy thresholds in brain mitochondria: potential involvement in neurodegeneration. J. Biol. Chem. 273, 12753–12757.

De, S.D., Raffaello, A., Teardo, E., Szabo, I., Rizzuto, R., 2011. A forty-kilodalton protein of the inner membrane is the mitochondrial calcium uniporter. Nature 476, 336–340.

De Brito, O.M., Scorrano, L., 2008. Mitofusin 2 tethers endoplasmic reticulum to mitochondria. Nature 456, 605–610.

Deas, E., Wood, N.W., Plun-Favreau, H., 2011. Mitophagy and Parkinson's disease: the PINK1-parkin link. Biochim. Biophys. Acta 1813, 623–633.

Deisenhofer, J., Michel, H., 1989. Nobel lecture. The photosynthetic reaction centre from the purple bacterium *Rhodopseudomonas viridis*. EMBO J. 8, 2149–2170.

Deisenhofer, J., Epp, O., Sinning, I., Michel, H., 1995. Crystallographic refinement at 2.3 Å resolution and refined model of the photosynthetic reaction centre from *Rhodopseudomonas viridis*. J Mol. Biol. 246, 429–457.

Denton, R.M., 2009. Regulation of mitochondrial dehydrogenases by calcium ions. Biochim. Biophys. Acta 1787, 1309–1316.

Diana, F.F., Silva Esteves, A.R., Oliveira, C.R., Cardoso, S.M., 2011. Mitochondria: the common upstream driver of amyloid-β and tau pathology in Alzheimer's disease. Curr. Alzheimer Res. 8, 563–572.

Diaz, F., Moraes, C.T., 2008. Mitochondrial biogenesis and turnover. Cell Calcium 44, 24–35.

Diaz-Ruiz, R., Rigoulet, M., Devin, A., 2011. The warburg and crabtree effects: on the origin of cancer cell energy metabolism and of yeast glucose repression. Biochim. Biophys. Acta 1807, 568–576.

Dickinson, B.C., Srikun, D., Chang, C.J., 2010. Mitochondrial-targeted fluorescent probes for reactive oxygen species. Curr. Opin. Chem. Biol. 14, 50–56.

Dimroth, P., Jockel, P., Schmid, M., 2001. Coupling mechanism of the oxaloacetate decarboxylase Na^+ pump. Biochim. Biophys. Acta 1505, 1–14.

Drago, I., Pizzo, P., Pozzan, T., 2011. After half a century mitochondrial calcium in- and efflux machineries reveal themselves. EMBO J. 30, 4119–4125.

Droge, W., 2002. Free radicals in the physiological control of cell function. Physiol. Rev. 82, 47–95.

D'Souza, G.G., Wagle, M.A., Saxena, V., Shah, A., 2011. Approaches for targeting mitochondria in cancer therapy. Biochim. Biophys. Acta 1807, 689–696.

Du, H., Guo, L., Yan, S., Sosunov, A.A., McKhann, G.M., Yan, S.S., 2010. Early deficits in synaptic mitochondria in an Alzheimer's disease mouse model. Proc. Natl. Acad. Sci. USA 107, 18670–18675.

Duchen, M.R., Szabadkai, G., 2010. Roles of mitochondria in human disease. Essays Biochem. 47, 115–137.

Duchen, M.R., Surin, A., Jacobson, J., 2003. Imaging mitochondrial function in intact cells. Methods Enzymol. 361, 353–389.

Dudkina, N.V., Oostergetel, G.T., Lewejohann, D., Braun, H.P., Boekema, E.J., 2010. Row-like organization of ATP synthase in intact mitochondria determined by cryo-electron tomography. Biochim. Biophys. Acta 1797, 272–277.

Duffy, L.M., Chapman, A.L., Shaw, P.J., Grierson, A.J., 2011. The role of mitochondria in the pathogenesis of amyotrophic lateral sclerosis. Neuropathol. Appl. Neurobiol. 37, 336–352.

Dukanovic, J., Rapaport, D., 2011. Multiple pathways in the integration of proteins into the mitochondrial outer membrane. Biochim. Biophys. Acta 1808, 971–980.

Eckert, A., Schmitt, K., Gotz, J., 2011. Mitochondrial dysfunction: the beginning of the end in Alzheimer's disease? Separate and synergistic modes of tau and amyloid-beta toxicity. Alzheimer's Res. Ther. 3, 15–25.

Edlich, F., Banerjee, S., Suzuki, M., Cleland, M.M., Arnoult, D., Wang, C., et al., 2011. Bcl-x(L) retrotranslocates Bax from the mitochondria into the cytosol. Cell 145, 104–116.

Efremov, R.G., Sazanov, L.A., 2011. Structure of the membrane domain of respiratory complex I. Nature 476, 414–420.

Endo, T., Yamano, K., Kawano, S., 2011. Structural insight into the mitochondrial protein import system. Biochim. Biophys. Acta 1808, 955–970.

Ericson, N.G., Kulawiec, M., Vermulst, M., Sheahan, K., O'Sullivan, J., Salk, J.J., et al., 2012. Decreased mitochondrial DNA mutagenesis in human colorectal cancer. PLoS Genet. 8, e1002689.

Ermler, U., Fritzsch, G., Buchanan, S.K., Michel, H., 1994. Structure of the photosynthetic reaction centre from *Rhodobacter sphaeroides* at 2.65Å resolution: cofactors and protein–cofactor interactions. Structure 2, 925–936.

Ernst, S., Duser, M.G., Zarrabi, N., Dunn, S.D., Borsch, M., 2012. Elastic deformations of the rotary double motor of single F(o)F(1)-ATP synthases detected in real time by Forster resonance energy transfer. Biochim. Biophys. Acta 1817, 1722–1731.

Erusalimsky, J.D., Moncada, S., 2007. Nitric oxide and mitochondrial signaling. From physiology to pathophysiology. Arterioscler. Thromb. Vasc. Biol. 27, 2524–2531.

Evans, D.S., Kapahi, P., Hsueh, W.C., Kockel, L., 2011. TOR signaling never gets old: aging, longevity and TORC1 activity. Ageing Res. Rev. 10, 225–237.

Exner, N., Lutz, A.K., Haass, C., Winklhofer, K.F., 2012. Mitochondrial dysfunction in Parkinson's disease: molecular mechanisms and pathophysiological consequences. EMBO J. 31, 3038–3062.

Faccenda, D., Campanella, M., 2012. Molecular regulation of the mitochondrial F(1)F(o)-ATP synthase: physiological and pathological significance of the inhibitory factor 1 (IF(1)). Int. J. Cell Biol 2012, 367934.

Falkenberg, M., Larsson, N.G., Gustafsson, C.M., 2007. DNA replication and transcription in mammalian mitochondria. Annu. Rev. Biochem. 76, 679–699.

Fan, M.M., Raymond, L.A., 2007. N-methyl-D-aspartate (NMDA) receptor function and excitotoxicity in Huntington's disease. Prog. Neurobiol. 81, 272–293.

Feniouk, B.A., Kozlova, M.A., Knorre, D.A., Cherepanov, D.A., Mulkidjanian, A.Y., Junge, W., 2004. The proton-driven rotor of ATP synthase: ohmic conductance (10 fS), and absence of voltage gating. Biophys. J. 86, 4094–4109.

Fercher, A., O'Riordan, T.C., Zhdanov, A.V., Dmitriev, R.I., Papkovsky, D.B., 2010. Imaging of cellular oxygen and analysis of metabolic responses of mammalian cells. Methods Mol. Biol. 591, 257–273.

Ferguson, S.J., 2000. ATP synthase: what dictates the size of the ring? Curr. Biol. 10, R804–R808.

Ferguson, S.J., Ingledew, W.J., 2008. Energetic problems faced by micro-organisms growing or surviving on parsimonious energy sources and at acidic pH: I. Acidithiobacillus ferrooxidans as a paradigm. Biochim. Biophys. Acta 1777, 1471–1479.

Fernandez-Marcos, P.J., Auwerx, J., 2011. Regulation of PGC-1α, a nodal regulator of mitochondrial biogenesis. Am. J. Clin. Nutr. 93, 884S–890S.

Finley, L.W., Haigis, M.C., 2009. The coordination of nuclear and mitochondrial communication during aging and calorie restriction. Ageing Res. Rev. 8, 173–188.

Folmes, C.D., Nelson, T.J., Dzeja, P.P., Terzic, A., 2012. Energy metabolism plasticity enables stemness programs. Ann. N. Y. Acad. Sci. 1254, 82–89.

Fonteriz, R.I., de la Fuente, S., Moreno, A., Lobatin, C.D., Montero, M., Alvarez, J., 2010. Monitoring mitochondrial [Ca^{2+}] dynamics with rhod-2, ratiometric pericam and aequorin. Cell Calcium 48, 61–69.

Foretz, M., Hebrard, S., Leclerc, J., Zarrinpashneh, E., Soty, M., Mithieux, G., et al., 2010. Metformin inhibits hepatic gluconeogenesis in mice independently of the LKB1/AMPK pathway via a decrease in hepatic energy state. J. Clin. Invest. 120, 2355–2369.

Frank, H.A., Cogdell, R.J., 2012. Light capture in photosynthesis. Comp. Biophys. 8, 94–114.

Frey, T.G., Mannella, C.A., 2000. The internal structure of mitochondria. Trends Biochem. Sci. 25, 319–324.

Friedman, J.R., Lackner, L.L., West, M., Dibenedetto, J.R., Nunnari, J., Voeltz, G.K., 2011. ER tubules mark sites of mitochondrial division. Science 334, 358–362.

Galter, D., Pernold, K., Yoshitake, T., Lindqvist, E., Hoffer, B., Kehr, J., et al., 2009. MitoPark mice mirror the slow progression of key symptoms and L-DOPA response in Parkinson's disease. Genes Brain Behav. 9, 173–181.

Garlid, K.D., Halestrap, A.P., 2012. The mitochondrial K(ATP) channel—Fact or fiction? J. Mol. Cell. Cardiol. 52, 578–583.

Gautier, C.A., Kitada, T., Shen, J., 2008. Loss of PINK1 causes mitochondrial functional defects and increased sensitivity to oxidative stress. Proc. Natl. Acad. Sci. USA 105, 11364–11369.

Gerencser, A.A., Nicholls, D.G., 2008. Measurement of instantaneous velocity vectors of organelle transport: mitochondrial transport and bioenergetics in hippocampal neurons. Biophys. J. 95, 3079–3099.

Gibasiewicz, K., Pajzderska, M., Karolczak, J., Dobek, A., 2009. Excitation and electron transfer in reaction centers from Rhodobacter sphaeroides probed and analyzed globally in the 1-nanosecond temporal window from 330 to 700 nm. Phys. Chem. Chem. Phys. 11, 10484–10493.

Gil, J.M., Rego, A.C., 2008. Mechanisms of neurodegeneration in Huntington's disease. Eur. J. Neurosci. 27, 2803–2820.

Giorgi, C., De, S.D., Bononi, A., Rizzuto, R., Pinton, P., 2009. Structural and functional link between the mitochondrial network and the endoplasmic reticulum. Int. J. Biochem. Cell Biol. 41, 1817–1827.

Giralt, A., Villarroya, F., 2012. SIRT3, a pivotal actor in mitochondrial functions: metabolism, cell death and aging. Biochem. J. 444, 1–10.

Giuditta, A., Chun, J.T., Eyman, M., Cefaliello, C., Bruno, A.P., Crispino, M., 2008. Local gene expression in axons and nerve endings: the glia-neuron unit. Physiol. Rev. 88, 515–555.

Goehring, I., Gerencser, A.A., Schmidt, S., Brand, M.D., Mulder, H., Nicholls, D.G., 2012. Plasma membrane potential oscillations in insulin secreting ins-1 832/13 cells do not require glycolysis and are not initiated by fluctuations in mitochondrial bioenergetics. J. Biol. Chem. 287, 15706–15717.

Gomez-Duran, A., Pacheu-Grau, D., Martinez-Romero, I., Lopez-Gallardo, E., Lopez-Perez, M.J., Montoya, J., et al., 2012. Oxidative phosphorylation differences between mitochondrial DNA haplogroups modify the risk of Leber's hereditary optic neuropathy. Biochim. Biophys. Acta 1822, 1216–1222.

Griffiths, E.J., 2009. Mitochondrial calcium transport in the heart: physiological and pathological roles. J. Mol. Cell. Cardiol. 46, 789–803.

Griffiths, E.J., 2012. Mitochondria and heart disease. Adv. Exp. Med. Biol. 942, 249–267.

Guan, L., Kaback, H.R., 2006. Lessons from lactose permease. Annu. Rev. Biophys. Biomol. Struct. 35, 67–91.

Gutscher, M., Pauleau, A.L., Marty, L., Brach, T., Wabnitz, G.H., Samstag, Y., et al., 2008. Real-time imaging of the intracellular glutathione redox potential. Nat. Methods 5, 553–559.

Hajnoczky, G., Csordas, G., Das, S., Garcia-Perez, C., Saotome, M., Sinha, R.S., et al., 2006. Mitochondrial calcium signalling and cell death: approaches for assessing the role of mito-chondrial Ca^{2+} uptake in apoptosis. Cell Calcium 40, 553–560.

Haldrup, A., Jensen, P.E., Lunde, C., Scheller, H.V., 2001. Balance of power: a view of the mech-anism of photosynthetic state transitions. Trends Plant. Sci. 6, 301–305.

Halestrap, A.P., 2009. What is the mitochondrial permeability transition pore? J. Mol. Cell. Cardiol. 46, 821–831.

Halestrap, A.P., Pasdois, P., 2009. The role of the mitochondrial permeability transition pore in heart disease. Biochim. Biophys. Acta Bioenerg. I1787, 1402–1415.

Hancock, C.R., Han, D.H., Higashida, K., Kim, S.H., Holloszy, J.O., 2011. Does calorie restric-tion induce mitochondrial biogenesis? A reevaluation. FASEB J. 25, 785–791.

Handy, D.E., Loscalzo, J., 2011. Redox regulation of mitochondrial function. Antioxid. Redox Signal. 16, 1323–1367.

Hardie, D.G., 2011. AMP-activated protein kinase: a cellular energy sensor with a key role in metabolic disorders and in cancer. Biochem. Soc. Trans. 39, 1–13.

Harman, D., 1956. Aging: a theory based on free radical and radiation chemistry. J. Gerontol. 11, 298–300.

Harner, M., Korner, C., Walther, D., Mokranjac, D., Kaesmacher, J., Welsch, U., et al., 2011. The mitochondrial contact site complex, a determinant of mitochondrial architecture. EMBO J. 30, 4356–4370.

Hauser, D.N., Hastings, T.G., 2012. Mitochondrial dysfunction and oxidative stress in Parkinson's disease and monogenic parkinsonism. Neurobiol. Dis. 51, 35–42.

He, W., Newman, J.C., Wang, M.Z., Ho, L., Verdin, E., 2012. Mitochondrial sirtuins: regulators of protein acylation and metabolism. Trends Endocrinol. Metab. 23, 467–476.

Heathcote, P., Jones, M.R., 2012. The structure -function relationships of photosynthetic reaction centres. Comp. Biophys. 8, 116–144.

Hekimi, S., Lapointe, J., Wen, Y., 2011. Taking a "good" look at free radicals in the aging process. Trends Cell Biol. 21, 569–576.

Henderson, P.J., 2012. Membrane proteins for secondary active transport and their molecular mechanism. Compr. Biophys. 8, 265–288.

Herrmann, J.M., 2011. MINOS is plus: a mitofilin complex for mitochondrial membrane contacts. Dev. Cell 21, 599–600.

Herrmann, J.M., Riemer, J., 2010. The intermembrane space of mitochondria. Antioxid. Redox Signal. 13, 1341–1358.

Herzig, S., Raemy, E., Montessuit, S., Veuthey, J.L., Zamboni, N., Westermann, B., et al., 2012. Identification and functional expression of the mitochondrial pyruvate carrier. Science 337, 93–96.

Hiller, S., Garces, R.G., Malia, T.J., Orekhov, V.Y., Colombini, M., Wagner, G., 2008. Solution structure of the integral human membrane protein VDAC-1 in detergent micelles. Science 321, 1206–1210.

Hirai, T., Subramaniam, S., Lanyi, J.K., 2009. Structural snapshots of conformational changes in a seven-helix membrane protein: lessons from bacteriorhodopsin. Curr. Opin. Struct. Biol. 19, 433–439.

Hirst, J., 2010. Towards the molecular mechanism of respiratory complex I. Biochem. J. 425, 327–339.

Hirst, J., Carroll, J., Fearnley, I.M., Shannon, R.J., Walker, J.E., 2003. The nuclear encoded subunits of complex I from bovine heart mitochondria. Biochim. Biophys. Acta Bioenergetics 1604, 135–150.

Hoek, J.B., Nicholls, D.G., Williamson, J.R., 1980. Determination of the mitochondrial proton-motive force in isolated hepatocytes. J. Biol. Chem. 255, 1458–1464.

Hoffman, D.L., Brookes, P.S., 2009. Oxygen sensitivity of mitochondrial reactive oxygen species generation depends on metabolic conditions. J. Biol. Chem. 284, 16236–16245.

Hohmann-Marriott, M.F., Blankenship, R.E., 2011. Evolution of photosynthesis. Annu. Rev. Plant. Biol. 62, 515–548.

Hunte, C., Zickermann, V., Brandt, U., 2010. Functional modules and structural basis of conformational coupling in mitochondrial complex I. Science 329, 448–451.

Isaev, P.I., Liberman, E.A., Samuilov, V.D., Skulachev, V.P., Tsofina, L.M., 1970. Conversion of biomembrane-produced energy into electric form: 3. Chromatophores of *Rhodospirillum rubrum*. Biochim. Biophys. Acta 216, 22–29.

Ishmukhametov, R., Hornung, T., Spetzler, D., Frasch, W.D., 2010. Direct observation of stepped proteolipid ring rotation in *E. coli* F(0)F(1)-ATP synthase. EMBO J. 29, 3911–3923.

Iverson, T.M., 2012. Catalytic mechanisms of complex II enzymes: a structural perspective. Biochim. Biophys. Acta [Epub ahead of print]

Iwabu, M., Yamauchi, T., Okada-Iwabu, M., Sato, K., Nakagawa, T., Funata, M., et al., 2010. Adiponectin and AdipoR1 regulate PGC-1α and mitochondria by Ca(2+) and AMPK/SIRT1. Nature 464, 1313–1319.

Iwai, M., Takizawa, K., Tokutsu, R., Okamuro, A., Takahashi, Y., Minagawa, J., 2010. Isolation of the elusive supercomplex that drives cyclic electron flow in photosynthesis. Nature 464, 1210–1213.

Iwata, M., Lee, Y., Yamashita, T., Yagi, T., Iwata, S., Cameron, A.D., et al., 2012. The structure of the yeast NADH dehydrogenase (Ndi1) reveals overlapping binding sites for water- and lipid-soluble substrates. Proc. Natl. Acad. Sci. USA 109, 15247–15252.

Jackson, J.B., 2012. A review of the binding-change mechanism for proton-translocating transhydrogenase. Biochim. Biophys. Acta 1817, 1839–1846.

Jacobson, J., Duchen, M.R., 2002. Mitochondrial oxidative stress and cell death in astrocytes: requirement for stored Ca^{2+} and sustained opening of the permeability transition pore. J. Cell Sci. 115, 1175–1188.

Jagendorf, A.T., 2002. Photophosphorylation and the chemiosmotic perspective. Photosynth. Res. 73, 233–241.

Jang, Y.C., Remmen, V.H., 2009. The mitochondrial theory of aging: insight from transgenic and knockout mouse models. Exp. Gerontol. 44, 256–260.

Janssen, R.J., Nijtmans, L.G., Heuvel, L.P., Smeitink, J.A., 2006. Mitochondrial complex I: structure, function and pathology. J. Inherit. Metab. Dis. 29, 499–515.

Jekabsons, M.B., Nicholls, D.G., 2004. *In situ* respiration and bioenergetic status of mitochondria in primary cerebellar granule neuronal cultures exposed continuously to glutamate. J. Biol. Chem. 279, 32989–33000.

Jiang, D., Zhao, L., Clapham, D.E., 2009. Genome-wide RNAi screen identifies Letm1 as a mitochondrial Ca^{2+}/H^+ antiporter. Science 326, 144–147.

Jitrapakdee, S., Wutthisathapornchai, A., Wallace, J.C., MacDonald, M.J., 2010. Regulation of insulin secretion: role of mitochondrial signalling. Diabetologia 53, 1019–1032.

Johnson-Cadwell, L.I., Jekabsons, M.B., Wang, A., Polster, B.M., Nicholls, D.G., 2007. "Mild uncoupling" does not decrease mitochondrial superoxide levels in cultured cerebellar granule neurons but decreases spare respiratory capacity and increases toxicity to glutamate and oxidative stress. J. Neurochem. 101, 1619–1631.

Jokinen, R., Battersby, B.J., 2012. Insight into mammalian mitochondrial DNA segregation. Ann. Med. 45, 149–155.

Jones, M.R., 2009. The petite purple photosynthetic powerpack. Biochem. Soc. Trans. 37, 400–407.

Jose, C., Bellance, N., Rossignol, R., 2010. Choosing between glycolysis and oxidative phosphorylation: a tumor's dilemma? Biochim. Biophys. Acta 1807, 552–561.

Junge, W., Sielaff, H., Engelbrecht, S., 2009. Torque generation and elastic power transmission in the rotary F_oF_1-ATPase. Nature 459, 364–370.

Kaila, V.R., Verkhovsky, M.I., Wikstrom, M., 2010. Proton-coupled electron transfer in cytochrome oxidase. Chem. Rev. 110, 7062–7081.

Kajimura, S., Seale, P., Spiegelman, B.M., 2010. Transcriptional control of brown fat development. Cell Metab. 11, 257–262.

Kapahi, P., et al., 2010. With TOR, Less is more: a key role for the conserved nutrient-sensing TOR pathway in aging. Cell Metab. 11, 453–465.

Karbowski, M., Neutzner, A., 2012. Neurodegeneration as a consequence of failed mitochondrial maintenance. Acta Neuropathol. 123, 157–171.

Kawakami, K., Umena, Y., Kamiya, N., Shen, J.R., 2011. Structure of the catalytic, inorganic core of oxygen-evolving photosystem II at 1.9Å resolution. J. Photochem. Photobiol. B 104, 9–18.

Kellosalo, J., Kajander, T., Kogan, K., Pokharel, K., Goldman, A., 2012. The structure and catalytic cycle of a sodium-pumping pyrophosphatase. Science 337, 473–476.

Kemp, G.J., Brindle, K.M., 2012. What do magnetic resonance-based measurements of $P_i \rightarrow ATP$ flux tell us about skeletal muscle metabolism? Diabetes 61, 1927–1934.

Kim, J.S., Jin, Y., Lemasters, J.J., 2006. Reactive oxygen species, but not Ca^{2+}, trigger pH- and mitochondrial permeability transition-dependent death of adult rat myocytes after ischemia/reperfusion. Am. J. Physiol. Heart Circ. Physiol. 290, H2024–H2034.

Klingenberg, M., 2008. The ADP and ATP transport in mitochondria and its carrier. Biochim. Biophys. Acta 1778, 1978–2021.

Komary, Z., Tretter, L., Adam-Vizi, V., 2010. Membrane potential-related effect of calcium on reactive oxygen species generation in isolated brain mitochondria. Biochim. Biophys. Acta 1797, 922–928.

Korkhov, V.M., Mireku, S.A., Locher, K.P., 2012. Structure of AMP–PNP-bound vitamin B12 transporter BtuCD-F. Nature 490, 367–372.

Korshunov, S.S., Skulachev, V.P., Starkov, A.A., 1997. High protonic potential actuates a mechanism of production of reactive oxygen species in mitochondria. FEBS Lett. 416, 15–18.

Krishnan, K.J., Turnbull, D.M., 2010. Mitochondrial DNA and genetic disease. Essays Biochem. 47, 139–151.

Kunji, E.R., 2012. Structural and mechanistic aspects of mitochondrial transport proteins. Compr. Biophys. 8, 174–205.

Lange, C., Hunte, C., 2002. Crystal structure of the yeast cytochrome bc_1 complex with its bound substrate cytochrome c. Proc. Natl. Acad. Sci. USA 99, 2800–2805.

Lanyi, J.K., 2004. Bacteriorhodopsin. Annu. Rev. Physiol. 66, 665–688.

Lanyi, J.K., 2012. Light capture and energy transduction in bacterial rhodopsins and related proteins. Comp. Biophys. 8, 206–227.

Lax, N.Z., Turnbull, D.M., Reeve, A.K., 2011. Mitochondrial mutations: newly discovered players in neuronal degeneration. Neuroscientist 17, 645–658.

Lee, H.C., Wei, Y.H., 2012. Mitochondria and aging. Adv. Exp. Med. Biol. 942, 311–327.

Lee, J., Giordano, S., Zhang, J., 2012. Autophagy, mitochondria and oxidative stress: cross-talk and redox signalling. Biochem. J. 441, 523–540.

Lee, W.K., Thevenod, F., 2006. A role for mitochondrial aquaporins in cellular life-and-death decisions? Am. J. Physiol. Cell Physiol. 291, C195–C202.

Lenaers, G., Reynier, P., Elachouri, G., Soukkarieh, C., Olichon, A., Belenguer, P., et al., 2009. OPA1 functions in mitochondria and dysfunctions in optic nerve. Int. J. Biochem. Cell Biol. 41, 1866–1874.

Lewin, R., 1987. The unmasking of mitochondrial eve. Science 238, 24–26.

Liao, J., Li, H., Zeng, W., Sauer, D.B., Belmares, R., Jiang, Y., 2012. Structural insight into the ion-exchange mechanism of the sodium/calcium exchanger. Science 335, 686–690.

Lin, S., Jaschke, P.R., Wang, H., Paddock, M., Tufts, A., Allen, J.P., et al., 2009. Electron transfer in the *Rhodobacter sphaeroides* reaction center assembled with zinc bacteriochlorophyll. Proc. Natl. Acad. Sci. USA 106, 8537–8542.

Lin, S.M., Tsai, J.Y., Hsiao, C.D., Huang, Y.T., Chiu, C.L., Liu, M.H., et al., 2012. Crystal structure of a membrane-embedded H^+-translocating pyrophosphatase. Nature 484, 399–403.

Liu, T., O'Rourke, B., 2009. Regulation of mitochondrial Ca^{2+} and its effects on energetics and redox balance in normal and failing heart. J. Bioenerg. Biomembr. 41, 127–132.

Liu, Z., Yan, H., Wang, K., Kuang, T., Zhang, J., Gui, L., et al., 2004. Crystal structure of spinach major light-harvesting complex at 2.72 Å resolution. Nature 428, 287–292.

Locher, K.P., Lee, A.T., Rees, D.C., 2002. The *E. coli* BtuCD structure: a framework for ABC transporter architecture and mechanism. Science 296, 1091–1098.

Locke, R.M., Rial, E., Nicholls, D.G., 1982. Fatty acids as acute regulators of the proton conductance of hamster brown fat mitochondria. Eur. J. Biochem. 129, 373–380.

Loiseau, D., Chevrollier, A., Verny, C., Guillet, V., Gueguen, N., Pou De Crescenzo, M.A., et al., 2007. Mitochondrial coupling defect in Charcot–Marie–Tooth type 2A disease. Ann. Neurol. 61, 315–323.

Lu, M., Fu, D., 2012. Structure–function relationships in P-type ATPases. Science 317, 1746–1748.

Lü, W., Du, J., Schwarzer, N.J., Andrade, S.L.A., Einsle, O., 2013. The Formate/Nitrite Transporter family of anion channels. Biol. Chem. 394, 715–727.

Luecke, H., Schobert, B., Richter, H.T., Cartailler, J.P., Lanyi, J.K., 1999. Structure of bacteriorhodopsin at 1.55 Å resolution. J. Mol. Biol. 291, 899–911.

Lyons, J.A., Aragao, D., Slattery, O., Pisliakov, A.V., Soulimane, T., Caffrey, M., 2012. Structural insights into electron transfer in caa_3-type cytochrome oxidase. Nature 487, 514–518.

Maechler, P., Wollheim, C.B., Bentzen, C.L., Niesor, E., 1992. Role of the intestinal acyl-CoA:cholesterol acyltransferase activity in the hyperresponse of diabetic rats to dietary cholesterol. J. Lipid Res. 33, 1475–1484.

Maechler, P., Li, N., Casimir, M., Vetterli, L., Frigerio, F., Brun, T., 2010. Role of mitochondria in β-cell function and dysfunction. Adv. Exp. Med. Biol. 654, 193–216.

Maklashina, E., Cecchini, G., 2010. The quinone-binding and catalytic site of complex II. Biochim. Biophys. Acta 1797, 1877–1882.

Malnoe, A., Wollman, F.-A., de Vitry, C., Rappaport, F., 2011. Photosynthetic growth despite a broken Q-cycle. Nat. Commun. 2 (301), 1–6.

Martin, L.J., 2012. Biology of mitochondria in neurodegenerative diseases. Prog. Mol. Biol. Transl. Sci. 107, 355–415.

Martinez, T.N., Greenamyre, J.T., 2012. Toxin models of mitochondrial dysfunction in Parkinson's disease. Antioxid. Redox Signal. 16, 920–934.

Matthies, D., Haberstock, S., Joos, F., Dotsch, V., Vonck, J., Bernhard, F., et al., 2011. Cell-free expression and assembly of ATP synthase. J. Mol. Biol. 413, 593–603.

McCarty, M.F., 2005. Up-regulation of PPARgamma coactivator-1α as a strategy for preventing and reversing insulin resistance and obesity. Med. Hypotheses 64, 399–407.

McKenzie, M., Liolitsa, D., Akinshina, N., Campanella, M., Sisodiya, S., Hargreaves, I., et al., 2007. Mitochondrial ND5 gene variation associated with encephalomyopathy and mitochondrial ATP consumption. J. Biol. Chem. 282, 36845–36852.

McManus, M.J., Murphy, M.P., Franklin, J.L., 2011. The mitochondria-targeted antioxidant MitoQ prevents loss of spatial memory retention and early neuropathology in a transgenic mouse model of Alzheimer's disease. J. Neurosci. 31, 15703–15715.

Menalled, L.B., Chesselet, M.F., 2002. Mouse models of Huntington's disease. Trends Pharmacol. Sci. 23, 32–39.

Menz, R.I., et al., 2001. Structure of bovine mitochondrial F1-ATPase with nucleotide bound to all three catalytic sites: implications for the mechanism of rotary catalysis. Cell 106, 331–341.

Merritt, E.A., Stout, G.H., Turley, S., Sieker, L.C., Jehsen, L.H., Orme-Johnson, W.H., 1993. Structure at pH 6.5 of ferredoxin I from *Azotobacter vinelandii* at 2.3Å resolution. Acta Crystallogr. D Biol. Crystallogr. 49, 272–281.

Mimaki, M., Wang, X., McKenzie, M., Thorburn, D.R., Ryan, M.T., 2011. Understanding mitochondrial complex I assembly in health and disease. Biochim. Biophys. Acta 1817, 851–862.

Mitchell, P., 1961. Coupling of phosphorylation to electron and hydrogen transfer by a chemiosmotic type of mechanism. Nature 191, 144–148.

Mitchell, P., 1966. Chemiosmotic Coupling in Oxidative and Photosynthetic Phosphorylation. Glynn Research, Bodmin, UK.

Mitchell, P., 2011. Chemiosmotic coupling in oxidative and photosynthetic phosphorylation. Biochim. Biophys. Acta 1807, 1507–1538.

Mitchell, P., Moyle, J., 1967. Respiration-driven proton translocation in rat liver mitochondria. Biochem. J. 105, 1147–1162.

Mitchell, P., Moyle, J., 1969. Estimation of membrane potential and pH difference across the cristae membrane of rat liver mitochondria. Eur. J. Biochem. 7, 471–484.

Mootha, V.K., Lindgren, C.M., Eriksson, K.F., Subramanian, A., Sihag, S., Lehar, J., et al., 2003. PGC-1α-responsive genes involved in oxidative phosphorylation are coordinately downregulated in human diabetes. Nat. Genet. 34, 267–273.

Morgan, B., Sobotta, M.C., Dick, T.P., 2011. Measuring E(GSH) and H_2O_2 with roGFP$_2$-based redox probes. Free Radic. Biol. Med. 51, 1943–1951.

Morgan, J.E., Gennis, R.B., Maeda, A., 2008. A role for internal water molecules in proton affinity changes in the Schiff base and Asp85 for one-way proton transfer in bacteriorhodopsin. Photochem. Photobiol. 84, 1038–1045.

Morino, K., Petersen, K.F., Shulman, G.I., 2006. Molecular mechanisms of insulin resistance in humans and their potential links with mitochondrial dysfunction. Diabetes 55 (**Suppl 2**), S9–S15.

Morino, M., Natsui, S., Ono, T., Swartz, T.H., Krulwich, T.A., Ito, M., 2010. Single site mutations in the hetero-oligomeric Mrp antiporter from alkaliphilic *Bacillus pseudofirmus* OF4 that affect Na^+/H^+ antiport activity, sodium exclusion, individual Mrp protein levels, or Mrp complex formation. J. Biol. Chem. 285, 30942–30950.

Moroni, F., 2008. Poly(ADP-ribose)polymerase 1 (PARP-1) and postischemic brain damage. Curr. Opin. Pharmacol. 8, 96–103.

Moser, C.C., Farid, T.A., Chobot, S.E., Dutton, P.L., 2006. Electron tunneling chains of mitochondria. Biochim. Biophys. Acta 1757, 1096.

Muench, S.P., Trinick, J., Harrsion, M.A., 2011. Structural divergence of rotary ATPases. Quart. Rev. Biophys. 44, 311–356.

Mulder, H., Ling, C., 2009. Mitochondrial dysfunction in pancreatic β-cells in type 2 diabetes. Mol. Cell. Endocrinol. 297, 34–40.

Muller, M., Mentel, M., Van Hellemond, J.J., Henze, K., Woehle, C., Gould, S.B., et al., 2012. Biochemistry and evolution of anaerobic energy metabolism in eukaryotes. Microbiol. Mol. Biol. Rev. 76, 444–495.

Murakami, S., Nakashima, R., Yamashita, E., Matsumoto, T., Yamaguchi, A., 2006. Crystal structures of a multidrug transporter reveal a functionally rotating mechanism. Nature 443, 173–179.

Murphy, M.P., 2009. How mitochondria produce reactive oxygen species. Biochem. J. 417, 1–13.

Murphy, M.P., 2011. Mitochondrial thiols in antioxidant protection and redox signalling: distinct roles for glutathionylation and other thiol modifications. Antioxid. Redox Signal. 16, 476–484.

Murphy, M.P., Holmgren, A., Larsson, N.G., Halliwell, B., Chang, C.J., Kalyanaraman, B., et al., 2011. Unraveling the biological roles of reactive oxygen species. Cell Metab. 13, 361–366.

Narendra, D.P., Youle, R.J., 2011. Targeting mitochondrial dysfunction: role for PINK1 and Parkin in mitochondrial quality control. Antioxid. Redox Signal. 14, 1929–1938.

Nedergaard, J., Cannon, B., 2003. The 'novel' 'uncoupling' proteins UCP2 and UCP3: what do they really do? Pros and cons for suggested functions. Exp. Physiol. 88, 65–84.

Neupert, W., Herrmann, J.M., 2007. Translocation of proteins into mitochondria. Annu. Rev. Biochem. 76, 723–749.

Newsholme, P., Gaudel, C., Krause, M., 2012. Mitochondria and diabetes: an intriguing pathogenetic role. Adv. Exp. Med. Biol. 942, 235–247.

Newstead, S., Drew, D., Cameron, A.D., Postis, V.L., Xia, X., Fowler, P.W., et al., 2011. Crystal structure of a prokaryotic homologue of the mammalian oligopeptide-proton symporters, PepT1 and PepT2. EMBO J. 30, 417–426.

Nicholls, D.G., 1974. The influence of respiration and ATP hydrolysis on the proton electrochemical potential gradient across the inner membrane of rat liver mitochondria as determined by ion distribution. Eur. J. Biochem. 50, 305–315.

Nicholls, D.G., 2005a. Commentary on: 'Old and new data, new issues: the mitochondrial Δψ' by H. Tedeschi. Biochim. Biophys. Acta 1710, 63–65.

Nicholls, D.G., 2005b. Mitochondria and calcium signalling. Cell Calcium 38, 311–317.

Nicholls, D.G., 2006a. Simultaneous monitoring of ionophore- and inhibitor-mediated plasma and mitochondrial membrane potential changes in cultured neurons. J. Biol. Chem. 281, 14864–14874.

Nicholls, D.G., 2006b. The physiological regulation of uncoupling proteins. Biochim. Biophys. Acta 1757, 459–466.

Nicholls, D.G., 2008a. Oxidative stress and energy crises in neuronal dysfunction. Ann. N. Y. Acad. Sci. 1147, 53–60.

Nicholls, D.G., 2008b. The Peter Pitchell medal lecture: forty years of Pitchell's proton circuit: from little grey books to little grey cells. Biochim. Biophys. Acta 1777, 550–556.

Nicholls, D.G., 2012. Fluorescence measurement of mitochondrial membrane potential changes in cultured cells. Methods Mol. Biol. 810, 119–133.

Nicholls, D.G., Chalmers, S., 2004. The integration of mitochondrial calcium transport and storage. J. Bioenerg. Biomembr. 36, 277–281.

Nield, J., Barber, J., 2006. Refinement of the structural model for the photosystem II supercomplex of higher plants. Biochim. Biophys. Acta 1757, 353–361.

Nogueiras, R., Habegger, K.M., Chaudhary, N., Finan, B., Banks, A.S., Dietrich, M.O., et al., 2012. Sirtuin 1 and sirtuin 3: physiological modulators of metabolism. Physiol. Rev. 92, 1479–1514.

Noinaj, N., Guillier, M., Barnard, T.J., Buchanan, S.K., 2010. TonB-dependent transporters: regulation, structure, and function. Ann. Rev. Microbiol. 64, 43–60.

Nowikovsky, K., Schweyen, R.J., Bernardi, P., 2009. Pathophysiology of mitochondrial volume homeostasis: potassium transport and permeability transition. Biochim. Biophys. Acta 1787, 345–350.

Okita, K., Yamanaka, S., 2011. Induced pluripotent stem cells: opportunities and challenges. Philos. Trans. R. Soc. London B Biol. Sci. 366, 2198–2207.

Oliveira, J., Ellerby, L.M., Rego, A.C., Nicholls, D.G., 2007. Mitochondrial dysfunction in Huntington's disease: the bioenergetics of isolated and *in-situ* mitochondria from transgenic mice. J. Neurochem. 101, 241–249.

O'Rourke, B., Blatter, L.A., 2009. Mitochondrial Ca^{2+} uptake: tortoise or hare? J. Mol. Cell. Cardiol. 46, 767–774.

Pagani, L., Eckert, A., 2011. Amyloid-β interaction with mitochondria. Int. J. Alzheimer's Dis. 2011, 925050.

Palmer, T., Berks, B.C., 2012. The twin-arginine translocation (Tat) protein export pathway. Nat. Rev. Microbiol. 10, 483–496.

Palmieri, F., Pierri, C.L., 2010. Mitochondrial metabolite transport. Essays Biochem. 47, 37–52.

Palty, R., Sekler, I., 2012. The mitochondrial Na^+/Ca^{2+} exchanger. Cell Calcium 52, 9–15.

Palty, R., Silverman, W.F., Hershfinkel, M., Caporale, T., Sensi, S.L., Parnis, J., et al., 2010. NCLX is an essential component of mitochondrial Na^+/Ca^{2+} exchange. Proc. Natl. Acad. Sci. USA 107, 436–441.

Papa, S., Rasmo, D.D., Technikova-Dobrova, Z., Panelli, D., Signorile, A., Scacco, S., et al., 2012. Respiratory chain complex I, a main regulatory target of the cAMP/PKA pathway is defective in different human diseases. FEBS Lett. 586, 568–577.

Park, E., Rapoport, T.A., 2012. Mechanisms of Sec61/SecY-mediated protein translocation across membranes. Annu. Rev. Biophys. 41, 21–40.

Parsons, M.J., Green, D.R., 2010. Mitochondria in cell death. Essays Biochem. 47, 99–114.

Pebay-Peyroula, E., Dahout-Gonzalez, C., Kahn, R., Trézéguet, V., Lauquin, G.J.M., Brandolin, R., 2003. Structure of mitochondrial ADP/ATP carrier in complex with carboxyatractyloside. Nature 426, 39–44.

Pedrini, S., Sau, D., Guareschi, S., Bogush, M., Brown Jr., R.H., Naniche, N., et al., 2010. ALS-linked mutant SOD1 damages mitochondria by promoting conformational changes in Bcl-2. Hum. Mol. Genet. 19, 2974–2986.

Pellegrini, L., Scorrano, L., 2007. A cut short to death: parl and opa1 in the regulation of mitochondrial morphology and apoptosis. Cell Death Differ. 14, 1275–1284.

Perez-Pinzon, M.A., Stetler, R.A., Fiskum, G., 2012. Novel mitochondrial targets for neuroprotection. J. Cereb. Blood Flow Metab. 32, 1362–1376.

Perkins, G.A., Tjong, J., Brown, J.M., Poquiz, P.H., Scott, R.T., Kolson, D.R., et al., 2010. The micro-architecture of mitochondria at active zones: electron tomography reveals novel anchoring scaffolds and cristae structured for high-rate metabolism. J. Neurosci. 30, 1015–1026.

Perry, G.M., Tallaksen-Greene, S., Kumar, A., Heng, M.Y., Kneynsberg, A., van, G.T., et al., 2010. Mitochondrial calcium uptake capacity as a therapeutic target in the R6/2 mouse model of Huntington's disease. Hum. Mol. Genet. 19, 3354–3371.

Phillips, D., Aponte, A.M., Covian, R.G., Balaban, R.S., 2011. Intrinsic protein kinase activity of mitochondrial oxidative phosphorylation complexes. Biochemistry 50, 2515–2529.

Pi, J., Bai, Y., Daniel, K.W., Liu, D., Lyght, O., Edelstein, D., et al., 2009. Persistent oxidative stress due to absence of uncoupling protein 2 associated with impaired pancreatic beta-cell function. Endocrinology 150, 3040–3048.

Pilsl, A., Winklhofer, K.F., 2011. Parkin, PINK1 and mitochondrial integrity: emerging concepts of mitochondrial dysfunction in Parkinson's disease. Acta Neuropathol. 123, 173–188.

Pogoryelov, D., Yildiz, O., Faraldo-Gomez, J.D., Meier, T., 2009. High-resolution structure of the rotor ring of a proton-dependent ATP synthase. Nat. Struct. Mol. Biol. 16, 1068–1073.

Pogoryelov, D., Klyszejko, A.L., Krasnoselska, G.O., Heller, E.M., Leone, V., Langer, J.D., et al., 2012. Engineering rotor ring stoichiometries in the ATP synthase. Proc. Natl. Acad. Sci. USA 109, E1599–E1608.

Popov, V., Medvedev, N.I., Davies, H.A., Stewart, M.G., 2005. Mitochondria form a filamentous reticular network in hippocampal dendrites but are present as discrete bodies in axons: a three-dimensional ultrastructural study. J. Comp. Neurol. 492, 50–65.

Prebble, J., 2002. Peter Mitchell and the ox phos wars. Trends Biochem. Sci. 27, 209–212.

Prentki, M., Nolan, C.J., 2006. Islet beta cell failure in type 2 diabetes. J. Clin. Invest. 116, 1802–1812.

Price, N.C., Dwek, R.A., Ratcliffe, R.G., Wormald, M., 2001. Principles and Problems in Physical Chemistry for Biochemists. Oxford University Press, New York.

Pryde, K.R., Hirst, J., 2011. Superoxide is produced by the reduced flavin in mitochondrial complex I: a single, unified mechanism that applies during both forward and reverse electron transfer. J. Biol. Chem. 286, 18056–18065.

Quinlan, C.L., Gerencser, A.A., Treberg, J.R., Brand, M.D., 2011. The mechanism of superoxide production by the antimycin-inhibited mitochondrial Q-cycle. J. Biol. Chem. 286, 31361–31372.

Quinlan, C.L., Treberg, J.R., Perevoshchikova, I.V., Orr, A.L., Brand, M.D., 2012. Native rates of superoxide production from multiple sites in isolated mitochondria measured using endogenous reporters. Free Radic. Biol. Med. 53, 1807–1817.

Rabl, R., Soubannier, V., Scholz, R., Vogel, F., Mendl, N., Vasiljev-Neumeyer, A., et al., 2009. Formation of cristae and crista junctions in mitochondria depends on antagonism between Fcj1 and Su e/g. J. Cell Biol. 185, 1047–1063.

Radi, R., Cassina, A., Hodara, R., Quijano, C., Castro, L., 2002. Peroxynitrite reactions and formation in mitochondria. Free Radic. Biol. Med. 33, 1451–1464.

Ralph, S.J., Rodriguez-Enriquez, S., Neuzil, J., Saavedra, E., Moreno-Sanchez, R., 2010. The causes of cancer revisited: "Mitochondrial malignancy" and ROS-induced oncogenic transformation—Why mitochondria are targets for cancer therapy. Mol. Aspects Med. 31, 145–170.

Rapoport, S.I., 2003. Coupled reductions in brain oxidative phosphorylation and synaptic function can be quantified and staged in the course of Alzheimer's disease. Neurotox. Res. 5, 385–398.

Ravnskjaer, K., Boergesen, M., Rubi, B., Larsen, J.K., Nielsen, T., Fridriksson, J., et al., 2005. Peroxisome proliferator-activated receptor α (PPARα) potentiates, whereas PPARγ attenuates, glucose-stimulated insulin secretion in pancreatic beta-cells. Endocrinology 146, 3266–3276.

Ravussin, E., Kozak, L.P., 2009. Have we entered the brown adipose tissue renaissance? Obes. Rev. 10, 265–268.

Ray, P.D., Huang, B.W., Tsuji, Y., 2012. Reactive oxygen species (ROS) homeostasis and redox regulation in cellular signaling. Cell Signal. 24, 981–990.

Rees, D.M., Montgomery, M.G., Leslie, A.G., Walker, J.E., 2012. Structural evidence of a new catalytic intermediate in the pathway of ATP hydrolysis by F_1-ATPase from bovine heart mitochondria. Proc. Natl. Acad. Sci. USA 109, 11139–11143.

Rial, E., Poustie, E.A., Nicholls, D.G., 1983. Brown adipose tissue mitochondria: the regulation of the 32,000 Mr uncoupling protein by fatty acids and purine nucleotides. Eur. J. Biochem. 137, 197–203.

Rich, P.R., Marechal, A., 2012. Electron transfer chains: structures, mechanisms and energy coupling. Comp. Biophys. 8, 73–93.

Richardson, D.J., Butt, J.N., Fredrickson, J.K., Zachara, J.M., Shi, L., Edwards, M.J., et al., 2012. The "porin-cytochrome" model for microbe-to-mineral electron transfer. Mol. Microbiol. 85, 201–212.

Ristow, M., Zarse, K., 2010. How increased oxidative stress promotes longevity and metabolic health: the concept of mitochondrial hormesis (mitohormesis). Exp. Gerontol. 45, 410–418.

Robinson, A.J., Overy, C., Kunji, E.R., 2008. The mechanism of transport by mitochondrial carriers based on analysis of symmetry. Proc. Natl. Acad. Sci. USA 105, 17766–17771.

Rochet, J.C., Hay, B.A., Guo, M., 2012. Molecular insights into Parkinson's disease. Prog. Mol. Biol. Transl. Sci. 107, 125–188.

Rollauer, S.E., et al., 2012. Structure of the TatC core of the twin-arginine protein transport system. Nature 492, 210–214.

Rutter, G.A., 2004. Visualising insulin secretion: the minkowski lecture 2004. Diabetologia 47, 1861–1872.

Sack, G.H., 2011. Introduction to the minireviews series on mitochondrial matters in amyotrophic lateral sclerosis, Lou Gehrig's disease. J. Bioenerg. Biomembr. 43, 565–567.

Safiulina, D., Kaasik, A., Seppet, E., Peet, N., Zharkovsky, A., Seppet, E., 2004. Method for in situ detection of the mitochondrial function in neurons. J. Neurosci. Methods 137, 87–95.

Saft, C., Zange, J., Andrich, J., Muller, K., Lindenberg, K., Landwehrmeyer, B., et al., 2005. Mitochondrial impairment in patients and asymptomatic mutation carriers of Huntington's disease. Mov. Disord. 20, 674–679.

Sahin, E., DePinho, R.A., 2012. Axis of ageing: telomeres, p53 and mitochondria. Nat. Rev. Mol. Cell Biol. 13, 397–404.

Santos, R., Lefevre, S., Sliwa, D., Seguin, A., Camadro, J.M., Lesuisse, E., 2010. Friedreich's ataxia: molecular mechanisms, redox considerations and therapeutic opportunities. Antioxid. Redox Signal. 13, 651–690.

Sanz, A., Fernandez-Ayala, D.J., Stefanatos, R.K., Jacobs, H.T., 2010. Mitochondrial ROS production correlates with, but does not directly regulate lifespan in Drosophila. Aging 2, 200–223.

Saraste, M., 1999. Oxidative phosphorylation at the fin de siècle. Science 283, 1488–1493.

Saroussi, S., Schushan, M., Ben-Tal, N., Junge, W., Nelson, N., 2012. Structure and flexibility of the C-ring in the electromotor of rotary F_oF_1-ATPase of pea chloroplasts. PLoS ONE 7, e43045.

Satrustegui, J., Pardo, B., Del, A.A., 2007. Mitochondrial transporters as novel targets for intracellular calcium signaling. Physiol. Rev. 87, 29–67.

Sazanov, L.A., Hinchliffe, P., 2006. Structure of the hydrophilic domain of respiratory complex I from Thermus thermophilus. Science 311, 1430–1436.

Scaduto, R.C., Grotyohann, L.W., 1999. Measurement of mitochondrial membrane potential using fluorescent rhodamine derivatives. Biophys. J. 76, 469–477.

Scarpulla, R.C., 2008. Nuclear control of respiratory chain expression by nuclear respiratory factors and PGC-1-related coactivator. Ann. N. Y. Acad. Sci. 1147, 321–334.

Scarpulla, R.C., 2011. Metabolic control of mitochondrial biogenesis through the PGC-1 family regulatory network. Biochim. Biophys. Acta 1813, 1269–1278.

Schafer, F.Q., Buettner, G.R., 2001. Redox environment of the cell as viewed through the redox state of the glutathione disulfide/glutathione couple. Free Radic. Biol. Med. 30, 1191–1212.

Schapira, A.H., 2012. Mitochondrial diseases. Lancet 379, 1825–1834.

Schieke, S.M., Phillips, D., McCoy Jr., J.P., Aponte, A.M., Shen, R.F., Balaban, R.S., et al., 2006. The mTOR pathway regulates mitochondrial oxygen consumption and oxidative capacity. J. Biol. Chem. 281, 27643–27652.

Scott, I., Webster, B.R., Li, J.H., Sack, M.N., 2012. Identification of a molecular component of the mitochondrial acetyl transferase program; a novel role for GCN5L1. Biochem. J. 443, 655–661.

Sears, I.B., MacGinnitie, M.A., Kovacs, L.G., Graves, R.A., 1996. Differentiation-dependent expression of the brown adipocyte uncoupling protein gene: regulation by peroxisome proliferator-activated receptor gamma. Mol. Cell Biol. 16, 3410–3419.

Semenza, G.L., 2007. Oxygen-dependent regulation of mitochondrial respiration by hypoxia-inducible factor 1. Biochem. J. 405, 1–9.

Serviddio, G., Sastre, J., 2010. Measurement of mitochondrial membrane potential and proton leak. Methods Mol. Biol. 594, 107–121.

Shiba, T., Kido, Y., Sakamoto, K., et al., 2013. Structure of the trypanosome cyanide-insensitive alternative oxidase. Proc. Natl. Acad. Sci. U.S.A. 110, 4580–4585.

Shibata, N., Inoue, T., Nagano, C., Nishio, N., Kohzuma, T., Onodera, K., et al., 1999. Novel insight into the copper-ligand geometry in the crystal structure of *Ulva pertusa* plastocyanin at 1.6-A resolution: structural basis for regulation of the copper site by residue 88. J. Biol. Chem. 274, 4225–4230.

Shigeto, M., Katsura, M., Matsuda, M., Ohkuma, S., Kaku, K., 2006. First phase of glucose-stimulated insulin secretion from MIN 6 cells does not always require extracellular calcium influx. J. Pharmacol. Sci. 101, 293–302.

Shimamura, T., Weyand, S., Beckstein, O., Rutherford, N.G., Hadden, J.M., Sharples, D., et al., 2010. Molecular basis of alternating access membrane transport by the sodium-hydantoin transporter Mhp1. Science 328, 470–473.

Shimomura, K., Galvanovskis, J., Goldsworthy, M., Hugill, A., Kaizak, S., Lee, A., et al., 2009. Insulin secretion from beta-cells is affected by deletion of nicotinamide nucleotide transhydrogenase. Methods Enzymol. 457, 451–480.

Shin, J.H., Ko, H.S., Kang, H., Lee, Y., Lee, Y.I., Pletinkova, O., et al., 2011. PARIS (ZNF746) repression of PGC-1α contributes to neurodegeneration in Parkinson's disease. Cell 144, 689–702.

Shoshan-Barmatz, V., De, P.V., Zweckstetter, M., Raviv, Z., Keinan, N., Arbel, N., 2010. VDAC, a multi-functional mitochondrial protein regulating cell life and death. Mol. Aspects Med. 31, 227–285.

Shoubridge, E.A., Wai, T., 2007. Mitochondrial DNA and the mammalian oocyte. Curr. Top. Dev. Biol. 77, 87–111.

Shoubridge, E.A., Wai, T., 2008. Medicine. Sidestepping mutational meltdown. Science 319, 914–915.

Shuttleworth, C.W., 2010. Use of NAD(P)H and flavoprotein autofluorescence transients to probe neuron and astrocyte responses to synaptic activation. Neurochem. Int. 56, 379–386.

Smith, J.C., 1990. Potential-sensitive molecular probes in membranes of bioenergetic relevance. Biochim. Biophys. Acta 1016, 1–28.

Smith, R.A., Murphy, M.P., 2011. Mitochondria-targeted antioxidants as therapies. Discov. Med. 11, 106–114.

Soares, P., Ermini, L., Thomson, N., Mormina, M., Rito, T., Rohl, A., et al., 2009. Correcting for purifying selection: an improved human mitochondrial molecular clock. Am. J. Hum. Genet. 84, 740–759.

Solaini, G., Harris, D.A., 2005. Biochemical dysfunction in heart mitochondria exposed to ischaemia and reperfusion. Biochem. J. 390, 377–394.

Solcan, N., Kwok, J., Fowler, P.W., Cameron, A.D., Drew, D., Iwata, S., et al., 2012. Alternating access mechanism in the POT family of oligopeptide transporters. EMBO J. 31, 3411–3421.

Soubannier, V., McBride, H.M., 2009. Positioning mitochondrial plasticity within cellular signaling cascades. Biochim. Biophys. Acta 1793, 154–170.

Sowa, Y., Berry, R.M., 2012. The rotary bacterial flagella motor. Comp. Biophys. 8, 50–71.

Spetzler, D., Ishmukhametov, R., Hornung, T., Day, L.J., Martin, J., Frasch, W.D., 2009. Single molecule measurements of F1-ATPase reveal an interdependence between the power stroke and the dwell duration. Biochemistry 48, 7979–7985.

Stanika, R.I., Winters, C.A., Pivovarova, N.B., Andrews, S.B., 2010. Differential NMDA receptor-dependent calcium loading and mitochondrial dysfunction in CA1 vs. CA3 hippocampal neurons. Neurobiol. Dis. 37, 403–411.

Stavrovskaya, I.G., Kristal, B.S., 2005. The powerhouse takes control of the cell: is the mitochondrial permeability transition a viable therapeutic target against neuronal dysfunction and death? Free Radic. Biol. Med. 38, 687–697.

Stewart, J.B., Freyer, C., Elson, J.L., Wredenberg, A., Cansu, Z., Trifunovic, A., et al., 2008. Strong purifying selection in transmission of mammalian mitochondrial DNA. PLoS Biol. 6, e10.

Supale, S., Li, N., Brun, T., Maechler, P., 2012. Mitochondrial dysfunction in pancreatic beta cells. Trends. Endocrinol. Metab. 23, 477–487.

Swerdlow, R.H., 2011. Does mitochondrial DNA play a role in Parkinson's disease? A review of cybrid and other supportive evidence. Antioxid. Redox Signal. 16, 950–964.

Swierczek, M., Cieluch, E., Sarewicz, M., Borek, A., Moser, C.C., Dutton, P.L., et al., 2010. An electronic bus bar lies in the core of cytochrome bc_1. Science 329, 451–454.

Symersky, J., Osowski, D., Walters, D.E., Mueller, D.M., 2012a. Oligomycin frames a common drug-binding site in the ATP synthase. Proc. Natl. Acad. Sci. USA 109, 13961–13965.

Symersky, J., Pagadala, V., Osowski, D., Krah, A., Meier, T., Faraldo-Gomez, J.D., et al., 2012b. Structure of the c_{10} ring of the yeast mitochondrial ATP synthase in the open conformation. Nat. Struct. Mol. Biol. 19, 485–491. S1

Szendroedi, J., Phielix, E., Roden, M., 2012. The role of mitochondria in insulin resistance and type 2 diabetes mellitus. Nat. Rev. Endocrinol. 8, 92–103.

Takeda, K., Matsui, Y., Kamiya, N., Adachi, S., Okumura, H., Kouyama, T., 2004. Crystal structure of the M intermediate of bacteriorhodopsin: allosteric structural changes mediated by sliding movement of a transmembrane helix. J. Mol. Biol. 341, 1023–1037.

Taylor, C.T., 2008. Mitochondria and cellular oxygen sensing in the HIF pathway. Biochem. J. 409, 19–26.

Tedeschi, H., 2005. Old and new data, new issues: the mitochondrial DeltaPsi 17. Biochim. Biophys. Acta 1709, 195–202.

Terzioglu, M., Galter, D., 2008. Parkinson's disease: genetic versus toxin-induced rodent models. FEBS J. 275, 1384–1391.

Thauer, R.K., Kaster, A.K., Seedorf, H., Buckel, W., Hedderich, R., 2008. Methanogenic archaea: ecologically relevant differences in energy conservation. Nat. Rev. Microbiol. 6, 579–591.

Toyabe, S., Watanabe-Nakayama, T., Okamoto, T., Kudo, S., Muneyuki, E., 2011. Thermodynamic efficiency and mechanochemical coupling of F_1-ATPase. Proc. Natl. Acad. Sci. USA 108, 17951–17956.

Treberg, J.R., Quinlan, C.L., Brand, M.D., 2011. Evidence for two sites of superoxide production by mitochondrial NADH-ubiquinone oxidoreductase (complex I). J. Biol. Chem. 286, 27103–27110.

Trifunovic, A., Wredenberg, A., Falkenberg, M., Spelbrink, J.N., Rovio, A.T., Bruder, C.E., et al., 2004. Premature ageing in mice expressing defective mitochondrial DNA polymerase. Nature 429, 417–423.

Trouillard, M., Meunier, B., Rappaport, F., 2011. Questioning the functional relevance of mitochondrial supercomplexes by time-resolved analysis of the respiratory chain. Proc. Natl. Acad. Sci. U.S.A. 108, E1027–E1034.

Tsika, E., Moore, D.J., 2012. Mechanisms of LRRK2-mediated neurodegeneration. Curr. Neurol. Neurosci. Rep. 12, 251–260.

Twig, G., Graf, S.A., Wikstrom, J.D., Mohamed, H., Haigh, S.E., Elorza, A.G., et al., 2006. Tagging and tracking individual networks within a complex mitochondrial web using photoactivatable GFP. Am. J. Physiol. Cell Physiol. 291, C176–C184.

Twig, G., Elorza, A., Molina, A.J., Mohamed, H., Wikstrom, J.D., Walzer, G., et al., 2008a. Fission and selective fusion govern mitochondrial segregation and elimination by autophagy. EMBO J. 27, 433–446.

Twig, G., Hyde, B., Shirihai, O.S., 2008b. Mitochondrial fusion, fission and autophagy as a quality control axis: the bioenergetic view. Biochim. Biophys. Acta 1777, 1092–1097.

Umena, Y., Kawakami, K., Shen, J.R., Kamiya, N., 2011. Crystal structure of oxygen-evolving photosystem II at a resolution of 1.9Å. Nature 473, 55–60.

Usukura, E., Suzuki, T., Furuike, S., Soga, N., Saita, E., Hisabori, T., et al., 2012. Torque generation and utilization in motor enzyme F_0F_1-ATP synthase: half-torque F_1 with short-sized pushrod helix and reduced ATP synthesis by half-torque F_0F_1. J. Biol. Chem. 287, 1884–1891.

van Oven, M., Kayser, M., 2009. Updated comprehensive phylogenetic tree of global human mitochondrial DNA variation. Hum. Mutat. 30, E386–E394.

van Spanning, R.J., Richardson, D.J., Ferguson, S.J., 2012. Introduction to the biochemistry and molecular biology of denitrification. In: Bothe, H., Ferguson, S.J., Newton, W.E. (Eds.), Biology of the Nitrogen Cycle. Elsevier, New York, pp. 3–20.

Vander Heiden, M.G., Cantley, L.C., Thompson, C.B., 2009. Understanding the warburg effect: the metabolic requirements of cell proliferation. Science 324, 1029–1033.

Vaseva, A.V., Moll, U.M., 2009. The mitochondrial p53 pathway. Biochim. Biophys. Acta 1787, 414–420.

Vaubel, R.A., Isaya, G., 2012. Iron–sulfur cluster synthesis, iron homeostasis and oxidative stress in Friedreich ataxia. Mol. Cell Neurosci. [Epub ahead of print]

Vendelbo, M.H., Nair, K.S., 2011. Mitochondrial longevity pathways. Biochim. Biophys. Acta 1813, 634–644.

Venter, J.C., Remington, K., Heidelberg, J.F., Halpern, A.L., Rusch, D., Eisen, J.A., et al., 2004. Environmental genome shotgun sequencing of the Sargasso Sea. Science 304, 66–74.

Verdin, E., Hirschey, M.D., Finley, L.W., Haigis, M.C., 2010. Sirtuin regulation of mitochondria: energy production, apoptosis, and signaling. Trends Biochem. Sci. 35, 669–675.

Viollet, B., Guigas, B., Sanz, G.N., Leclerc, J., Foretz, M., Andreelli, F., 2012. Cellular and molecular mechanisms of metformin: an overview. Clin. Sci.(London) 122, 253–270.

Vithayathil, S.A., Ma, Y., Kaipparettu, B.A., 2012. Transmitochondrial cybrids: tools for functional studies of mutant mitochondria. Methods Mol. Biol. 837, 219–230.

Vives-Bauza, C., Zhou, C., Huang, Y., Cui, M., de Vries, R.L., Kim, J., et al., 2010. PINK1-dependent recruitment of Parkin to mitochondria in mitophagy. Proc. Natl. Acad. Sci. USA 107, 378–383.

von Ballmoos, C., Brunner, J., Dimroth, P., 2004. The ion channel of F-ATP synthase is the target of toxic organotin compounds. Proc. Natl. Acad. Sci. USA 101, 11239–11244.

Walker, J.E., 2013. The ATP synthase: the understood, the uncertain and the unknown. Biochem. Soc. Trans. 41, 1–16.

Wang, C., Youle, R.J., 2009. The role of mitochondria in apoptosis. Annu. Rev. Genet. 43, 95–118.

Wang, X., Petrie, T.G., Liu, Y., Liu, J., Fujioka, H., Zhu, X., 2012. Parkinson's disease-associated DJ-1 mutations impair mitochondrial dynamics and cause mitochondrial dysfunction. J. Neurochem. 121, 830–839.

Warburg, O., Wind, F., Negelein, E., 1927. The metabolism of tumors in the body. J. Gen. Physiol. 8, 519–530.

Ward, A., Reyes, C.L., Yu, J., Roth, C.B., Chang, G., 2007. Flexibility in the ABC transporter MsbA: alternating access with a twist. Proc. Natl. Acad. Sci. USA 104, 19005–19010.

Ward, M.W., Rego, A.C., Frenguelli, B.G., Nicholls, D.G., 2000. Mitochondrial membrane potential and glutamate excitotoxicity in cultured cerebellar granule cells. J. Neurosci. 20, 7208–7219.

Wasilewski, M., Scorrano, L., 2009. The changing shape of mitochondrial apoptosis. Trends Endocrinol. Metab. 20, 287–294.

Watmough, N.J., Frerman, F.E., 2010. The electron transfer flavoprotein: ubiquinone oxidoreductases. Biochim. Biophys. Acta 1797, 1910–1916.

Watt, I.N., Montgomery, M.G., Runswick, M.J., Leslie, A.G., Walker, J.E., 2010. Bioenergetic cost of making an adenosine triphosphate molecule in animal mitochondria. Proc. Natl. Acad. Sci. USA 107, 16823–16827.

Wei, A.C., Liu, T., Winslow, R.L., O'Rourke, B., 2012. Dynamics of matrix-free Ca^{2+} in cardiac mitochondria: two components of Ca^{2+} uptake and role of phosphate buffering. J. Gen. Physiol. 139, 465–478.

Weisova, P., Anilkumar, U., Ryan, C., Concannon, C.G., Prehn, J.H., Ward, M.W., 2012. "Mild mitochondrial uncoupling" induced protection against neuronal excitotoxicity requires AMPK activity. Biochim. Biophys. Acta 1817, 744–753.

Weyand, S., Shimamura, T., Yajima, S., Suzuki, S., Mirza, O., Krusong, K., et al., 2008. Structure and molecular mechanism of a nucleobase-cation-symport-1 family transporter. Science 322, 709–713.

Wiederkehr, A., Park, K.S., Dupont, O., Demaurex, N., Pozzan, T., Cline, G.W., et al., 2009. Matrix alkalinization: a novel mitochondrial signal for sustained pancreatic beta-cell activation. EMBO J. 28, 417–428.

Wilson, C.Y., Rubinstein, J.L., 2012. Subnanometre-resolution structure of the intact Thermus thermophilus H^+-driven ATP synthase. Nature 481, 214–218.

Winklhofer, K.F., Haass, C., 2010. Mitochondrial dysfunction in Parkinson's disease. Biochim. Biophys. Acta 1802, 29–44.

Wraight, C.A., 2004. Proton and electron transfer in the acceptor quinone complex of photosynthetic reaction centers from Rhodobacter sphaeroides. Front Biosci. 9, 309–337.

Yao, J., Irwin, R.W., Zhao, L., Nilsen, J., Hamilton, R.T., Brinton, R.D., 2009. Mitochondrial bioenergetic deficit precedes Alzheimer's pathology in female mouse model of Alzheimer's disease. Proc. Natl. Acad. Sci. USA 106, 14670–14675.

Yoshikawa, S., 1999. X-ray structure and reaction mechanism of bovine heart cytochrome c oxidase. Biochem. Soc. Trans. 27, 351–362.

Zeuthen, T., 2001. How water molecules pass through aquaporins. Trends Biochem. Sci. 26, 77–79.

Zhang, C.Y., Baffy, G., Perret, P., Krauss, S., Peroni, O., Grujic, D., et al., 2001. Uncoupling protein-2 negatively regulates insulin secretion and is a major link between obesity, beta cell dysfunction, and type 2 diabetes. Cell 105, 745–755.

Zhou, N., Gordon, G.R., Feighan, D., MacVicar, B.A., 2010. Transient swelling, acidification, and mitochondrial depolarization occurs in neurons but not astrocytes during spreading depression. Cereb. Cortex 20, 2614–2624.

Zoccarato, F., Nicholls, D.G., 1982. The role of phosphate in the regulation of the Ca efflux pathway of liver mitochondria. Eur. J. Biochem. 127, 333–338.

Zouni, A., Witt, H.T., Kern, J., Fromme, P., Krauss, N., Saenger, W., et al., 2001. Crystal structure of photosystem II from Synechococcus elongatus at 3.8Å resolution. Nature 409, 739–743.

INDEX

Note: Page numbers followed by "*f*" and "*t*" refer to figures and tables, respectively.

A

A23187, 19–20
Acr B protein, 239–240
Adaptive thermogenesis, 355
Adenine nucleotide transporter (ANT),
    7–8, 82, 222–223, 223–225, 224*f*,
    229, 270–273, 296, 347–348
Ag/AgCl reference electrode, 62*f*
Aging, mitochondrial theories of, 383–386
    dietary restriction and impact on TOR
        pathway, 384–385
    free radical theory, 383–384
    'mitohormesis,' concept of, 384
Alternative oxidase (AOX), 133–134
    physiological function of, 134–136
    structure of, 134, 135*f*
Alzheimer's disease (AD), 373–376
Ammonium swelling technique, 24–25, 25*f*
AMP-activated protein kinase (AMPK),
    331–333, 332*f*
Amt proteins, 248–249
Anion exchange systems, 246
Antimycin, 122–123
Antiport, 16, 228–229, 236–238
Apoptosis. *See* Programmed cell death
    (apoptosis)
Aralar (AGC1), 273
Arcobacter butzleri, 244
Aspartate/glutamate carriers (AGCs), 273
ATP:ADP ratio, 65
ATP-binding cassette (ABC) proteins, 240–243
ATP content of a cell, 286–287
ATP-dependent multidrug resistance (MDR)
    transporters, 281
ATP hydrolysis, 204–205, 213, 235–236,
    279–280
    bacterial transport by, 240–244
    conformational changes at the catalytic site
        during, 206–209, 208*f*
    by SecA protein, 249
ATP hydrolysis reaction, equilibrium of, 32
ATP-Mg/P$_i$ carrier, 270–271
ATP synthase, 4, 11–12, 42*f*, 47, 64, 64–65,
    229
    binding change mechanism, 204–206, 207*f*
    by bovine heart, 85–87
    catalytic components of, 7–8
    chloroplast, 9–10
    c ring torque generation, 211–213
    c subunits, 210–211, 218–219
    DELSEED region, 57*f*
    F$_0$, 209–215
    F$_1$, 200–209
    F$_1$ and F$_0$, 197–198
    H$^+$/ATP stoichiometry of thylakoid, 187–188
    IF$_1$ inhibitor protein, 301–302, 302*f*
    inhibitor proteins of, 216–217, 302*f*
    mechanisms of torque generation, 213–215
    molecular structure, 198–200
    oligomycin, role of, 198
    peripheral stalk, 209
    and respiratory control, 77–83
    reversal, 75–76
    rotation mechanism in, 198–200, 201*f*
    vs flagellum motor, 220
ATP synthesis, 20, 45, 188–189, 272
    binding change mechanism, 204–206, 207*f*
    chemical coupling mechanism, 35
    driven by an artificial protonmotive force,
        83–84
    light-dependent, 85–87
    by *M. barkeri*, 152

ATP synthesis (*Continued*)
  by photosynthetic energy-transducing
    membranes, 159
  proton utilisation in, 84–85
  reconstitution of, 94
  vesicles and, 8–9
  voltage gating and, 74–75
ATP turnover, 80–83, 351
Atractylis gummifera, 270
A-type ATPases, 217–218
Autofluorescence, 285–286
Avogadro's constant, 44
Axonal mitochondria, 313
Azotobacter vinelandii, 146

B

Bacterial energy transduction, 8–9
Bacterial flagellum, 218–220
Bacterial membrane, 6
Bacterial photosynthetic reaction centre,
    165–171
  charge movements, 170–171
  functioning of, 170–171
  generation of light, 172–174
  in green sulfur bacteria and heliobacteria,
    174
  photosystem I-type reaction centre, 174
  quinone binding sites, 169
  structural correlations, 169–170
Bacterial respiratory chains, 136–157
  bioenergetics of methane synthesis by
    bacteria, 152–156
  electron transfer into and out of bacterial
    cells, 150–151
  electron transport systems of *P. denitrificans*
    and *E. coli*, 145–146
  Escherichia coli, 142–145
  Helicobacter pylori, 146–147
  Nitrobacter, 147–148
  Paracoccus denitrificans, 137–142
  Propionigenium modestum, 156–157
  reversed electron transfer, 151
  Thiobacillus ferrooxidans, 149–150
Bacterial transport, 228–251
  Acr B protein, 239–240
  by ATP-binding cassette (ABC) proteins,
    240–243
  by bacterial symporters, 235–236
  CorA family, 245–246
  driven by anion exchange, 246
  driven by phosphoryl transfer from
    phosphoenolpyruvate, 246–248

driven directly by ATP hydrolysis, 240–244
  FocA and NirC-type proteins, 245
  LeuT family, 236–238, 237*f*
  of macromolecules across the bacterial
    cytoplasmic membrane, 249–251
  by major facilitator superfamily (MFS)
    proteins, 229–236
  by multidrug effluxers, 239–240
  multidrug resistance (MDR) protein, 242
  by proton-dependent oligopeptide
    transporters, 234–235
  protonmotive force-driven transport across
    outer membrane, 238–240
  proton symport and antiport systems,
    228–229
  by P-type ATPases, 243–244
  role of a Na$^+$ circuit, 236
  sodium–hydantoin symporter (Mhp1),
    236–237
  sodium symport and antiport systems,
    236–238
  TonB system, 238–239
  YiiP, 246
Bacteriochlorophyll (Bchl), 162
Bacteriorhodopsin (BR), 87, 191–196
  conformational changes in, 192*f*–193*f*
  photocycle, structure and function, 191–194
Basal proton leak, 74–75
Basal respiration, 279
B850 chlorophylls, 163–165
Bilayer-mediated transport, 13–15, 14*f*
Bilayers
  ionophore-induced permeability properties
    of, 17–18
  K$^+$ permeability of, 15–16
  natural permeability properties of, 17
Bisoxonols, 70
Brain and mitochondrial dysfunction, 361–377
  Alzheimer's disease (AD), 373–376
  β-amyloid effects on, 374–376
  amyotrophic lateral sclerosis (ALS), 376
  Friedreich's ataxia, 373
  glutamate excitotoxic neuronal cell death,
    362–366, 364*f*
  Huntington's disease, 370–373, 372*f*
  neurodegenerative disorders, 362
  Parkinson's disease, 366–370
  PARP and NAD$^+$ depletion, 363–366
  'spreading depression' (SD), 366
  stroke (focal ischaemia), 362–366
Brown fat mitochondria, 64
BtuCDF protein, 242

C

Ca^{2+}-ATPase, 257
Ca^{2+}/2H$^+$ antiport ionophores, 280–281
Calvin cycle, 9–10, 160, 186–189
Camgaroos, 260
Ca^{2+}/3Na$^+$ exchange, 264
Cancer, 377–381
  mtDNA mutations, contribution of, 380
  therapy targeting mitochondria and
    glycolysis, 381
  transcription factors and metabolic
    reprogramming, 379–380
  Warburg and Crabtree effects, 378–379
Cardiolipin, 15
Carotenoid band shift, 85
Carotenoids of photosynthetic energy-
  transducing membranes, 71
Cell biology of mitochondrion
  architecture of, 303–306
  electron transport chain and ATP synthase
    complexes, 305
  intermembrane space (IMS), 305–306
  outer mitochondrial membrane (OMM),
    305–306
  structure of inner membrane, 304–305
  ultrastructure, 304*f*
  voltage-dependent anion channel (VDAC),
    305–306, 306, 306*f*
Cell-free vesicular systems, 8–9
β cell mitochondria, 356
Cell respiratory control (CRC), concept of,
  277–278, 278*f*
  experiments, 278–280
Cell respirometer, 276, 277*f*
Cellular bioenergetics, stages in
  ATP turnover, 279
  basal respiration, 279
  maximal respiration, 279–280
  non-mitochondrial respiration, 279
  proton leak, 279
  spare respiratory capacity, 280
Chameleons, 260
Chemiosmotic coupling of thylakoid
  membrane, 5*f*
Chemiosmotic hypothesis, 11–12
Chemiosmotic mechanism, 1–2
Chemiosmotic theory, 6, 20, 65
  brief history, 11–12
Chloroplasts, 3, 9–10
  ATP synthase, 9–10
  preparation of, 10
  thylakoids of, 9–10

Chromatophores, 11
Citrate carrier, 21
Citrin (AGC2), 273
Clark oxygen electrode, 61, 62*f*, 63
Complex I (NADH-UQ oxidoreductase), 93,
  108–115, 137
  action of, 112
  bacterial nomenclature, 109
  core subunits, 109
  electron transfers, 115–118
  Fe–S centre of, 112
  flux control coefficients, 80
  hydrophobic domain of, 111–113
  mitochondrial, 115
  N2 centre of, 112
  proton pumping by subunits, 113–115
  structure of, 110*f*–111*f*
  of *Thermus thermophilus*, 109–110
Complex I polypeptides (ND1–ND6 and
  ND4L), 328–329
Complex II (succinate dehydrogenase), 93,
  116, 137, 146
  electron transfers, 115–116
  Fe–S centres, 116, 117*f*
  inhibition of, 125
  as a Δ*p* consumer, 116
  quinone binding site, 116
Complex III (UQH 2–cyt *c* oxidoreductase/*bc*$_1$
  complex), 93, 137
  cytochrome *c* interaction with, 125–126
  haems of, 104–105
  'Q-cycle' of electron transfer, 98–100, 119,
    120*f*
  Rieske protein 2Fe–2S cluster, 123
  structure of, 123–125
  transfer of electrons, 118–119
Complex IV (cytochrome *c* oxidase), 93,
  126–131
  copper A (Cu$_A$) centre of, 125
  cycle of catalytic activity, 129, 130*f*
  D and K channels of, 128
  electron transfer and reduction of oxygen,
    129–131
  haem groups of, 128
  H-channel mechanism in, 128–129
  reaction between cytochrome oxidase and
    oxygen, 130*f*
  structure of, 127–129, 127*f*
Connectivity theorem, 80
CorA family, 245–246
Creatine/creatine-phosphate (Cr/CrP) pathway,
  271*f*, 272–273

^{14}C-sucrose, 68
CTFR (cystic fibrosis transmembrane regulator) protein, 243
C-type cytochromes, 145–146, 150–151, 173
Cyanobacteria, 11
Cyanohydroxycinnamate, 21
Cybrids, 347–350, 348f
Cyclic electron transport, 188–189
Cyclophilin D (CyP-D), 267
Cyclosporine A (CsA), 267
Cyt f, 183
Cytochrome c, 45, 125–126
  copper A (Cu$_A$) centre of, 125–126
  lysine patch on, 125
Cytochrome oxidase. See Complex IV (cytochrome c oxidase)
Cytochrome redox couple, 38
Cytochromes, 95
  electron transfers in, 103
Cytoplasmic environment, 256–257
  free cytoplasmic Ca^{2+} concentration in, 257
  interrelations between mitochondrial and plasma membrane transport processes in, 258f
  K$^+$ conductance of plasma membrane, 256
  Na$^+$/Ca^{2+} exchanger, 256
  Na$^+$ conductance of plasma membrane, 256

D

Dicarboxylate translocator, 82
Dicyclodihexylcarbodiimide (DCCD), 211
Diffusion potentials, 50
Dihydroorotate dehydrogenase, 115–116
Dimethylsulfoxide, 145
D-malate dehydrogenase, 147
Donnan potentials, 50–51
Dual-wavelength spectrophotometry, 105–106
Dynein motors, 313

E

Elasticity coefficient, 80
Electrical circuit analogy, 5–6, 6f
Electrochromism, 71
Electroneutral K$^+$/H$^+$ antiport, 7
Electroneutral transport, 16–17
Electron transfer, mechanism of, 101–106
  in B. subtilis, 116
  E$_h$ values for respiratory chain component, 106, 107f
  electron transport and proton pumping, 113–115
  electron tunnelling process, 101–103

Em values for respiratory chain component, 105–106, 107f
  in flavoprotein–ubiquinone oxidoreductase, 116–117
  Gibbs activation energy, 101–102
  inverted region in, 102
  mediators, role of, 104–105
  midpoint redox potentials, 103–104
  in mitochondria of non-mammalian cells, 133–136
  pathways in green plants, algae, and cyanobacteria, 174–191, 175f
  redox potentiometry of, 104–106
  reorganisation energy, 102
  in Rhodopseudomonas viridis, 104f
  separation distance, effect of, 103
Encephalomyopathies, 346
Endosymbiotic theory, 328
Energy-conserving membranes, 1, 15
  lipid compositions, 15
Energy-transducing membranes, 3
  in bacteria, 8–9
  carotenoids of, 71
  chloroplasts, 9–10
  chromatophores, 11
  inner membrane, 7–8
  mitochondrial cross-section, 7–8, 7f
  proton pumps of, 3, 4f
  submitochondrial particles, 7–8, 7f
Entropy of a system, 29–30
Equilibrium distribution of permeant ions and species, 47–50
  of charged species, 48
  of electroneutrally permeant weak acids and bases, 48–50, 49f
  equilibrium Nernst equation, 48
  membrane potential, 48
  by passive uniport across a membrane, 49t
Escherichia coli, 21, 47, 140–146, 216, 240–241, 246
  aerobic and anaerobic respiratory systems, 143f
  anaerobic metabolism in, 144–145
  ATP synthase, 301–302
  chloride channel/transporter in, 244–245
  cyt bd complex, 142–143
  d-type haem, 142–143
  electron transport chain of, 142f, 143, 143–144, 146
  F$_o$ F$_1$-ATP synthases, 198, 198, 199f, 200
  glycerol-3-phosphate/phosphate exchanger (GlpT) from, 233

MotB Asp32 in, 219–220
Na$^+$/H$^+$ antiporter, 236
oxidases in, 142
oxidation of D-lactate, 143–144
phosphotransferase system (PTS) for
  mannitol in, 247f
synthase rotation in, 200
ETF–ubiquinone oxidoreductases, 93, 94,
  94, 115–116, 116–117, 117, 118, 118f,
  139–140, 151
Eukaryotic cytoplasmic ATP/ADP + P$_i$ pool,
  46–47
Eukaryotic plasma membrane potentials, 50

F

F$_0$-ATP synthase, 209–215, 229
  structure and proposed rotary mechanism of
    yeast, 212f
F$_1$-ATP synthase, 200–209, 301–302
  organisation of α, β, and γ chains in, 200,
    202–203, 202f–203f
  P-loop ('phosphate-binding loop') region,
    202
  rotation of γ subunit, 203
FCCP protonophore, 22–24, 74, 79, 279–280
FepA, 238–239
Ferredoxins, 43
Ferrocytochrome c:O$_2$ oxidoreductase, 126
Fe–S centres, 92, 96–97, 98f, 109–110, 112,
  116–117, 117f, 123, 144–145, 159, 176f,
  184, 185f, 250–251, 288–289, 290, 343,
  373
Flavin adenine dinucleotide (FAD), 97–98,
  116–118
Flavin mononucleotide (FMN), 97–98
Flavins, 97–100, 99f
Flavoproteins, 285–286
Flavoprotein–ubiquinone oxidoreductase,
  electron transfer in, 116–117
Fluorescent lipophilic cations, 21
Flux control coefficient, 79–80
FocA-type protein, 245
F$_o$F$_1$-ATP synthases, 197–198, 248–249,
  304–305
  c ring torque generation, 211–213
  c subunits, 210–211
  electron cryomicroscopy of bovine, 210f
Free energy changes. *See* Gibbs energy
  changes
Free-living trypanosomes, 135–136
Friedreich's ataxia, 373
F-type ATP synthases, 197–198

FucP (fucose-proton symporter), 232–233,
  233f

G

Gibbs energy changes, 29–30, 205, 331–332
  for ATP hydrolysis reaction, 32–33
  derivation of equilibrium constants, 34–35
  different manifestations of, 29–30
  and displacement from equilibrium, 30–36
  equilibrium constant, 35
  equilibrium distribution of permeant ions
    and species, 47–50
  for the formation of transition state, 101–102
  hydrolysis of ATP to ADP, 33, 33t, 35
  and oxidation–reduction (redox) potentials,
    41–43
  for PEP hydrolysis, 248
  phosphorylation potential, 33
  and photons, 44
  in photosynthetic systems, 29
  relevant to bioenergetics, 34–35
  in reversible regions of the respiratory chain,
    58
  standard, 33–35
Gibbs–Helmholtz equation, 29
Glucose-6-phosphate (G6P), 246
Glucose-stimulated insulin secretion (GSIS),
  356–358, 357f
Glutathione, 292
Goldman equation, 50
Gramicidin, 19
Gram-negative bacteria, energy transduction
  in, 8, 9f
Green bacteria, 11
Green plant photosystems, 179
Green sulfur bacteria, 186
  photosynthesis in, 174

H

Halobacteria, 11
Halorhodopsin, 194–196
  crystal structure of, 195
  Schiff's base nitrogen, 195–196
Haplotypes, 330–331
H+/ATP stoichiometry, 215–216
HCCS (holocytochrome c synthase), 316
Heart and mitochondrial dysfunction, 350–354
  altered workload, 351–352
  cardiac ischaemia/reperfusion injury,
    352–354, 353f
Heat shock protein mtHsp70, 315
Helicase (snail-gut enzyme), 8

Helicobacter pylori, 146–147
  electron transfer chain of, 146, 147*f*
  oxidation of a flavodoxin in, 147
Heliobacteria, photosynthesis in, 174
Henderson–Hasselbalch equation, 48–50
H^+ extrusion, non-steady-state determinations
  of, 58
High-energy phosphate bond, 36
HiPiP (high potential iron–sulfur protein),
  173–174
Histidine phosphocarrier protein (HPr), 247
H^+/O stoichiometry, 45
  experimental determination of, 59
Human mtDNA migrations, 331*f*
Huntington's disease, 80, 370–373, 372*f*
Hydrogen electrode, potential difference
  of, 37
Hypoxia, 334–337
  AMPK activation during, 334–336
  HIF-1α, 336*f*
  hypoxia-inducible factor, 336–337

I

Indirect proton translocation, 20
INDY proteins, 236
Intact cells, 53–54
  ATP content of, 286–287
  flavoproteins, 285–286
  membrane potential ($\Delta\psi$), 281–284, 282*f*
  NAD(P)H fluorescence, 285–286
  protonmotive force (pmf) (Δp), 281–287
  protonophores, 280–281
In vivo bioenergetics, 288
Ion electrochemical potential differences,
  43–44
Ionomycin, 19–20
Ionophores, 17, 18*f*, 280–281
  A23187, 19–20
  as carriers of protons and charge, 20
  channel-forming, 17–19
  gramicidin, 19
  in intact cells, 20
  ionomycin, 19–20
  as mobile carriers, 17–18
  nigericin, 16–17, 19
  valinomycin, 19
Ion transport, across energy-conserving
    membranes
  antiport, 16
  bilayer-mediated, 17–21
  bilayer-mediated *vs* protein-catalysed
    transport, 13–15

classification of, 13–17, 14*f*
coordinate movement of ions, 22
directly coupled to metabolism *vs* passive
    transport, 15
electroneutral *vs* electrical, 16–17
and fluid-mosaic model of membrane
    structure, 13
primary active transport, 15
protein-catalysed, 21–22
secondary active transport, 15
swelling mechanism, 22–25
symport, 15–16
uniport, 15–16
Isolated mitochondria, 53–54

K

K^+/H^+ antiport, 16–17
Kinesin-1 (KIF5), 312–313
Kinesin motors, 312–313
Klebsiella pneumoniae, 146
K^+ permeability of bilayers, 15–16

L

Lactose (galactoside)/H^+ symporter, 230–232,
  231*f*
Legionella pneumophila CopA Cu^+-ATPase,
  244
LeuT family, 236–238, 237*f*
Light-capture mechanism, 174–191
Light-dependent energy transduction, 11, 160*f*.
    See also Bacterial photosynthetic reaction
    centre
  cyclic electron transport, 188–189
  generation of light, 172–174
  pathways in green plants, algae, and
    cyanobacteria, 174–191, 175*f*
Light-driven proton pump, 11
Light-harvesting complexes, LH1 and LH2,
  165, 177–178, 189*f*
Lipid compositions of energy-conserving
    membranes, 15
Lipophilic cations and anions, 21
L-malate dehydrogenase, 147
Luciferase, 286–287

M

Magnesium transporters, 245–246
Major facilitator superfamily (MFS) proteins,
  229–236
  in bacterial symporters, 235–236
  EmrD (putative multidrug efflux pump),
    233–234

FucP (fucose-proton symporter), 232–233, 233f

lactose (galactoside)/H$^+$ symporter, 230–232, 231f

in proton-dependent oligopeptide transporters, 234–235

Malate–aspartate shuttle (MAS), 273

Mammalian mitochondria, 316–317, 317–318, 317f

Mammalian mitochondrial DNA (mtDNA), 328, 330. *See also* MtDNA mutations

Membrane potential ($\Delta\psi$), 43–44, 48, 50, 256–257

across energy-transducing membranes, 48

defined, 44

estimation by permeant ion distribution, 66–67

extrinsic optical indicators of, 69–70

factors influencing, 72–73

factors influencing partition of, 72–73, 72f

fluorescence response at single-cell level, 283

in intact cells, 281–284, 282f

intrinsic optical indicators of, 71

monitoring using safranine fluorescence, 70f

Menaquinone, 98–100

MERRF (myoclonic epilepsy ragged red fibres), 346

Metabolic control analysis, 77–83

bottom-up analysis, 80

elasticity coefficient, 80

flux control coefficient, 79–80

modules for, 81f

top-down (modular) analysis, 80–83

Methane synthesis by bacteria, 152–156

acetate, growth on, 155–156

CH$_3$OH, growth by disproportionation of, 155

CH$_3$OH, reduction of, 152–153

CO$_2$, reduction of, 153–155

Methanogenesis, 156

Methanosarcina barkeri, 152

Methanosarcina mazei, 152

Methylmalonyl CoA, decarboxylation of, 156–157

Mhp1, 236–238

Microbacterium liquefaciens, 236–238

MICU1, 263–264

Mild uncoupling, concept of, 296

Mitochondria and submitochondrial particles, 3, 7–8, 7f

brown adipose tissue (BAT), current/voltage relationship of, 75f

brown fat, 64

Ca^{2+} accumulation, 15–16

creatine/creatine-phosphate (Cr/CrP) pathway, 272–273

determination of matrix volume, 67, 67f

electron transfer, in non-mammalian cells, 133–136, 134f

glutathione pool (GSH) of, 292

inner mitochondrial membrane (IMMs), 197–198, 257–259, 269

isolated, 53–54

mitochondrial membrane, 6

oxidative phosphorylation in, 47

proportion of protein in, 15

proton leak across, 61

redox potentials for NAD$^+$/NADH couple, 41

respiratory control, 77–83

respiring, 45

and reversed electron transfer (RET), 76–77, 77f

swelling mechanism, 22, 23f

ultrasonic disintegration of, 8

from yeast cells, 8

Mitochondrial ATP synthase, 57–58, 76, 132

Mitochondrial bc_1 complex, 122–123, 147–148, 177

conformational changes in, 124f

crystal structures of, 125

Mitochondrial biogenesis, 313–318

assembly of mitochondrial complexes, 316–318

of brown adipose tissue (BAT), 338–339

mTOR (mammalian target of rapamycin), role of, 338–339

protein import, 314–316, 315f

protein phosphorylation, mitochondrial, 337–338

redox signalling and oxidative stress, 342–344

sirtuins, role of, 340–342

transcription factors and transcriptional coregulators in, 333–334

Mitochondrial Ca^{2+} cycling, 264

Mitochondrial calcium transport, 259–268

accumulation and steady-state cycling of Ca^{2+}, 262f

Ca^{2+} buffering, 263–268

kinetics of Ca^{2+} uptake, 264–265

matrix free Ca^{2+} concentrations, 265–267

and permeability transition, 267–268

Mitochondrial calcium transport (*Continued*)
  techniques for monitoring Ca^{2+}, 260,
    260–261, 261*f*
  total matrix Ca^{2+}, 260–261, 262*f*
Mitochondrial Ca^{2+} uniporter, 263–264
Mitochondrial complex I, 115
Mitochondrial complex III, 42–43
Mitochondrial disease, 345–350
  autoimmune destruction of β cells and
    diabetes, 355–361
  brain and mitochondrial dysfunction,
    361–377. *See also* Brain and
    mitochondrial dysfunction
  cancer, 377–381
  definition, 345–346
  heart and mitochondrial dysfunction,
    350–354
  Kearns–Sayre syndrome, 346
  mitochondrial myopathy and
    encephalopathy with lactic acidosis and
    stroke-like episodes (MELAS), 346
  mtDNA mutations, 346
  nuclear mutations, 350
Mitochondrial dynamics, 306–312
  discrete mitochondria *vs* integrated
    reticulum, 307–308
  fission/fusion cycle, 309–310, 309*f*, 310*f*
  interactions with endoplasmic reticulum,
    311, 311*f*
Mitochondrial dysfunction, 255
Mitochondrial Eve, 331
Mitochondrial genome, 327–331, 329*f*
  haplotypes, 330–331
  mammalian mitochondrial DNA (mtDNA),
    328, 330
  single nucleotide polymorphisms (SNPs),
    330
Mitochondrial metabolite carriers (MCs),
    268–276
  acylcarnitine/carnitine carrier (SLC25A20,
    29), 275
  adenine nucleotide carrier (or translocator),
    270–273
  deoxynucleotide carrier (SLC25A19), 275
  dicarboxylate carrier (SLC25A21), 275
  electron import from cytoplasm, 273–274
  glutamate carrier (SLC25A18, 22),
    274–275
  isoforms, 268
  mammalian, 269*t*
  metabolite equilibria across the inner
    mitochondrial membrane, 275–276

ornithine/citrulline carriers (SLC25A15, 2),
    275
2-oxoglutarate carrier (SLC25A11), 275
phosphate carrier (PiC), 270–273
of pyruvate, 275
subfamilies, 268
tricarboxylate carrier (SLC25A1), 275
uncoupling proteins, role of, 269
Mitochondrial monovalent ion transport,
    257–259
  ATPsensitive K channel (KATP), 259
  K$^+$/H$^+$ exchanger in yeast, 259
Mitochondrial permeability transition pore
    (mPTP), 267–268
Mitochondrial respiratory chain, 56*f*
Mitochondrial transport, 312–313, 312*f*
Mitochondrial transport protein family,
    222–228
  adenine nucleotide transporter (ANT),
    222–223, 223–225, 224*f*
  Ca^{2+} uniporter, 227
  glutamate uniporter, 226
  phosphate carrier, 225–226
  pyruvate, 226–227
  SLC25A family, 222
  uncoupling protein 1 (UCP1, SLC25A7),
    226
Mitophagy, 318–321, 319*f*
Mitoquinone (MitoQ), 295
Mobile carrier ionophores, 17–18
  gramicidin, 19
  valinomycin, 19
MsbA, 242
MtAlpHi (mitochondrial alkaline pH
    indicator), 284
MtDNA mutations, 346–347
  cybrids and, 347–350, 348*f*
  oocytes and generational quality control,
    347
MTOR (mammalian target of rapamycin),
    338–339
  major signalling pathways via, 340*f*
MtrA, 150–151
MtrB, 150–151
Multidrug effluxers, 239–240
Multidrug resistance (MDR) protein, 242
Myxothiazol, 122–123

N

Na$^+$ /Ca^{2+} exchanger, 256, 266
NADH–cyt *c* oxidoreductase activity, 94
NADH oxidation, 94–95, 151

NAD$^+$/NADH couple
  electron flow in, 47
  in mitochondrion, 41
  redox potential for, 41t
NAD$^+$/NADPH couple, 132–133, 273
NADP$^+$, 10
NADPH, 9–10
NAD(P)H fluorescence, 285–286
NADP$^+$/NADPH couples, 132–133, 175
Na$^+$/H$^+$ antiport activity, 16
Ndi1, 136
Nernst equation, 50, 67, 71
Nernst equilibrium across energy-transducing
    membranes, 69–70
Nernst potential, 48
N-ethylmaleimide, 59
Nicotinamide nucleotide transhydrogenase,
    132–133
Nigericin, 19, 22–23
NirC-type protein, 245
Nitric oxide reductase, 141, 146
$Nitrobacter$, 42–43, 147–148
  growth of, 147–148
  reversed electron transport in, 148f
Nitrous oxide reductase, 141
NMDA receptor activation, consequences of,
    363, 364f
Non-mitochondrial respiration, 279
Non-ohmic current/voltage relationship, 74–75
Novel uncoupling proteins (nUCPs), 300–301
NuoLMN subunit, 113

O
Observed mass action ratio, 30
Occluded state, 237–238
Ohm's law, 65
Oligomycin, 76, 279
Oligomycin-sensitivity conferral (conferring)
    protein, 209
Outer mitochondrial membrane (OMM),
    272, 305–306
  OMM-located TOB/SAM complex, 316
$Oxalobacter formigenes$, 248, 249f
Oxidation–reduction (redox) potentials, 36–43
  actual redox potentials, 39, 39, 40t
  determination of, 37–38
  and Gibbs energy change, 41–43
  glutathione couple, 40–41
  and pH, 38–39
  redox couples, 36–37
  relative concentrations of oxidised and
      reduced species, 38

Oxidative phosphorylation
  'efficiency' of, 46–47
  metabolic control analysis applied to, 80–83
  protonophores and, 79
Oxidative stress, 342–344, 363
Oxido-reduction reactions, 29
Oxygen electrode experiments, design and
    interpretation of, 63–64, 256
Oxygen-evolving complex, 179–182
Oxygen-sensitive electrode, 60–61

P
P870, 161
$Paracoccus denitrificans$, 137–142, 160, 216
  anaerobic electron acceptors in, 138
  c- and d_1-type haem centres, 140–141
  cbb_3 oxidase, 138
  c-type cytochromes, 137
  denitrification process in, 139–141, 140f
  electron carriers in, 138
  electron transport components in, 137f
  membrane-bound NO$_3^-$ reductase, 139
  nitric oxide reductase, 141
  nitrite reductase, 140–141
  nitrous oxide reductase, 141
  oxidation of compounds with one carbon
      atom, 138
  periplasmic nitrate reductase of, 139–140
  routes of electron transfer, 138
Parkinson's disease, 80, 310, 366–370
  familial, 369
  sporadic, 368
  α-synuclein, DJ-1, and LRRK2, role of,
      369–370
Passive transport of ions, 15
Pasteur effect, 331–332, 356
P/2e$^-$ ratio, 64–65
Peripheral stalk, 209
Periplasmic binding proteins, 240–241, 242,
    243f
Permeabilised cells, 287
Peroxiredoxins, 292
PGC (peroxisome proliferator-activated
    receptor γ coactivator) family, 334, 335f
pH gradient (ΔpH), 5, 18f, 24–25, 44
  across energy-transducing membranes,
      48–50
  across thylakoid membrane, 5
  electron transport-coupled proton extrusion
      and, 59, 65–66
  extrinsic indicators of, 71
  factors influencing partition of, 72–73, 72f

pH gradient (*Continued*)
  in intact cells, 284–285, 284*f*
  nigericin and, 19–20
  in steady state, 190–191
  swelling mechanism and, 23–24
Phosphoenolpyruvate (PEP), 247
Phosphonium cations, 67–69
Phosphotransferase system (PTS), 246–248
  for mannitol in *E.coli*, 247*f*
Photons, 44
  energy in 1 mol (or einstein) of, 44
Photosynthetic activity, factors influencing, 159
Photosynthetic bacteria, 3
Photosynthetic state transitions, 190
Photosynthetic thylakoids, 12
Photosystem II (PSII), 177–183
  b_6f complex, 183–184
  carotenoid orientation, 186–187
  2,5-dibromo-3-methyl-6-isopropylbenzoquinone (DBMIB), 183
  organisation of core polypeptides and cofactors in, 180*f*
  oxygen evolving complex (OEC), 180, 181*f*, 182, 187
  redox centre in plastocyanin, 183–184
Photosystem I (PSI), 177, 184–186
  Chl molecules, function of, 184–186
  PsaA and PsaB subunits, 184–186
  structure of, 184, 185*f*
PINK1 (PTEN-induced kinase 1), 320–321
Plastoquinol (PQH_2), 177, 187
Plastoquinone (PQ), 177, 187
P/O ratio, 64–65
$P870^+/P870$ redox couple, 161
Preconditioning, 333
Programmed cell death (apoptosis), 321–325
  cristae remodelling and, 325
  extrinsic and intrinsic pathways of, 322*f*, 323, 323–324
  by outer membrane permeabilisation, 324
Prokaryotic cells, 3
Propionigenium modestum, 156–157, 236
Protein-catalysed transport, 13–15, 14*f*, 21–22
  characteristics of, 21
Protein-catalysed transport properties, 15
Protein phosphorylation, mitochondrial, 337–338, 339*f*
Proteorhodospsin, 194–196
Proton circuit, 5–6, 6*f*, 54–58. *See also* Proton current

basic states of, 61
kinetic and thermodynamic competence of Δp in, 83–87
with light-driven proton pump, 85–87, 86*f*
mitochondrial respiratory chain, 56*f*
proton current, 58–65
secondary transport processes, 57*f*
uncoupling in, 55
'useful' (ATP-synthesising) and 'wasteful' (proton leak) pathways, 55–57
voltage and current measurements, 57–58
Proton conductance, 73–75
  basal proton leak, 74–75
Proton current, 58–65. *See also* Proton circuit
  $H^+/2e^-$ ratios, 59–60
  H^+/O stoichiometry, experimental determination of, 59
  in intact cells, quantification of, 276–281
  monitoring of, 60–61
  oxygen electrode experiments, design and interpretation of, 63–64, 68–69, 68*f*
  P/O ratio and $P/2e^-$ ratio, 64–65
  practical determination of, 61–63
  stoichiometry of proton extrusion, 58
Proton-dependent oligopeptide transporters, 234–235
Proton electrochemical gradient, 4, 44, 47
Proton electrochemical potential difference across membrane, 51
Proton leak, 61, 64, 279, 296–301
Protonmotive force (pmf) (Δp), 4, 20, 44–45, 64, 159, 229
  bacterial transport by, 238–240
  and binding affinity of ATP, 205–206
  distribution of phosphonium cations, 67–69
  early estimates of, 65–66
  energy transduction between respiratory chain and, 78
  factors influencing partition of, 72–73, 72*f*
  generation by bacterial transport, 248–249
  generation by Z-scheme, 186–188
  generation of light, role in, 173
  in intact cells, 281–287
  kinetic and thermodynamic competence in proton circuit, 83–87
  measurement in isolated organelles, 65–73
  metabolic control analysis and, 80–82
  quantification of, 65
  respiration-dependent, 245
  RET and, 76–77
  in steady state, 190–191

Protonophores, 6, 20, 79
   in intact cells, 280–281
Proton–polypeptide antiport system, 249–250
Proton pumping, 48
   by ATP synthase reaction, 46
   in complex IV, 128–129, 130f
   direct, 107
   electron transport and, 113–115
   by respiratory chain complexes, 45
Proton symport, 15–16
   and antiport systems, 228–229, 230f
Proton-translocating pyrophosphatase, 218
Proton translocation by respiratory chains,
      106–108, 108f
   conformational pump model, 108, 109f
   loop mechanism, 106–107
Proton translocators, 20, 75–76
Pseudoazurin, 140–141
Pseudomonas cocovenenans, 270
P-type ATPases, 243–244
Purple bacteria, 11
Pyrroloquinoline quinone (PQQ), 138
Pyruvate, 21

Q

Q-cycle mechanism, 98–100, 119, 120f, 172
   inhibitors of, 122–123
   thermodynamics of, 122
Quinols, 97–100, 99f
Quinones, 97–100, 99f

R

Ratiometric pericam, 260
Reactive nitrogen species (RNS),
      mitochondrial, 295
Reactive oxygen species (ROS), mitochondrial,
      288–295, 289f
   in complex I, 290
   in complex III, 291
   measurement of, 292–294
   superoxide metabolism, 291–292
   thiol redox potentials, 294–295
Red drop, 175–177
Redox-sensitive GFPs (roGFPs), 294–295
Redox signalling, 342–344
Reorganisation energy, 102
Respiration-dependent swelling, 23–24
Respiratory bacteria, energy transduction in,
      8–9
Respiratory chains of mitochondria
   bacterial, 136–157

complex I:II:III:IV ratio, 94
components of, 91–100
E_h values for, 106, 107f
electron transfer activity of each complex,
      93–94, 101–106
Em values for, 105–106, 107f
fractionation of, 91–100
loop mechanism, 106–107
organisation of, 91–100
overall proton and charge movements
      catalysed by, 131–132, 132f
proton translocation by, 106–108
reconstitution of, 91–100
redox centres, 95–100
redox potentiometry of, 104–106
sequence of redox carriers in, 100
spectroscopic techniques for, 96f–97f
Respiratory control ratio (RCR), 64
Reversed electron transfer (RET), 76–77, 77f,
      290
Revised Cambridge Reference Sequence
      (rCRS), 330
Rhodobacter capsulatus, 173–174, 235–236
Rhodobacter sphaeroides, 160, 173
   bacterial photosynthetic reaction centre of,
      162, 165–171, 166f
   delocalised exciton coupling, 162
   fluorescence resonance energy transfer
      (FRET), 162, 164–165
   light-harvesting complexes, LH1 and LH2,
      163, 163f, 165
   light reaction of photosynthesis in, 161–172
   pathways of electron transfer in, 161f
   purified reaction centres, 165
   use of antennae for absorption of light,
      162–165
Rhodopseudomonas acidophila, 163
Rhodopseudomonas viridis, 165
   reaction centre, 171, 172f
Rhodospirillaceae, 11
Rhodospirillum rubrum, 218
Rieske protein Fe–S centre, 123
Rnf system, 151
Rotenone-insensitive NADH dehydrogenase,
      136
Rusticyanin, 149–150

S

S, n-glycerophosphate dehydrogenase,
      115–116
Saccharomyces cerevisiae, 136

Salicylhydroxamic acid, 133–134
Salmonella typhimurium, 240–241
Seahorse extracellular flux analyser, 63
Sec system, 249–250
Semiquinone, 121
Shewanella species, 150–151
Single nucleotide polymorphisms (SNPs), 330
Sirtuins, 340–342, 341*f*
S,n -glycerophosphate dehydrogenase, 93
S,n-glycerophosphate shuttle, 273–274
Sodium-translocating pyrophosphatase, 218
Standard redox potential, 38
   at a pH other than zero, 39
Staphylococcus aureus, 246
Steady-state proton translocation, 48
Stem cells, 381–382, 382*f*
Stigmatellin, 122–123
Streptococcus faecalis, 76
Streptococcus lactis, 246
Streptomyces lividans, 244
Submitochondrial particles (SMPs), 7*f*, 8
Superoxide dismutases (SODs), 288–289,
   291–292
Superoxide metabolism, 291–292
Surface potentials, 51
Swelling mechanism, 22–25, 23*f*
   ammonium, 24–25, 25*f*
   Ca^{2+} overload, 24
   respiration-dependent, 23–24
Symport, 15–16, 221
Symporter proteins, 229
Synechococcus elongatus, 184

T
TatABC proteins, 250–251
Tetramethylrhodamine methyl and ethyl esters
   (TMRM and TMRE), 281–282
Tetraphenylborate anion (TPB$^-$), 21
Tetraphenyl phosphonium cation (TPP$^+$), 21,
   66, 74
Thermodynamics, 2, 204–205
   contribution to bioenergetics, 27–29
   entropy, 29–30
   equilibrium constants (*K*), 31, 34–35
   Gibbs energy change, 29–30
   mass action ratio equations, 35
   systems, 27–29
Thermodynamic stoichiometry, 46
Thermosynechococcus vulcanus, 179
Thermus thermophilus, 145–146, 217–218
Thiobacillus ferrooxidans, 149–150

bioenergetics of, 150
electron transfer and ATP synthesis by, 149*f*
oxidation of Fe^{2+}, 149–150
Thioredoxin (Trx), 292
Thylakoid membrane, 5*f*, 9–10, 63, 84–85,
   216–217
   of chloroplasts, 9–10, 160
   H$^+$/ATP stoichiometry of, 187–188
   lumen space, 9–10
   photosynthetic, 12
   physiological cyclic electron transport in,
      188*f*
   reaction centres in, 175–177
   thylakoid light-harvesting complex II, 178*f*
TIM40 'disulfide relay' system, 316
TonB/ExbB/ExbD complex, 238–239
TonB system, 238–239
Trafficking of mitochondria, 312–313
Transfer RNAs (tRNAs), 328–329
TRAP (*tri*partite *A* TP-independent
   *peri*plasmic) transporters, 235–236
Trialkyl tin compounds, 213
Tricarboxylic acid (TCA) cycle, 115–116
Trimethylamine-*N*-oxide, 145
Triphenylmethyl phosphonium (TPMP$^+$), 66
Tunnelling, 101
Type 2 diabetes, 358–361
   β cell failure and, 360–361
   UCP2, role of, 361

U
Ubiquinol, 139–140
   NADH oxidation by, 151
Ubiquinone, 173, 295
   and complex III, 118–125
   delivering electron to, 115–118
Ubisemiquinone, 98
Uncouplers, mechanism of, 20
Uncoupler-stimulated ATPase activity,
   75–76
Uncoupling protein 1 in brown fat (UCP1),
   268, 354–355
Uncoupling protein 1 (UCP1), 297–300, 298*f*
Uniport, 15–16
Ureaplasma urealyticum, 248–249

V
Valinomycin, 19, 22–23, 50–51, 71
   K$^+$ ionophore, 67
Vesicle preparations, 8–9, 9*f*
V-type ATPases, 217–218

W

Water-splitting reaction, 179–180
    of a $Mn_4 CaO_5$ cluster, 180
    of photosystem II, 181f

X

Xanthorhodopsin, 194–195

Y

Yarrowia lipolytica, 115
Yeast mitochondria, 136
YiiP, 246

Z

Zinc transporters, 245–246
Z-scheme, 177, 186–188

Printed in the United States
By Bookmasters